IUTAM Symposium on One Hundred Years of Boundary Layer Research

SOLID MECHANICS AND ITS APPLICATIONS
Volume 129

Aims and Scope of the Series

The fundamental questions arising in mechanics are: *Why?*, *How?*, and *How much?* The aim of this series is to provide lucid accounts written by authoritative researchers giving vision and insight in answering these questions on the subject of mechanics as it relates to solids.

The scope of the series covers the entire spectrum of solid mechanics. Thus it includes the foundation of mechanics; variational formulations; computational mechanics; statics, kinematics and dynamics of rigid and elastic bodies: vibrations of solids and structures; dynamical systems and chaos; the theories of elasticity, plasticity and viscoelasticity; composite materials; rods, beams, shells and membranes; structural control and stability; soils, rocks and geomechanics; fracture; tribology; experimental mechanics; biomechanics and machine design.

The median level of presentation is the first year graduate student. Some texts are monographs defining the current state of the field; others are accessible to final year undergraduates; but essentially the emphasis is on readability and clarity.

For a list of related mechanics titles, see final pages.

IUTAM Symposium on

One Hundred Years of Boundary Layer Research

Proceedings of the IUTAM Symposium held at DLR-Göttingen, Germany, August 12–14, 2004

Edited by

G.E.A. MEIER

DLR, Göttingen, Germany

and

K.R. SREENIVASAN

ICTP, Trieste, Italy

Managing Editor:

H.-J. Heinemann

DLR, Göttingen, Germany

Springer

A C.I.P. Catalogue record for this book is available from the Library of Congress.

ISBN-10 1-4020-4149-7 (HB)
ISBN-13 978-1-4020-4149-5 (HB)
ISBN-10 1-4020-4150-0 (e-book)
ISBN-13 978-1-4020-4150-1 (e-book)

Published by Springer,
P.O. Box 17, 3300 AA Dordrecht, The Netherlands.

www.springer.com

Printed on acid-free paper

Printed in the Netherlands.

CONTENTS

Poster-Presentation

PREFACE

Prandtl's famous lecture with the title "Über Flüssigkeitsbewegung bei sehr kleiner Reibung" was presented on August 12, 1904 at the Third Internationalen Mathematischen Kongress in Heidelberg, Germany. This lecture invented the phrase "Boundary Layer" (Grenzschicht). The paper was written during Prandtl's first academic position at the University of Hanover. The reception of the academic world to this remarkable paper was at first lukewarm. But Felix Klein, the famous mathematician in Göttingen, immediately realized the importance of Prandtl's idea and offered him an academic position in Göttingen. There Prandtl became the founder of modern aerodynamics. He was a professor of applied mechanics at the Göttingen University from 1904 until his death on August 15, 1953. In 1925 he became Director of the Kaiser Wilhelm Institute for Fluid Mechanics. He developed many further ideas in aerodynamics, such as flow separation, base drag and airfoil theory, especially the law of the wall for turbulent boundary layers and the instability of boundary layers en route to turbulence.

During the fifty years that Prandtl was in the Göttingen Research Center, he made important contributions to gas dynamics, especially supersonic flow theory. All experimental techniques and measurement techniques of fluid mechanics attracted his strong interest. Very early he contributed much to the development of wind tunnels and other aerodynamic facilities. He invented the soap-film analogy for the torsion of noncircular material sections; even in the fields of meteorology, aeroelasticity, tribology and plasticity his basic ideas are still in use. Aside from the boundary layer and the boundary layer equations for which Prandtl rightly occupies an immortal place, his name lives through the Prandtl number, Prandtl's momentum transport theory and the mixing length, the Prandtl-Kolmogorov formula in turbulence closure, the Prandtl-Lettau equation for eddy viscosity, the Prandtl-Karman law of the wall, Prandtl's lifting line theory, Prandtl's minimum induced drag, the Prandtl-Meyer expansion, the Prandtl-Glauert rule, and so forth. The string of young men he mentored is nothing short of remarkable. Among them we easily recognize Ackert, Betz, Blasius, Flachsbart, Karman, Nikuradse, Schiller, Schlichting, Tietjens, Tollmien and Wieselsberger. The list could, of course, be larger.

The hundredth anniversary of Prandtl's invention was the first reason for us to apply for an IUTAM Symposium "One Hundred Years of Boundary Layer Research". The other reason was to summarize the progress in the field by inviting the best known specialists for related contributions. The overwhelming response led to the many interesting lectures and contributions collected in these proceedings.

We thank F. Smith, R. Narasimha, H. Hornung, T. Kambe, I. Wygnanski, A. Roshko, P. Huerre, E. Reshotko, K. R. Sreenivasan for the revision of the manuscripts and helpful advice.

We especially appreciate Dr. Hans-Joachim Heinemann's organisation of the meeting and his work managing the edition of the proceedings, without which the task would have been impossible. Monika Hannemann provided our internet presentation, Oliver Fries was responsible for finances, Helga Feine, Catrin Rosenstock and Monika Hannemann managed the conference office, and Karin Hartwig assisted in the preparation of the symposium.

All the technical organization and support was provided by the Institute of Aerodynamics and Flow Technology, DLR Göttingen, directed by Prof. Dr. Andreas Dillmann. We appreciate this support very much.

The Editors and the Managing-Editor are very grateful to Mrs. Anneke Pot, Senior Assistant to the Publisher, and Springer, Dordrecht, The Netherlands, for the excellent support and help in publishing this book.

It is our hope that the readers of this book will find it as pleasant as we do and discover new views on boundary layers and the related research which flows from Ludwig Prandtl's work in 1904.

Göttingen, August 2004

G.E.A.Meier and K.R.Sreenivasan
(Cochairmen)

Scientific Committee:

D.H. van Campen	Eindhoven University of Technology; IUTAM
P. Huerre	Ecole Polytechnique; Palaiseau
T. Kambe	Science Council of Japan, Tokyo
G.E.A. Meier	DLR Göttingen - Chairmen
H.K. Moffatt	Center for Mathematical Sciences, Cambridge, IUTAM
A. Roshko	CALTEC, Pasadena
F. Smith	University College London
K.R. Sreenivasan	International Center for Theoretical Physics, Trieste - Chairman
I.J. Wygnanski	The University of Arizona

Sponsors of Symposium

German Research Foundation, DFG, Bonn
International Union of Theoretical and Applied Mechanics (IUTAM)
Bundesland Niedersachsen, Hannover
Deutsches Zentrum für Luft- und Raumfahrt e.V. (DLR), Köln
Kluwer Academic Publishers B.V., Dordrecht

PRANDTL'S BOUNDARY LAYER CONCEPT AND THE WORK IN GÖTTINGEN

A historical view on Prandtl's scientific life

Gerd E. A. Meier
Institut für Strömungsmaschinen, Universität Hannover und
DLR–Institut für Aerodynamik und Strömungstechnik, Göttingen, Germany

Abstract: The invention of the "Boundary Layer" by Ludwig Prandtl goes back to his famous lecture in August 8, 1904 with the title "Über Flüssigkeitsbewegung bei sehr kleiner Reibung" which was held at the "III. International Mathematischen Kongreß" in Heidelberg. These proceedings and the related IUTAM Symposium celebrate the 100th anniversary of this event. The following historical remarks will be a short record of Prandtl's scientific life with emphasis on his "Boundary Layer" work.

Key words: Ludwig Prandtl, history, scientific work, fluid mechanics, boundary layer.

1. PRANDTL'S EDUCATION AND HIS EARLY PROFESSIONAL CAREER

Ludwig Prandtl was born February 4, 1875 in Freising, Bavaria. His father was a professor at an agricultural school in Weihenstephan. He spent his school years in Freising and lived later in Munich until 1894. After graduation from school he studied eight semesters of "Maschinentechnik" (mechanical engineering) at the Technical High School in Munich where he was awarded the degree of a "Maschineningenieur" (mechanical engineer) in 1898. Professor August Föppl was his teacher in Technical Mechanics and became his mentor later on. Prandtl spent an additional year in Föppl's laboratory for his dissertation at the University of Munich as a doctor of

G.E.A. Meier and K.R. Sreenivasan (eds.), IUTAM Symposium on One Hundred Years of Boundary Layer Research, 1-18,

philosophy, because the Technical High School was not allowed to provide a doctoral thesis in those days. His Dissertation with the title "Kipperscheinungen, ein Fall von instabilem elastischem Gleichgewicht" was the foundation of his scientific carrier.

In the beginning of the year 1900 he was affiliated as an engineer at the "Maschinenfabrik Augsburg-Nürnberg" (MAN) in Augsburg. There he was involved with work on diffusers for wood cutting machines. When designing for this company a device for sucking dust and splices, Prandtl noticed that the pressure recovery he expected from a divergent nozzle was not realized. Soon he detected the still famous rule that half the divergence angle of a diffuser may not be larger than about 7° in order to avoid separation of the decelerating flow. In those experiments his ideas of a special behavior of the near wall parts of the flow field have been born obviously. Already there, he was confronted with the phenomenon of flow separation and this consequently was the initiation of his interest in flow phenomena and the real reason of his invention of the boundary layer concept [**1,2**]. Later as a professor at the University of Hanover he showed the compatibility of his boundary layer approximations with the Navier-Stokes Equations which led to a development of historical dimensions.

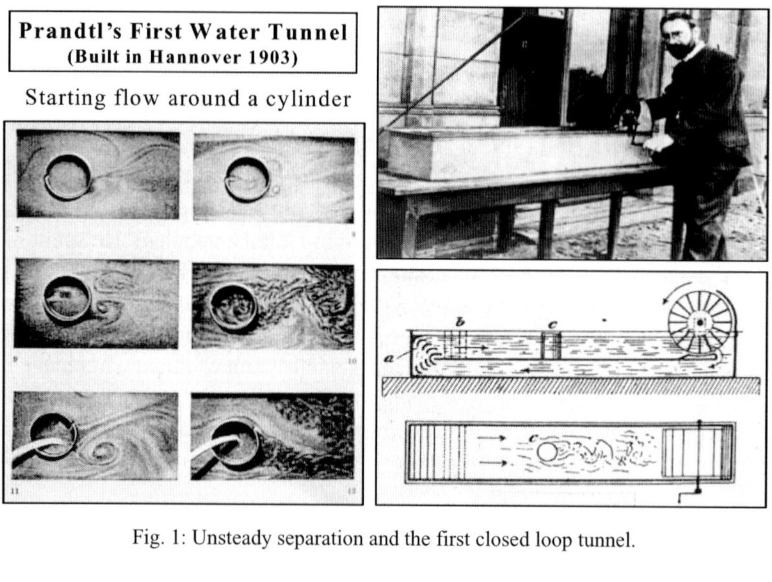

Fig. 1: Unsteady separation and the first closed loop tunnel.

Already in October 1901 Prandtl became a full professor of mechanics at the Technical High School of Hanover. There he built his first hand driven water tunnel for his elucidating experiments (Fig. 1). The flow was seeded on the free surface for visualizing the separation and vortices. He spent only three years in this place and position, but he published several important papers and finally also the famous lecture at the "III. Internationaler Mathematischer Kongress in Heidelberg 1904" with the title "Über Flüssigkeitsbewegung bei sehr kleiner Reibung", which publication nowadays is seen as the publication presenting the discovery of the boundary layer concept and as the beginning of the related research (Fig. 2).

$$\varrho\left(\frac{\partial v}{\partial t} + v \circ \nabla v\right) + \nabla(V + p) = k\nabla^2 v$$

Neglection of "small" Terms

$$\varrho\left(u\frac{\partial u}{\partial x} + v\frac{\partial u}{\partial y}\right) + \frac{dp}{dx} = k\frac{\partial^2 u}{\partial y^2};$$

$$\frac{\partial u}{\partial x} + \frac{\partial v}{\partial y} = 0.$$

Pressure is imposed
Set is numerically integrable
Self similarity of solutions

BL-thickness goes like square root of viscosity and x.

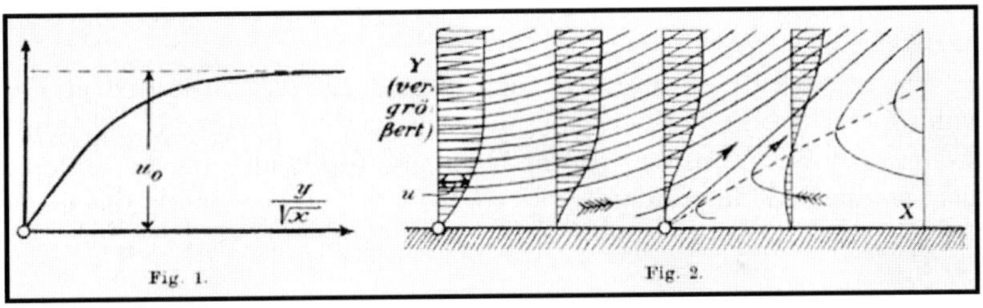

Fig. 2: Prandtl's Discovery of the Boundary Layer.
An asymptotic approach to the full momentum equation.

This lecture in Heidelberg was also the reason for the famous mathematician Felix Klein, who was a professor of mathematics in the University of Göttingen, to offer Prandtl a university position in Göttingen as an Extra Ordinarius. Although Prandtl had to step back this way from a full professorship, he finally took the position to change into an environment with his own laboratory and to contact the famous scientists in the University of Göttingen **[1,2,3,5]**.

2. THE EARLY WORK IN GÖTTINGEN

In 1905 Felix Klein also motivated the mathematician Carl Runge to come from Hanover to Göttingen, with the three later founding the "Institut für Angewandte Mathematik und Mechanik" and this became a very fruitful scientific environment for themselves and their students in the following years. Following the common enthusiasm about aeronautics together with Runge in 1907, Prandtl held his first seminar on aerodynamics in the University.

Prandtl directed in this institute the PhD works of Blasius, Boltze and Hiemenz covering boundary layer problems. Prandtls former own work in boundary layer theory has been continued with the thesis of Blasius in 1908 on laminar boundary layer development on a flat plate. Blasius solved Prandtls boundary layer equations in his PhD thesis for the flat plate successfully. Boltze solved in 1908 the laminar boundary layer for a body of revolution and Hiemenz in 1910 solved the laminar boundary layer for a cylinder in cross flow.

In 1906 the young Theodor von Karman from Hungary was asking Prandtl for a PhD opportunity and was promoted in 1908 to Göttingen with a topic in the field of elasticity. Later, in connection with the work of Hiemenz and Rubach, he invented the "Wirbelstraße" (vortex street). Already in April 1913 von Karman became a professor at the Technical High School of Aachen.

The organised boundary layer and turbulence research started in 1909 with the PhD works of Hochschild, Rubach, Kröner, Nikuradse and Dönch. In this early work, one can see the beginning of Prandtls interest in turbulence research and flow control (Fig. 3). Later this research work was intensified by his contributions to the problem of the drag of a sphere.

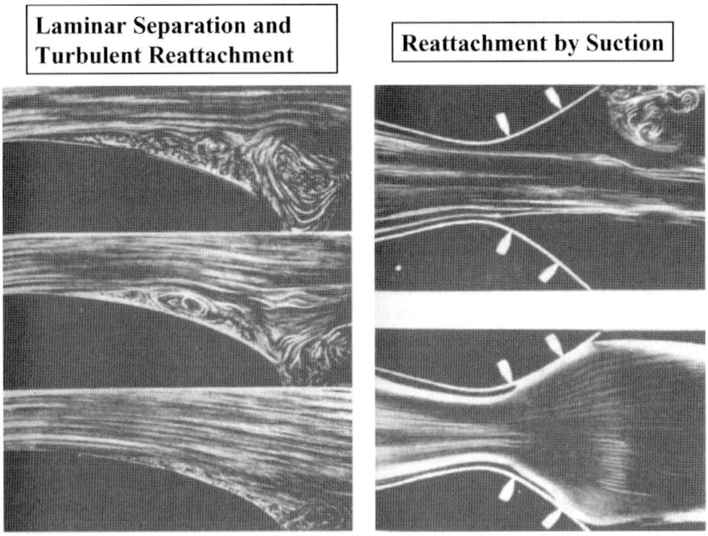

Fig. 3: Reattachment of boundary layers by turbulence and suction.

In measuring the drag on spheres, scientists like Prandtl and Eiffel from Paris were very surprised about large differences in the drag coefficients measured in their wind tunnels. The contradiction in drag coefficients for spheres, which differed by 50%, finally could be explained by the different separation at different Reynolds numbers. It was Prandtl who explained these discrepancies with an "experimentum crucis" where he introduced for the first time a trip wire at the wall to change the state of the boundary layer from laminar to turbulent. Prandtl made this special experiment with the trip wire to demonstrate that also in case of lower Reynolds numbers, the drag figures of the supercritical regime could be achieved.

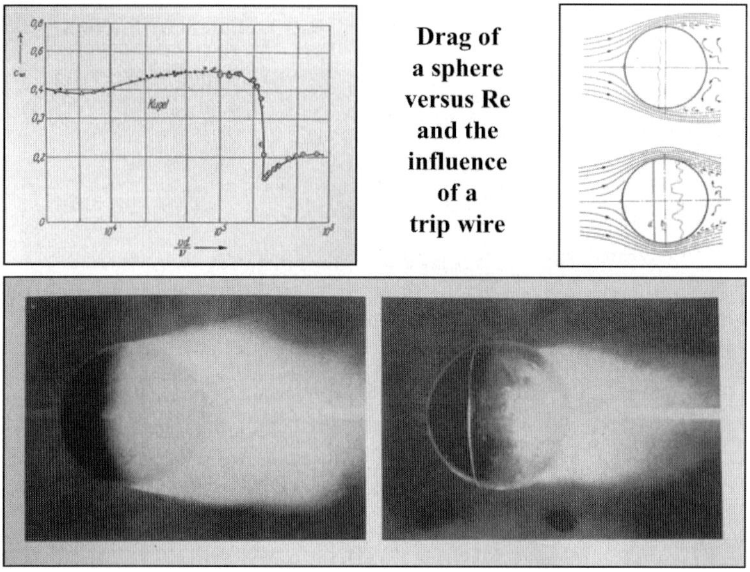

Fig. 4: Influence of a trip wire on laminar separation.

Using the trip wire with the wind tunnel set at a constant speed, the drag could be reduced considerably. It was once again the different separation location which led to this phenomenon. He clearly pointed out that due to the more downstream separation in case of a turbulent boundary layer, the pressure drag is reduced substantially. The test results could finally be understood by Prandtl's boundary layer theory with the introduction of the critical Reynolds number for transition (Fig. 4) [**1,4**].

Inspired by the experiments in the habilitation thesis of W. Nusselt, Prandtl discovered in 1910 the analogy between heat convection and friction in fluid boundary layers. His idea was based on the analogy between the differential equations of heat convection and flow in the vicinity of the wall. In connection with his boundary layer theory, he solved some problems for laminar and turbulent flows on plates and through tubes. Later in 1928 he improved this simulation by introducing more precise properties of the turbulent flow. Honouring his work in this field, the ratio of cinematic viscosity and temperature conductivity was called the "Prandtl Number" later on. But, since Nusselt had used this ratio in his former work, Prandtl was never very accepting of this honour (Fig. 5).

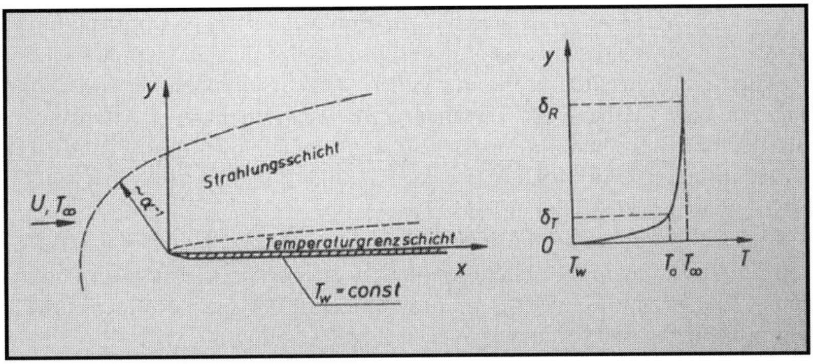

Fig. 5: By the analogy of the equations of flow and heat convection, Prandtl established a mapping of heat exchange in flows over walls.

In the years after 1912, Carl Wieselsberger was one of the important scientists in Prandtl's "Aerodynamische Versuchsanstalt (AVA)" (aerodynamic research establishment). Wieselsberger mainly conducted drag measurements for airships and airfoils (Fig. 6). Also the drag of sails was measured in the wind tunnel and the results have been compared with those of Gustave Eiffel from Paris, France.

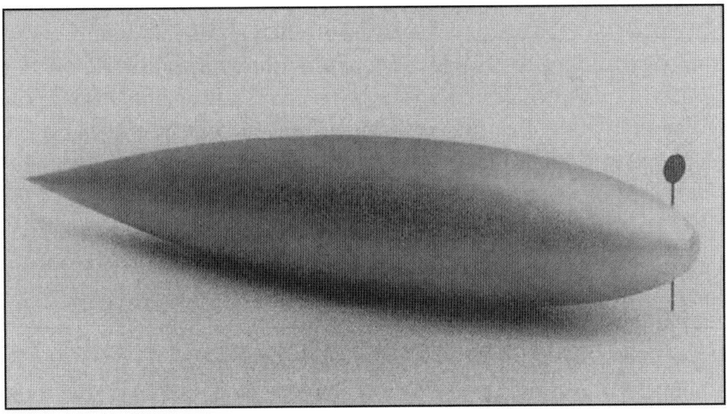

Fig. 6: Comparable drag of a small disk and a streamlined body.

In 1920 Prandtl realized that the drag of a flat plate is closely related to the drag of a straight pipe by considering that only the flow field close to the wall (the boundary layer) is important for the friction effects. This also implies that the velocity distribution near the wall is determined only by the law of friction. So he concluded that the flow velocity is proportional to the square root of wall distance as previously shown by Blasius for pipe flow.

This finally resulted in the law called "Universelles Wandgesetz" - the universal law of the wall. The theory for the friction on a flat plate by Blasius for the laminar case from 1908 and Prandtls own theory from 1921 for the turbulent case were verified for Reynolds Numbers close to a million by Liepmann and Dhawan in 1951 (Fig. 8).

By later experiments with high Reynolds numbers, Prandtl in parallel to von Karman came to the conclusion that by a logarithmic formulation introducing the shear stress velocity, a fully universal law for the velocity distributions near the wall could be achieved (Fig. 9).

With respect to the description of the fully developed turbulent flow Prandtl had introduced in 1924 the term "Mischungsweg" (mixing length). His idea was that fully developed turbulence is characterized by some characteristic length, after which the eddies loose their individuality. He mainly used this idea to understand the momentum exchange between the turbulent eddies and to explain the turbulent shear stress this way. The mixing length formulation for the turbulent shear stress which is in essence identical to the earlier formulation by Reynolds was independently invented by Prandtl in 1926. His formulation had the advantage of introducing the wall distance y and a typical constant which later by von Karman was found to be k=0.4 (Fig. 7). The mixing length concept led to some useful theoretical considerations for the mixing of a free jet by W.Tollmien and some interpretations of the velocity profiles in ducts with rectangular cross sections by Dönch.

In 1926, Prandtl discovered on the basis of measurements of Nikuradse in rectangular and triangular ducts, turbulent secondary flows which had not been observed in the laminar case. Prandtl understood these phenomena as a consequence of the momentum exchange in the three dimensional turbulent flow. This was far from any possible theoretical treatment in those days. In contrast to the secondary flows in curved ducts he named these phenomena secondary flows of the second kind.

In 1907, Prandtl rejected an offer of the Technical High School of Stuttgart to become a full professor of Technical Mechanics; he preferred to stay in Göttingen to finish his plans for a "Modellversuchsanstalt" and to stay in the fruitful scientific environment of the Alma Mater there **[3,5,6,7]**.

3. THE "KAISER WILHELM INSTITUT FÜR AERODYNAMIK"

When in Berlin 1910, the plans for the founding of the "Kaiser Wilhelm Gesellschaft" (KWG) became virulent, Felix Klein had the idea to propose a

"Kaiser Wilhelm Institute for Aerodynamics". The purpose was mainly to keep Prandtl in Göttingen by providing him with an institute for all problems of aerodynamics and hydrodynamics. Prandtl himself later wrote a proposal for this research institute which was consisting of a "Kanal-Haus" with all kinds of test tubes and water test facilities for flow experiments, a machine house, a calibration chamber, shops and finally a flying station for measurement in open air. In recognition of Prandtl's merits in sciences and especially in aerodynamics and hydrodynamics, this institute was granted by the "Kaiser Wilhelm Gesellschaft" in June 1913.

But in 1914 the First World War began and so the plans for the founding of the Kaiser Wilhelm Institute were postponed. Only the wind tunnel project, which was important for the aircraft industry could be completed in 1917. Also in these difficult times, Prandtl could only use about one third of the wind tunnel time for research purposes. Special reports, the so called "Technische Berichte", dealt with problems of airfoil sections, drag of fans and coolers, and design of fuselage and propellers. In cooperation with Monk and Betz, Prandtl also made remarkable progress in his airfoil theory.

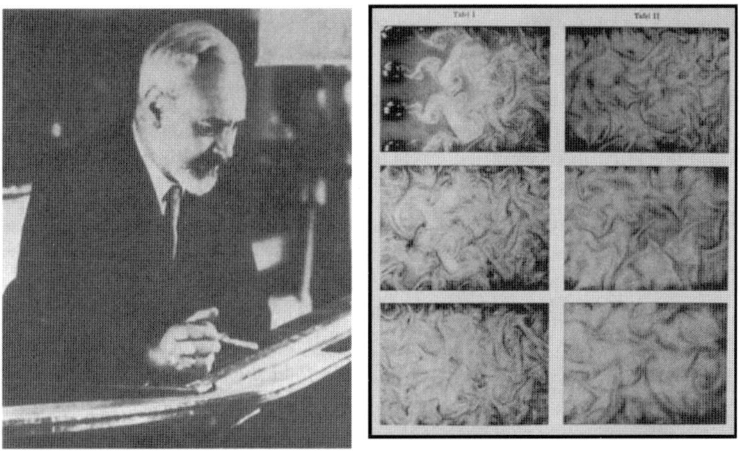

Fig. 6: Left: Prandtl studying turbulence. Right: Grid turbulence.

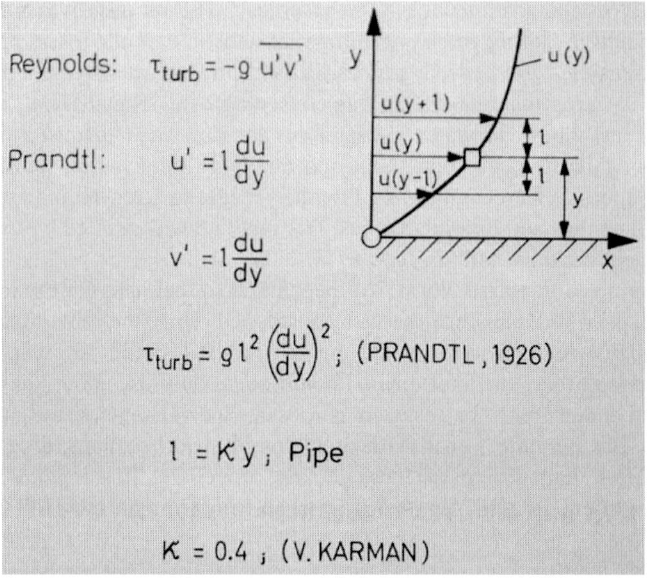

Fig. 7: Prandtls mixing length concept.

In August 1920, Prandtl was offered to become successor of his father in law August Föppl on a full chair for mechanics at the Technical High School in Munich. This was very attractive for him because many of his supporters in Göttingen like von Böttinger and Felix Klein faded away and the situation of the "Versuchsanstalt" was not very good.

So after this offer, a time of difficult negotiations started to keep Prandtl in Göttingen. His intention to switch from the more applied research in the "AVA" to a more scientific research in the frame of a fully developed "Kaiser Wilhelm Institute" and to get rid of the lectures at the university was a difficult problem in those days, since the financial situation of the government and the "Kaiser Wilhelm Gesellschaft" was poor. But finally, also with the help of his friends in the administration and in industry, he was granted a directorship in a "Kaiser Wilhelm Institute" and could keep his full professorship for Technical Physics in the University of Göttingen as well. The main reason that these negotiations came to a successful end was that the scientific community and also the administration realized that there was nobody else who could replace Prandtl at Göttingen in those days.

All the work of Prandtl in the years after the First World War was devoted to the aerodynamics of transport vehicles. Mainly, the aerodynamic problems of civil aircraft but also the drag and smoke emissions of railway steam engines and the drag of racing cars and automobiles were studied.

Prandtl could start in 1924 building his new institute which had a laboratory for gas dynamic experiments and later also a rotating laboratory which was designed for studies of atmospheric flows. The rotating laboratory was at first operated by the young Busemann studying the influence of Corriolis forces on the flows in an open water tank. Beside the scientific results, he got all information about dealing with sea sickness.

For the new "Kaiser-Wilhelm-Institut", which was physically built in 1924, Prandtl named beside, gas dynamics and cavitation, mainly boundary layers, vortices, and viscid flows as the targets of research. Among the experimental facilities were two towing tanks for boundary layer and wake studies. The bigger one had a length of 13 meters.

In this way, two institutes existed since 1925 in parallel, as Prandtl was the director of the "Kaiser Wilhelm Institute für Strömungsforschung" and the AVA, which was in fact directed by the deputy director Albert Betz.

Already in 1924 Prandtl became honorary member of the London Mathematical society and in 1927 he was invited for the Wilbur Wright Memorial Lecture by the Royal Aeronautical Society. In those years, he also got honorary PhD's from the Universities of Danzig and Zürich, Switzerland. Later he was honoured in the same way in Bukarest, Cambridge, Istambul, Prag and Trondheim.

In the twenties, Prandtl's work was devoted mainly to the problems of the origin of turbulence and the properties of turbulent flows (Fig. 6). The first studies of instability of laminar boundary layers had been conducted by Tietjens in his dissertation. In 1925, Prandtl published his results about the drag in pipes and the first ideas about his mixing length model for turbulent flows.

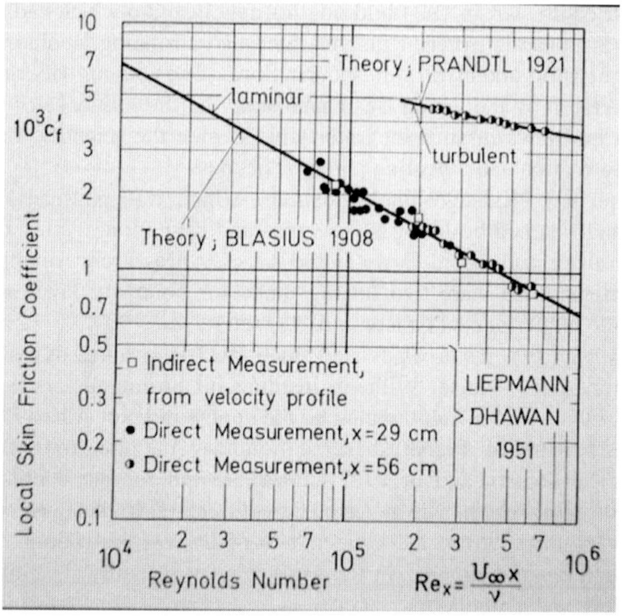

Fig. 8: The skin friction predicted by Prandtl's theory and its experimental verification.

In the early twenties, Prandtl started intensive considerations about the origin of turbulence. He built a special tunnel about six meters long with a seeding possibility to observe the flow on the surface by floating particles. The intermittent vortices and waves he observed were not what he expected, because small amplitude distortions were considered to be stable in those days. Together with Tietjens, he found in theoretical considerations instability of the laminar flow with respect to small distortions. But these simplified theoretical considerations did not explained the stability of the boundary layer for small Reynolds numbers. From this experience he concluded that the understanding and quantitative treatment of turbulence was a futile task [**1,3,5**].

4. THE WORK OF PRANDTL IN THE THIRTIES

The reason why Prandtl was so important for the Research Centre in Göttingen was mainly due to his work in the field of boundary layers. By consideration that friction in flows with small viscosity is only important in the vicinity of walls, the whole range of complex flow phenomena in vehicles and engines became transparent. Another field was airfoil theory which mainly, by the introduction of the induced drag, provided a foundation for all kind of airfoil designs. Since many other researchers and institutions were in those days doing successful research in this field, one can understand Prandtl's idea to switch to new horizons in the newly built institute.

In the new institute for "Strömungsforschung", Prandtl gathered a lot of young students, who became famous researchers later on, like J. Ackeret, H. Blenk, A. Busemann, H. Goertler, H. Ludwieg, J. Nikuradze, K. Oswatitsch, H. Schlichting, R. Seifert, W. Tollmien, O. Tietjens, W. Wuest, and others. Counting the number of the resulting PhD thesis's and his own publications, about one quarter of Prandtls work was devoted to boundary layer and turbulence research.

Prandtl had understood in the twenties with his initial ideas from the beginning of the century the main properties of the laminar boundary layer, the reasons for separation and also the consequences for pressure drag. Additionally, he also found the possibility of reducing the pressure drag by shifting the separation point downstream by diminishing the area of separated flow. But in the thirties he was still excited about the problem of instability of the boundary layers and the route to turbulence (Fig. 6). Around 1930, Prandtl studied the influence of stabilizing effects on turbulence especially by curved surfaces and stratified fluids.

An important step to understand the mechanisms of instability was the asymptotic theory, which was put in final form by W. Tollmien. This theory for first time provided the stability limit for the flat plate accurately. Contributions in this field had been made by Prandtl and Tietjens before but also Lord Rayleigh and W. Heisenberg had contributed in this field. With Tollmiens method, Schlichting and Pretsch solved the problem for other geometries, especially for curved walls. But Prandtl was always a little bit skeptical about this theory because the predicted instability waves could not be seen in his simple experiments. So Prandtl built a new water tunnel, better designed for studying laminar flow, but after his own words it was impossible to avoid all the distortions from the intake so that here and there a "turbulence herd" appeared. This indicates that Prandtl observed turbulent spots, which was later introduced in the literature by Emmons, Schubauer and Klebanoff.

It took another fifteen years until the end of the Second World War that Schubauer and Skramstad in the NBS under the supervision of H. L. Dryden conducted experiments in a tunnel with very low turbulence to prove the concept of Tollmien-Schlichting instability waves.

But also the mechanism of transition of the boundary layers and the persistant turbulence, which were not really understood until now, were still

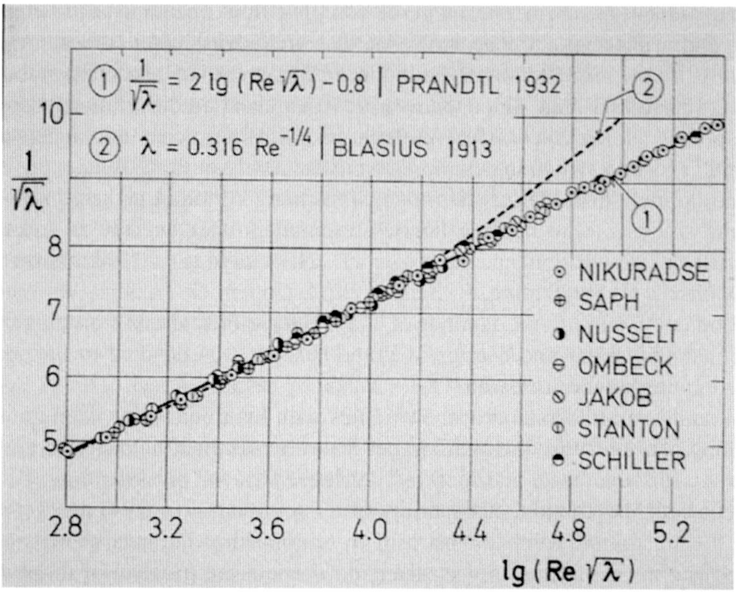

Fig. 9: The logarithmic law of the wall

Prandtl's concern and he proposed a semi empirical approach to use the momentum equation of stationary boundary layers with an input of turbulent velocity distributions. With an additional empirical approach for the shear stress at the wall he could calculate the velocity profiles of the turbulent boundary layer.

In 1936, Prandtl built a new "Wall Roughness Tunnel" which was a wooden construction with a 6m long test section where the pressure gradient could be varied. Many interesting papers about turbulent boundary layers by famous authors like Ludwieg, Schultz-Grunow, Wieghardt and Tillmann are originating from there. In this context for Prandtl, the work of Ludwieg and Tillmann was very helpful. They made the most accurate measurements of the shear stress in turbulent boundary layers in those days. This way, the universal "Law of the wall" which had been proposed by Prandtl and also von Karman in the days of considerations about Prandtl's earlier power law hypothesis could be confirmed in a more precise way as by the early measurements in the thirty's performed by Nikuradse (Fig. 9).

Nikuradse later mainly contributed under the supervision of Prandtl with some striking experiments on the influence of wall roughness on the drag in pipe flow. These were important data for the industry, especially chemical engineering. These data are still in use today and have been extended to all kinds of flow geometries (Fig. 10). Based on Nikuradses experiments, Prandtl and Schlichting published in 1934 a paper about the drag of plates with roughness. Schlichting worked with Prandtl until 1939 when he became a full professor in Braunschweig. In 1957, he followed Betz as director of the AVA in Göttingen.

But it was also in the thirties that Prandtl's interest changed and the work in the "Kaiser Wilhelm Institute für Strömungsforschung" shifted to other fundamental problems which made use of his former research experiments in boundary layer flows. For instance together with H. Reichert he studied the influence of heat layers on the turbulent flow and he spent as well some activity in meteorology. Prandtl also wrote in those years a contribution to "Aerodynamic Theory" which was edited by W. F. Durand. In this book, Prandtl described all the work which had been done up to that time in Göttingen. The "Aerodynamic Theory" became standard literature in the field and was really the breakthrough for Prandtl's ideas and his fame in the international community **[3,8]**.

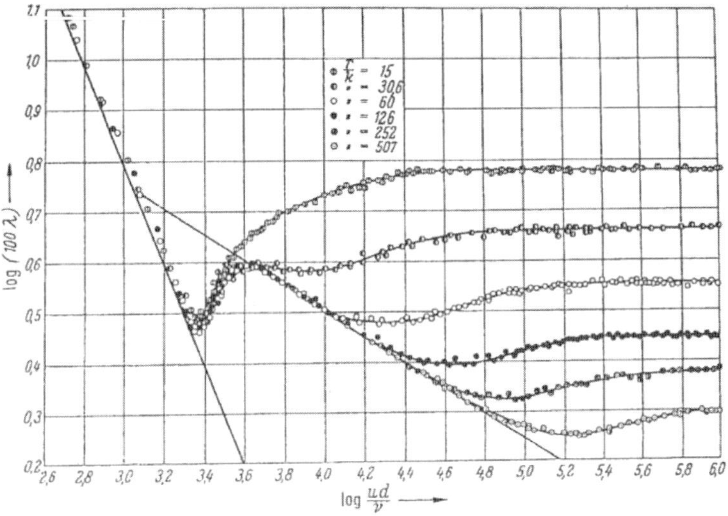

Fig. 10: Nikuradses drag measurements for pipes.

5. PRANDTL'S WORK IN THE FORTIES

Even in the war in 1941 Prandtl built a small wind tunnel for the study of laminar to turbulent transition studies. Here the work of H. Reichardt and W. Tillmann about turbulence structure has to be mentioned. In 1945, Prandtl published two papers: One on the transport of turbulent energy and the other one on three dimensional boundary layers. The question where in the boundary layer turbulent energy is created and how it is propagated into the flow was still addressed by Prandtl and some co workers up to his death in 1956.

After the Second World War, the "Kaiser Wilhelm Institute für Strömungsforschung" was transformed into the "Max Planck Institut für Strömungsforschung" (MPI) in Goettingen. In 1946, Prandtl retired from the directorship of the new MPI where Betz was his successor. After his retirement he had still a small group until 1951 where he studied the theory of tropical cyclones with E. Kleinschmidt. The main parts of the MPI were the two departments headed by Betz and Tollmien. In 1957 the AVA (Aerodynamische Versuchsanstalt) was established and the MPI-department of Betz was the core of the new AVA headed by Schlichting. Prandtl also gave up his chair in the University of Göttingen which was granted to Tollmien in 1947. Under Prandtls direction and by his initiative, 85 PhD theses have been conducted in the years from 1905 to 1947 at the University of Göttingen [**6**]. About 30 of these publications are devoted to problems of boundary layer and turbulence.

6. BOUNDARY LAYER WORK AFTER PRANDTL

The "Max Planck Institut für Strömungsforschung" was Prandtl's scientific home for his last years and was always devoted to research on boundary layers and turbulence. Under Tollmien who followed Prandtl in 1956 as a director, the work in boundary layer instability, intermittency and turbulent structures was promoted in many doctoral theses. Also the work of Reichardt, Herbeck and Tillmann was directed on the structure and statistics of intermittent and turbulent flows. The work of Eckelmann and his co-workers with a newly built oil channel for extremely low Reynolds Numbers contributed to the ideas about the structure of sublayer instabilities and intermittency development.

In the seventies, pipe flow experiments found the locations where fluctuation energy is mainly generated and how it is propagating from this well defined location of generation into the boundary layer: Downstream with flow velocity and perpendicular to the wall with shear stress velocity. So something like a certain propagation angle for turbulent energy propagation is defined by the two velocities locally. This is similar to the Mach angle in acoustics, defined by the flow velocity and the velocity of sound [**12**]. It is interesting that Prandtl's question about the turbulent energy propagation was answered with the help of the shear stress velocity, which he introduced for his logarithmic law of the wall.

With the same pipe flow tunnel, Dinckelacker made interesting experiments on the influence of riblets on the boundary layer and friction. He was able to reduce the drag of turbulent pipe flow by more than 10%.

Until the end of fluid mechanics research in the Max-Planck-Institute, when it's last director E.-A. Müller retired in 1998, a lot of work was done in vortex dynamics, turbulence control and the structure of turbulent boundary layers.

The successor of the former AVA in Göttingen, the "DFVLR-Institute für Strömungsmechanik" was headed since 1957 by Schlichting and had with Becker, Ludwieg, Riegels, Rotta and many others an excellent team for boundary layer research in the many wind tunnels of the institute but also in numerical and theoretical research projects.

Later, the "DLR Institut für Aerodynamik und Strömungstechnik", also did a lot of work on boundary layers. The mysterious transition scenarios and the mechanisms of instability have been a major target in the years of improved experimental and numerical methods. Many interesting results for boundary layer instability have been received by solving the Navier Stokes equations numerically and also by experiments, using new optical tools, which have confirmed these results. The main finding was that the well known Tollmien-Schlichting-waves and other new instability forms undergo higher order instability processes which lead to new special wave forms and vortices which finally disintegrate in chaotic interaction [**10,11**].

One can say that from the initiative of Ludwig Prandtl as a scientist and organizer, boundary layer research was connected to the research centre in Göttingen for over 100 years from its reception and that we are proud to have hosted the related IUTAM symposium for the celebration in Göttingen.

ACKNOWLEDGEMENTS

The author gratefully acknowledges the support of the "DLR Institut für Aerodynamic und Strömungstechnik" in preparing this article especially the figures which stem from the institute's archives. Mrs. Karin Hartwig assisted in typing the text.

REFERENCES

1. Prandtlt L, Oswatitsch K, Wieghardt K, *Führer durch die Strömungslehre*, Braunschweig, Vieweg, 1984.
2. Görtler H, Tollmien W, (Eds), *50 Jahre Grenzschichtforschung*, Braunschweig, Vieweg, 1955.
3. Rotta JC. *Die Aerodynamische Versuchsanstalt in Göttingen, ein Werk Ludwig Prandtls*, Göttingen, Vandenhoeck und Ruprecht, 1990.
4. Meier GEA, Viswanath PR, (Eds), *Mechanics of Passive and Active Flow Control*, Dordrecht, Kluwer, 1999.
5. Meier GEA, (Ed), *Ludwig Prandtl, ein Führer in der Strömungslehre*, Braunschweig, Vieweg, 2000.
6. Fütterer H, Weingarten K, *Ludwig Prandtl und sein Werk, Ausstellung zu seinem 125. Geburtstag*, Deutsches Zentrum für Luft- und Raumfahrt und Max-Planck-Institut für Strömungsforschung, Göttingen, 2002.
7. Busemann A, *Ludwig Prandtl, 1875-1953,* Biographical Memories of Fellows of the Royal Society, Vol. 5, Feb. 1960 1960, p.193.
8. Flügge-Lotz I, Flügge W, *Ludwig Prandtl in the nineteen-thirties: reminiscences*, Ann. Rev. Fluid Mech., Vol. 5, 1973, p. 1.
9. Oswatitsch K, Wieghardt K, *Ludwig Prandtl and his Kaiser-Wilhelm–Institut*, Ann. Rev Fluid Mech. 19, 1987, p. 1.
10. *50 Jahre Max-Planck Institut für Strömungsforschung Göttingen 1925-1975*, Göttingen, 1975, Hubert & Co.
11. Meier GEA, *35 Jahre Aerodynamik und Aeroelastik in Göttingen*, in: 35 Jahre Deutsches Zentrum für Luft- und Raumfahrt e. V., Köln, Sept. 2004, DLR.
12. Schildknecht M, Miller JA, Meier GEA, *The influence of suction on the structure of turbulence in a fully developed pipe flow*, pp. 67-107, vol. 90, part 1, JFM, 1979.

THE FULL LIFESPAN OF THE BOUNDARY-LAYER AND MIXING-LENGTH CONCEPTS[1]

Philippe R. Spalart
Boeing Commercial Airplanes. P.O. Box 3707, Seattle, WA 98124, USA. (425) 234 1136
philippe.r.spalart@boeing.com

Abstract: Ludwig Prandtl's most penetrating contributions are approximations to the dynamics of fluids. As such, they are liable to be superseded, at the time it becomes possible to solve the original equations analytically or, more probably, to routinely obtain numerical solutions so accurate they solve the problem without explicit use of the approximations. The engineering value of the theories is distinguished from their educational and intuitive value. The purpose here is to envision when and how this shift will happen for the boundary-layer and mixing-length concepts, with an aside on lifting-line theory, thus defining in some sense the lifespan of Prandtl's ideas.

Key words: Boundary layer, CFD, grid, mixing length, logarithmic layer, turbulence model, lifting line

1. BOUNDARY-LAYER THEORY

Engineering increasingly relies on Computational Fluid Dynamics. Few CFD codes use the boundary-layer equations today. They tend to be special-purpose codes, applied to the repeatable topologies and nearly-attached flow typical of airplanes in cruise, as opposed to vehicles, houses, factories, and airplanes landing. Examples of viscous-inviscid coupling are Boeing's Tranair full-aircraft code and Drela's MSES (Multiple-Element Streamline Euler Solver) airfoil code. Cruise and slightly off-design conditions for an airliner are an excellent application; the lower computing cost relative to Navier-Stokes codes allows multi-point, multi-disciplinary optimisation.

[1] In tribute to Dr. W.-H. Jou

G.E.A. Meier and K.R. Sreenivasan (eds.), IUTAM Symposium on One Hundred Years of Boundary Layer Research, 19-28,

Transition prediction also involves the boundary layer as an entity instead of local quantities, for physical reasons, and in fact depends on fine details of it. Often, the Navier-Stokes solution fields are unfortunately not "clean" enough to accurately provide these details, so that the rather awkward state of the art is to run a separate boundary-layer solution using the Navier-Stokes pressure distribution. Nevertheless, very few codes offer transition prediction and, in broad terms, the boundary-layer equations have been displaced from CFD, victims of the complexity of coupling methodologies and of the Goldstein singularity, added to computing-power increases that facilitate Navier-Stokes solutions and give access to more complex geometries.

On the other hand, when the Navier-Stokes equations are solved, it is most often on grids with an obvious boundary-layer structure. Over a smooth surface, the grid is clustered at the wall in the normal direction only, clearly following the boundary-layer approximation. This is valid for laminar solutions and for the Reynolds-Averaged Navier-Stokes (RANS) equations with turbulence. Several major codes even use the "thin-layer Navier-Stokes equations", thus dropping cross-direction viscous terms, which pre-supposes the grid is aligned with a thin shear layer. All grid generators are attuned to wall units and to the grid-stretching ratios acceptable in the logarithmic layer. These accuracy requirements derive from the physics of the wall layer and are easy to implement before any solution is obtained, the friction velocity needed to express wall units being fairly predictable. The true difficulty is to predict the boundary-layer thickness, in order to switch from a "viscous grid" inside the boundary-layer to an "Euler grid" outside it with both good accuracy and economy. Therefore, careful RANS users design grids to match boundary layers. Flows such as a wing with high-lift system also benefit from anisotropic grid clustering in the off-body thin shear layers. These layers are essentially unmanageable with viscous-inviscid coupling, at least in 3D, because the shape and topology of the free wakes, which would need to be explicitly described and inserted as velocity jumps in the inviscid solution, become too complex. They may also thicken far beyond the range of the thin-layer approximation, especially over a flap. On the other hand, ensuring grid convergence in every shear layer in a 3D high-lift RANS solution is also very difficult when the grid is user-designed; thus, the Navier-Stokes equations do not make this problem trivial in any sense.

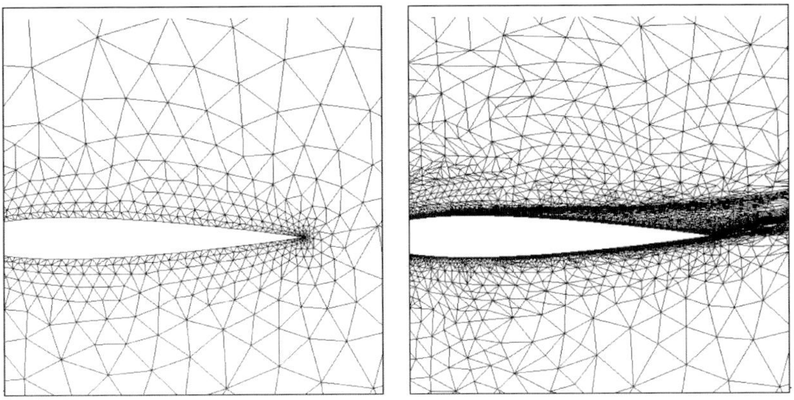

Figure 1. Initial and final grids for RANS airfoil calculation.

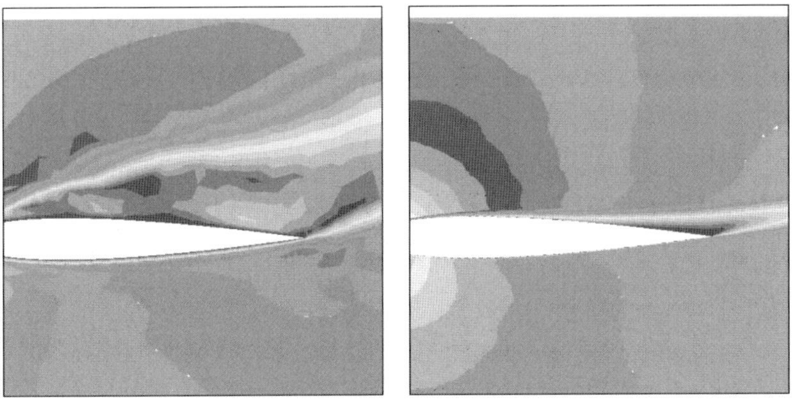

Figure 2. Initial and final Mach-number distributions for RANS airfoil calculation.

It is a clear goal for the next generation of Navier-Stokes codes to remove this imposition, through grid generation concurrent with the solution. Figures 1-3 were provided by S. Allmaras for the Boeing General-Geometry Navier-Stokes team, based in Seattle and Moscow. A NACA 0012 airfoil is at 15° angle of attack, at Mach 0.2 and Reynolds number 10^6 with fully-turbulent boundary layers. The solution begins with a coarse "Euler" grid (128 points on the airfoil, 907 in total) and mild isotropic clustering near the wall (1a). The solution on this Grid 1 is only partially iteration-converged and is very inaccurate, since none of the viscous effects are captured well (2a). The lack of turbulent viscosity gives essentially zero skin friction, and causes spurious separation. Through the cycles, the solver identifies the boundary layer and other shear layers, provides grid points, establishes the turbulence model, and iterates as the shear layers find their place. The refinement approach is fairly empirical at this point, using derivatives of the Mach number. The remediation of spurious separation requires the ability to de-refine the grid, as would the motion of shocks during convergence. Here, "de-refining" means that the next grid iteration can be coarser than the last one, in some region; in other words, the iterative grid generation does not only involve the addition of grid points (which would be easier). The final grid, Grid 11 (698 points on the airfoil, 33438 total), is in figure 1b, and the solution in 2b. The grid refinement naturally produces anisotropic cells in the boundary layer and other thin shear layers, and eventually respects wall units for the wall-normal spacing. This is seen in figure 3, which shows that the wall-parallel spacing was merely halved, and also makes the interface between boundary-layer and Euler regions evident. The wall-normal clustering is seen to occur in steps, in this early version of the code. This is not optimal, and "hand-made" grids are smoother. On the other hand, such grids can never match the boundary-layer thickness all along the wall at all angles of attack. As a result, either they extend the viscous spacing into the Euler region, which is somewhat wasteful or, worse, they begin the Euler spacing inside the viscous region, which is inaccurate.

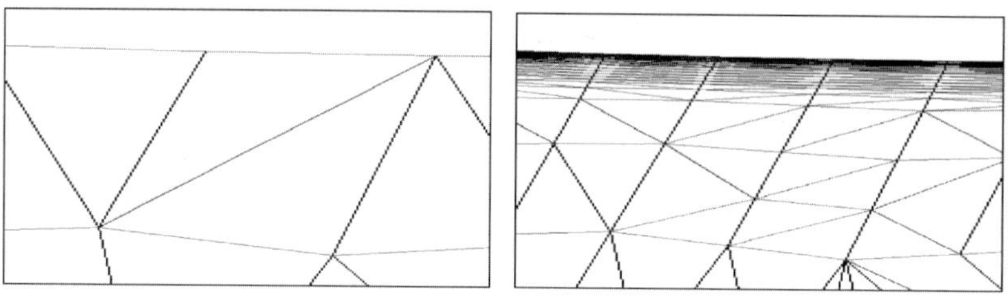

Figure 3. Initial and final grids for RANS airfoil calculation. Detail near lower surface.

The figures vividly illustrate how a boundary-layer structure imposes itself with automatic adaptation in a steady RANS case. Unsteady RANS is left for future work. Large-Eddy Simulation (LES) and Detached-Eddy Simulation (DES) present additional challenges to grid adaptation, but none that are insurmountable. Such simulations naturally lead to nearly isotropic grid cells, away from the wall. Very near the wall, an effective LES relies on wall modelling, again requiring anisotropic cells, and so does any DES.

Once this global strategy of concurrent grid generation and solution succeeds and spreads, which is in high demand and is likely within a decade, CFD users (and automatic optimisers) will proceed without knowing boundary-layer theory.

This will not apply to designers, or to those who wish to understand aerodynamics, in engineering or in nature. Most flows of interest contain boundary layers, which often control the rest of the flow, and intuition will not be effective without the boundary-layer idea and a grasp of the complex interplay between pressure gradient, transition and separation. This is to obtain high-quality predictions, as well as to design control of the flows, active or passive. In the design of an airliner wing, under intense competitive pressure to reduce drag, the concept of "pushing the boundary layer" is central. Significant gains ensue from bringing the boundary layer close to separation at the trailing edge and in other regions of "stress," but the risks related to unforeseen separation are also very large, and neither the wind tunnel nor CFD can be completely trusted to predict flight.

Another intellectual attraction is that the underlying mathematical technique of matched asymptotic expansions is more general than boundary-layer theory. It enters lifting-line theory [1], also due to Prandtl with the influence of Lanchester, which has similarly been displaced from CFD codes but not as a fundamental tool to understand and design wings. This lasting value creates much interest in extracting the induced drag from CFD solutions and wind-tunnel surveys, as opposed to lifting-line solutions. Unfortunately, years of effort have not led to a definition of induced drag in a general viscous flow, even assuming complete access to the flow field. A practical method would address finite loading (which lifting-line theory does not) on a non-planar geometry, and multiple surfaces (the induced drag of the wing and horizontal tail need to be treated together, and the high-lift system is more complex still).

Within CFD, a related argument has been made that forces would be better extracted from far-field quantities than from wall quantities (pressure and skin friction). This has always seemed dubious to the author; the boundary layer can be accurate and the wake inaccurate, but not the converse. An additional argument is made that far-field extraction will separate induced drag, wave drag, and viscous or "parasite" drag. It echoes the fact that within lifting-line theory, many results can be expressed "at the wing" or "in the wake" through elementary manipulations of integrals, and also that viscous drag has been added to induced drag successfully in practical design methods for simple wings. However, conclusive results are lacking for these far-field extraction strategies. The current methods based on wake surveys, experimental or numerical, suffer from rather poor accuracy. Furthermore, surveys at different stations give a different split between apparent induced drag and apparent parasite drag, which defeats the purpose.

The permanence of the boundary-layer concept can be attributed to the high values of the Reynolds number in human-size and larger flows. More precisely, it is due to the fact that even *turbulent* skin-friction coefficients are much smaller than unity, with 0.002 being typical; "bei sehr kleiner Reibung" in Prandtl's 1904 words ("with very small friction"). Small values of constants such as 0.0168 in the Cebeci-Smith turbulence model are another illustration (a point made by Melnik). Could this be predicted by thought alone, without experiment or direct simulations?

2. MIXING-LENGTH THEORY

The nature of mixing-length theory is different from that of boundary-layer theory. Instead of being a mathematical approximation with proved formal validity in a limit, it is a physical argument that the turbulence at a given location can be described from a small number of parameters, provided that it is fully developed. In fact, only *one* feature of the turbulence, namely the Reynolds shear stress, can be described (coupled with the mean shear rate). Even the other Reynolds stresses do not conform when the global Reynolds number of the boundary layer varies [2], a fact which essentially all turbulence models are unable to duplicate. On the other hand, the dissipation rate follows an equivalent model very closely, possibly because it adapts to the turbulent-energy production, which is well-behaved [2]. Mixing-length theory is strongly tied to the logarithmic "law" for the velocity profile of a turbulent boundary layer, and the concepts will be treated as nearly interchangeable.

Mixing-length theory has been applied to simple free shear flows, but needs different constants and is slightly less accurate than the assumption of uniform eddy viscosity (also due to Prandtl) [3], whereas in wall-bounded flows it rests on only one primary constant and a secondary one, and has been dominant. Both approaches (mixing length and log law) have been described as "amounting only to dimensional analysis," unfairly. They make the sweeping assumption that the only length scale needed to build a potent model of the turbulence is proportional to the distance from the wall, with the ratio a *universal* constant named after von Kármán. Once this is posited, dimensional analysis is used. However, sweeping assumptions can be wrong, and this one is successful.

The Kármán constant κ which sets the mixing length has received attention of a mixed kind in the last five years. While it had seemed safely confined to the bracket [0.40, 0.41] for decades, serious experimental papers have given values as different as 0.436 [4] and 0.383 [5]. This impacts extrapolations to high Reynolds numbers; a difference of 0.025 in κ changes the skin friction at length Reynolds number $Re_x = 10^8$ by 2%, and therefore the drag of an airplane by 1%. This is significant in terms of guarantees in the airline industry. It is also disappointing for a presumed universal constant to be challenged by ±5%, and it is hoped that the differences are not eventually traced to different instrumentation (Pitot tube versus hot wire). The impact of the subtle corrections for finite probe size on experimental values for κ has also been disturbing. Conversely, Direct Numerical Simulation (DNS) is still far from powerful enough to conclusively set this constant.

It remains that essentially all authors view the Kármán constant as universal, not entertaining the idea that it could differ in a pipe and in a boundary layer, for instance, or depend on Reynolds number and pressure gradient. The concept itself is not under attack here. Similarly, challenges to the log law itself and proposals to replace it with a power law are, in the author's opinion, without merit [5, 6]. They are incompatible with the Galilean invariance that is implied in much of the thinking in turbulence, and is built into all transport-equation turbulence models.

Mixing-length theory and log law are equivalent only when the turbulent shear stress is independent of the position. Experiments and simulations suggest that when it is not the case, because of a pressure gradient, the log law is closer to being preserved. This applies to channel or pipe flow and boundary layers in pressure gradients, even with the stress as far as ±40% from its wall value. In addition, it was argued in [2] that even in the flat-plate turbulent boundary layer, the stress is not constant to leading order in the outer expansion, contrary to the common view. Its slope over the range of validity of the log law is finite, about –0.6 when normalized with the skin friction and the boundary-layer thickness δ (the near-equivalent slope in a channel or pipe is –1). The argument in [2] is based on the mean momentum equation, simple, and supported by DNS results.

This near-consensus preference for the log law is regarded as fortuitous, physically, and in some sense unfortunate. The reason is that the mixing length has more intuitive meaning and relates local quantities (making it useable in a RANS model), whereas the log law involves the wall value of shear stress. In other words, many "motivations" for the log law fail when the stress is not constant; their logic evaporates. The word "motivation" is a reminder that these are not actual derivations, based on any valid governing equation. A definitive generalisation for pressure gradients and suction/blowing now appears unachievable.

The mixing-length theory is essential in algebraic turbulence models, which have also lost much ground in CFD, again because of coding complexity, loss of meaning after separation, and incompatibility with unstructured grids. The turbulence models in wide use today are built on between one and seven partial differential equations, and even the simplest ones can claim somewhat better physics than algebraic models when the turbulence travels from boundary layer to free shear layers, or from one type of free shear layer to another. Among the common models, some use the wall distance as an essential parameter in the log layer, very much in the spirit of mixing-length theory; this includes those of Secundov *et al.* [7] and later Spalart-Allmaras [8]. Others such as Menter's [9] use it in a different manner, in the upper region of the boundary layer, and yet others do not use it at all. In fact, some authors consider the use of wall distance as a serious flaw, both for reasons of CFD convenience and for more "philosophical" reasons. This controversy over local and non-local influences is not about to end, especially in a field as arbitrary as RANS modelling. It is unlikely that the distance-using models will be surpassed and retired for quite a few years, plausibly for two decades. In that case, the heritage of the 1925 mixing-length theory will have lived for at least a century in pure RANS models.

The mounting threat to mixing-length theory, and to RANS in general, comes from DNS and LES. However, even if Moore's rate for the growth of computing power is sustained, DNS of a full-size wing will be possible only by 2080, and then only as a "grand challenge" [10]. LES will be possible far earlier, near 2045, but this will be "true" LES. By this we mean that the grid spacing, at least parallel to the wall, can take unlimited values in wall units. Instead of being of the order of 10 to 20, the lateral spacing Δz^+ can be 10,000, for instance. Such a capability is far from standard, and much LES work sadly still takes place at very modest Reynolds numbers, of little practical value and where clear scientific conclusions cannot be drawn either. This leaves both engineers and theoreticians rather un-impressed. DES was applied as a wall model at very high Reynolds number, with fair results [11].

An important point is that wall modelling is empirical and akin to RANS modelling, although narrower in purpose and often given to simple algebraic or one-equation models. Many researchers wish to escape from empiricism, with good reason, but rarely with much success in the field of turbulence. A litmus test when a new approach claims *not* to be empirical is to ask, "Does this approach imply a value for κ?" All the effective approaches to wall modelling do imply a value and therefore are empirical, so that only full DNS will eventually displace κ. The Kármán constant will remain a crucial empirical constant in engineering, and the most pivotal one in CFD, essentially until the end of the 21st century.

3. OUTLOOK

Prandtl's boundary-layer, mixing-length and lifting-line approximations have been extremely fruitful, and their place in engineering fluid dynamics is only slowly being eroded a century or almost a century after they were imagined. Their educational value is permanent. The mixing length, although it is the least elegant of the three, will live the longest: roughly, for another century, in superficially modified form and confined to the very-near-wall regions. This is remarkable especially in view of the "acceleration" of science.

It seems unlikely that Prandtl would be surprised with the eventual "victory" of computing power over his intuitive approximations, since he did believe in the Navier-Stokes equations, and appreciated the one-dimensional numerical solutions that were possible in his days, for instance that due to Blasius. It is likely he would enjoy the formal mathematics that were used to support and expand his ideas, although only marginally. Note how higher-order extensions of Prandtl's theories have not proven very useful, or even been available. Repeated attempts at systematic improvements have remained very debatable, both in the boundary-layer and mixing-length arenas. Some have been simply erroneous [1], and the others are dependent on additional assumptions that are far from being supported strongly enough by data. It appears Prandtl had the wisdom not to attempt extensions of his approximations, formal or not, that would be too fragile.

REFERENCES

1. Van Dyke M., *Perturbation methods in fluid mechanics*. Stanford, Parabolic Press, 1975.
2. Spalart P. R. "Direct simulation of a turbulent boundary layer up to $R_\theta = 1410$." *J. Fluid Mech.* **187**, pp. 61-98, 1988.
3. Schlichting H., *Boundary-layer theory*. New York, McGraw-Hill, 1979.
4. Zagarola, M. V., Perry, A. E., Smits, A. J. "Log laws or power laws: the scaling in the overlap region." *Phys. Fluids,* **9**, pp. 2094-2100, 1997.
5. Nagib, H. M., Christophorou, C., Monkewitz, P. A. "High Reynolds number turbulent boundary layers subjected to various pressure-gradient conditions". IUTAM 2004: 100 years of boundary-layer research. Aug. 12-14. Göttingen, Germany.
6. Barenblatt, G. I., Chorin, A. J. "Scaling in the intermediate region in wall-bounded turbulence: the power law." *Phys. Fluids,* **10**, pp. 1043-1046.
7. Gulyaev, A., Kozlov, V., Secundov, A. "A universal one-equation turbulence model for turbulent viscosity." *Fluid Dyn.*, **28**, 4, pp. 485-494, 1994.
8. Spalart, P. R., Allmaras, S. R. "A one-equation turbulence model for aerodynamic flows." *Rech. Aérospatiale*, **1**, pp. 5-21, 1994.
9. Menter, F. "Two-equation eddy-viscosity turbulence models for engineering applications." *AIAA J.*, **32** (8), pp. 269-289, 1994.

10. Spalart P. R. "Strategies for turbulence modelling and simulations." *Int. J. Heat & Fluid Flow,* **21**, pp. 252-263, 2000.
11. Nikitin, N. V., Nicoud, F., Wasistho, B., Squires, K. D., Spalart, P. R. "An Approach to Wall Modeling in Large-Eddy Simulations". *Phys. Fluids,* **12** (7), pp. 7-10, 2000.

RATIONAL BASIS OF THE INTERACTIVE BOUNDARY LAYER THEORY

J. Cousteix[a] and J. Mauss[b]

[a] *Département Modèles pour l'Aérodynamique et l'Énergétique, ONERA, and École Nationale Supérieure de l'Aéronautique et de l'Espace, 2 avenue Édouard Belin, 31055 Toulouse - France. Tél : 05 62 25 25 80 - Fax : 05 62 25 25 83 - Email : Jean. Cousteix@onecert.fr*

[b] *Institut de Mécanique des Fluides de Toulouse UMR-CNRS and Université Paul Sabatier, 118 route de Narbonne, 31062 Toulouse Cedex, France. Tél : 05 61 55 67 94 - Fax : 05 61 55 83 26 - Email : mauss@imft.fr*

Abstract: The interactive boundary layer theory has been used successfully for a long time but the theory received no formal justification. Flows at high Reynolds number are analyzed here with an asymptotic method in which generalized expansions are used and applied to a laminar or a turbulent boundary layer.

1. INTRODUCTION

The boundary layer theory proposed by Prandtl [12] was a major step in the understanding of the flow behaviour in aerodynamics and became an extremely useful practical tool for predicting aerodynamic flows. A great difficulty has been encountered in applications for flows subject to an adverse pressure gradient strong enough to lead to separation. Goldstein [5] analyzed the behaviour of the boundary layer solution—for a given pressure distribution— close to the point of separation. He showed that the solution is singular if the prescribed velocity profile has a zero derivative at the wall (zero shear stress) and pointed out that the pressure distribution around the separation point cannot be taken arbitrarily. Goldstein also suggested that the use of inverse methods could be a way to overcome the singularity. In these inverse techniques, the external velocity distribution is not prescribed but is a part of the calculation method; the input is for example the distribution of the displacement thickness. Catherall and Mangler [2] showed numerically that separated flow can be calculated in this way without any sign of singularity.

Another major contribution is due to Lighthill [7] who analyzed the upstream influence in supersonic flow. When an oblique shock wave impacts a two-dimensional flat plate boundary layer, it is observed that the boundary

G.E.A. Meier and K.R. Sreenivasan (eds.), IUTAM Symposium on One Hundred Years of Boundary Layer Research, 29-38,

layer grows more than expected well upstream of the shock wave and possibly separates also upstream of the shock wave. The theoretical difficulty was that perturbations cannot travel upstream neither in a supersonic flow nor in an attached boundary layer. The explanation that perturbations can travel upstream in the subsonic part of the boundary layer is not valid either because the length of upstream influence would not be properly predicted.

A key point in Lighthill's analysis is the *mutual interaction* between the outer inviscid flow and the near wall viscous layer. This feature supplants the hierarchy of the Prandtl theory in which the inviscid flow imposes the pressure distribution to the viscous layer. Another important result is the calculation of a measure of the length of upstream influence; this length is determined as the distance in which the disturbance is reduced by a factor e^{-1}. From this result, the streamwise length scale of interaction is $LRe^{-3/8}$ where L is the distance of disturbance from the boundary layer origin and Re is the Reynolds number based on L.

A breakthrough occurred with the triple deck theory (TD) attributed to Stewartson and Williams [13, 14] and to Neyland [11]; Messiter [9] analyzed the flow near the trailing edge of a flat plate and also arrived, independently, at the triple deck structure. Stewartson and Williams considered their theory as an extension of Lighthill's theory to nonlinear interactions. The triple deck structure is a degeneracy of the Navier-Stokes equations which describes certain separated boundary layers without singularity.

In engineering calculation methods, the viscous-inviscid interaction is addressed by solving the Navier-Stokes equations or by using the interactive boundary layer theory (IBL). In this theory, the hierarchy between the inviscid flow equations and the boundary layer equations is replaced by a strong coupling of the equations. The IBL theory was used and applied successfully for some time [1, 3, 6, 17, 18]. The best justification, provided by Veldman, is that IBL contains all terms that are relevant in TD. However, Sychev et al. [15] commented that: "No rational mathematical arguments (based, say, on asymptotic analysis of the Navier-Stokes equations) have been given to support the model approach". In this paper, this problem is examined by using the successive complementary expansions method (SCEM) described in section 2. This method is used to obtain the IBL model for laminar (section 3) and turbulent flows (section 4).

2. SUCCESSIVE COMPLEMENTARY EXPANSIONS METHOD

Consider a singular perturbation problem where the function $\Phi(x,\varepsilon)$ is defined in a domain D and ε is the small parameter. Assume that two significant domains have been identified—an outer domain where the relevant variable is x and an inner domain where the boundary layer variable is X.

According to the Successive Complementary Expansions Method (SCEM), the starting point requires a *uniformly valid generalized approximation*:

$$\bar{\Phi}_a = \sum_{i=1}^{n} \bar{\delta}_i(\varepsilon)\left[\bar{\varphi}_i(x,\varepsilon) + \bar{\psi}_i(X,\varepsilon)\right]$$

where $\bar{\delta}_i$ is an *order function*. This approximation is constructed step by step without requiring any matching principles. The boundary conditions are sufficient for calculating the successive approximations. More detailed information about the SCEM is given in Ref. [8].

By using asymptotic expansions, the function $\bar{\Phi}_a$ can be written as:

$$\bar{\Phi}_a = \hat{\Phi}_a + \mathrm{o}(\delta_m) \text{ with } \hat{\Phi}_a = \sum_{i=1}^{m} \delta_i(\varepsilon)\left[\varphi_i(x) + \psi_i(X)\right] \; ; \; \bar{\delta}_n = \mathrm{O}(\delta_m)$$

where $\hat{\Phi}_a$ is a *regular approximation*—the sum of two regular expansions— and $\delta_i(\varepsilon)$ are *gauge functions*, i.e. δ_i is a suitable representative order function chosen in the corresponding equivalence class defined from the relation of strict order. It is not necessary that the set $\bar{\delta}_i$ is the same as the set δ_i since new terms can appear and since the functions δ_i are gauge functions.

The difference between the generalized and regular expansions is that $\bar{\varphi}_i$ is a function of x *and* ε whereas φ_i is a function of x only; in the same way, $\bar{\psi}_i$ is a function of X *and* ε whereas ψ_i is a function of X only.

3. IBL MODEL

3.1 Second order IBL Model

For a laminar incompressible two-dimensional steady flow, the Navier-Stokes equations can be written in dimensionless form as

$$\frac{\partial \mathcal{U}}{\partial x} + \frac{\partial \mathcal{V}}{\partial y} = 0 \tag{1a}$$

$$\mathcal{U}\frac{\partial \mathcal{U}}{\partial x} + \mathcal{V}\frac{\partial \mathcal{U}}{\partial y} = -\frac{\partial \mathcal{P}}{\partial x} + \frac{1}{\mathcal{R}}\left(\frac{\partial^2 \mathcal{U}}{\partial x^2} + \frac{\partial^2 \mathcal{U}}{\partial y^2}\right) \tag{1b}$$

$$\mathcal{U}\frac{\partial \mathcal{V}}{\partial x} + \mathcal{V}\frac{\partial \mathcal{V}}{\partial y} = -\frac{\partial \mathcal{P}}{\partial y} + \frac{1}{\mathcal{R}}\left(\frac{\partial^2 \mathcal{V}}{\partial x^2} + \frac{\partial^2 \mathcal{V}}{\partial y^2}\right) \tag{1c}$$

The Reynolds number $\mathcal{R}$ is high compared to unity and a small parameter ε is

$$\varepsilon^2 = \frac{1}{\mathcal{R}} = \frac{\nu}{VL} \tag{2}$$

with V and L denoting reference quantities. The coordinate normal to the wall is y and the coordinate along the wall is x; the x- and y-velocity components are $\mathcal{U}(\equiv u/V)$ and $\mathcal{V}(\equiv v/V)$; the pressure is $\mathcal{P}(\equiv p/\rho V^2)$.

We first look for an outer generalized approximation beginning with the terms

$$\mathcal{U} = u_1(x,y,\varepsilon) + \cdots; \quad \mathcal{V} = v_1(x,y,\varepsilon) + \cdots; \quad \mathcal{P} = p_1(x,y,\varepsilon) + \cdots \tag{3}$$

Neglecting terms of order $\mathrm{O}(\varepsilon^2)$, Eqs. (1a-1c) reduce to the Euler equations. Uniform flow at infinity provides the usual boundary conditions for the Euler equations. At the wall, the no-slip conditions cannot be applied to the Euler equations but the wall condition is not known and will be given later. Away from the wall, the outer flow is certainly well described by the Euler equations but not near the wall. According to the SCEM, the outer approximation is complemented as shown in Fig. (1)

$$\mathcal{U} = u_1(x,y,\varepsilon) + U_1(x,Y,\varepsilon) + \cdots \tag{4a}$$

$$\mathcal{V} = v_1(x,y,\varepsilon) + \varepsilon V_1(x,Y,\varepsilon) + \cdots \tag{4b}$$

$$\mathcal{P} = p_1(x,y,\varepsilon) + \varepsilon^2 P_1(x,Y,\varepsilon) + \cdots \tag{4c}$$

where Y is the boundary layer variable $Y = \dfrac{y}{\varepsilon}$. The $\mathcal{V}$-expansion comes from the continuity equation which must be non-trivial and the $\mathcal{P}$-expansion comes from the analysis of the y-momentum equation.

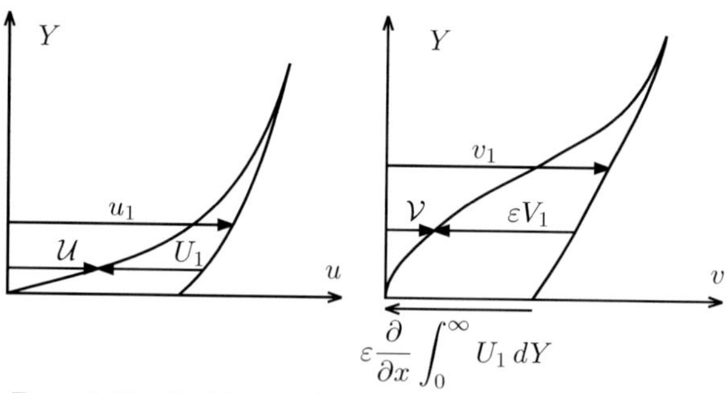

Figure 1: Sketch of the velocity components in the boundary layer.

The Navier-Stokes equations are rewritten with expansions (4a–4c). A first-order IBL model is obtained by neglecting terms of order $\mathrm{O}(\varepsilon)$ in the x-momentum equation and a second-order IBL model is obtained by neglecting terms of order $\mathrm{O}(\varepsilon^2)$.

By using the definitions

$$u = u_1 + U_1 \quad ; \quad v = v_1 + \varepsilon V_1 \tag{5}$$

the second order model leads to the following generalized boundary layer equations

$$\left.\begin{aligned} \frac{\partial u}{\partial x} + \frac{\partial v}{\partial y} &= 0 \\ u\frac{\partial u}{\partial x} + v\frac{\partial u}{\partial y} &= u_1\frac{\partial u_1}{\partial x} + v_1\frac{\partial u_1}{\partial y} + \frac{1}{\mathcal{R}}\frac{\partial^2 (u - u_1)}{\partial y^2} \end{aligned}\right\} \tag{6}$$

which must be solved together with the Euler equations for u_1, v_1 and p_1. The solution for u and v applies over the whole domain, thereby providing *a uniformly valid approximation*. Indeed, Eqs. (6) are valid in the whole field and not only in the boundary layer; the solution of these equations outside the boundary layer gives $u \to u_1$ and $v \to v_1$, which implies that we recover the solution of the Euler equations.

The boundary conditions are

$$\begin{array}{lllll} y = 0 & : & u = 0 & ; & v = 0 \\ y \to \infty & : & u - u_1 \to 0 & ; & v - v_1 \to 0 \end{array} \tag{7}$$

Boundary conditions at infinity are also prescribed for the Euler equations.

The condition $v - v_1 \to 0$ when $y \to \infty$ implies that the system of Eqs. (6) and the Euler equations must be solved together. It is not possible to solve the Euler equations independently from the boundary layer equations since *the two sets of equations interact*. The IBL theory has been proposed earlier heuristically or on the basis of the triple deck theory [1,3,6,17,18] and is fully justified here thanks to the use of *generalized expansions*.

3.2 Reduced Model for an Outer Irrotational Flow

When the outer flow is irrotational and if the validity of Eqs. (6) is restricted to the boundary layer only, it is shown that Eqs. (6) can be simplified into the standard boundary layer equations

$$\left.\begin{aligned} \frac{\partial u}{\partial x} + \frac{\partial v}{\partial y} &= 0 \\ u\frac{\partial u}{\partial x} + v\frac{\partial u}{\partial y} &= u_1(x,0)\frac{du_1(x,0)}{dx} + \frac{1}{\mathcal{R}}\frac{\partial^2 u}{\partial y^2} \end{aligned}\right\} \tag{8}$$

and the boundary conditions are

$$u(x,0,\varepsilon) = 0 \quad ; \quad v(x,0,\varepsilon) = 0 \tag{9}$$

$$\lim_{y\to\infty} u = u_1(x,0,\varepsilon) \quad ; \quad \lim_{y\to\infty}\left[v + y\frac{du_1(x,0,\varepsilon)}{dx}\right] = v_1(x,0,\varepsilon) \tag{10}$$

The last equation can be interpreted in terms of displacement thickness and may be written as

$$v_1(x,0,\varepsilon) = \frac{d}{dx}\left\{\int_0^\infty [u_1(x,0,\varepsilon) - u]\, dy\right\} \tag{11}$$

This reduced model is the usual model used in IBL calculations. It must be noted that the boundary layer and inviscid flow equations are strongly coupled due to condition (10). There is no hierarchy between the boundary layer and inviscid flow equations; *the two sets of equations interact.*

It is also interesting to note that the first order triple deck theory can be deduced from the IBL theory [4]. This completes the link with the method proposed by Veldman [17].

4. TURBULENT FLOW

4.1 Equations and Turbulent Scales

For a two-dimensional incompressible steady flow, the Reynolds averaged Navier-Stokes equations in dimensionless form can be written as

$$\frac{\partial \mathcal{U}}{\partial x} + \frac{\partial \mathcal{V}}{\partial y} = 0 \tag{12a}$$

$$\mathcal{U}\frac{\partial \mathcal{U}}{\partial x} + \mathcal{V}\frac{\partial \mathcal{U}}{\partial y} = -\frac{\partial \mathcal{P}}{\partial x} + \frac{\partial}{\partial x}\left(\mathcal{T}_{xx} + \frac{1}{\mathcal{R}}\frac{\partial \mathcal{U}}{\partial x}\right) + \frac{\partial}{\partial y}\left(\mathcal{T}_{xy} + \frac{1}{\mathcal{R}}\frac{\partial \mathcal{U}}{\partial y}\right) \tag{12b}$$

$$\mathcal{U}\frac{\partial \mathcal{V}}{\partial x} + \mathcal{V}\frac{\partial \mathcal{V}}{\partial y} = -\frac{\partial \mathcal{P}}{\partial y} + \frac{\partial}{\partial x}\left(\mathcal{T}_{xy} + \frac{1}{\mathcal{R}}\frac{\partial \mathcal{V}}{\partial x}\right) + \frac{\partial}{\partial y}\left(\mathcal{T}_{yy} + \frac{1}{\mathcal{R}}\frac{\partial \mathcal{V}}{\partial y}\right) \tag{12c}$$

where the turbulent stresses $\mathcal{T}_{ij}$ are defined from the correlations between velocity fluctuations :

$$\mathcal{T}_{ij} = - < \mathcal{U}_i'\mathcal{U}_j' >$$

Usually, the boundary layer is described by two layers: an outer layer the thickness of which is δ and an inner layer the thickness of which is of the order of ν/u_τ where u_τ is the friction velocity. In the outer and inner layers the turbulent velocity scale $\boldsymbol{u}$ is of the order of u_τ. In the outer layer, the turbulent length scale $\boldsymbol{\ell}$ is of the order of δ and in the inner layer, the turbulent length scale is $\nu/\boldsymbol{u}$.

In the outer layer, it is assumed that the turbulent time scale is of the order of the time scale of the mean motion, i.e.

$$\frac{\boldsymbol{\ell}}{\boldsymbol{u}} = \frac{L}{V} \tag{13}$$

The asymptotic analysis introduces two small parameters ε and $\hat{\varepsilon}$ which represent the order of the thicknesses of the outer and inner layers

$$\varepsilon = \frac{\boldsymbol{\ell}}{L} \quad ; \quad \hat{\varepsilon} = \frac{\nu}{\boldsymbol{u}L} \tag{14}$$

Taking into account the relation (13), we have

$$\varepsilon\hat{\varepsilon}\mathcal{R} = 1 \tag{15}$$

Using the strict order notation $\mathrm{O_S}$, it can be shown that

$$\varepsilon = \mathrm{O_S}\left(\frac{1}{\ln \mathcal{R}}\right) \tag{16}$$

and, using the symbol $\succ$ which means "asymptotically larger than", it is deduced that for all $n \geq 0$

$$\varepsilon^n \succ \hat{\varepsilon} \succ \frac{1}{\mathcal{R}} \tag{17}$$

The variables η and $\hat{y}$ adapted to the study of the outer and inner layers are

$$\eta = \frac{y}{\varepsilon} \quad ; \quad \hat{y} = \frac{y}{\hat{\varepsilon}} \tag{18}$$

4.2 Second order IBL Model

According to the SCEM, we look for a uniformly valid approximation in the form

$$\mathcal{U} = u_1(x,y,\varepsilon) + \varepsilon U_1(x,\eta,\varepsilon) + \varepsilon \widehat{U}_1(x,\hat{y},\varepsilon) + \cdots \tag{19a}$$
$$\mathcal{V} = v_1(x,y,\varepsilon) + \varepsilon^2 V_1(x,\eta,\varepsilon) + \varepsilon\hat{\varepsilon} \widehat{V}_1(x,\hat{y},\varepsilon) + \cdots \tag{19b}$$
$$\mathcal{P} = p_1(x,y,\varepsilon) + \varepsilon^2 P_1(x,\eta,\varepsilon) + \varepsilon^2 \widehat{P}_1(x,\hat{y},\varepsilon) + \cdots \tag{19c}$$
$$\mathcal{T}_{ij} = \varepsilon^2 \tau_{ij,1}(x,\eta,\varepsilon) + \varepsilon^2 \hat{\tau}_{ij,1}(x,\hat{y},\varepsilon) + \cdots \tag{19d}$$

The flow defined by u_1, v_1 and p_1 is governed by the Euler equations and the second order generalized boundary layer equations are

$$\frac{\partial U_1}{\partial x} + \frac{\partial V_1}{\partial \eta} = 0 \tag{20a}$$

$$U_1\frac{\partial u_1}{\partial x} + u_1\frac{\partial U_1}{\partial x} + \varepsilon U_1\frac{\partial U_1}{\partial x} + \varepsilon V_1\frac{\partial u_1}{\partial y} + \frac{v_1}{\varepsilon}\frac{\partial U_1}{\partial \eta} + \varepsilon V_1\frac{\partial U_1}{\partial \eta} = \frac{\partial \tau_{xy,1}}{\partial \eta} + \varepsilon\left(\frac{\partial \tau_{xx,1}}{\partial x} - \frac{\partial \tau_{yy,1}}{\partial x}\right) \tag{20b}$$

$$\frac{\partial \widehat{U}_1}{\partial x} + \frac{\partial \widehat{V}_1}{\partial \hat{y}} = 0 \tag{20c}$$

$$\frac{\varepsilon^2}{\hat{\varepsilon}}\frac{\partial \hat{\tau}_{xy,1}}{\partial \hat{y}} + \frac{1}{\varepsilon\mathcal{R}}\frac{\partial^2 U_1}{\partial \eta^2} + \frac{\varepsilon}{\hat{\varepsilon}^2\mathcal{R}}\frac{\partial^2 \widehat{U}_1}{\partial \hat{y}^2} = 0 \tag{20d}$$

The boundary conditions are

$$\eta \to \infty \quad : \quad U_1 \to 0 \quad ; \quad V_1 \to 0 \tag{21a}$$

$$\hat{y} \to \infty \quad : \quad \widehat{U}_1 \to 0 \quad ; \quad \widehat{V}_1 \to 0 \tag{21b}$$

$$\hat{y} = 0 \quad : \quad u_1 + \varepsilon U_1 + \varepsilon \widehat{U}_1 = 0 \tag{21c}$$

$$\hat{y} = 0 \quad : \quad v_1 + \varepsilon^2 V_1 + \hat{\varepsilon}\varepsilon \widehat{V}_1 = 0 \tag{21d}$$

At infinity, conditions of uniform flow are usually applied to u_1 and v_1.

4.3 Global Model for the Boundary Layer

Defining

$$\begin{aligned} u &= u_1 + \varepsilon U_1 + \varepsilon \widehat{U}_1 \\ v &= v_1 + \varepsilon^2 V_1 + \varepsilon \hat{\varepsilon} \widehat{V}_1 \\ -< u_i' u_j' > &= \varepsilon^2 \tau_{ij,1} + \varepsilon^2 \hat{\tau}_{ij,1} \end{aligned}$$

it is possible to write a global model which *contains* Eqs. (20a-20d)

$$\frac{\partial u}{\partial x} + \frac{\partial v}{\partial y} = 0 \tag{22a}$$

$$u\frac{\partial u}{\partial x} + v\frac{\partial u}{\partial y} = u_1\frac{\partial u_1}{\partial x} + v_1\frac{\partial u_1}{\partial y} + \frac{\partial}{\partial y}\left(- < u'v' >\right) + \frac{1}{\mathcal{R}}\frac{\partial^2 (u - u_1)}{\partial y^2} + \frac{\partial}{\partial x}(< v'^2 > - < u'^2 >) \tag{22b}$$

The above equations must be completed by the Euler equations for u_1 and v_1. The boundary conditions are

$$y \to \infty \quad : \quad u - u_1 \to 0 \quad ; \quad v - v_1 \to 0 \tag{23a}$$

$$\text{at the wall} \quad : \quad u = 0 \quad ; \quad v = 0 \tag{23b}$$

The global model reduces to the standard turbulent boundary layer equations for an irrotational inviscid flow if the term with $(< v'^2 > - < u'^2 >)$ is neglected; the boundary layer equations are similar to Eqs. (8) except that the viscous stress is replaced by the sum of the viscous and turbulent stresses. However, it is stressed that the strong coupling with the inviscid flow is maintained due to the condition (23a) on the velocity normal to the wall.

4.4 Uniformly Valid Approximation of the Velocity Profile in the Boundary Layer

For an irrotational inviscid flow, Eq. (20d) can be written as

$$\frac{\tau}{\tau_w} = \frac{\tau_{\text{outer}}}{\tau_w} \tag{24}$$

In this equation, τ represents the *total stress* $\tau = -<u'v'> + \frac{1}{\mathcal{R}}\frac{\partial u}{\partial y}$ in the whole boundary layer whereas τ_{outer} represents the *turbulent stress* in the outer part of the boundary layer and τ_w is the wall shear stress. It must be noted that Eq. (24) is obtained with generalized expansions. With *regular expansions*, Eq. (24) would reduce to the inner layer equation $\tau/\tau_w = 1$, the solution of which is the standard law of the wall valid only in the inner layer.

Eq. (24) has been solved by using a mixing length model and $\tau_{\text{outer}}/\tau_w$ has been obtained from similarity solutions valid in the outer part of the boundary layer [10]. In this way, a *uniformly valid approximation* of the velocity profile in the whole boundary layer is obtained. Fig. (2) shows the results for a flat plate boundary layer. It is observed in particular that the logarithmic evolution of the velocity disappears at the lower Reynolds numbers.

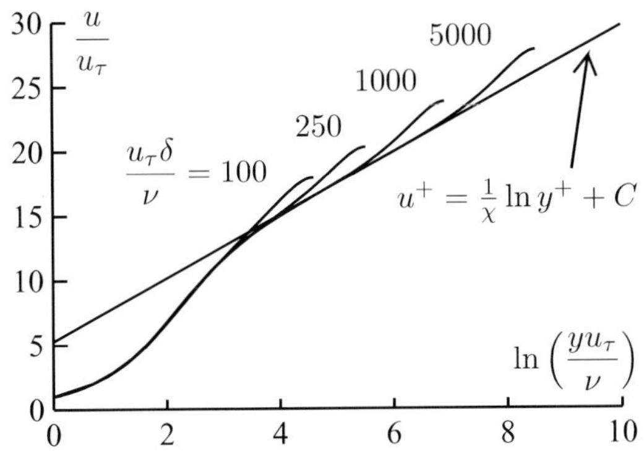

Figure 2: Uniformly valid approximation of velocity profiles in a flat plate turbulent boundary layer at different Reynolds numbers.

5. CONCLUSION

The interactive boundary layer theory (IBL) is fully justified by applying to the analysis of high Reynolds number flows the successive complementary expansions method (SCEM) with generalized expansions.

The key is the condition on the velocity normal to the wall between the external outer flow and the boundary layer. In the triple deck theory, thanks to an appropriate choice of the scales, the matching on the velocity normal to the wall between the decks produces an equivalent characteristic. In fact, it is shown that the first order triple deck theory can be deduced from the IBL model. The Prandtl boundary layer model and the second order Van Dyke model [16] can also be deduced from the second order IBL model.

ACKNOWLEDGEMENTS

The authors want to thank T. Cebeci who read the paper very carefully and made valuable comments.

REFERENCES

[1] J.E. Carter. A new boundary layer inviscid iteration technique for separated flow. In *AIAA Paper 79-1450*. 4th Computational fluid dynamics conf., Williamsburg, 1979.

[2] D. Catherall and W. Mangler. The integration of a two-dimensional laminar boundary-layer past the point of vanishing skin friction. *J. Fluid. Mech.*, 26(1):163–182, 1966.

[3] T. Cebeci. *An Engineering Approach to the Calculation of Aerodynamic Flows*. Horizons Publishing Inc, Long Beach, Ca - Springer-Verlag, Berlin, 1999.

[4] J. Cousteix and J. Mauss. Approximations of the Navier-Stokes equations for high Reynolds number flows past a solid wall. *Jour. Comp. and Appl. Math.*, 166(1):101–122, 2004.

[5] S. Goldstein. On laminar boundary-layer flow near a position of separation. *Quarterly J. Mech. and Appl. Math.*, 1:43–69, 1948.

[6] J.C. Le Balleur. Couplage visqueux–non visqueux : analyse du problème incluant décollements et ondes de choc. *La Rech. Aérosp.*, 6:349–358, 1977.

[7] M.J. Lighthill. On boundary–layer and upstream influence: II. Supersonic flows without separation. *Proc. R. Soc.*, Ser. A 217:478–507, 1953.

[8] J. Mauss and J. Cousteix. Uniformly valid approximation for singular perturbation problems and matching principle. *C. R. Mécanique*, 330, issue 10:697–702, 2002.

[9] A.F. Messiter. Boundary–layer flow near the trailing edge of a flat plate. *SIAM J. Appl. Math.*, 18:241–257, 1970.

[10] R. Michel, C. Quémard, and R. Durant. Application d'un schéma de longueur de mélange à l'étude des couches limites turbulentes d'équilibre. N.T. 154, ONERA, 1969.

[11] V.YA. Neyland. Towards a theory of separation of the laminar boundary–layer in supersonic stream. *Izv. Akad. Nauk. SSSR, Mekh. Zhid. Gaza.*, 4, 1969.

[12] L. Prandtl. Űber Flűßigkeitsbewegung bei sehr kleiner Reibung. Proceedings 3rd Intern. Math. Congr., Heidelberg, pages 484–491, 1904.

[13] K. Stewartson. Multistructured boundary–layers of flat plates and related bodies. *Adv. Appl. Mech.*, 14:145–239, 1974.

[14] K. Stewartson and P.G. Williams. Self induced separation. *Proc. R. Soc.*, A 312:181–206, 1969.

[15] V.V. Sychev, A.I. Ruban, Vic.V. Sychev, and G.L. Korolev. *Asymptotic theory of separated flows*. Cambridge University Press, Cambridge, U.K., 1998.

[16] M. Van Dyke. Higher approximations in boundary-layer theory. Part 2. Application to leading edges. *J. of Fluid Mech.*, 14:481–495, 1962.

[17] A.E.P. Veldman. New, quasi–simultaneous method to calculate interacting boundary layers. *AIAA Journal*, 19(1):79–85, January 1981.

[18] A.E.P. Veldman. Viscoous-Inviscid Interaction: Prandtl's Boundary Layer challenged by Goldstein's Singularity. In J. Cousteix and J. Mauss, editors, *Proc. BAIL2004 Conf. on Boundary and Interior Layers*, 2004.

SYMMETRY METHODS IN TURBULENT BOUNDARY LAYER THEORY

New wake region scaling laws and boundary layer growth

M. Oberlack, G. Khujadze
Fluid Mechanics Group, Technische Universitat Darmstadt,
Petersenstr. 13, 64287 Darmstadt, Germany
oberlack@hyhy.tu-darmstadt.de, khujadze@hyhy.tu-darmstadt.de

Abstract The Lie group or symmetry approach developed by Oberlack (see e.g. Oberlack 2001 and references therein) is used to derive *new scaling laws* for various quantities of a zero pressure gradient (ZPG) turbulent boundary layer flow. In an extension of the earlier work a third scaling group was found in the two-point correlation (TPC) equations for the one-dimensional turbulent boundary layer. This is in contrast to the Navier-Stokes and Euler equations which respectively admits one and two scaling groups. The present focus is on the exponential law in the outer region of turbulent boundary layer and corresponding new scaling laws for one- and two-point correlation functions. Theoretical results are compared to direct numerical simulation (DNS) data of a flat plate turbulent boundary layer at ZPG and at two different Reynolds numbers $Re_\theta = 750, 2240$ with up to 140 million grid points. DNS data show good agreement with the theoretical results though due to the moderate Reynolds number for a limited range of applicability. Finally it is shown that the boundary layer growth is linear.

Keywords: Lie group method, turbulent scaling law, wake law

1. Introduction

The classical logarithmic law of the wall

$$\bar{u}_1^+ = \frac{1}{\kappa}\ln(x_2^+) + C. \tag{1}$$

is still considered as one of the corner stones of turbulence theory.

In recent years there has been a variety of publications describing alternative functional forms of the mean velocity distribution in this region (Barenblatt, *et al.* 2000, George and Castillo 1997, Zagarola *et al.* 1997) some of which were rather controversial. Nevertheless, high quality data such as by Österlund *et al.* (2000a), (2000b) show that the classical theory gives the most accurate

G.E.A. Meier and K.R. Sreenivasan (eds.), IUTAM Symposium on One Hundred Years of Boundary Layer Research, 39-48,

description of the data. The logarithmic scaling obtained by Oberlack (2001) using first principles by employing Lie group methods only once again confirmed the validity of the law though in slightly extended form

$$\bar{u}_1^+ = \frac{1}{\kappa} \ln({x_2}^+ + A^+) + C^+. \tag{2}$$

This scaling law was nicely confirmed by Lindgren *et al.* (2004) using the experimental data of the KTH data-base for turbulent boundary layers for a wide range of Reynolds numbers. They found that with the extra constant A^+, numerically fixed to $A^+ \approx 5$, the modified law describes the experimental data down to $x_2^+ \approx 100$ instead of $x_2^+ \approx 200$ for the classical logarithmic law.

A second law important for ZPG boundary layer flow was derived in Oberlack (2001), because it describes the mean velocity distribution in the wake region (outer region) of the turbulent boundary layer flow. For the outer region of the turbulent boundary layer experimental results have shown that the consideration of the velocity difference $(U_\infty - \bar{u}_1)$, gives a scaling law for its distribution, if this difference is rescaled by u_τ and the distance is normalized by the boundary layer thickness Δ. Thus, in the outer part of the boundary layer flow the mean velocity is represented by the equation

$$\frac{\bar{u}_\infty - \bar{u}_1}{u_\tau} = f\left(\frac{x_2}{\Delta}\right). \tag{3}$$

In Oberlack (2001) it was shown that the exponential law

$$\frac{\bar{u}_\infty - \bar{u}_1}{u_\tau} = \alpha \exp\left(-\beta \frac{x_2}{\Delta}\right). \tag{4}$$

is in fact an explicit form of the classical velocity defect law (3).

Subsequently we derive new and validate the scaling laws discussed above by employing data of a direct numerical simulation (DNS) of the Navier-Stokes equations (for details see Khujadze and Oberlack 2004).

At this point the presented ZPG turbulent boundary layer DNS ($Re_\theta = 2240$) almost doubles the Reynolds number of the classical benchmark of Spalart (1988) for the same flow.

2. Lie group analysis and new scaling laws

The approach developed in Oberlack (2001) based on the fluctuation equations have in Oberlack and Busse (2002) been applied to the TPC equations to find their symmetry groups. In its most general form the TPC equations read

$$
\begin{aligned}
&\left.\frac{\bar{D}R_{ij}}{\bar{D}t} + R_{kj}\frac{\partial \bar{u}_i(\boldsymbol{x},t)}{\partial x_k} + R_{ik}\frac{\partial \bar{u}_j(\boldsymbol{x},t)}{\partial x_k}\right|_{\boldsymbol{x}+\boldsymbol{r}} \\
&+ \left[\bar{u}_k(\boldsymbol{x}+\boldsymbol{r},t) - \bar{u}_k(\mathbf{x},t)\right]\frac{\partial R_{ij}}{\partial r_k} \\
&+ \frac{\partial \overline{p'u'}_j}{\partial x_i} - \frac{\partial \overline{p'u'}_j}{\partial r_i} + \frac{\partial \overline{u'_i p'}}{\partial r_j} + \nu\left[\frac{\partial^2 R_{ij}}{\partial x_k \partial x_k} - 2\frac{\partial^2 R_{ij}}{\partial x_k \partial r_k} + \frac{\partial^2 R_{ij}}{\partial r_k \partial r_k}\right] \\
&+ \frac{\partial R_{(ik)j}}{\partial x_k} - \frac{\partial}{\partial r_k}\left[R_{(ik)j} - R_{i(jk)}\right] = 0,
\end{aligned}
\tag{5}
$$

where correlation vectors and tensors are defined in the usual way (see e.g. Oberlack and Busse 2002) and $\frac{\bar{D}}{\bar{D}t} = \left(\frac{\partial}{\partial t} + \bar{u}_k \frac{\partial}{\partial x_k}\right)$ is the mean substantial derivative. $\boldsymbol{x}$ and $\boldsymbol{r} = \boldsymbol{x}' - \boldsymbol{x}$ are coordinates in the physical and the correlation spaces respectively. The present analysis is based on the TPC equations with the limited mean velocity profile $\bar{u}_1 \equiv \bar{u}_1(x_2)$ i.e. the parallel flow assumption, where x_1, x_2 and x_3 are respectively the streamwise, the wall normal and the spanwise direction. The TPC equations will be considered in the outer part of boundary layer flow i.e. sufficiently apart from the viscous sublayer, and hence viscosity is negligible.

In order to distinguish between viscosity dominated small-scale quantities and inertia dominated large-scale quantities a singular asymptotic expansion was introduced in Oberlack and Peters (1993) for isotropic turbulence and extended to inhomogeneous flows in Oberlack and Busse (2002). Therein the two sets of equations for the large and small scales are derived. Of the two given scales, for the former (outer layer in correlation space) we have TPC equations with $\nu = 0$. The outer part of the asymptotic expansion in r-space is obtained by taking the limit $\nu \rightarrow 0$ in equation (5).

It is apparent that this reduced equation is not valid in the limit $r \rightarrow 0$ since no dissipation is contained which becomes important when r is in the order of the Kolmogorov length-scale η_k.

It is important to note that in the subsequent analysis only the large-scale equations are investigated and hence only large-scale quantities such as the mean velocity or the Reynolds stresses are determined. Small scale quantities such as dissipation can formally obtained once the large scale quantities are derived.

In the present study Lie's procedure is used to find symmetry transformations and self-similar solutions of equation (5). In the first step the infinitesimal generators $\boldsymbol{\xi}(\boldsymbol{x},\boldsymbol{y})$ and $\boldsymbol{\eta}(\boldsymbol{x},\boldsymbol{y})$ must be determined from equation (5) in the large-scale limit $\nu = 0$. As a result, an over-determined set of ~ 700 linear partial differential equations are obtained using the Lie group software package by Carminati and Vu (2000). Imposing the reduction of a parallel flow, the solution of the system for the desired symmetries is given below

$$\xi_{x2} = c_1 x_2 + c_4, \tag{6}$$
$$\xi_{r1} = c_1 r_1, \tag{7}$$
$$\xi_{r2} = c_1 r_2, \tag{8}$$
$$\xi_{r3} = c_1 r_3, \tag{9}$$
$$\eta_{\bar{u}_1} = (c_1 - c_2)\bar{u}_1 + c_5, \tag{10}$$
$$\eta_{R_{ij}} = [2(c_1 - c_2) + c_3]\, R_{ij}, \tag{11}$$
$$\eta_{\overline{u'_i p'}} = [3(c_1 - c_2) + c_3]\overline{u'_i p'} \tag{12}$$
$$\eta_{\overline{p' u'}_i} = [3(c_1 - c_2) + c_3]\overline{p' u'_i}, \tag{13}$$
$$\eta_{Z_{ij}} = [2c_1 - 3c_2 - c_3]\, Z_{ij}. \tag{14}$$

Z_{ij} is the sum of derivatives of the triple correlation functions in (5) and c_i are group parameters. Beside the symmetry groups given in (6)-(14) other symmetries were obtained some of which are unphysical. This has been first reported in Oberlack (2000).

For the present problem we focus on the scaling symmetries, Galilean invariance and the translation groups in (6)-(14).

For a better understanding we may employ Lie's theory, to derive the global transformations

$$\mathbf{G}_{s1} : \tilde{x}_2 = x_2 e^{a_1},\ \tilde{r}_i = r_i e^{a_1},\ \tilde{\bar{u}}_1 = \bar{u}_1 e^{a_1},\ \tilde{R}_{ij} = R_{ij} e^{2a_1}, \cdots \tag{15}$$
$$\mathbf{G}_{s2} : \tilde{x}_2 = x_2,\ \tilde{r}_i = r_i,\ \tilde{\bar{u}}_1 = \bar{u}_1 e^{-a_2},\ \tilde{R}_{ij} = R_{ij} e^{-2a_2}, \cdots \tag{16}$$
$$\mathbf{G}_{s3} : \tilde{x}_2 = x_2,\ \tilde{r}_i = r_i,\ \tilde{\bar{u}}_1 = \bar{u}_1,\ \tilde{R}_{ij} = R_{ij} e^{-a_3}, \cdots \tag{17}$$
$$\mathbf{G}_{transl} : \tilde{x}_2 = x_2 + a_4,\ \tilde{r}_i = r_i,\ \tilde{\bar{u}}_1 = \bar{u}_1,\ \tilde{R}_{ij} = R_{ij}, \cdots \tag{18}$$
$$\mathbf{G}_{galil} : \tilde{x}_2 = x_2,\ \tilde{r}_i = r_i,\ \tilde{\bar{u}}_1 = \bar{u}_1 + a_5,\ \tilde{R}_{ij} = R_{ij}, \cdots . \tag{19}$$

The variables $a_1 - a_5$ are the group parameters of the corresponding parameter $c_1 - c_5$ in the infinitesimals (6)-(14).

The most interesting fact with respect to the latter groups is that *three independent scaling groups* $\mathbf{G}_{s1}$, $\mathbf{G}_{s2}$, $\mathbf{G}_{s3}$ have been computed. Two symmetry groups correspond to the scaling symmetries of the Euler equations. The first one is the scaling in space, the second one scaling in time. The third group ($\mathbf{G}_{s3}$) is a new scaling group that is a characteristic feature of the one-dimensional turbulent boundary layer flow. This is in striking contrast to the Euler and Navier-Stokes equations, which only admit two and one scaling groups, respectively. $\mathbf{G}_{transl}$ and $\mathbf{G}_{galil}$ represent the translation symmetry in space and the Galilean transformation, respectively.

The corresponding characteristic equations for the invariant solutions read

$$\frac{dx_2}{c_1 x_2 + c_4} = \frac{dr_i}{c_1 r_i} = \frac{d\bar{u}_1}{(c_1 - c_2)\bar{u}_1 + c_5} = \frac{dR_{ij}}{[2(c_1 - c_2) + c_3]R_{ij}} = \ldots. \quad (20)$$

In Oberlack (2001) it has been shown that the symmetry breaking of the scaling of space leads to a new exponential scaling law corresponding to the outer part of a boundary layer flow, the wake region.

Thus, imposing the assumption of symmetry breaking of the scaling of space ($c_1 = 0$) here as well and integrating the corresponding equations we obtain an extended set of scaling laws for the mean velocity and the TPC functions as follows

$$\bar{u}_1(x_2) = k_1 + k_2 e^{-k_3 x_2}, \quad \text{with} \quad k_1 \equiv \frac{c_5}{c_2} \quad \text{and} \quad k_3 \equiv \frac{c_2}{c_4} \quad (21)$$

and k_2 is a constant of integration and

$$R_{ij}(x_2, \boldsymbol{r}) = e^{-k_4 x_2} B_{ij}(\mathbf{r}), \ldots \quad (22)$$

where $k_4 \equiv \frac{2c_2 - c_3}{c_4}$ is a constant and B_{ij} is a function of $\boldsymbol{r}$ only.

For positive k_3 the velocity law (21) converges to a constant velocity for $x_2 \to \infty$ ($k_1 = \bar{u}_\infty$). For a plane shear flow this may only be applicable to a boundary layer type of flow. In this flow the symmetry breaking length scale is the boundary layer thickness.

In normalized and non-dimensional variables the exponential scaling law (21) may be re-written in the form given above in equation (4).

One can derive the scaling laws for the Reynolds stresses from the (22) by taking the limit $\boldsymbol{r} = 0$:

$$\frac{\overline{u_i' u_j'}(x_2)}{u_\tau^2} = b_{ij} \exp\left(-a\frac{x_2}{\Delta}\right). \quad (23)$$

b_{ij} and a are universal constants that should be taken from DNS or experimental data. Δ is the Rotta-Clauser length scale $\Delta \equiv \int_0^\infty \frac{\bar{u}_\infty - \bar{u}_1}{u_\tau} = \frac{\bar{u}_\infty}{u_\tau}\delta^*$, while δ^* is the boundary layer displacement thickness.

Recently (4) was validated in detail using high Reynolds number experimental data from the KTH data-base for ZPG turbulent boundary layers by Lindgren *et al.* (2004). The range of Reynolds number based on momentum-loss thickness was from 2500 to 27000. It was found that the exponential scaling law of the mean velocity defect in the outer or wake region fits well with the experimental data over a large part of the boundary layer thickness.

3. Two-dimensional turbulent boundary layer

Putting the strict limit of a fully parallel flow aside and consider the case of a spatially growing and hence two-dimensional boundary layer the number of symmetries we find is, on the one side, an extended one but, on the other side, a reduced set of symmetries. The scaling group $\mathbf{G}_{s3}$ disappears but new symmetries arise such as the usual rotation group as well as the translation and Galilean group in x_2 direction (for an extensive discussion see Oberlack 2000 or Oberlack and Busse 2002). For the present problem of a ZPG boundary layer flow rotation is incompatible and we obtain the characteristic equation

$$\frac{dx_1}{c_1x_1+c_3}=\frac{dx_2}{c_1x_2+c_4}=\frac{dr_i}{c_1r_i}=\frac{d\bar{u}_1}{(c_1-c_2)\bar{u}_1+c_5}=$$
$$\frac{d\bar{u}_2}{(c_1-c_2)\bar{u}_2+c_6}=\frac{d\bar{p}}{[2(c_1-c_2)]\bar{p}}=\frac{dR_{ij}}{[2(c_1-c_2)]R_{ij}}=\ldots. \tag{24}$$

For the present purpose of investigating the boundary layer growth only the first two terms are relevant which, once integrated lead to

$$\delta_{2D}=\frac{x_2+\frac{c_4}{c_1}}{x_1+\frac{c_3}{c_1}} \tag{25}$$

Apparently we find that in a ZPG turbulent boundary layer the boundary layer thickness grows like a linear function which is in contrast to a laminar boundary layer which growths like a square root. Note that any scaling constant, say c_s, in front of δ_{2D} in the form $c_s\frac{x_2+\frac{c_4}{c_1}}{x_1+\frac{c_3}{c_1}}$ would not change the result of δ_{2D} being a similarity variable. In fact we should mention that it means that the growth factor is not determined from the above analysis and appears to be some kind of "eigenvalue" of the flow.

4. DNS and scaling law validation

The DNS code was developed at KTH, Stockholm (for details see Lundbladh *et al.* 1999 and Skote 2001) using a spectral method with Fourier decomposition in the horizontal directions and Chebyshev discretization in the wall normal direction. Time integration is performed using a third order Runge-Kutta scheme for the advective and forcing terms and Crank-Nicolson for the viscous terms.

The first simulations were performed at two different resolutions (number of grid points 7.9 and 31.2 millions respectively). The simulations in this case were run for a total of 10000 time units (δ^*/u_∞) and the sampling for the statistics was performed during the last 5000 time units. The useful region was confined to $150-300\delta^*|_{x=0}$ which corresponds to Re_{δ^*} from 800 to 1100 or Re_θ from 540 to 750. The differences between the two resolutions were very small.

Thus we decided that the coarser grid is sufficient to resolve the flow with this Reynolds number and it has to be used as the starting point for the calculation of the number of grid points for the higher Reynolds number flows according to

$$N \approx N_0 \left(\frac{Re^1_{\delta^*}}{Re^0_{\delta^*}} \right)^{9/4} . \tag{26}$$

Assuming that $N_0 \approx 7.9 * 10^6$ was sufficient to resolve the flow at low Reynolds number ($Re_{\delta^*} \approx 1100$) we employed $Re_{\delta^*} = 3050$ into (26) to obtain $N \approx 138.8$ million grid points for the larger Reynolds number.

Again the useful region was confined to $150 - 300\delta^*|_{x=0}$ (total length of computational box is $450\delta^*|_{x=0}$) which corresponds to Re_{δ^*} from 2230 to 3050 or Re_θ from 1670 to 2240. Resolutions in plus units are $\Delta x^+ \approx 15$, $\Delta z^+ \approx 11$ and $\Delta y^+ \approx 0.13 - 18$.

Scaling law validation of one- and two-point quantities

The mean velocity of the turbulent boundary layer data is plotted in Figure 1 for $Re_\theta = 2240$. As it is observed from the figure, DNS and theoretical results (4) are in good agreement in the region $x_2/\Delta \approx 0.01 - 0.15$. Above this region the velocity defect law in the outer part of the boundary layer decreases more rapidly than what was derived from the theoretical result. There may be two reasons for this. First it might be the result of the low Reynolds number phenomenon in DNS while for the derivation of the theoretical results we assumed the large Reynolds number limit. Second as it is argued in Lindgren *et al.* (2004) that the non-parallel effects become dominant in the outer part of the wake region, which could cause the deviation from the exponen-

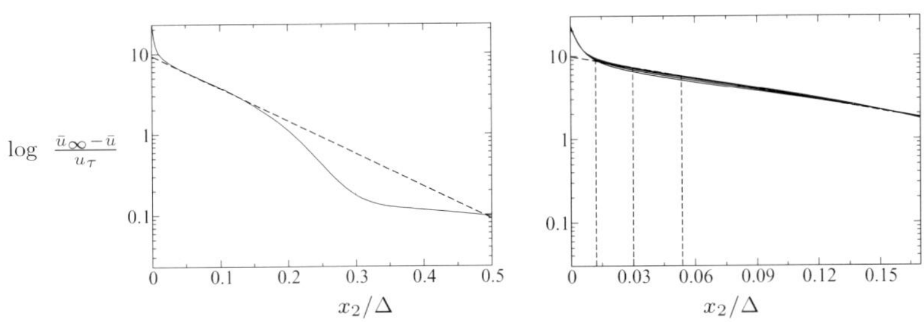

Figure 1. Mean velocity profile in log-linear scaling. Left figure: Dashed line corresponds to theoretical results from the law (4). Solid line represents DNS results at $Re_\theta = 2240$; Right Figure: Close up plot of mean velocity profile at different Reynolds numbers.

tial law, while the theoretical results were derived assuming a fully parallel flow. Because of these reasons, we have a relatively small "coincidence region" of the theoretical and DNS results. The close up of mean velocity profiles is presented in the same figure (right plot) at different Reynolds numbers

$Re_\theta = 1670, 1870, 2060, 2240$. Good collapse of profiles in the exponential region is seen from the plot.

In Figures 2 and 3 Reynolds normal stresses are presented from DNS results. The scaling law (23) (dashed lines on the figure) are compared to the DNS results. Constants in the exponential scaling law for Reynolds stresses (23) are different for the different Reynolds stress tensor components. The last means that the "coincidence region" for each component of Reynolds stress tensor is located in the different area of the outer part of the boundary layer flow which may be low Reynolds number phenomenon. TPC functions are represented

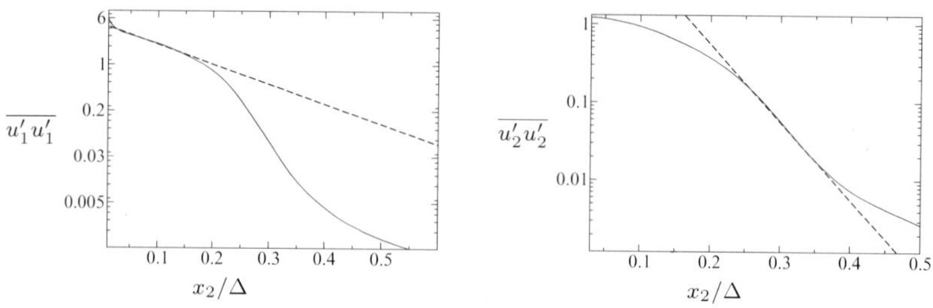

Figure 2. Reynolds normal stresses $\overline{u_1'u_1'}$ and $\overline{u_2'u_2'}$. —— DNS ($Re_\theta = 2240$), - - - theoretical result from equation (23).

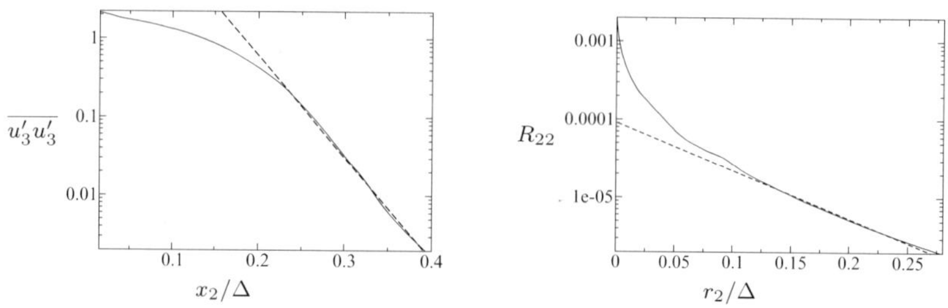

Figure 3. Left figure: $\overline{u_3'u_3'}$, —— DNS ($Re_\theta = 2240$), - - - theoretical result from equation (23). Right figure: R_{22}, - - - theoretical results, equation (22), —— DNS.

in Figures 3 (right plot) and 4. The region where the exponential law is valid, for different component of TPC function is different. For R_{22} the exponential law is valid in the interval $0.1 - 0.25$; For R_{12} and R_{21} these intervals are $0.02 - 0.06$ and $0.001 - 0.02$ respectively.

TPC functions were calculated for both cases of DNS. The results from $Re_\theta = 750$ and $Re_\theta = 2240$ are compared. $R_{12}(x_2, r_2)$, $R_{21}(x_2, r_2)$ and $R_{22}(x_2, r_2)$ are plotted against $r_2 = x_2' - x_2$ normalized by the Reynolds

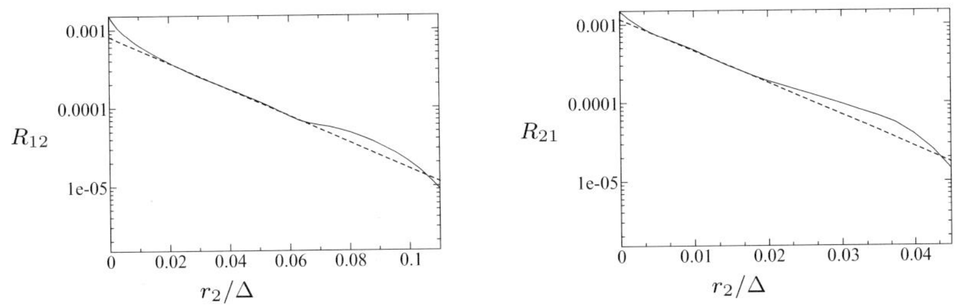

Figure 4. R_{12} and R_{21}, - - - theoretical results, equation (22), —— DNS for $Re_\theta = 2240$.

stresses

$$\mathcal{R}_{[ij]}(x_2, \mathbf{r}) = \frac{R_{ij}(x_2, \mathbf{r})}{u'_i(x_2)u'_j(x_2)}, \tag{27}$$

where [] is the index denoting componentwise ratios. Note that $\mathcal{R}_{[ij]}$ is not a tensor. One important result from the present analysis is equation (22) which, employed in (27), leads to the fact that $\mathcal{R}_{[ij]}$ is independent of the wall normal coordinate x_2 i.e. $\mathcal{R}_{[ij]} = F_{[ij]}(\mathbf{r})$. A validation of the latter may be taken from Figure 5 and 6, where TPC functions for different initial points in wall-normal direction collapse in one, as it is expected from equation (27).

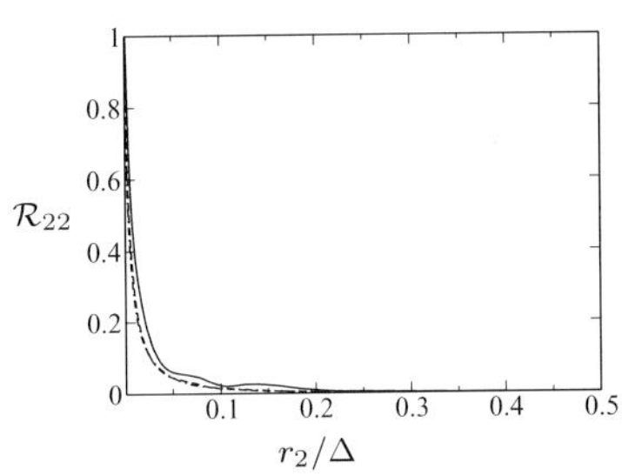

Figure 5. Normalized TPC for $Re_\theta = 750$ at different initial points. —— $x_2/\Delta = 0.013$, - - - $x_2/\Delta = 0.026$, —·— $x_2/\Delta = 0.052$

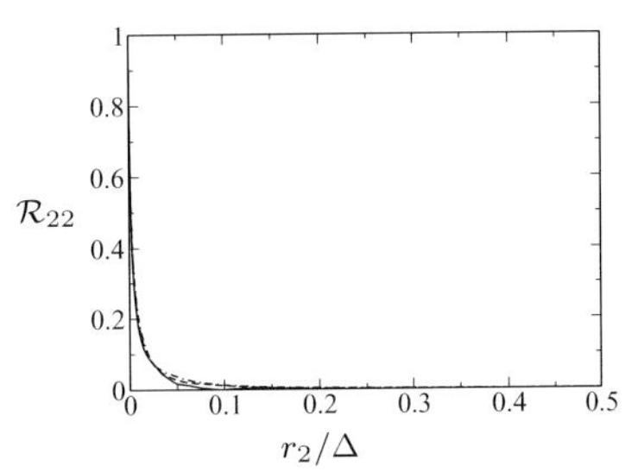

Figure 6. Normalized TPC for $Re_\theta = 2240$ at different initial points. —— $x_2/\Delta = 0.028$, - - - $x_2/\Delta = 0.057$, —·— $x_2/\Delta = 0.085$

For the low Reynolds number case ($Re_\theta = 750$) we computed TPC for three different initial points $x_2/\Delta = 0.013, 0.026, 0.052$ in the region where the exponential law is valid (see Figure 1, where the vertical dashed lines represent the initial points for TPC functions calculation). For the high Reynolds number case computations were done for $x_2/\Delta = 0.028, 0.057, 0.085$ as initial points.

5. Conclusions

The most interesting theoretical result is that new symmetries of the TPC equations are found which are not intrinsic to the Navier-Stokes or Euler equations. These symmetries have in turn been used to derive new scaling laws for the two-point and Reynolds stress quantities. A DNS of turbulent boundary layer flow was performed at $Re_\theta = 750, 2240$ to validate the theoretical results. The data show good collapse for one and two-point statistical quantities. DNS shows the validity of the scaling laws though due to the moderate Reynolds number only for a limited range of applicability. Further it is shown that the boundary layer growth is linear.

References

Barenblatt, G.I., Chorin, A.J. and Prostokishin, V.M. (2000). Self-similar intermediate structures in turbulent boundary layers at large Reynolds numbers *J.Fluid Mech.* **410**, 263–283.

Carminati, J. and Vu, K. (2000). Symbolic Computation and differential equations: Lie symmetries. *J.Sym. Comp.*, **29**, 95–116.

George, W.K. and Castillo, L. (1997). Zero-pressure-gradient turbulent boundary layer *Appl. Mech. Rev.*, **50**, 689–729.

Khujadze, G., Oberlack, M. (2004). DNS and scaling laws from new symmetry groups of ZPG turbulent boundary layer flow. *Theo. Comp. Fluid Dyn.*, (in press).

Lindgren, B., Österlund, J. M., Johansson, A. V. (2004). Evaluation of scaling laws derived from Lie group symmetry methods in zero pressure-gradient turbulent boundary layers. *J.Fluid Mech.* **502**, pp. 127–152.

Lundbladh A., Berlin, S., Skote, M., Hildings, C., Choi, J., Kim, J. and Henningson, D.S. (1999). An efficient spectral method for simulation of incompressible flow over a flat plate. *Tech. Rep.* 1999:11. KTH, Stockholm.

Oberlack, M. (2000). Symmetrie, Invarianz und Selbstähnlichkeit in der Turbulenz. *Habilitation thesis RWTH Aachen*, Shaker, Aachen.

Oberlack, M. (2001). A unified approach for symmetries in plane parallel turbulent shear flows. *J.Fluid Mech.*, **427**, pp. 299–328.

Oberlack, M. and Busse, F.H., (Editors) (2002) Theories of turbulence, *CISM Courses and Lectures*, **442**, Springer Wien, New-York.

Oberlack, M. and Peters, N., (1993) Closure of the two-point correlation equations as a basis of Reynolds stress models, in So, R., Speziale, C., and Launder, B., eds. *Near-Wall Turbulent flows*, 85–94. Elsevier Science Publisher.

Österlund, J., Johansson, A., Nagib, H. and Hites, M. (2000a). A note on the overlap region in turbulent boundary layers, *Phys. Fluids* **12**(1), 1–4.

Österlund, J., Johansson, A., Nagib, H. (2000b). Comment on 'a note on the intermediate region in turbulent boundary layers' [Phy. Fluids 12, 2159]. *Phys. Fluids* **12**, 2360–2363.

Skote, M. (2001). Studies of turbulent boundary layer flow through direct numerical simulation. Doctoral thesis, Royal Institute of Technology, Stockholm, Sweden.

Spalart, P.R. (1988). Direct simulation of a turbulent boundary layer up to $R_\theta = 1410$. *J. Fluid Mech.* **187**, 61–98.

Zagarola, M.V., Perry, A.E. and Smits, A.J. (1997). Log laws or power laws: the scaling in the overlap region *Phys. Fluids* **9**, 2094–2100.

Viscous/Inviscid Interaction Procedures for Compressible Aerodynamic Flow Simulations

Mohamed Hafez and Essam Wahba

University of California, Davis CA 95616, USA

Abstract. Steady transonic flows over a wing are calculated based on a hierarchical formulation where the potential flows are corrected due to the entropy and vorticity effects of shock waves and/or laminar boundary layers. Preliminary results are in agreement with standard Euler and Navier-Stokes calculations. The merits of the present approach are briefly discussed

1 Introduction

For a long time, D'Alembert's paradox remained unsolved until Prandtl introduced his boundary layer theory in 1904. Prandtl introduced also the concept of the displacement thickness to couple the inviscid and viscous flow calculations. The inviscid pressure over the augmented body is used in the boundary layer equations as a forcing function. An iterative procedure based on this strategy can be used for attached low speed flows (see [1]-[5]).

Prandtl's ideas and concepts were the bases for singular perturbation and matched asymptotic analyses. The purpose of the present work is to implement numerically some of the strategies of viscous/inviscid interaction procedures for the simulation of viscous transonic separated flows over wings.

For high Reynolds number external flows over wings, it is not necessary to solve Navier-Stokes equations everywhere in the flow field. The grid has to be stretched and due to truncation errors, artificial entropy and vorticity will be generated leading to global inaccuracy of the calculations. It is proposed, in this work, to solve a potential equation in the far field and limit the viscous flow calculations to a small region around the wing and in the wake. A conservative estimate of the size of this region is assumed to be known a priori. The potential flow formulation must be corrected in the case of strong shocks and a region of inviscid rotational flow is needed for these cases. The outer boundary condition is based on Prandtl's lifting line theory where the circulation at each cross-section, including downwash due to three dimensional effects, is calculated iteratively, thus reducing the domain of calculations considerably.

The present hierarchical formulation can be viewed as a viscous/inviscid interaction procedure, saving memory and time of calculations. The formulation does not suffer however from the convergence difficulties of classical methods of coupling boundary layer and potential flow calculations. It avoids also difficulties associated with reflection of the error from artificial interface boundaries in the standard zonal approach based on heterogeneous domain decomposition.

In the following, the details of the derivation are delineated and preliminary numerical results of some test cases are presented and compared to standard Navier-Stokes solutions.

G.E.A. Meier and K.R. Sreenivasan (eds.), IUTAM Symposium on One Hundred Years of Boundary Layer Research, 49-58,

2 Present Formulation

In [6], the authors used a Helmholtz type decomposition of the velocity vector into the gradient of a potential function plus a rotational component

$$\boldsymbol{q} = \nabla\phi + \boldsymbol{q}^* \tag{1}$$

Alternative decompositions are possible, see for example [8]. Based on (1), the vorticity $\boldsymbol{\omega}$ can be expressed in terms of $\boldsymbol{q}^*$ as follows

$$\boldsymbol{\omega} = \nabla \times \boldsymbol{q} = \nabla \times \boldsymbol{q}^* \tag{2}$$

Conservation of mass leads to

$$\nabla \cdot (\rho \nabla \phi) = -\nabla \cdot (\rho \boldsymbol{q}^*) \tag{3}$$

The right hand side term of (3) is a source term representing the effect of vorticity on the potential flow field. The density and pressure are written in terms of the entropy and total enthalpy

$$\rho = \rho_i e^{-\frac{\Delta S}{R}}, \;\; P = P_i e^{-\frac{\Delta S}{R}} \tag{4}$$

where

$$\rho_i = \left[(\gamma - 1) M_\infty^2 \left(H - \frac{1}{2} q^2 \right) \right]^{\frac{1}{\gamma-1}}, \;\; P_i = \frac{\rho_i^\gamma}{\gamma M_\infty^2} \tag{5}$$

The total enthalpy, H, can be obtained from the energy equation. In the present work, it is assumed that $Pr = 1$ and for high Reynolds number flows, following Buzmann, H is constant everywhere

$$H = \frac{\gamma}{\gamma - 1} \frac{P_i}{\rho_i} + \frac{1}{2} q^2 = H_\infty \tag{6}$$

The above simplification is not valid for the case of heat transfer and the energy equation must be solved instead. For the calculations of entropy and vorticity effects, three domains of the flow field are identified. The most inner one, is the viscous flow layer, followed by an inviscid rotational flow region, while the flow in the outer region is a potential flow.

In the viscous flow region, pressure is updated from the normal momentum equation, where denotes tensor product,

$$\frac{\partial(\delta P)}{\partial n} = -\boldsymbol{n} \cdot \left(\nabla \cdot (\rho \boldsymbol{q} \otimes \boldsymbol{q}) + \nabla P - \frac{1}{Re} \left[\nabla^2 \boldsymbol{q} + \frac{1}{3} \nabla (\nabla \cdot \boldsymbol{q}) \right] \right) \tag{7}$$

The entropy is then obtained from the pressure using (4). For the evaluation of $\boldsymbol{q}^*$, one component of $\boldsymbol{q}^*$ is chosen to vanish. The other two components are chosen to be tangential to the grid lines and are updated based on the momentum equations as follows

$$\rho \boldsymbol{q} \cdot \nabla (\boldsymbol{t_1} \cdot \boldsymbol{\delta q}^*) - \frac{1}{Re} \nabla^2 (\boldsymbol{t_1} \cdot \boldsymbol{\delta q}^*) = -R_{t_1 - Momentum} \tag{8}$$

$$\rho \boldsymbol{q} \cdot \nabla (\boldsymbol{t_2} \cdot \boldsymbol{\delta q}^*) - \frac{1}{Re} \nabla^2 (\boldsymbol{t_2} \cdot \boldsymbol{\delta q}^*) = -R_{t_2 - Momentum} \tag{9}$$

where $\boldsymbol{t_1}$ and $\boldsymbol{t_2}$ are unit vectors tangent to the body surface.

For the inviscid rotational flow region, a correction for the entropy function, $\left(\kappa = e^{-\frac{\Delta S}{R}}\right)$, is obtained from the tangential momentum equation,

$$P_i \boldsymbol{q}.\nabla(\delta\kappa) = -\boldsymbol{q}.\left[\nabla.\left(\rho\boldsymbol{q}\otimes\boldsymbol{q}\right) + \nabla P\right] \tag{10}$$

while $\boldsymbol{q}^*$ is updated from

$$\boldsymbol{q}\times\left[\left(\nabla\times\delta\boldsymbol{q}^*\right)\times\boldsymbol{q}\right] = -\boldsymbol{q}\times\left[\left(\nabla\cdot\left(\rho\boldsymbol{q}\otimes\boldsymbol{q}\right) + \nabla P\right)/\rho\right] \tag{11}$$

Notice that one of the components of $\boldsymbol{q}^*$ can be chosen to be identically zero. In the outer region, $\boldsymbol{q}^* = \boldsymbol{0}$ and $\Delta S = 0$, and the formulation reduces to the standard potential equation.

The present formulation has a built-in preconditioning for low Mach number flows. The incompressible limit ($M_\infty = 0$) is recovered as a special case where the density becomes constant and the entropy calculation is not needed in a similar way to the simulation of two dimensional flows of [7].

2.1 Boundary Conditions

At the solid surface, the no penetration and no slip conditions are imposed, i.e.

$$\boldsymbol{q} = \boldsymbol{0} \quad \Rightarrow \quad \boldsymbol{q}^* = -\nabla\phi \tag{12}$$

To recover potential flows in the absence of $\boldsymbol{q}^*$, we use $\frac{\partial\phi}{\partial n} = 0$. The component of $\boldsymbol{q}^*$ normal to the surface is zero, the boundary conditions for the other two components are

$$\boldsymbol{q}^*\cdot\boldsymbol{t_1} = -\frac{\partial\phi}{\partial t_1}\,, \quad \boldsymbol{q}^*\cdot\boldsymbol{t_2} = -\frac{\partial\phi}{\partial t_2} \tag{13}$$

The upstream conditions are assumed to be uniform, i.e. the flow is isentropic and irrotational ($\Delta S = 0, \boldsymbol{q}^* = \boldsymbol{0}$). The outflow boundary conditions are obtained via extrapolation. Outside the viscous and inviscid rotational flow regions, the flow is assumed to be isentropic and irrotational again.

The outer boundary conditions for the potential depends on the free stream Mach number, M_∞. For supersonic flows, upstream uniform conditions are imposed in an elongated domain such that the shocks intersect the exit boundary. For $M_\infty < 1$, the asymptotic behavior of the potential flow is needed. For two dimensional flows, the potential in the far field represents uniform flow plus an irrotational vortex. On the other hand, for three dimensional flows, following Prandtl lifting line theory, modified with compressibility effect via Prandtl-Glauert transformation, the wing shrinks to a line of bound vortices with a vortex sheet following the streamlines downstream to infinity. It is assumed that the angle of attack is small so the trailing vortices may be taken parallel to the wing planform. The trailing vortex system induces downwash velocities in the vicinity of the wing and hence decrease the angle of attack. Therefore, one can still use the two dimensional far field behavior, at several chords from the surface with a circulation modified by the downwash, at each cross section of the wing. More accurate formula can be derived using the Biot-Savart law for a horse-shoe vortex system. An alternative approach is to use a uniform flow boundary condition located at several spans from the wing. The saving from using an asymptotic far field solution is obvious.

To test the present outer boundary condition, numerical simulations are performed for inviscid subsonic flow ($M_\infty = 0.3$) over a wing of elliptic planform and NACA0012 cross sections, of high aspect ratio ($AR = 9.0$) at small angle of attack 2°. Results of the 3-D simulation are shown in Fig. 1 where the surface pressure distributions are in good agreement with 2-D calculations based on an effective angle of attack given by Prandtl. In Fig. 2, local lift coefficient, C_l, distributions are plotted for 2-D and 3-D calculations and are compared to standard formulas for 2-D lift coefficient of thick airfoils with and without Prandtl's correction. The pressure contours on the upper and lower surfaces of the wing are plotted in Fig. 3 and 4.

3 Numerical Methods

In the present work, a structured C-H grid of ($140 \times 40 \times 40$) points is generated around the wing using algebraic methods, where a C-grid is wrapped around each wing cross-section and a H-grid is used in the spanwise direction which collapses into a single plane after the wing tip. The generated grid is shown in Fig. 5.

Finite volume methods based on Gauss theorem are used to discretize the governing equations written in conservation form. The areas of the faces and the volume of the hexahedron are evaluated from the position vectors of the vertices. The surface integrals, representing the fluxes across the faces of the control volume, are expressed in terms of the direction cosines for the cell-face surface areas and the average of the flux at the corners. To evaluate derivatives, co-volumes are used in standard manner.

Upwinding schemes are used for the convective terms in the momentum equation for both the corrections and the residuals. Artificial time dependent terms are added to the equations and a line relaxation procedure, marching with the main flow direction, is implemented to update the flow field.

For the augmented potential equation, the flux biasing scheme of Hafez, Whitlow and Osher [9] is used in the present work, together with the Zebra-SLOR procedure, see [10]; where all the even planes are solved at the same time, followed by the odd ones. The Zebra scheme avoids the problems of marching along the spanwise direction, which may lead to convergence difficulties.

Multigrid acceleration techniques can be easily implemented at least in each plane as a quasi two dimensional problem following [11].

4 Numerical Results

Three dimensional inviscid and viscous flows over a NACA0012 wing are simulated using the present formulation. The NACA0012 wing is derived from ONERA M6 wing where the cross sections are replaced by NACA0012 airfoils. The results are compared with those of Overflow, NASA standard Navier-Stokes code [12], [13].

First, potential subsonic flows for a lifting wing at $M_\infty = 0.3$ and $\alpha = 2^o$ are calculated. Comparison with Euler codes show good agreement as expected. The surface pressure contours for upper and lower surfaces are plotted in Fig. 6 and 7.

For $M_\infty = 0.84$ and $\alpha = 3^o$, potential flow calculation has a shock wave on the upper surface. The shock location and strength are different if entropy and vorticity effects are included. The surface pressure contours and the surfaces pressure distributions for both calculations are shown in Fig. 8 to 12.

For $M_\infty = 0.9$ and $\alpha = 1^o$, the potential flow solution exhibit strong shocks on both upper and lower surfaces of the wing as shown in Fig. 13 and 14. The Mach contours at the symmetry plane is plotted in Fig. 17. On the other hand, in the viscous flow calculations, at $Re = 1000$, no shocks appear. The surface pressure contours and the Mach contours are plotted in Fig. 15, 16 and 18. Finally, the present calculations are in good agreement with Overflow results based on standard Navier-Stokes equations as shown in Fig. 19.

5 Concluding Remarks

A hierarchical formulation is presented for 3-D high Reynolds number flows over wings. Numerical results for some test cases are compared to standard Navier-Stokes solutions. It is argued that both the efficiency and the accuracy of the calculations are benefited from the velocity decomposition into the gradient of a potential plus a rotational component, where the latter is restricted to small regions. Further studies are required to assess the merits quantitatively.

References

1. L. Prandtl: *Essentials of Fluid Dynamics*, Blackie, London, 1952
2. L. Prandtl and O. G. Tietjens: *Fundamentals of Hydro and Aeromechanics*, Dover publications, 1957
3. L. Prandtl and O. G. Tietjens: *Applied Hydro and Aeromechanics*, Dover publications, 1957
4. H. Schlichting and K. Gersten: *Boundary Layer Theory*, Springer, 1999
5. H. Schlichting and E. Truckenbrodt: *Aerodynamics of the Airplane*, McGraw-Hill, 1979
6. M. Hafez and E. Wahba: *Numerical Simulations of Transonic Aerodynamic Flows based on a Hierarchical Formulation*, AIAA Paper 03-3564, 2003
7. M. Hafez, A. Shatalov and E. Wahba: *Numerical Simulations of Incompressible Aerodyanmic Flows using Viscous/Inviscid Interaction Procedures*, Comp. Methods Appl. Mech. Eng, To appear
8. R. E. Gordnier and S. G. Rubin: *Transonic Flow Solutions using a Composite Velocity Procedure for Potential, Euler and RNS Equations*, Comp. & Fluids, Vol. 17, pp 85-98, 1989
9. M. M. Hafez, W. Whitlow and S. Osher: *Improved finite difference schemes for transonic potential calculations*, AIAA J., Vol. 25, No. 11, pp 1456-1462, 1987
10. M. M. Hafez and D. Lovell: *Improved relaxation schemes for transonic potential calculations*, Int. J. Numer. Methods Fluids, Vol. 8, No. 1, pp 1-16, 1988
11. M. Hafez and E. Wahba: *Multigrid Acceleration of Transonic Aerodynamic Flow Simulations based on a Hierarchical Formulation*, ICCFD3 Proc., 2004, to appear
12. C. Tang and M. M. Hafez: *Numerical simulation of steady compressible flows using a zonal formulation. Part I:Inviscid Flows*, Comp. & Fluids, pp 898-1002, Vol. 30, 2001
13. C. Tang and M. M. Hafez: *Numerical simulation of steady compressible flows using a zonal formulation. Part II:Viscous Flows*, Comp. & Fluids, pp 1003-1016, Vol. 30, 2001

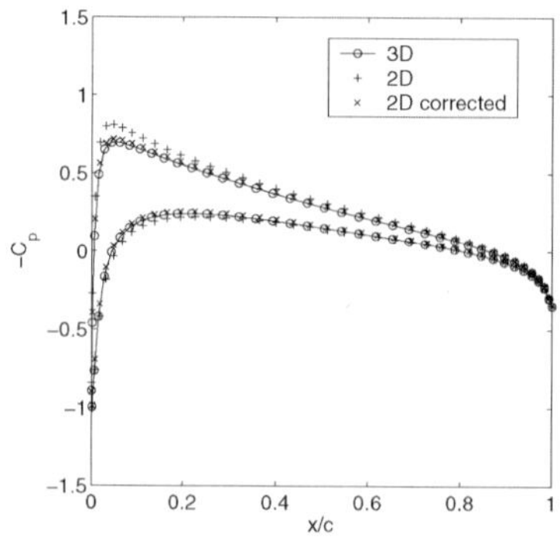

Fig. 1. Surface pressure distribution (Subsonic flow over an elliptic wing)

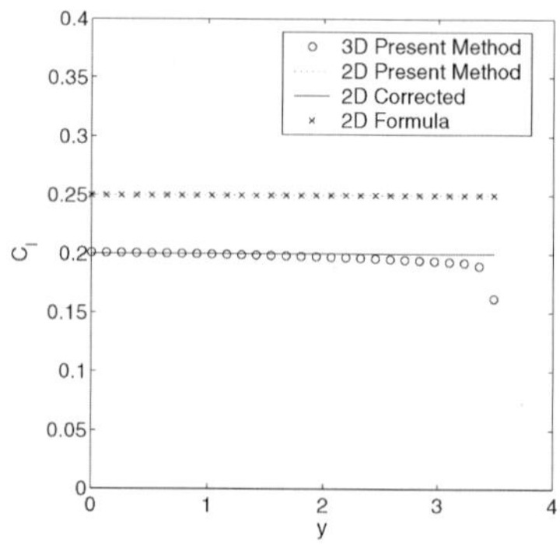

Fig. 2. Lift coefficient distribution (Subsonic flow over an elliptic wing)

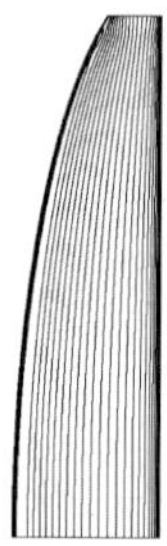

Fig. 3. Pressure contours on upper surface of elliptic wing (AR=9)

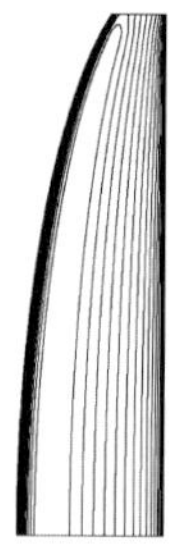

Fig. 4. Pressure contours on lower surface of elliptic wing (AR=9)

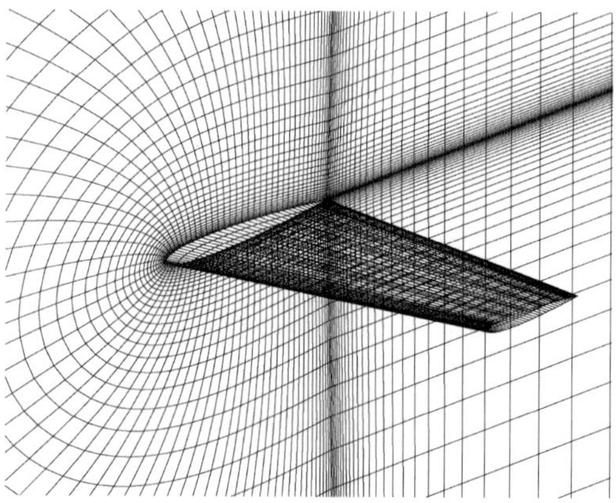

Fig. 5. Structured C-H Grid for NACA0012 Wing (140 × 40 × 40)

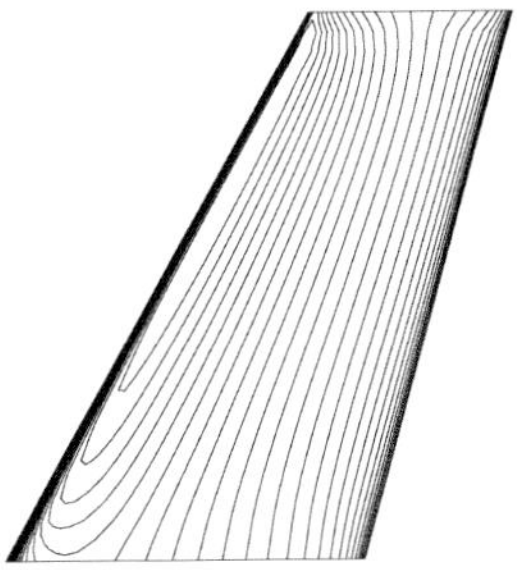

Fig. 6. Pressure contours on upper surface of NACA0012 wing ($M_\infty = 0.3$, $\alpha = 2^o$)

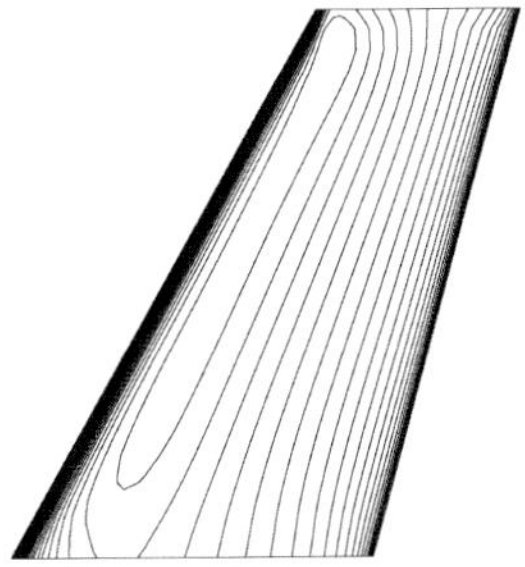

Fig. 7. Pressure contours on lower surface of NACA0012 wing ($M_\infty = 0.3$, $\alpha = 2^o$)

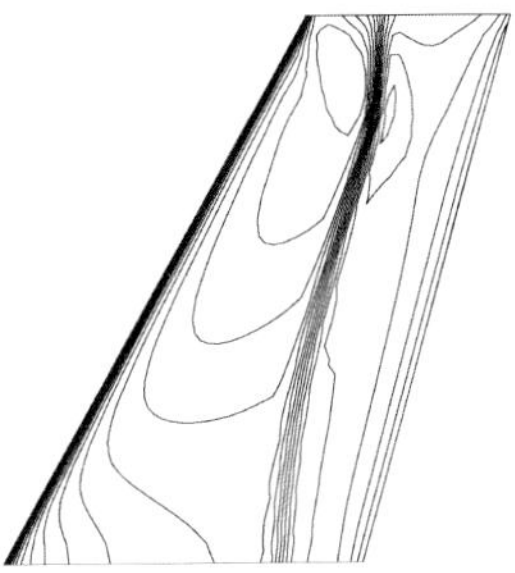

Fig. 8. Pressure contours on upper surface of NACA0012 wing ($M_\infty = 0.84$, $\alpha = 3^o$, Potential Flow)

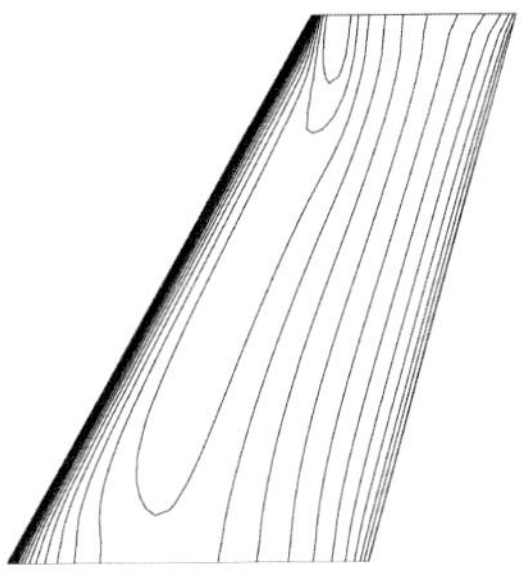

Fig. 9. Pressure contours on lower surface of NACA0012 wing ($M_\infty = 0.84$, $\alpha = 3^o$, Potential Flow)

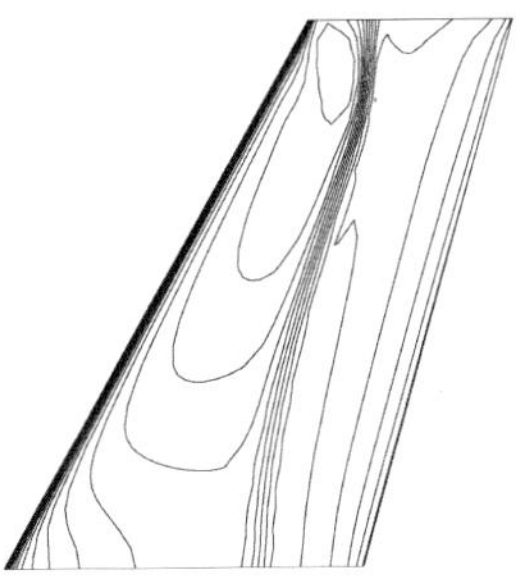

Fig. 10. Pressure contours on upper surface of NACA0012 wing ($M_\infty = 0.84$, $\alpha = 3^o$, Inviscid Rotational Flow)

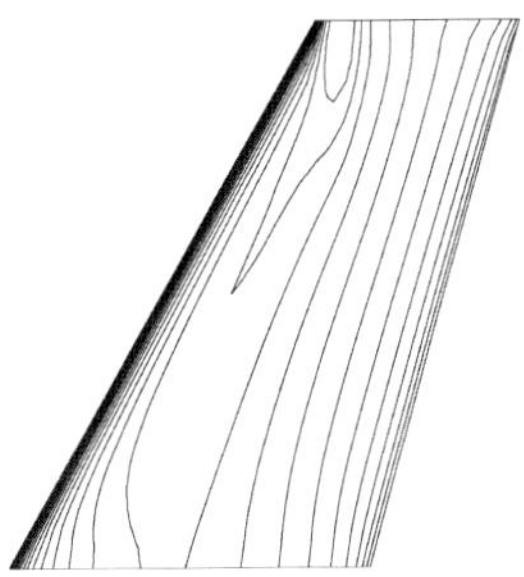

Fig. 11. Pressure contours on lower surface of NACA0012 wing ($M_\infty = 0.84$, $\alpha = 3^o$, Inviscid Rotational Flow)

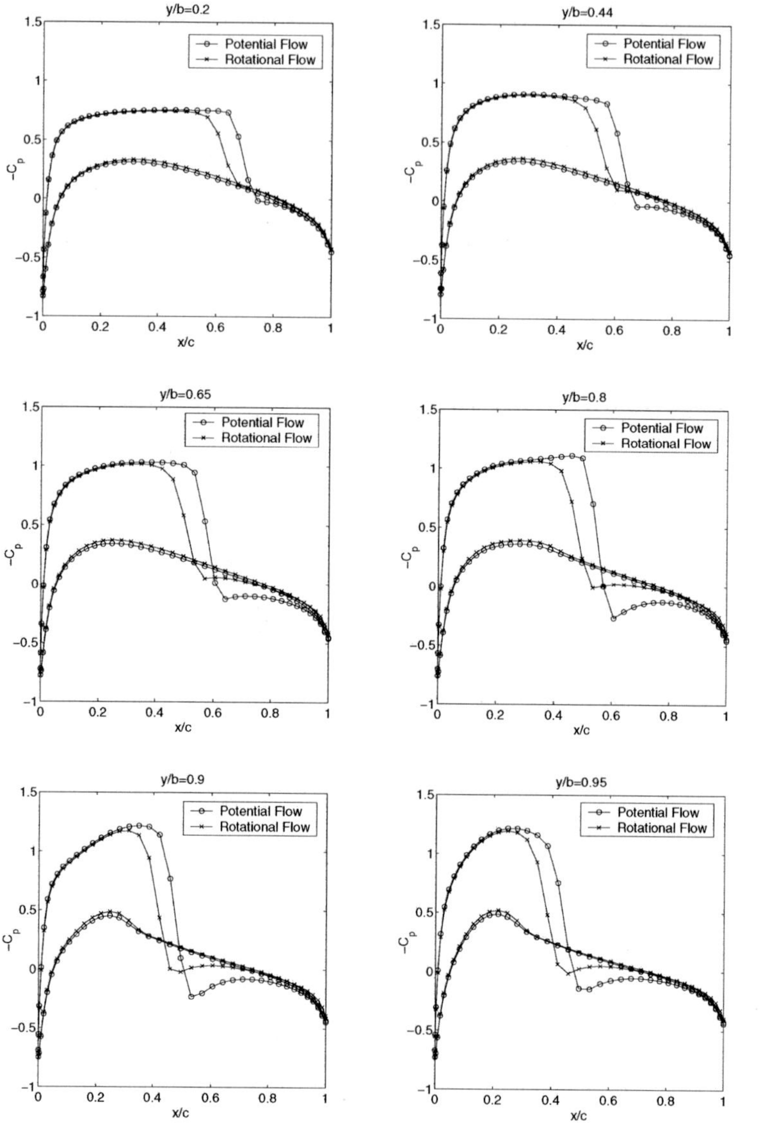

Fig. 12. Surface pressure distributions for inviscid flow over NACA0012 wing ($M_\infty = 0.84$, $\alpha = 3^o$)

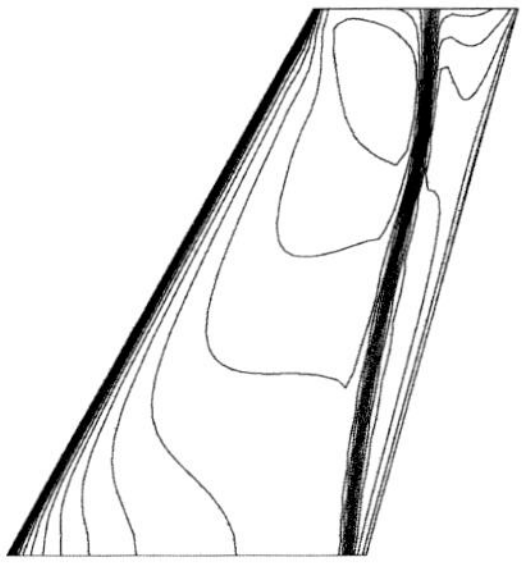

Fig. 13. Pressure contours on upper surface of NACA0012 wing ($M_\infty = 0.9$, $\alpha = 1^o$, Potential Flow)

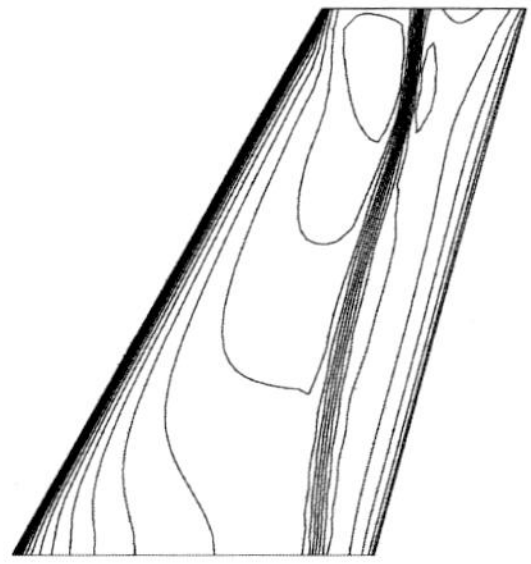

Fig. 14. Pressure contours on lower surface of NACA0012 wing ($M_\infty = 0.9$, $\alpha = 1^o$, Potential Flow)

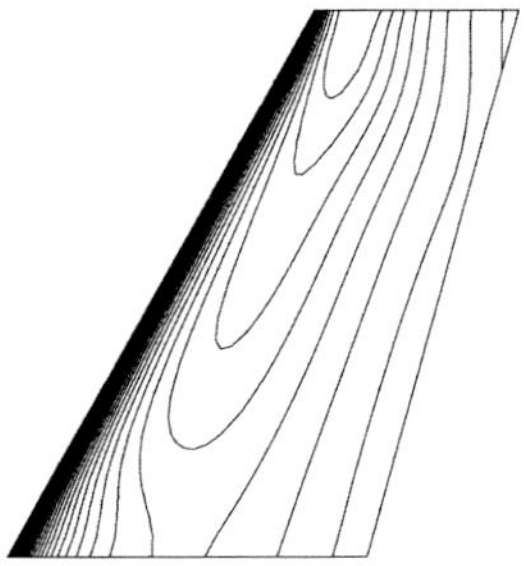

Fig. 15. Pressure contours on upper surface of NACA0012 wing ($M_\infty = 0.9$, Re=1000, $\alpha = 1^o$)

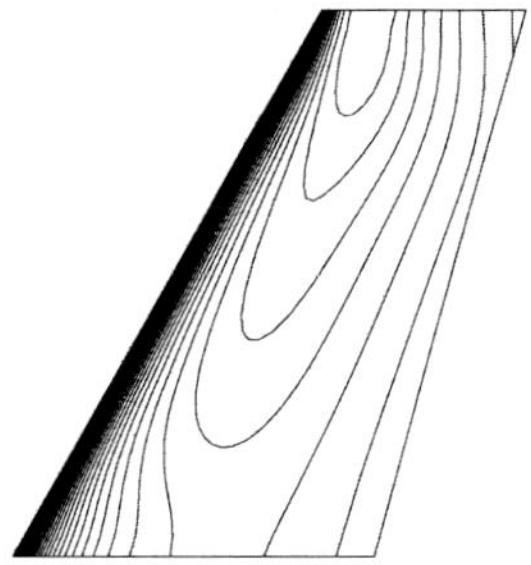

Fig. 16. Pressure contours on lower surface of NACA0012 wing ($M_\infty = 0.9$, Re=1000, $\alpha = 1^o$)

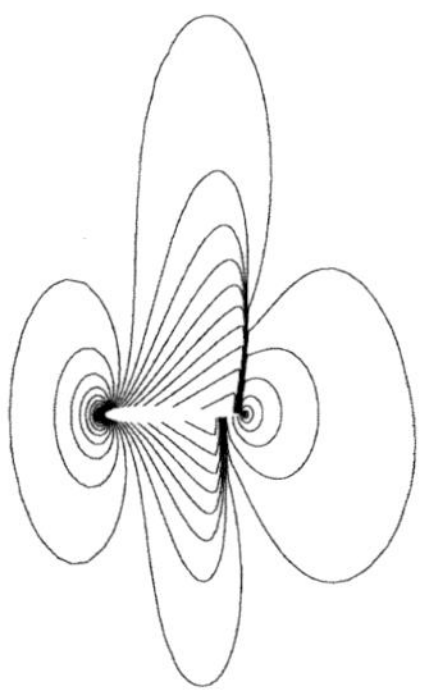

Fig. 17. Mach contours at symmetry plane of NACA0012 wing ($M_\infty = 0.9$, $\alpha = 1^o$, Potential Flow)

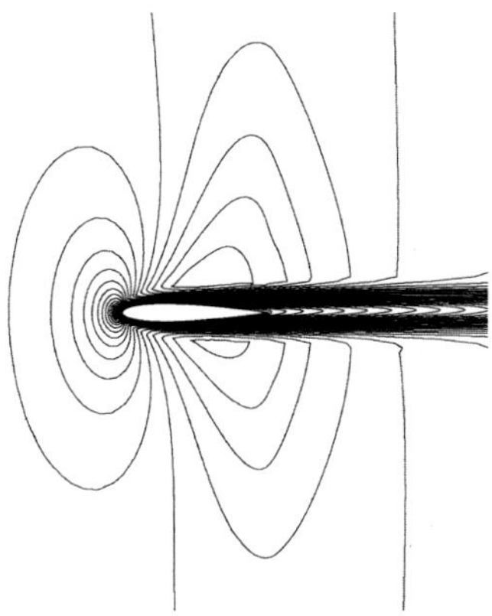

Fig. 18. Mach contours at symmetry plane of NACA0012 wing ($M_\infty = 0.9$, Re=1000, $\alpha = 1^o$)

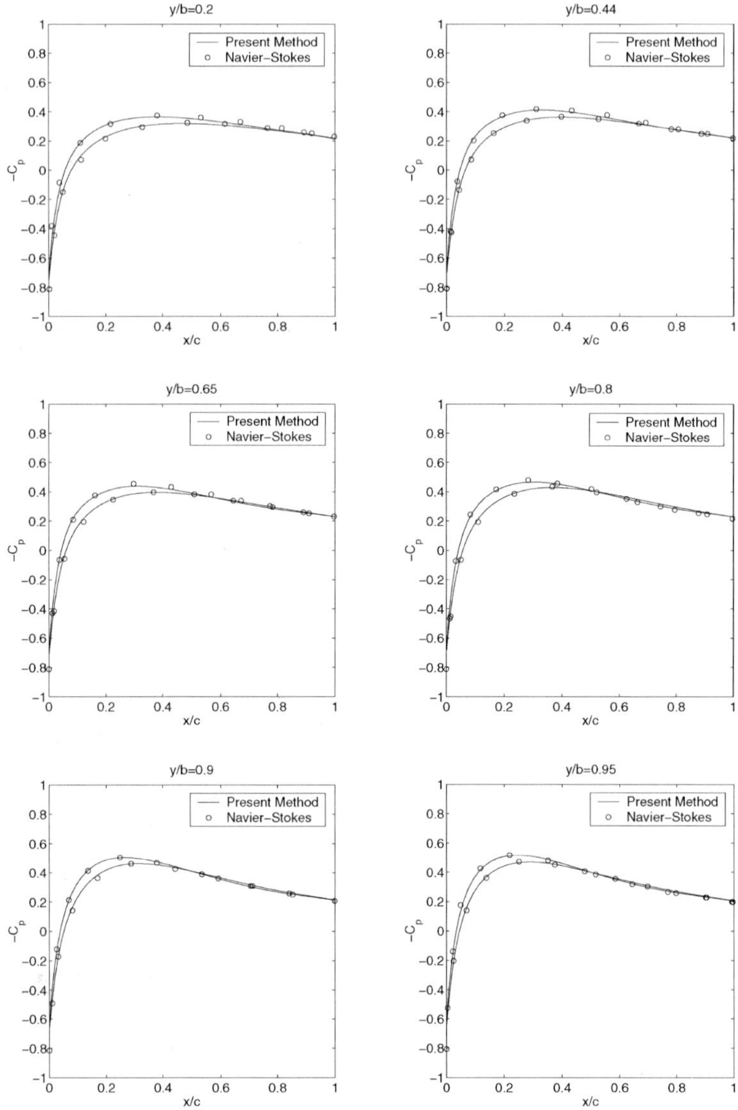

Fig. 19. Surface pressure distributions for viscous flow over NACA0012 wing ($M_\infty = 0.9$, Re=1000, $\alpha = 1^o$)

THE APPLICATION OF OPTIMAL CONTROL TO BOUNDARY LAYER FLOW

D.S. Henningson and A. Hanifi
Department of Mechanics, KTH and
Swedish Defence Research Agency (FOI)

Modern optimal control theory can be used to calculate the optimal steady suction needed to e.g. relaminarize the flow or to delay transition. This has been used to devise the best possible suction distributions for keeping the flow laminar, and applied for flat plate boundary layers as well as boundary layers on swept wings of airplanes. Optimal control theory can also be used to device the best possible measurement feedback control. Real time measurements of flow quantities at the wall is fed back to control the flow through wall actuation, using e.g. blowing and suction. We have applied modern control theory to channel flows as well as two- and three-dimensional boundary layers, and found that flow disturbances can be cancelled, transition delayed and low Reynolds number turbulence relaminarized.

1 INTRODUCTION

Professor Prandtl's epoch-making lecture on the boundary layer nature of near wall motion of fluid flows at high Reynolds numbers [22], also pointed out the possibilities this discovery implied for the control of flows with boundary layers. Prandtl made use of wall suction in the boundary layer on a circular cylinder to show that the global flow pattern could be greatly influenced by only influencing the boundary layer in an appropriate manner. In the past decade optimal control theory has been applied to boundary layer flow in order to harvest the maximum effects out of this discovery.

G.E.A. Meier and K.R. Sreenivasan (eds.), IUTAM Symposium on One Hundred Years of Boundary Layer Research, 59-71,

Reducing the viscous drag on a wing while maintaining operational properties such as e.g. lift, is of great interest and the research in this area is vast [16]. It is known that the viscous drag increases dramatically as the boundary layer flow changes from a laminar to a turbulent state. Transition in the boundary layer on aircraft wings is usually caused by break down of small disturbances which grow as they propagate down stream. The stabilization effect of steady boundary layer suction on growth of these disturbances is well known [25] and has been utilized for Laminar Flow Control (LFC) and Hybrid Laminar Flow Control (HLFC) [16]. However, in most cases the design of suction distributions rely on the experiences of the engineers which may not always give the optimal solution, i.e. giving the largest delay of laminar-turbulence transition for a given suction power. In the recent decade, the development of optimal control theory applied in fluid mechanics problems has been rapid and a number of attempts have been made to optimize the steady suction distribution in order to control growth of disturbances [1, 3, 7, 10, 21]. In all of these investigations the optimization methods are gradient based and they utilize the potential of adjoint methods to obtain the gradients of interest. A common approach [1, 3, 7, 21] is to minimize some measure of the disturbance growth, either the disturbance kinetic energy [1, 7, 21] or the so called N-factor [3].

The feedback control of fluid flow systems is a problem that has received growing attention in recent years and has been approached in a number of different ways. Here we will use linear control theory, basing the control algorithm on the linearised Navier–Stokes equations governing small perturbations to the flow system, a mathematical statement of the control objective, and models of the unknown disturbances acting on the system. Recent reviews of related flow control efforts can be found in, for instance, Bewley [4], Kim [17], and the introduction of Högberg, Bewley & Henningson [13].

The classical problem of linear model-based feedback control based on noisy measurements can be decomposed into two independent subproblems: first, the state-feedback (full-information) control problem, in which full state information is used to determine effective control feedback, and, second, the state estimation problem, in which measurements are continuously used to force a real-time calculation of the flow system in an appropriate manner such that the calculated flow state eventually approximates the actual flow state. Once both subproblems are solved, one can synthesize them into what is referred to as a compensator, and control a flow based on limited measurements of the flow system.

2 OPTIMAL CONTROL AND DESIGN OF LAMINAR WINGS

There are different approaches to establish a laminar boundary layer on the surface of a wing. The first one, which can be seen as a passive control approach, is the so called Natural Laminar Flow (NLF) design. Here, the shape of the wing is designed such that the resulting pressure distribution suppresses growth of disturbances inside the boundary layer. The second approach is an active control approach and is usually referred to as Hybrid Laminar Flow Control (HLFC). Here, the boundary layer is stabilized by means of suction at the wall. Both of these problems can be attacked using the optimal control theory. In NLF design the control parameter is the geometry of the wing, while in HLFC approach the control parameter is the suction rate at the surface of the wing. The latter can be translated to the static pressure in the so called pressure chambers. Recent works of Pralits [19] and Pralits et al. [21] are examples of optimal control theory applied to HLFC and Amoignon et al. [2] demonstrate its application to NLF design.

Consider a time and spanwise periodic disturbnace with amplitude function $\tilde{\mathbf{q}}$. The objective function $\mathcal{J}$ to be minimized is usually based on some norm of the disturbance which generally can be written

$$\mathcal{J} = \xi E(x_f) + (1-\xi)\int_{x_0}^{x_f} E\,dx + \ell^2 g^2,\ E = \frac{1}{2}\int_{z_0}^{z_1}\int_0^{\infty} \tilde{\mathbf{q}}^H M \tilde{\mathbf{q}}\,dy\,dz,$$

where x, y, z are streamwise, normal and spanwise coordinates, respectively. ξ and M are a weighting scalar and a weighting matrix, respectively. ℓ is a regularization parameter and g represents deviation of some aerodynamic properties, like pressure drag or lift coefficients, from their reference values. Using Lagrange multiplier technique, constraints on the geometry and control efforts can also easily be implemented. An important issue in the optimization procedure is the accuracy and efficiency of gradient of the objective function w.r.t. the control parameter. An efficient method for calculating this gradient, when degrees of freedom of the control parameter is large, is the adjoint method.

Let us have a closer look to the case of HLFC. Assume the disturbance evolution is modeled by Parabolized Stability Equations (PSE) and the mean flow is given by the solution of the boundary-layer equations (BLE). We write these equations in symbolic form as

$$L_P\,\hat{\mathbf{q}} = 0, \quad \int_0^\infty \hat{\mathbf{q}}^H \hat{\mathbf{q}}_x\,dx = 0, \quad L_B(\mathbf{Q})\,\mathbf{Q} = 0,$$

Here, $\tilde{\mathbf{q}}(x,y) = \hat{\mathbf{q}}(x,y)\exp(i\int_{x_0}^{x}\alpha(x')dx')$ with α being the streamwise wavenumber. The second of the equations above is the so called auxiliary condition which is needed to remove the ambiguity caused by x-dependency of both $\hat{\mathbf{q}}$ and α. It should be seen as a state equation when adjoints are derived. Then, the adjoint system is

$$L_P^*\,\mathbf{q}^* = S_P(\hat{\mathbf{q}}, r^*), \quad \frac{\partial}{\partial x}\int_0^\infty \mathbf{q}^{*H}\frac{\partial L_P}{\partial \alpha}\,\hat{\mathbf{q}}\,dy = -\,i\int_0^\infty \tilde{\mathbf{q}}^H M \tilde{\mathbf{q}}\,dy,$$

$$L_B^*\,\mathbf{Q}^* = S_B(\hat{\mathbf{q}}, \mathbf{q}^*, \mathbf{Q}),$$

Here, $\mathbf{Q}$ is the vector of mean flow quantities and superscript * refers to the adjoint quantities. The solution of the adjoint boundary layer equations, when appropriate initial and boundary conditions are used, gives the desired gradient of the objective function with respect to the mean flow quantities. Instead, its gradient w.r.t. mean mass flow at the wall (for $\ell = 0$) is given as (see e.g. [21])

$$\frac{\partial \mathcal{J}}{\partial \dot{m}_{\mathrm{w}}} = V_{\mathrm{w}}^*,$$

where V_{w}^* is the Lagrangian multiplier of the continuity equation for the mean momentum. Then, a gradient-based method can be used to find the optimal distribution of normal velocity at the wall. In figure 1 (left) a schematic of the design procedure for optimal suction distributions are given. In figure 1 (right) the derivative of the objective function, as a function of the streamwise position, calculated

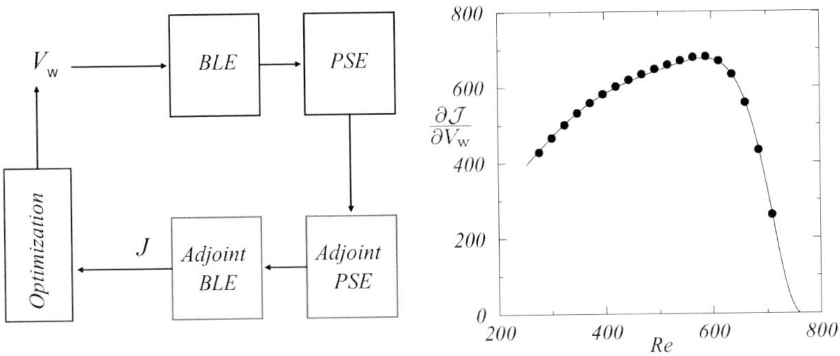

Figure 1: Left: Schematic of the optimization procedure for suction distribution. Right: Comparison of calculated derivatives of disturbance energy using finite differences and adjoint method. From Pralits et al. [21].

using the adjoint technique are compared with finite differences. As can be seen there the agreement is excellent. In real applications, the steady boundary-layer suction is usually done through a number of discrete pressure chambers [6, 8, 16, 23]. In this case, the size, position and the internal static pressure of each chamber, P_{c_j}, are the design variables. Here, the resulting suction velocity is a function of the surface porosity, hole geometry and the pressure difference between the pressure distribution on the wing and static pressure in the chambers [6]. Knowing this function, the gradient of the objective function w.r.t. the pressure in the chambers is found as

$$\frac{\partial \mathcal{J}}{\partial P_{c_j}} = -\int_{\Gamma_j} V_w^* \frac{\partial \dot{m}_w}{\partial P_{c_j}}\, dx,$$

where Γ_j denotes streamwise extension of each chamber. Pralits & Hanifi [20] used this expression to design an optimal distribution of pressure for different combination of chambers. The results were presented for an airfoil designed for medium range commercial aircraft. In figure 2 a comparison between results for uncontrolled, optimal continuous suction and optimal pressure chambers cases is given.

Amoignon et al. [2] advanced the technique discussed above to derive the expression for gradient of disturbance growth w.r.t. the parameters defining the geometry of an airfoil. Here, the procedure given in figure 1 also includes solution of the Euler equations and

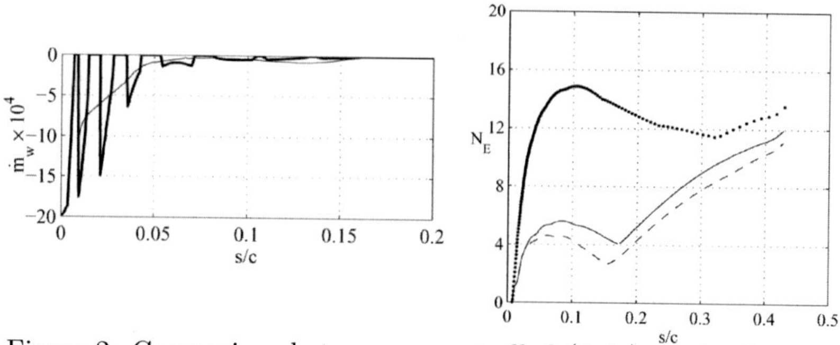

Figure 2: Comparison between uncontrolled (dots), optimal continuous suction (dashed) and optimal pressure chambers (solid). Left: mass flow distribution. Right: envelope of envelopes. Adapted from Pralits & Hanifi [20]

their adjoints. They demonstrated that suppression of disturbance amplification with restriction on changes in geometry and aerodynamic coefficients was possible.

3 LQG COMPENSATION

The full linear feedback control problem can be put in the following standard form

$$\frac{du}{dt} = Au + B\phi + B_1 f \qquad (1)$$

$$y = Cu + g \qquad (2)$$

$$\frac{du_e}{dt} = Au_e + B\phi + \hat{L}(y - y_e) \qquad (3)$$

$$y_e = Cu_e \qquad (4)$$

$$\phi = \hat{K}u_e \qquad (5)$$

Equation (1) describes the evolution of the flow state u, where A is the linear evolution operator, ϕ is the control, B is the input operator for the control, f is the disturbances acting on the system and B_1 is the input operator for the disturbances. In equation (2) y is the measurement, C is the output operator for the measurement and g is the measurement noise.

Equation (3) represents the estimated flow state, which is needed since the full flow state is not assumed to be accessible to measurement. The estimator is forced proportional to the difference in the measurement between the actual flow and the estimated flow, i.e. $\hat{L}(y - y_e)$, where L is the measurement gain. Finally, in equation (5) the estimated state is used to find the control ϕ using the control gain $\hat{K}$. This dynamical systems representation is used in the LQG (Linear Quadratic Gaussian) control problem. "Linear" refers to the linearized representation of the flow evolution equation, "Quadratic" to the fact that we will be minimizing the energy, a quadratic function of the flow state and "Gaussian" to the assumed distribution of disturbances and noise in the flow. The optimal choice of gains $\hat{K}$ and $\hat{L}$ can be found by the solution of Riccati equations which are derived e.g. in Lewis & Syrmos [18]. We will follow this book in the optimal control material presented here and the description is a shortened version of that presented in Henningson [9].

3.1 The linearized Navier-Stokes equations in state space form

The horizontally Fourier transformed Navier-Stokes equations linearized around the streamwise velocity U and spanwise velocity W can be written in normal velocity $\hat{v}$, normal vorticity $\hat{\eta}$ form as follows (see Schmid & Henningson [24])

$$\frac{\partial \hat{v}}{\partial t} = \underbrace{(\hat{\nabla}^2)^{-1}\left[-(i\alpha U + i\beta W)\hat{\nabla}^2 + i\alpha U'' + i\beta W'' + \frac{1}{Re}\hat{\nabla}^4\right]}_{\mathcal{L}_{OS}} \hat{v}$$

$$\frac{\partial \hat{\eta}}{\partial t} = \underbrace{\left[-(i\alpha U + i\beta W) + \frac{1}{Re}\hat{\nabla}^2\right]}_{\mathcal{L}_{SQ}} \hat{\eta} + \underbrace{(i\alpha W' - i\beta U')}_{\mathcal{L}_C} \hat{v}$$

where (α, β) is the wavenumber vector, $'$ denotes the normal derivative, $\hat{\nabla}^2$ the horizontally Fourier transformed Laplacian and Re is the Reynolds number. We have also defined the Orr-Sommerfeld ($\mathcal{L}_{OS}$), Squire ($\mathcal{L}_{SQ}$) and coupling ($\mathcal{L}_C$) operators. The boundary condition is assumed to be given by the blowing and suction velocity at the wall $v_{wall} = \varphi$.

This system can be put in state space form by a lifting procedure which transfers the inhomogeneous boundary condition to a forcing term. If we introduce

$$\begin{pmatrix} \hat{v} \\ \hat{\eta} \end{pmatrix} = \begin{pmatrix} \hat{v}_h \\ \hat{\eta}_h \end{pmatrix} + \begin{pmatrix} \hat{v}_p \\ \hat{\eta}_p \end{pmatrix} \varphi$$

into the OS-SQ-system we find

$$\underbrace{\frac{d}{dt}\begin{pmatrix} \hat{v}_h \\ \hat{\eta}_h \\ \varphi \end{pmatrix}}_{\dot{u}} = \underbrace{\begin{pmatrix} \mathcal{L}_{OS} & 0 & 0 \\ \mathcal{L}_C & \mathcal{L}_{SQ} & 0 \\ 0 & 0 & 0 \end{pmatrix}}_{A} \underbrace{\begin{pmatrix} \hat{v}_h \\ \hat{\eta}_h \\ \varphi \end{pmatrix}}_{u} + \underbrace{\begin{pmatrix} -\hat{v}_p \\ -\hat{\eta}_p \\ 1 \end{pmatrix} \dot{\varphi}}_{B\phi}$$

where $(\hat{v}_h, \hat{\eta}_h)$ satisfy homogeneous boundary conditions at the wall and $(\hat{v}_p, \hat{\eta}_p)$ denotes a given particular solution with unit normal velocity at the wall. Note that it is now the time derivative of the control which naturally enters in the state space formulation.

3.2 The solution to the optimal control problem

We now have the linearized Navier-Stokes equations for the evolution of small disturbances in boundary layer flow in state space form, i.e.

$$\frac{du}{dt} = Au + B\phi \qquad u(0) = u_0 \tag{6}$$

We choose the following quadratic objective function to minimize

$$J = \frac{1}{2} \int_0^T \int_\Omega (u^* Q u + \phi^* M \phi)\, dt d\Omega \tag{7}$$

where the operator Q is chosen such that the expression $u^* Q u$ is the disturbance energy and the last term of the integral is a penalization of the control.

The gradient of the objective function with respect to the control is

$$\nabla_\phi J = B^* p + M\phi \tag{8}$$

where the adjoint variable p is found to satisfy the equation

$$-\frac{dp}{dt} = A^* p + Qu \qquad p(T) = 0 \tag{9}$$

Using expression for the gradient of the objective function, the forward and the adjoint equations, it can be shown that the optimal feedback gain can be found from the non-negative self-adjoint solution to the algebraic Riccati equation

$$\left(XA + A^*X - XBM^{-1}B^*X + Q\right) u = 0 \qquad \forall u \tag{10}$$

giving the resulting optimal control or Kalman gain $\hat{K}$ as

$$\phi \; = \; -M^{-1}B^*Xu \; = \; \hat{K}u \tag{11}$$

We have now found the following closed loop system for the full information control. For each wavenumber pair we have

$$\frac{d}{dt}u = Au + B\phi = (A + B\hat{K})u \tag{12}$$

It is possible to inverse Fourier transform the optimal Kalman gain $\hat{K}$ into physical space to obtain a convolution kernel K. We find that the time derivative of the optimal wall blowing and suction is given by the following convolution integral between the kernel K and state u

$$\dot{v}_{wall}(x, z) = \int\int\int K(x - \bar{x}, \bar{y}, z - \bar{z})u(\bar{x}, \bar{y}, \bar{z})d\bar{x}d\bar{y}d\bar{z} \tag{13}$$

3.3 The Kalman filter and the optimal measurement gain

There remains to determine the optimal measurement gain $\hat{L}$ in the estimator problem, in order to have the complete solution to the LQG problem. We will find that $\hat{L}$ can be found from a Riccati equation related to that used to find $\hat{K}$. Using equations (1)-(4) we can find an equation for the estimation error $\tilde{u} = u - u_e$ as

$$\frac{d\tilde{u}}{dt} = (A - \hat{L}C)\tilde{u} - \hat{L}g + B_1 f$$

Note that this is a stochastic differential equation since it is driven by the two noise terms in the right hand side. The noise is assumed to have a Gaussian distribution and can thus be completely characterized by its mean and covariance. We assume that the covariance of the state disturbances f and the measurement noise g have the following form

$$R_{ff}(\tau) = E\{f(t+\tau)f^*(t)\} = R\delta(\tau) \quad (14)$$
$$R_{gg}(\tau) = E\{g(t+\tau)g^*(t)\} = G\delta(\tau) \quad (15)$$

where $E\{\ \}$ denotes the expectation operator. Now we choose $\hat{L}$ to minimize covariance of error $P = E\{\tilde{u}\tilde{u}^*\}$. Lewis & Syrmos [18] gives the derivation and it is found that the non-negative self-adjoint solution to the algebraic Riccati equation

$$\left(AP + PA^* - PC^*G^{-1}CP + B_1RB_1^*\right)\tilde{u} = 0 \quad \forall\tilde{u} \quad (16)$$

gives the covariance of the estimation error P, which can be shown to give the optimal measurement gain $\hat{L}$ as

$$\hat{\psi} = -PC^*G^{-1}\Delta y = \hat{L}\Delta y \quad (17)$$

where $\Delta y = y - y_e$ is the measurement error.

The optimal $\hat{L}$ is also given in Fourier space and can be inverse transformed to give the following optimal state estimation forcing

$$\psi(x,y,z) = \int\int L(x-\bar{x}, y, z-\bar{z})\Delta y(\bar{x},\bar{z})d\bar{x}d\bar{z} \quad (18)$$

4 BRIEF REVIEW OF RECENT FEEDBACK CONTROL RESULTS

The type of state feedback control described here has been applied to boundary layer flows in a number of recent investigations. Here we will briefly mention a number of investigations performed by the collaborating groups at KTH and UCSD, although there are also others pursuing this avenue, most notably the UCLA group. The latter activities, primarily regarding control and estimation of turbulent channel flows, were recently reviewed by Kim [17].

Control and estimation in channel flow have been performed by Bewley & Liu [5] for small amplitude disturbances and applied to transition by Högberg, Bewley & Henningson [13], were the disturbance amplitudes needed for transition to occur increased drastically when feedback control was applied. In fact, using a gain scheduling technique, low Reynolds number turbulence has been relaminarized

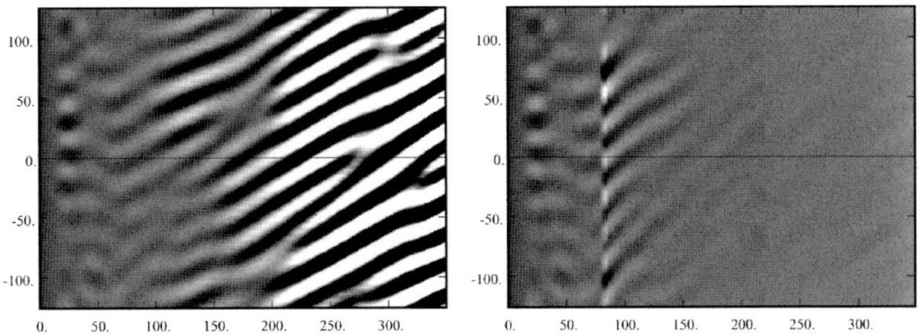

Figure 3: Snapshots of normal velocity in an horizontal plane without and with feedback control. Black and white indicates negative and positive velocity, respectively, and the blowing and suction control is applied between 75 and 225 in x.

by Högberg, Bewley & Henningson [14] using full state information control. Recent work aims at increasing the performance of the estimator by better modelling the covariance of the state disturbances and measurement noise, see equations (14) and (15), and are presented by Hoepffner, Chevalier, Bewley & Henningson [11].

This theory has also been applied to spatially developing boundary layers. Högberg & Henningson [12] studied the control of linear and non-linear disturbances in two- and three-dimensional boundary layers and Högberg, Chevalier & Henningson [15] the full compensation problem in Falkner-Skan-Cooke flow.

References

[1] Airiau, C., Bottaro, A., Walther, S. and Legendre, D., 2003, A methodology for optimal laminar flow control: Application to the damping of tollmien-schlichting waves in a boundary layer. *Phys. Fluids*, **15**, 1131-1145.

[2] Amoignon, O., Pralits, J.O., Hanifi, A., Berggren, M. and Henningson, D.S., 2004, Shape optimization for delay of laminar-turbulent transition, *submitted to AIAA J.*

[3] Balakumar, P. and Hall, P., 1999, Optimum suction distribution for transition prediction. *Theor. Comput. Fluid Dyn.*, **13**, 1-19.

[4] Bewley, T.R., 2001, Flow control: new challenges for a new Renaissance, *Progress in Aerospace Sciences* **37**, 21-58.

[5] Bewley, T.R. and Liu, S., 1998, Optimal and Robust Control and Estimation of Linear Paths to Transition, *J. Fluid Mech* **365**, 305-349.

[6] Bieler, H. and Preist, J., 1992 HLFC for commercial aircraft. In *First european forum on laminar flow technology*, Hamburg, 193-199.

[7] Cathalifaud, P. and Luchini, P., 2000, Algebraic growth in a boundary layer: optimal control by blowing and suction at the wall. *Eur. J. Mech. B/Fluids*, **19**(4), 469-490.

[8] Ellis, J.E. and Poll, D.I.A., 1996, Laminar and laminarizing boundary layers by suction through perforated plates. In *Second european forum on laminar flow technology*, Bordeaux, 8.17-8.26.

[9] Henningson, D.S., 2004, Optimal feedback control applied to boundary layer flow, *Advances in Turbulence X*, Proceedings of the Tenth European Turbulence Conference H. I. Andersson & P.-Å. Krogstad (Eds.), CIMNE, Barcelona, ??-??.

[10] Hill, D.C., 1997, Inverse design for laminar three-dimensional boundary layers. *Bull. Am. Phys. Soc.*, **42**, : 2120.

[11] Hoepffner, J., Chevalier, M., Bewley, T.R. and Henningson, D.S, 2004, State estimation in wall-bounded systems, Submitted to *J. Fluid Mech.*

[12] Högberg, M. and Henningson, D.S., 2002, Linear optimal control applied to instabilities in spatially developing boundary layers, *J. Fluid Mech* **470**, 151-179.

[13] Högberg, M., Bewley, T.R. and Henningson, D.S., 2003, Linear feedback control and estimation of transition in plane channel flow, *J. Fluid Mech* **481**, 149-175.

[14] Högberg, M., Bewley, T.R. and Henningson, D.S., 2003, Relaminarization of $Re_\tau = 100$ turbulence using gain scheduling and linear state-feedback-control, *Phys. Fluids* **15**, 3572-3575.

[15] Högberg, M., Chevalier, M. and Henningson, D.S., 2003, Linear compensator control of a point source induced perturbation in a Falkner-Skan-Cooke boundary layer, *Phys. Fluids* **15**, 2449-2452.

[16] Joslin, R.D., 1998, Overview of laminar flow control. Technical Report 1998-208705, NASA, Langley Research Center, Hampton, Virginia.

[17] Kim, J., 2003, Control of turbulent boundary layers, *Phys. Fluids* **15**, 1093-1105.

[18] Lewis, F.L. and Syrmos, V.L., 1995, Optimal Control, *Wiley-Interscience* Second edition.

[19] Pralits, J. 2003, Optimal Design of Natural and Hybrid Laminar Flow Control on Wings, Ph.D. Thesis, KTH Mechanics.

[20] Pralits, J.O. and Hanifi, A. 2003, Optimization of steady suction for disturbance control on infinite swept wings. *Phys. Fluids* **15**, 2756-2772.

[21] Pralits, J.O., Hanifi, A. and Henningson, D.S. 2002, Adjoint-based optimization of steady suction for disturbance control in incompressible flows. *J. Fluid Mech.* **467**, 129-161.

[22] Prandtl, L., 1904, Uber Flüssigkeitsbewegung bei sehr Kleiner Reibung, *Verh. d. III. Intern. Mathem. Kongresses*, Heidelberg (Neudruck in Prandtl-Betz, *Vier Abhandlungen zur Hydro- und Aerodynamik*, Göttingen, 1927). Auslieferung durch Springer.

[23] Reneaux, J. and Blanchard, A., 1992, The design and testing of an airfoil with hybrid laminar flow control. In *First european forum on laminar flow technology*, Hamburg, 164-174.

[24] Schmid, P.J. and Henningson, D.S., 2001, Stability and transition in shear flows, *Springer-Verlag*.

[25] Schlichting, H., 1943/44, Die beeinflussung der grenzschicht durch absaugung und ausblasen. *Jb. dt. Adad. d. Luftfahrtforschung*, 90-108.

LEADING-EDGE BOUNDARY LAYER FLOW

Prandtl's vision, current developments and future perspectives

V. Theofilis*, A.V. Fedorov** and S.S. Collis***

* *E.T.S.I. Aeronáuticos, U. Politécnica de Madrid, E-28040 Madrid, SPAIN (vassilis@torroja.dmt.upm.es)*

** *Moscow Institute of Physics and Technology, 141700 Moscow Region, RUSSIA (fedorov@falt.ru)*

*** *Sandia National Laboratories†, P.O. Box 5800, Albuquerque, NM 87185-0370, U.S.A (sscoll@sandia.gov)*

Abstract: The first viscous compressible three-dimensional BiGlobal linear instability analysis of leading-edge boundary layer flow has been performed. Results have been obtained by independent application of asymptotic analysis and numerical solution of the appropriate partial-differential eigenvalue problem. It has been shown that the classification of three-dimensional linear instabilities of the related incompressible flow [**13**] into symmetric and anti-symmetric mode expansions in the chordwise coordinate persists for compressible, subsonic flow-régime at sufficiently large Reynolds numbers.

Key words: Compressible Hiemenz flow, BiGlobal linear instability analysis

1. INTRODUCTION

In the context of external aerodynamics, the flow near the windward stagnation line of a swept cylinder serves as a canonical model of a leading-edge boundary layer. Research into boundary-layer flows over an unswept cylinder immediately followed the discovery of the boundary-layer concept itself. Indeed, it was Prandtl's interest in measuring the pressure on the surface of a cylinder that led to the discovery, by Hiemenz [**4**], of an exact solution of the incompressible Navier-Stokes equations that describes the stagnation point flow and bears his name. As the performance benefits of

† Sandia is a multiprogram laboratory operated by Sandia Corporation, a Lockheed Martin Company, for the United States Department of Energy's National Nuclear Security Administration under contract DE-AC04-94AL85000.

G.E.A. Meier and K.R. Sreenivasan (eds.), IUTAM Symposium on One Hundred Years of Boundary Layer Research, 73-82,

swept wings were realized and demonstrated in Göttingen [**5-6**] this solution was extended to the well-known three-dimensional stagnation-line flow [**10**].

Concurrently with the extension of the swept Hiemenz solution to the compressible régime by Reshotko and Beckwith [**9**], investigations into the instability of leading-edge boundary layer commenced. The first instability results were presented 50 years ago with the contributions of Görtler [**1**] and Hämmerlin [**2**] to the meeting "Fifty Years Boundary-Layer Research," celebrating Prandtl's boundary-layer idea. Both those contributions dealt with incompressible *stagnation point* (unswept Hiemenz) flow and put forward what has become known as the Görtler-Hämmerlin (GH) Ansatz, whereby linear disturbances of the leading-edge boundary layer inherit the functional dependence of the basic flow itself. Accordingly, the streamwise and wall-normal disturbance velocity components are functions of the wall-normal coordinate and, in addition, the streamwise velocity component depends linearly on the chordwise spatial coordinate. This Ansatz was later extended [**3**] and verified [**11**] for the incompressible *stagnation line* (swept Hiemenz) flow. These studies demonstrated that the most unstable eigenmode of the leading-edge boundary layer, denoted as the GH-mode, compares well with experiment and direct numerical simulation under linear conditions.

Recent advances in computing hardware and algorithms have permitted generalizations of the GH Ansatz in the context of BiGlobal linear theory based on solution of partial-differential eigenvalue problems (EVP). Lin and Malik [7] discovered new eigenmodes besides the GH-mode, and Theofilis, Fedorov, Obrist and Dallmann [**13**] demonstrated that the instability of incompressible three-dimensional swept leading-edge boundary-layer flow is amenable to analysis. The latter authors identified all (BiGlobal) eigenmodes as having a polynomial structure along the chordwise direction and reduced the partial-differential EVP to a system of one-dimensional ordinary-differential EVPs of the Orr-Sommerfeld class. The solution of this system delivers the *complete three-dimensional instability characteristics* for incompressible swept leading-edge boundary-layer flow.

The present contribution demonstrates that this reduction is also possible for compressible flows, albeit restricted to certain ranges of Reynolds and Mach numbers. We proceed along lines analogous to [**13**] and arrive at the instability characteristics of viscous compressible three-dimensional leading-edge boundary-layer flows by independent application of asymptotic analysis [**13**] and a novel numerical solution of the compressible BiGlobal EVP [**14**]. In Section 2 the fundamentals of our theoretical approach are discussed. Section 3 presents details of the basic flow followed by instability analysis results obtained using both theoretical approaches. A brief discussion of our ongoing efforts closes our present contribution.

2. THEORY

2.1 The basic state

The leading-edge flow in the vicinity of the attachment line of a swept wing is treated as a compressible stagnation line flow, with a non-zero velocity component along the attachment line. If the viscous boundary layer thickness is small compared with the leading-edge radius then the surface near the attachment line can be approximated as locally flat. Under these conditions, the Reynolds number is defined as

$$R = W_e^* \Delta^* / \nu_e^*, \qquad \Delta^* = \sqrt{\nu_e^* / (\partial U_e^* / \partial x^*)_{x=0}} \tag{1}$$

where W_e^* is the spanwise component of the velocity vector (U_e^*, W_e^*) at the boundary-layer edge — a scale consistent with that adopted in [**12-13**]. In the Cartesian coordinate system $(x,y,z) = (x^*, y^*, z^*)/\Delta^*$ (asterisk denotes dimensional quantities), the basic flow quantities are expressed in the form

x-component velocity: $U_s^*(x,y,z) = W_e^* x U_0(y)/R$ (2)

y-component velocity: $V_s^*(x,y,z) = W_e^* V_0(y)/R$ (3)

z-component velocity: $W_s^*(x,y,z) = W_e^* W_0(y)$ (4)

temperature: $T_s^*(x,y,z) = T_e^* T_0(y)$ (5)

pressure: $P_s^*(x,y,z) = \rho_e^* W_e^{*2}\left(\dfrac{1}{\gamma M^2} - \dfrac{x^2}{2R^2}\right)$ (6)

density: $\rho_s^*(x,y,z) = \rho_e^* \rho_0(y) = \rho_e^* / T_0(y)$ (7)

viscosity: $\mu_s^*(x,y,z) = \mu_e^* \mu(T_0(y))$. (8)

The profiles $U_0(y)$, $V_0(y)$, $W_0(y)$ and $T_0(y)$ are solutions of the ordinary-differential-equation system [**9**]

$$\frac{1}{T_0}\left(U_0^2 + V_0 U_0'\right) = 1 + \frac{d\mu}{dT_0} T_0' U_0' + \mu U_0'' \tag{9}$$

$$\frac{1}{T_0} V_0 W_0' = \frac{d\mu}{dT_0} T_0' W_0' + \mu W_0'' \tag{10}$$

$$U_0 - \frac{V_0}{T_0} T_0' + V_0' = 0 \tag{11}$$

$$\frac{d\mu}{dT_0} \frac{1}{\Pr} T_0'^2 + \frac{\mu}{\Pr} T_0'' - \frac{T_0' V_0}{T_0} + (\gamma - 1) M^2 \mu W_0'^2 = 0\,, \tag{12}$$

subject to the boundary conditions

$$U_0(0) = W_0(0) = 0\,,\ V_0(0) = -C_q T_w \tag{13}$$
$$U_0(\infty) = W_0(\infty) = T_0(\infty) = 1\,.$$

In these expressions, $C_q = -R\cdot(V_s^*(0)\rho_s^*(0))/(W_e^*\rho_e^*)$ is a normalized suction parameter, $T_w = T_0(0)$ is the wall temperature, $T_0'(0) = 0$ on adiabatic walls, and $M = W_e^*/a_e^*$ is the local Mach number.

2.2 Asymptotic analysis

The spatial homogeneity of the basic state along z permits the introduction of the decomposition

$$\boldsymbol{Q}(x,y,z,t) = \boldsymbol{Q}_b(x,y) + \boldsymbol{q}(x,y)\exp\left[\,i\,(\,\beta z - \omega t\,)\,\right] \tag{14}$$

into the governing three-dimensional viscous compressible equations of motion. Here, $\boldsymbol{Q}_b$ is the steady basic state, constructed using equations (2-6) after solving the system (9-13), and $\boldsymbol{q} = (u, v, w, \theta, p)^T$ are the *two-dimensional* amplitude functions of the velocity components, temperature and pressure. In the temporal framework considered here, β is a real wavenumber parameter related with a periodicity length $L_z = 2\pi/\beta$ along the spanwise direction, while the frequency ω is the sought eigenvalue. Additional free parameters are the Reynolds and Mach numbers, R and M, respectively.

The extended disturbance vector-function is specified as $\mathbf{F} \equiv (u, \frac{\partial u}{\partial y}, v, p, \theta, \frac{\partial \theta}{\partial y}, w, \frac{\partial w}{\partial y})^T$. Under the assumption of large Reynolds number R, a small parameter $\varepsilon = R^{-1}$ and slow variables $x_1 = \varepsilon x$, $t_1 = \varepsilon t$ are introduced and the vector function is given by the asymptotic expansion

$$\mathbf{F} = \mathbf{Z}_0(y; x_1, t_1, \beta, \omega) + \varepsilon \mathbf{Z}_1(y; x_1, t_1, \beta, \omega) + \ldots \tag{15}$$

The zero-order term is expressed as $\mathbf{Z}_0 = C(x_1 t_1)\xi(y; x_1)$, where ξ is a solution of the eigenvalue problem

$$\frac{\partial \xi}{\partial y} = \mathbf{A}\xi \tag{16}$$

$$\xi_1 = \xi_3 = \xi_5 = \xi_7 = 0\,, \quad y = 0$$
$$\xi_1 = \xi_3 = \xi_5 = \xi_7 = 0\,, \quad y = \infty$$

which delivers the eigenvalue $\omega = \omega_0(\beta, R)$. Here $\mathbf{A}$ is an 8×8 matrix of the stability problem. For the compressible Hiemenz flow, the eigenvalue ω_0 does not depend on x_1 and the eigenvector has the explicit form $\xi = (x_1\xi_{01}(y), x_1\xi_{02}(y), \xi_{03}(y), \xi_{04}(y), \xi_{05}(y), \xi_{06}(y), \xi_{07}(y))^T$. This fact allows for substantial simplifications of further analysis.

The second-order approximation leads to the inhomogeneous problem

$$\frac{\partial \mathbf{Z}_1}{\partial y} = \mathbf{A}\mathbf{Z}_1 + \mathbf{G}_t \frac{\partial \mathbf{Z}_0}{\partial t_1} + \mathbf{G}_x \frac{\partial \mathbf{Z}_0}{\partial x_1} + \mathbf{G}\mathbf{Z}_0 \tag{17}$$

$$Z_{11} = Z_{13} = Z_{15} = Z_{17} = 0, \quad y = 0$$

$$Z_{11} = Z_{13} = Z_{15} = Z_{17} = 0, \quad y = \infty$$

where $\mathbf{G}_t = i\partial\mathbf{A}/\partial\omega$, $\mathbf{G}_x = -i\partial\mathbf{A}/\partial\alpha$ with $\mathbf{A}$ being derived for the disturbance: $\exp[i(\alpha x + \beta z - \omega t)]$; the matrix $\mathbf{G}$ includes the basic-flow terms associated with nonparallel effects and higher-order terms of the stability problem of the parallel flow. The problem (17) has a non-trivial solution if the inhomogeneous part is orthogonal to the correspondent solution ζ of the adjoint problem. This leads to the equation for the amplitude function $C(x_1, t_1)$

$$\langle \mathbf{G}_t\xi,\zeta\rangle \frac{\partial C}{\partial t_1} + \langle \mathbf{G}_x\xi,\zeta\rangle \frac{\partial C}{\partial x_1} + C\left[\left\langle \mathbf{G}_x \frac{\partial \xi}{x\partial_1},\zeta\right\rangle + \langle \mathbf{G}\xi,\zeta\rangle\right] = 0 \tag{18}$$

where the scalar products are defined as $\langle \mathbf{G}\xi,\zeta\rangle \equiv \int_0^\infty \left(\sum_{j,k=1}^{8} G_{jk}\xi_k, \zeta_j \right) dy$.

Significantly, (18) can be written in a form with structure similar to the incompressible case [**13**],

$$S_1 \frac{\partial C}{\partial t_1} + S_2 x_1 \frac{\partial C}{\partial x_1} + S_3 = 0, \tag{19}$$

where S_1, S_2, S_3 are constants. This equation admits the set of solutions

$$C_n(x_1, t_1) = x_1^n \exp(-i\omega_{1n} t_1), \tag{20}$$

$$\omega_{1n} = -i(nS_2 + S_3)/S_1, \; n = 0, 1, \ldots \tag{21}$$

which give the modes with

$$\mathbf{F}_n(x, y, z, t) = x_1^n y(\xi) + O(\varepsilon), \tag{22}$$

$$\omega_n = \omega_0 + \varepsilon\omega_{1n} + O(\varepsilon^2)\,. \tag{23}$$

Here $n = 0, 2, ...$ corresponds to the symmetric modes S1, S2, ..., and $n = 1, 3, ...$ corresponds to the antisymmetric modes A1, A2, ... The first symmetric mode S1 is equivalent to the GH mode.

In summary, the following algorithm is formulated for the calculation of symmetric and antisymmetric modes: 1) Solve the zero-order problem (16) at $x_1 = 0$, which is simply a 2-D stability problem for the parallel boundary layer with the profiles $W_0(y)$ and $T_0(y)$; 2) Solve the corresponding adjoint problem and calculate the coefficients S_1, S_2, S_3 of (19); 3) Calculate the eigenvalues ω_n and the disturbance vector $\mathbf{F}_n$ using the formulae (21)-(23).

2.3 The BiGlobal EVP

Without resort to the explicit dependence of the disturbance quantities on the chordwise coordinate, as done in (15), the chordwise, x and wall-normal, y, directions are resolved in a coupled manner. Linearization and subtraction of the basic-flow related quantities lead to a generalized eigenvalue problem that may be converted into a matrix EVP, amenable to numerical solution, once numerical prescriptions for the differential operators (here spectral collocation) and appropriate boundary conditions are provided. The general form of the viscous compressible three-dimensional BiGlobal eigenvalue problem is

$$\Lambda\, \boldsymbol{q} = \omega\, \mathrm{P}\, \boldsymbol{q}, \tag{24}$$

where the entries of the matrices Λ and P may be found in [**14**]. By contrast to the latter work, in the open flow system considered here, the boundary conditions are no-slip at the wall, $y = 0$; homogeneous Dirichlet at the free-stream, $y = y_\infty$; and linear extrapolation from the interior of the computational domain at the endpoints $x = \pm x_\infty$ of the truncated domain along the chordwise direction. The related parameters were taken as $y_\infty = 100$ and $x_\infty = 25$ in all computations performed here. Finally, the EVP (24) was solved using an Arnoldi iteration for the recovery of the relevant part of the eigenspectrum.

3. RESULTS

3.1 The compressible swept Hiemenz basic flow

The basic flow is considered on an adiabatic wall, $T_0'(0) = 0$, the suction parameter is taken as 0 and a perfect gas with specific heat ratio $\gamma = 1.4$ and Prandtl number $Pr = 0.72$ is considered. The viscosity coefficient is calculated using Sutherland's formula at the local temperature $T_e^* = 300\,\mathrm{K}$. The problem (9-13) has been solved using a shooting method. Table 1 shows the dependence of the shear-stress and the wall-temperature on Mach number. Note that a characteristic overshoot in $U_0(y)$ appears at high Mach numbers [**9**], creating an inflection point in this velocity component. This, in turn, may give rise to inviscid instabilities of the compressible leading-edge flow in addition to those known to exist in the compressible flat-plate boundary-layer [**8**].

Table 1. Dependence of basic flow shear-stresses and wall-temperature on *M*.

M	$U_0'(0)$	$W_0'(0)$	$T_0(0)$
10^{-5}	1.232588	0.570465	1.000000
0.25	1.228483	0.566095	1.010746
0.50	1.216689	0.553511	1.042990
0.75	1.198610	0.534132	1.096741
1.00	1.176141	0.509881	1.172016
1.50	1.125582	0.454480	1.387198
2.00	1.076630	0.399360	1.688629
2.50	1.034609	0.350344	2.076344
3.00	1.000514	0.308888	2.550336

3.2 BiGlobal instability of the compressible Hiemenz flow

Results for the symmetric GH (S1), S2 and anti-symmetric A1, A2 modes are presented at the Reynolds number $R = 800$ and subsonic Mach numbers $M = 0.02, 0.5, 0.9$. The case $M = 0.02$ is in good agreement with the incompressible flow discussed in [**13**]. Distributions of the amplification rate $c_i = \mathrm{Im}(\omega_i/\beta)$ at the first two Mach numbers considered are shown in Figure 1. Full lines correspond to results of the analysis of Section 2.2 while symbols denote results obtained by numerical solution of the BiGlobal EVP discussed in Section 2.3. Good agreement may be seen between the results for the most unstable GH-mode predicted by the two approaches. This reinforces both methodologies as valid research tools to predict instability characteristics within appropriate parameter ranges. It should be noted here that, unlike the case of [**13**] where an explicit closed-form one-dimensional model could be written for the description of three-dimensional eigenmodes, deriving such a model for the compressible leading-edge boundary-layer flow is of limited usage, due to the error in the second-order asymptotic expansion presented, which scales as

$$O(M^2/R^2). \tag{25}$$

Nevertheless, in the appropriate parameter ranges, the spatial structure of the compressible analogues of the modes GH, A1, S2, A2, ... closely resemble those of incompressible flow. A demonstration can be seen in the amplitude functions of the GH mode at $M = 0.5$, $R = 800$, $\beta = 0.255$ shown in Figure 2, where the linear dependence of $u(x,y)$ on x and the independence of $w(x,y)$ (as well as v, θ and p not shown here) can be clearly seen. While this functional dependence is only asymptotically valid in the analysis of Section 2.2, the numerical solution of the BiGlobal EVP, without *a-priori* imposition of the GH Ansatz, delivers the first demonstration of the form of the most unstable linear eigenmode in compressible leading-edge boundary layer flow at these parameters.

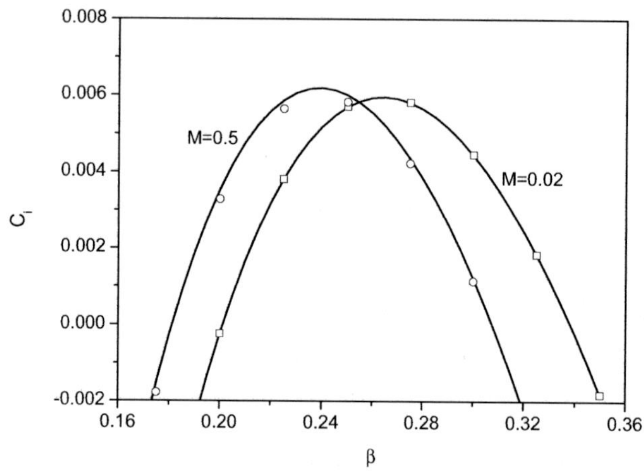

Figre 1. Distributions over β of the amplification-rate c_i of mode GH (S1) at $R = 800$ and $M = 0.02$, 0.5. Solid lines obtained by the asymptotic analysis of Section 2.2, symbols by numerical solution of the BiGlobal eigenvalue problem presented in Section 2.3.

However, as the Mach number increases (keeping all other parameters fixed) the two approaches deviate substantially in their instability predictions. Figure 3 shows that at $M = 0.9$ and $R = 800$ the BiGlobal EVP predicts a slightly wider and substantially stronger amplified flow compared with the asymptotic solution. The agreement improves with increasing Reynolds number, as can be seen in the case of $R = 1500$ shown in the same figure. This is in line with (25), while in both cases Branch I is predicted in a consistent manner. Work is currently underway to identify the stability boundaries using both approaches.

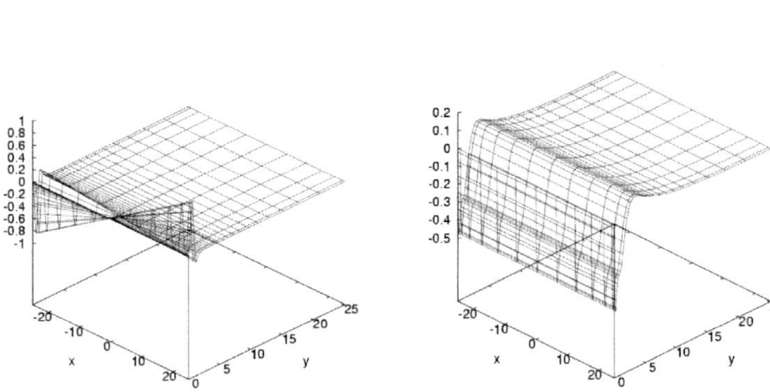

Figre 2. Amplitude functions of the disturbanc e velocity components of the leading eigenmode at $R = 800$, $\beta = 0.255$, $M = 0.5$. Left: $u(x,y)$, Right: $w(x,y)$.

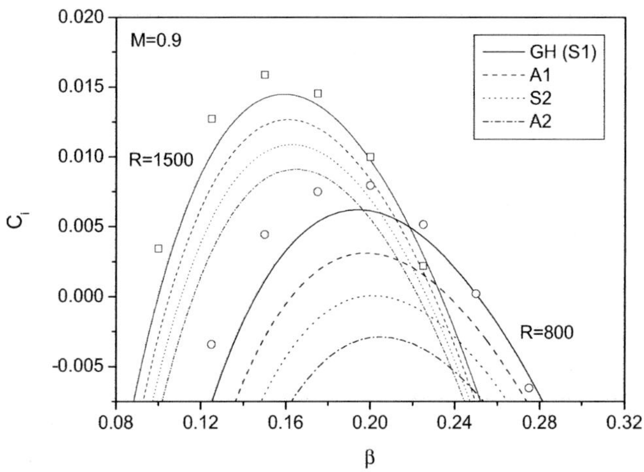

Figure 3. Dependence of c_i on β for modes GH, A1, S2, A2 at $M = 0.9$, $R = 800$ and 1500.

4. DISCUSSION

The first BiGlobal instability analysis of viscous compressible swept Hiemenz flow has been performed. Good agreement between asymptotic analysis and numerical solution of the partial-differential eigenvalue problem has been obtained within appropriate parameter ranges. It has been demonstrated that the three-dimensional "polynomial" eigenmodes of incompressible flow **[13]** persist in the subsonic flow regime. However, differences of the two approaches are found to occur at moderate Reynolds and high Mach numbers. This underlines both the efficiency of the asymptotic approach at the high-Reynolds number subsonic regime and the need for accurate numerical methodologies in order to provide reliable predictions of instability characteristics of this flow at all speed ranges.

ACKNOWLEDGEMENTS

The work of V. Theofilis was partly supported by the European Office of Aerospace Research and Development, the Air Force Research Laboratory, and the Air Force Office of Scientific Research, under Grant No. FA8655-03-1-3059 monitored by Dr. John D. Schmisseur (AFOSR) and Mr. Wayne Donaldson (EOARD). Additional partial support was provided by a Ramón y Cajal research fellowship of the Spanish Ministry of Science and Technology.

REFERENCES

1. Görtler, H. "Dreidimensionale Instabilität der ebenen Staupunktströmung gegenüber wirbelartigen Störungen". In *50 Jahre Grenzschichtforschung* (ed. H. Görtler, W. Tollmien), Vieweg und Sohn., pp. 304-314, 1955.
2. Hämmerlin, G. "Zur instabilitätstheorie der ebenen Staupunktströmung". In 50 Jahre Grenzschichtforschung (ed. H. Görtler, W. Tollmien), Vieweg und Sohn, pp. 315-327, 1955.
3. Hall, P, Malik, MR, Poll, DIA. "On the stability of an infinite swept attachment-line boundary layer". *Proc. R. Soc. Lond. A* **395**, pp. 229-245, 1984.
4. Hiemenz, K. "Die Grenzschicht an einem in den gliechförmigen Flüssigkeitsstrom eingetauchten geraden Kreiszylinder. Thesis, Göttingen. Also *Ding l. Polytechn J.* **326**, pp. 321-324, 1911.
5. Horten, R., Selinger, P.F. "Nurflügel - Die Geschichte der Horten-Flugzeuge 1933-1960". Weishaupt., 1983.
6. Kármán, T., Edson, L. "The wind and beyond. Theodore von Kármán: Pioneer in aviation and pathfinder in space". Little, Brown & Co. Boston, 1967.
7. Lin, RS, Malik, MR. "On the stability of attachment-line boundary layers. Part 1.The incompressible swept Hiemenz flow". *J. Fluid Mech.* **311**, pp. 239-255, 1996.
8. Mack, LM. "Boundary-layer linear stability theory". *AGARD Rep 709*, 1984.
9. Reshotko, E. and Beckwith, IE. "Compressible laminar boundary layer over a yawed infinite cylinder with heat-transfer and arbitrary Prandtl number". NACA TR1379, 1958.
10. Schlichting, H. "*Grenzschichttheorie*", Braun, 1951.
11. Spalart, PR. "Direct numerical study of leading-edge contamination". AGARD CP-438, pp. 5-1 - 5-13, 1988.
12. Theofilis, V. "On linear and nonlinear instability of the incompressible swept attachment-line boundary layer". *J. Fluid Mech.* **355**, pp. 193-227, 1998.
13. Theofilis, V., Fedorov, A. Obrist, D., Dallmann, UCh. "The extended Görtler–Hämmerlin model for linear instability of three-dimensional incompressible swept attachment-line boundary layer flow". *J. Fluid Mech.* **487**, pp. 271-313, 2003.
14. Theofilis, V., Colonius, T. "Three-dimensional instabilities of compressible flow over open cavities: direct solution of the BiGlobal eigenvalue problem". AIAA Paper 2004-2544, 2004.

APPLICATION OF TRANSIENT GROWTH THEORY TO BYPASS TRANSITION

Eli Reshotko
Case Western Reserve University
Cleveland, Ohio
U.S.A.

Anatoli Tumin
University of Arizona
Tucson, Arizona
U.S.A.

Abstract: Transient growth arises through the coupling between slightly damped, highly oblique (nearly streamwise) T-S and Squire modes leading to algebraic growth followed by exponential decay in a region that is subcritical with respect to the T-S neutral curve. A weak transient growth can also occur for two dimensional or axisymmetric modes since the Orr-Sommerfeld operator and its compressible counterpart are not self-adjoint, therefore their eigenfunctions are not strictly orthogonal. So transient growth is a candidate mechanism for many examples of bypass transition. The relevance to bypass transition is examined through the example of the hypersonic blunt body paradox.

Key words: Transient growth, bypass transition

1. INTRODUCTION

Prandtl [**1**] explained the early anomaly that flows such as the Blasius boundary layer, stable at infinite Reynolds number, could be unstable at finite Reynolds number. Thus, until about a dozen years ago, the predominant view of laminar-turbulent transition was centered around the slow linear amplification of exponentially growing disturbances (the familiar T-S waves), preceded by a receptivity process to the disturbance environment and followed by secondary instabilities, further non-linearity and finally a breakdown to a recognizable turbulent flow.

However, there are transition phenomena in flows that are linearly stable and so could not be attributed to the aforementioned "T-S path." These were labeled by Morkovin [**2**] as bypass transition. The general feeling then expressed by Morkovin as well as the present first author was that bypass transition was inherently non-linear, having bypassed the linear T-S processes. We often joked that bypass transition either bypassed the T-S

G.E.A. Meier and K.R. Sreenivasan (eds.), IUTAM Symposium on One Hundred Years of Boundary Layer Research, 83-93,

processes or bypassed our knowledge, or both. This picture had to be urgently reconsidered in the early 1990's with the emergence of a literature on transient growth.

Transient growth arises through the non-orthogonal nature of the Orr-Sommerfeld and Squire eigenfunctions. The largest effects come from the non-orthogonal superposition of slightly damped, highly oblique (near streamwise) T-S and Squire modes. These modes are subcritical with respect to the T-S neutral curve. The transient growth signature is essentially algebraic growth followed by exponential decay. A weak transient growth can also occur for two-dimensional or axisymmetric modes. So transient growth is therefore a candidate mechanism for many examples of bypass transition.

The early developments in transient growth are describe and summarized in the book by Schmid and Henningson [**3**]. Butler and Farrell [**4**] determined optimal disturbance parameters for maximum transient growth in plane Couette, plane Poiseuille and Blasius flows. These optimal disturbances have a decided three-dimensionality. In most cases, the optimal disturbances are stationary. They are for zero frequency and a particular spanwise wavenumber. It is important to emphasize that the transient growth theory is linear.

All of the above papers use a temporal formulation of the disturbance equations, that is, that the disturbances grow in time rather than in space. The spatial formulation described by Reshotko and Tumin [**5**] derives from the work of Ashpis and Reshotko [**6**] who studied the spatial response to a vibrating ribbon in a Blasius boundary layer. For a given real frequency, they showed that the upper half of the complex α-plane, properly indented to include the growing discrete modes, contains the eigenvalues that apply to the domain downstream of the vibrating ribbon while the lower half of the complex α-plane contains the eigenvalues that apply to the upstream domain. Thus for the case of the downstream response to a vibrating ribbon or any other disturbance source, one need consider only those eigenvalues in the upper half plane of the spatial spectrum properly indented to contain the growing discrete modes. The balance of the analysis parallels that done for the temporal case.

The consequence of these arguments is that transient growth can be a significant factor in the transition to turbulent flow for flows that are T-S stable. A summary of the early application of transient growth theory to cases of bypass transition is by Reshotko [**7**].

2. PATHS TO TURBULENCE IN WALL LAYERS

Consideration of transient growth has led to an enlargement and clarification of the paths to transition by Morkovin, Reshotko and Herbert [**8**] and is shown in Fig. 1.

Five paths to transition, A through E, are shown in this figure. A discussion of each of these paths follows. Examples are given particularly where related to transient growth and bypass transition.

Path A - Path A corresponds to the situation where transient growth is insignificant and transition is due to traditional T-S or Görtler mechanisms. This is the traditional path to transition for low disturbance environments where modal growth is significant. Summaries of all aspects of this path - disturbance environment, receptivity, linear and nonlinear instability, transition prediction and transition control - are available by Reshotko [**9**], Reed, Saric and Arnal [**10**] and Saric, Reed and White [**11**].

Path B - As described by the authors of Ref. [**8**], the Path B scenario indicates some transient growth providing a higher initial amplitude to the eigenmode growth upon crossing into an exponentially unstable region. There are no obvious examples in the literature of this scenario. It is somewhat troublesome because of the following:

Transient growth (nonmodal) is largest for stationary streamwise disturbances. Modal growth is largest for transverse disturbances at low speeds, or oblique disturbances at supersonic speeds. How a streamwise disturbance would couple to a transverse disturbance is not clear. It may be that a traveling nonmodal disturbance can couple with an oblique modal disturbance. It is more likely that the nonmodal and modal disturbances will develop independently. A good test case for Path B would be transient growth preceding a Görtler instability. Both involve streamwise disturbances of comparable wavelength.

In an interesting set of experiments, Kosorygin and Polyakov [**12**] report that for Tu < 0.1%, they observe T-S bands in their spectra, u' profiles that conform to T-S eigenfunctions, and amplitude growth in accordance with T-S theory. For Tu > 0.7%, low frequency disturbances are strong and display the Klebanoff mode [**13**]. The u' spectra fall monotonically with frequency. The fluctuations are three-dimensional with lateral scales of the order of a boundary layer thickness. For intermediate turbulence levels (0.1% < Tu < 0.7%) they report both T-S growth and Klebanoff mode growth to be

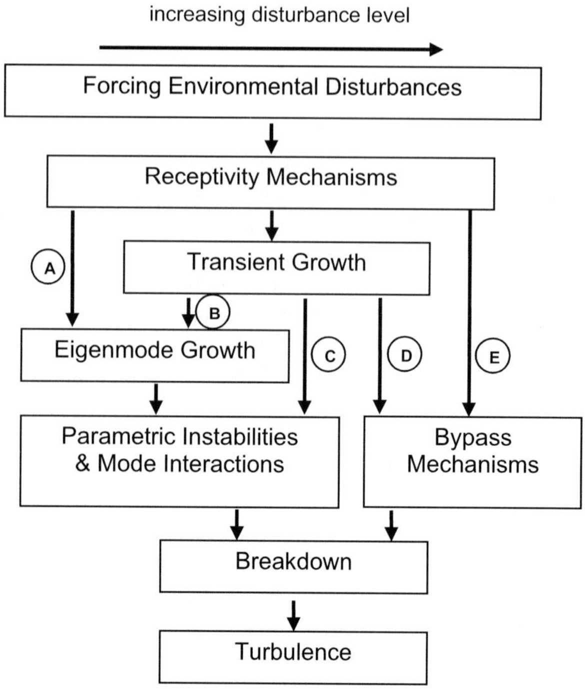

Figure 1. Paths to turbulence in wall layers

concurrent and that "transition has been determined by coexistence and interaction of two kinds of eddy motion whose lateral scales differ strongly from one another. The growth of the T-S wave changes weakly as compared to the estimate by the linear stability theory." Related experiments are by Suder et al [**14**] and Sohn and Reshotko [**15**]. For Tu = 0.3% - 0.4%, T-S bands are observed in the hot-wire spectra. For Tu > 0.9%, there is no clear evidence of T-S bands. Unfortunately, no attempts were made in those experiments to measure a spanwise scale.

On the other hand, Cossu and Brandt[**16**] suggest that the Blasius boundary layer can be stabilized by streamwise streaks of sufficiently large amplitude to cause a nonlinear distortion of the basic flow.

Path C - Path C is the case where eigenmode growth is absent. This is the transient growth path that has received the most attention because it covers the most salient cases of bypass transition. The optimal disturbances of Butler and Farrell [**4**] show large transient energy growth factors for plane

Couette flow as well as for plane Poiseuille flow below the Branch I Reynolds numbers. The particular case of Poiseuille pipe flow is described in detail by Reshotko and Tumin[**5**]. The "blunt body paradox" is discussed in in sections **3** of this paper.

Path D - In Path D, the result of the transient growth is that the spectrum of disturbances in the boundary layer is full – it looks like a turbulent spectrum (even while the basic flow profiles are still laminar). The spectra decrease monotonically with increase in frequency while the intensity level increases with distance downstream. Examples of Path D are in the experimental results of Suder et al [**14**] and of Sohn and Reshotko [**15**] for Tu > 1%. Based on transient growth theory, Andersson et al [**17**] have developed a very plausible correlation for flat plate transition at elevated freestream turbulence levels.

Path E - Path E represents the case of very large amplitude forcing where there is no linear regime. Such large amplitude forcing might come from chopping the free stream to obtain very large disturbance levels. The resulting freestream spectra do not resemble wind tunnel or grid turbulence spectra.

3. THE BLUNT BODY PARADOX

The "blunt body paradox" refers to the early transition on spherical forebodies (even those that are highly polished) observed at supersonic and hypersonic freestream speeds both in flight and in wind tunnels. This transition occurs usually in the subsonic portion of the flow behind the bow shock wave, a region of highly favorable pressure gradient that is stable to T-S waves. Surface cooling leads to even earlier transition. This phenomenon, identified in the mid-1950's, has defied clear explanation. The tentative suggestions are generally roughness related since stagnation point boundary layers are very thin. But no connection has been made between the microscopic roughness on the surface and the features of the observed early transition such as location, sensitivity to surface temperature level, etc. This has led to a search for an explanation through transient growth. This problem was first examined by Reshotko and Tumin [**18**]. A more recent work, resulting in a nosetip transition correlation, is by Reshotko and Tumin [**19**] and this section summarizes their findings.

Extensive calculations have also been carried out for axisymmetric stagnation point flows, $\beta_H = 0.5$,where β_H is the Hartree pressure gradient parameter. These are relevant to the spherical nosetip of hypersonic sphere-cone configurations for which there is an extensive experimental data base and significant transition correlations (Batt & Legner [**20 – 21**], Reda [**22 – 23**]). Fig. 2 shows the spatial transient growth factor G as a function of downstream distance for three surface temperature levels. For adiabatic wall conditions, the maximum growth factor is about 1100. For cooled walls, the growth factors are significantly larger reaching about 13500 for T_w / T_{aw}= 0.2. Since, most of the experimental data base is centered around T_w / T_{aw}= 0.5, the results for this latter case will be emphasized here. Further, in contrast to the flat plate, curvature is a significant factor for the sphere. Curvature is included in our calculations and in the results to be presented.

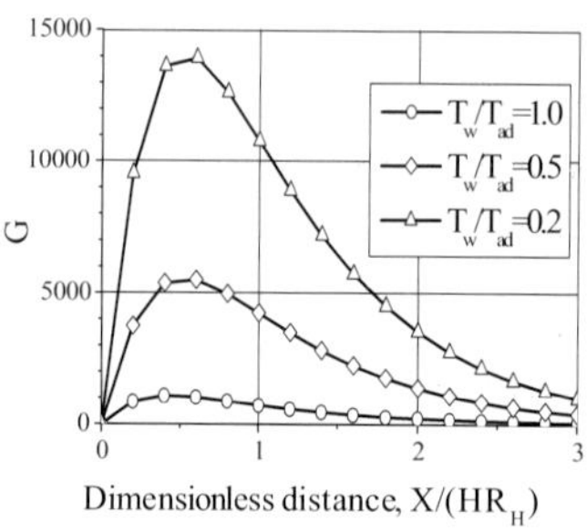

Figure 2. Spatial transient growth amplification factor for axisymmetric stagnation point flow at different surface temperature levels, $M_e = 0.6$, $Re_H = 1137$, $\omega = 0$, $\beta H = 0.762$

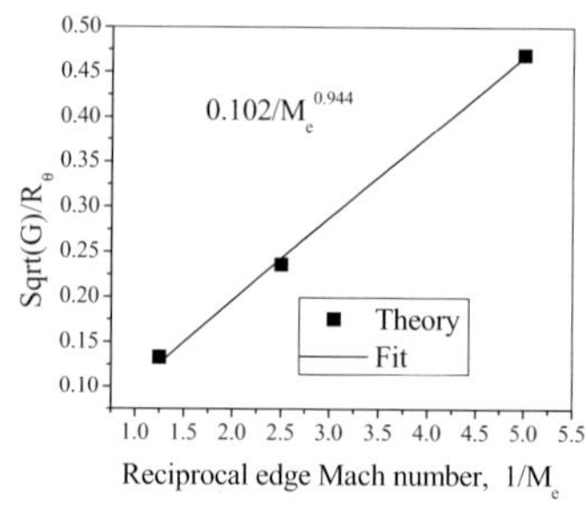

Figure 3. Optimal growth factors for axisymmetric stagnation point flow, $T_w / T_{aw} = 0.5$

For surface temperatures in the vicinity of T_w / T_{aw}= 0.5, the transient growth results without curvature effects are shown in Figure 3. Since the flow is compressible, the energy growth factors are based on the Mack [**24**] energy norm. The optimal spanwise wavenumber is essentially constant over the Mach number range at $\beta_{opt}\theta = 0.28$ which corresponds to about 3.2 boundary layer thicknesses. The curvature effects were included through the ratio G / G_0 of the energy growth factors with and without the curvature associated terms [**19**]. Since most of the experimental runs had surface temperature level variations during the run, and since the growth factors are sensitive to surface temperature level, it has further been determined from a least squares fit of the peak values in Fig. 2 that $G^{1/2}$ varies as $(T_w / T_{aw})^{-0.77}$. in the vicinity of $T_w / T_{aw} = 0.5$. The above can be summarized by:

$$G^{1/2} = \frac{0.1 \dfrac{\mathrm{Re}_\theta}{M_e^{0.944}} \left(\dfrac{2T_w}{T_{aw}} \right)^{-0.77}}{1 + 14.9 \,\mathrm{Re}_\theta \dfrac{\theta}{R_N}} \tag{1}$$

where the denominator in Eq. (1) represents the curvature effect. The larger the curvature is (smaller nose radius, R_N), the smaller the growth factor. This stabilizing effect agrees with previous experimental observations [**20**].

Influence of roughness on transition - The influence of roughness on transition can be modeled in a manner similar to that used by Andersson et al [**17**] for free-stream turbulence effects on transition. We assume that an energy norm at transition is related to an input energy through the transient growth factor *G*.

$$E_{tr} = GE_{in} \tag{2}$$

The input energy is in the form of a density times velocity squared where the roughness-induced disturbance velocities are assumed proportional to the roughness height, *k*. The momentum thickness, θ, is chosen as the reference length since it is the least sensitive to surface temperature level of any of the boundary layer scales. For stagnation-point flow, θ is also constant with distance from the stagnation point. The resulting expression for E_{in} is

$$E_{in} = \frac{\rho_w}{\rho_e} \left(\frac{k}{\theta} \right)^2 \tag{3}$$

For a boundary layer, Eq. (3) can be approximated by

$$E_{in} = \frac{T_e}{T_w} \left(\frac{k}{\theta} \right)^2 \tag{4}$$

Again, the growth factor *G* scales with the square of a thickness Reynolds number or with length Reynolds number to the first power. Thus from Eqs. (2) and (4) we can write

$$(E_{tr})^{1/2} = \frac{G^{1/2}}{\mathrm{Re}_\theta} \mathrm{Re}_\theta \frac{k}{\theta} \left(\frac{T_e}{T_w}\right)^{1/2} \qquad (5)$$

where $G^{1/2}/\mathrm{Re}_\theta$ is obtained from the transient growth results for the particular geometry and flow parameters. Transition is assumed to occur when, for the given flow, E_{tr} reaches a specific value, here taken as a constant.

We have to extract the factor $G^{1/2}/\mathrm{Re}_\theta$ from the transient growth results. The calculations summarized by Eq. (1) are for parallel flows (M_e = const). However, for the stagnation point flow, the edge Mach number varies almost linearly with angle from the stagnation point. From Eq. (1), it is seen that $G^{1/2}/\mathrm{Re}_\theta \sim 1/M_e^{0.944}$. Thus the growth factor is largest near the stagnation point and diminishes rapidly as the edge Mach number increases. An integration of the differential growth factors from the stagnation point to any downstream location shows that the integrated growth factor is essentially constant beyond $M_e \sim 0.1$ and therefore the integrated $G^{1/2}/\mathrm{Re}_\theta$ is essentially independent of the local Mach number at the transition location.

Thus, for the stagnation point at $T_w/T_{aw} \approx 0.5$, the relation is:

$$\frac{\mathrm{Re}_{\theta,tr}(k/\theta)(T_e/T_w)^{1/2}\,(2T_w/T_e)^{-0.77}}{(1+14.9\,\mathrm{Re}_\theta \frac{\theta}{R_N})} = const \qquad (6)$$

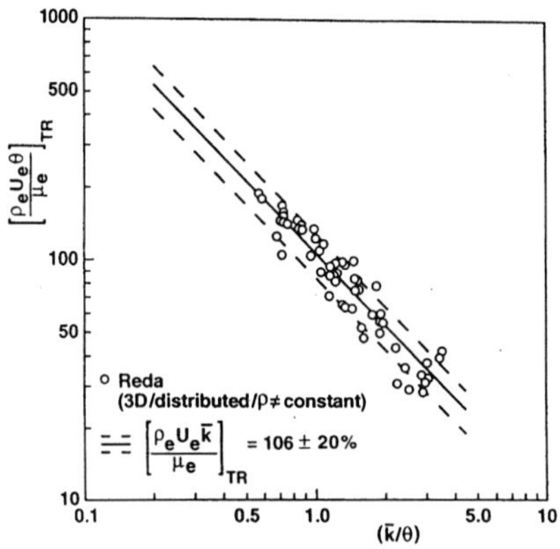

Figure 4. Nosetip transition data from ballistic-range experiments; 3D distributed roughness [22 – 23]

The last relation shows the trends of transition Reynolds number with roughness height and surface temperature level. For constant surface temperature level, $Re_{\theta,tr}$ should vary as $(k/\theta)^{-1}$. This is consistent with Reda's ballistic range data [**22 – 23**] as shown in Fig. 4.

The PANT wind-tunnel data [**20 – 21**] shown in Fig. 5a display this trend as well. In addition, some of the PANT data were taken for nearly constant (k/θ) but with varying surface temperature about $T_w/T_{aw} = 0.5$. For this case, Eq. (6) shows that $Re_{\theta,tr}$ should vary as $(T_w/T_{aw})^{1.27}$. This again is supported by the PANT data as shown by comparison of the data with lines of slope $n = 1.27$ in Fig. 5b. To be noted is that all of the nosetip transitions in the PANT data base take place well within the sonic point on the sphere ($0.2 < M_{e,tr} < 0.8$) and with $20 < Re_{\theta,tr} < 120$. The present summary relation for the PANT data base is

$$Re_{\theta,tr} = 180\ (k/\theta)^{-1}(T_e/2T_w)^{-1.27} \tag{7}$$

The curvature factor is ignored as it varies only within a narrow range for the whole data base. The numerical factor of 180 is for $T_w/T_{aw} = 0.5$ and comes

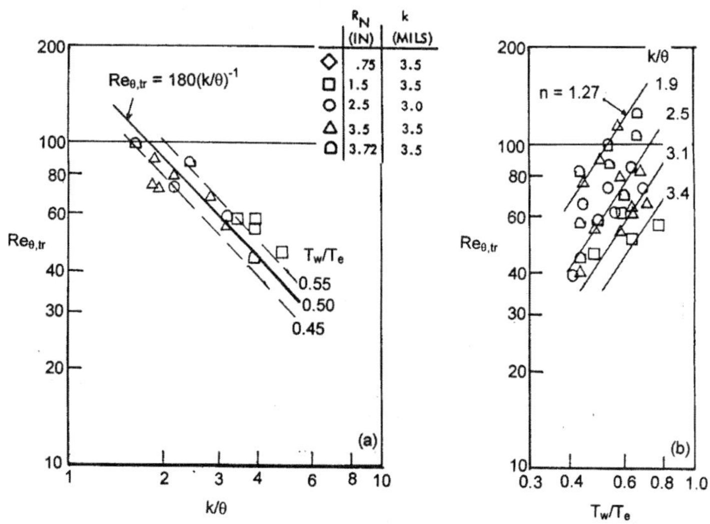

Figure 5. Transient growth based transition correlations of PANT Series J data

from averaging in only those points for which $0.45 < T_w / T_{aw} < 0.55$. The broken lines on either side of the solid line in Fig. 5a show the expected data spread for $0.45 < T_w/T_e < 0.55$ according to Eq. 7.

While the Reda correlation (Fig. 4) and the present correlation of the PANT data (Eq. 7) both vary as $(k/\theta)^{-1}$ for constant surface temperature, Reda's ballistic range data and the PANT wind tunnel data were taken at different temperature levels. Since θ appears in the numerator of both sides of Eq. 7, this relation can be rewritten as

$$U_e k / \nu_e = 180(2T_w / T_e)^{1.27} \tag{8}$$

The left side of Eq. (8) is the same as Reda's [**22 – 23**] $Re_{ke,tr}$. For T_w / T_{aw} = 0.33, Eq. (8) gives Reda's value of 106 (see Fig. 4). Reda estimates his surface temperature level to have been about 0.3. Thus the two data sets are also quantitatively comparable.

We propose therefore that the "blunt body paradox" is the result of transient growth due to surface roughness.

4. CONCLUDING REMARKS

A number of transition scenarios that involve transient growth have been reviewed and discussed. Transient growth does offer an explanation for a number of examples of bypass transition.

Transient growth is subject to a receptivity process that has not been here considered in any depth. It does imply however that the "optimal" disturbances, the focus of many transient growth studies in the literature, are not generally realized unless their parameters (frequency or wavenumber) are part of the disturbance input.

Perhaps most intriguing is transient growth as an explanation for roughness-induced transition. In the case of the blunt body paradox, it was shown that even small roughness could be important because of the large calculated transient growth factors over highly cooled surfaces. Since the transient growth factors are flow dependent, the sensitivity of transition to distributed roughness is very much flow dependent.

ACKNOWLEDGEMENTS

Support of this work by the U.S. Air Force Office of Scientific Research is gratefully acknowledged.

REFERENCES

1. Prandtl, L. "Bemerkungen uber die Enstehung der Turbulenz", ZAMM, **1**, pp. 431-435, 1921
2. Morkovin, MV. "Bypass transition to turbulence and research desiderata," in Graham, R, ed., *Transition in Turbines*, NASA Conf. Publ. 2386, 161, 1985
3. Schmid, PJ, Henningson,DS. *Stability and Transition in Shear Flows,* Springer-Verlag, New York, 2001
4. Butler, KM, Farrell, BF. "Three-dimensional optimal perturbations in viscous shear flow," *Phys. Fluids A*, **4**,**8**, pp. 1637-1650, 1992
5. Reshotko, E, Tumin, A. "Spatial theory of optimal disturbances in a circular pipe flow," *Phys. Fluids*, **13**,**4**, pp. 991-996, 2001
6. Ashpis, D, Reshotko, E. "The vibrating ribbon problem revisited," *J. Fluid Mech.*, **213**, pp. 513-547, 1990
7. Reshotko, E. "Transient Growth: A factor in bypass transition," *Phys. Fluids*, **13**,**5**, pp. 1067-1075, 2001
8. Morkovin, MV, Reshotko, E. Herbert, T. "Transition in open flow systems – a reassessment" *Bull. APS*, **39**,**9**, p. 1882, 1994
9. Reshotko, E. "Boundary Layer Instability, Transition and Control," Dryden Lecture in Research, AIAA Paper 94-0001, 1994
10. Reed, HL, Saric, WS, Arnal, D. "Linear Stability Theory Applied to Boundary Layers," *Ann. Rev. Fluid Mech.*, **28**, pp. 389-428, 1996
11. Saric, WS, Reed, HL, White, EB. "Stability and Transition of Three-Dimensional Boundary Layers," *Ann.Rev. Fluid Mech.*, **35**, pp. 413-440, 2003
12. Kosorygin, VS, Polyakov, NPh. "Laminar Boundary Layers in Turbulent Flows," in Arnal, D and Michel, R, eds, *Laminar-Turbulent Transition*, Springer-Verlag, pp. 573-578, 1990
13. Klebanoff, PS. *Bull. Am. Phys. Soc.*, **10**,**11**, p. 1323, 1971
14. Suder, KS, O'Brien, JE, Reshotko, E. "Experimental Study of Bypass Transition in a Boundary Layer," NASA Tech. Memorandum 100913, 1988
15. Sohn, KH, Reshotko, E. "Experimental Study of Boundary Layer Transition With Elevated Freestream Turbulence on a Heated Flat Plate," NASA Contractor Report 187068, 1991
16. Cossu, C, Brandt, L. "Stabilization of Tollmien-Schlichting waves by finite amplitude optimal streaks in the Blasius boundary layer," *Phys. Fluids*, **14**,**8**, pp. L57-L60, 2002
17. Andersson, P, Berggren, M. Henningson, D. "Optimal disturbances and bypass transition in boundary layers," *Phys. Fluids*, **11**, pp. 135-150, 1999
18. Reshotko, E, Tumin, A, "The Blunt Body Paradox – A Case for Transient Growth," in Fasel, HF, Saric, WS eds, *Laminar-Turbulent Transition*, Springer, pp. 403-408, 2000
19. Reshotko, E, Tumin, A. "Role of Transient Growth in Roughness-Induced Transition," *AIAA Journal*, **42**,**4**, pp. 766-770, 2004
20. Batt, RG, Legner, HL. "A Review and Evaluation of Ground Test Data on Roughness Induced Nosetip Transition," Report BMD-TR-81-58, 1980.
21. Batt, RG, and Legner, HL. "A Review of Roughness-Induced Nosetip Transition," *AIAA Journal*, **21**,**1**, pp. 7-22, 1983
22. Reda, DC. "Correlation of Nosetip Boundary Layer Transition Data Measured in Ballistic Range Experiments," *AIAA Journal* **19**,**3**, pp. 329-339, 1981
23. Reda, D.C. 2002, "Review and Synthesis of Roughness-Dominated Transition correlations for Reentry Applications," *Journal of Spacecraft and Rockets*, **39**,**2**, pp. 161- 167, 2002
24. Mack, L.M. "Boundary layer stability theory", JPL Report 900-277, 1969

ROUTES OF BOUNDARY-LAYER TRANSITION

Yury S. Kachanov
Institute of Theoretical and Applied Mechanics of Siberian Branch of the Russian Academy of Sciences, Institutskaya str. 4/1, 630090 Novosibirsk, Russia, phone: +7(3832)304278, fax: +7(3832)304278, e-mail: kachanov@itam.nsc.ru

Abstract: Based on multiple experimental observations it is well known at present that there are two most important kinds of fluid flow, in general, and boundary-layer flow in particular. One of them is called the laminar flow, another one — the turbulent flow. Meanwhile, there is a third kind of flow: *the transitional boundary layer*, which plays a role of a bridge between the other two main types of flow and represents a very special, intricate, and practically important physical phenomenon. The aim of the present paper is not to make a historical description of steps in boundary-layer transition research but rather to discuss the state of the art in the field and to give an overview of some important aspects and issues of the problem based on its *modern understanding*.

Key words: Boundary layer, receptivity, stability, non-linear disturbance interactions, laminar-turbulent transition, turbulence production.

1. INTRODUCTION

First purposive investigation of the laminar-turbulent transition problem was performed by O. Reynolds in 1883 [1] even earlier than the notion of the boundary layer was introduced by L. Prandtl in his famous work published in 1904 [2]. Despite the fact that experiments by O. Reynolds were carried out in a pipe flow, it turned out later that the transition processes occurring in various wall shear flows are often rather similar to each other. The investigations of the laminar-turbulent transition phenomenon *in boundary layers* were started very soon after publication of the Prandtl's work [2], the 100th anniversary of which we commemorate now.

G.E.A. Meier and K.R. Sreenivasan (eds.), IUTAM Symposium on One Hundred Years of Boundary Layer Research, 95-104,

2. TRANSITION AND INSTABILITIES

According to modern notions, boundary-layer transition is always associated with one or several *instabilities* of either the primary laminar base flow (in case of infinitely small perturbations) or the base flow disturbed by finite-amplitude perturbations. At present, three main kinds of instabilities have been found and investigated in boundary layers (Fig. 1): (i) convective instabilities, (ii) absolute instabilities, and (iii) global instabilities.

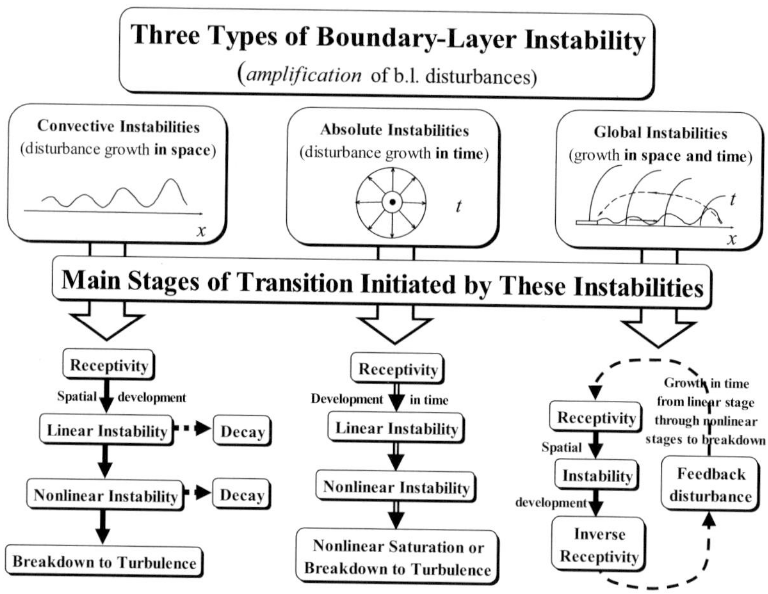

Figure 1. Boundary-layer instabilities and main stages of transition initiated by them.

The first of them is associated with a spatial growth of some steady or traveling disturbances. The second one assumes that the perturbations are amplified in time at some spatial locations. The third kind of instability occurs when a feedback coupling of perturbations is present and leads to appearance of a closed loop with a disturbance growth in time due to their convective (i.e. spatial) amplification in one or several parts of the loop.

These instabilities result in rather different general scenarios of transition. At the same time, every scenario includes four common elements (Fig. 1), corresponding to the most important aspects of the transition problem: (i) receptivity, (ii) linear instability, (iii) nonlinear instability, and (iv) final laminar-flow breakdown. (Nonlinear receptivity problems might be also important but not very often.) The term 'boundary-layer receptivity' stands for various mechanisms of transformation of external (with respect to the boundary layer) perturbations into boundary-layer perturbations. Note that instability mechanisms can only *amplify* boundary-layer disturbances, while the receptivity mechanisms are able to *produce* them.

In principle, every one of the instabilities (or a combination of several instabilities) can result in laminar flow breakdown. However, transition may not occur due to several reasons. First, *convectively* developing instability modes (waves or vortices) can start to decay in so far as properties of any real boundary layer *generally change downstream* (either its thickness, or shape factor, or the value of edge velocity). Second, disturbance growth associated with *absolute* or *global* instability can saturate due to nonlinear effects (including base flow distortions) but the disturbances cannot start to decay if the instability has begun. In this case some self-oscillating or steadily distorted secondary-flow regimes can appear and exist for a long time. Note that due to streamwise variation of boundary-layer properties, *convectively* amplified instability modes cannot in general become neutrally stablc, i.e. their growth cannot saturate for a long streamwise distance even at nonlinear stages of convective instability.

In contrast to some closed and separated flows, usual stationary boundary layers are convectively unstable in a *great majority* of situations due to: (i) rather large group velocities of amplified perturbations making impossible absolute instability and (ii) a very low efficiency of upstream feedbacks (usually acoustic ones) making impossible global instability. Therefore, we mainly concentrate further on the transition scenarios associated with convective instabilities. Let us consider different stages of such types of transition more systematically.

3. RECEPTIVITY ASPECT OF TRANSITION PROBLEM

The problem of boundary layer *receptivity* to external perturbations is complementary to the instability phenomenon (Fig. 2). The receptivity mechanisms are responsible for transformation of various external perturbations (including those originating from the wall) into boundary-layer perturbations.

Due to the predominantly convective character of boundary-layer instabilities, the role of the receptivity problem in boundary-layer transition is difficult to overestimate. Indeed, the transition enters a nonlinear stage when the disturbance amplitude reaches a certain threshold value A_t (that is very different for different types of instability and base flow). In a typical case of exponential disturbance growth and a localized type of linear receptivity (see below), this value can be expressed as $A_t = A_0 e^{\sigma \Delta x} = G A_e e^{\sigma \Delta x}$. Here A_0 is an initial amplitude of boundary-layer disturbance, A_e is an amplitude of external (with respect to the boundary layer) perturbation, G is the *receptivity coefficient* characterizing the efficiency of transformation of external perturbations into boundary-layer disturbances, σ is the disturbance growth rate, and $\Delta x = [\ln A_t - \ln(G A_e)]/\sigma$ is the location of the transition beginning. It is seen that when the efficiency of the receptivity mechanism tends to zero (i.e. $G \to 0$), Δx tends to infinity and the transition can never occur even if external perturbations have very high amplitudes A_e and the flow is strongly unstable (i.e. σ is very large).

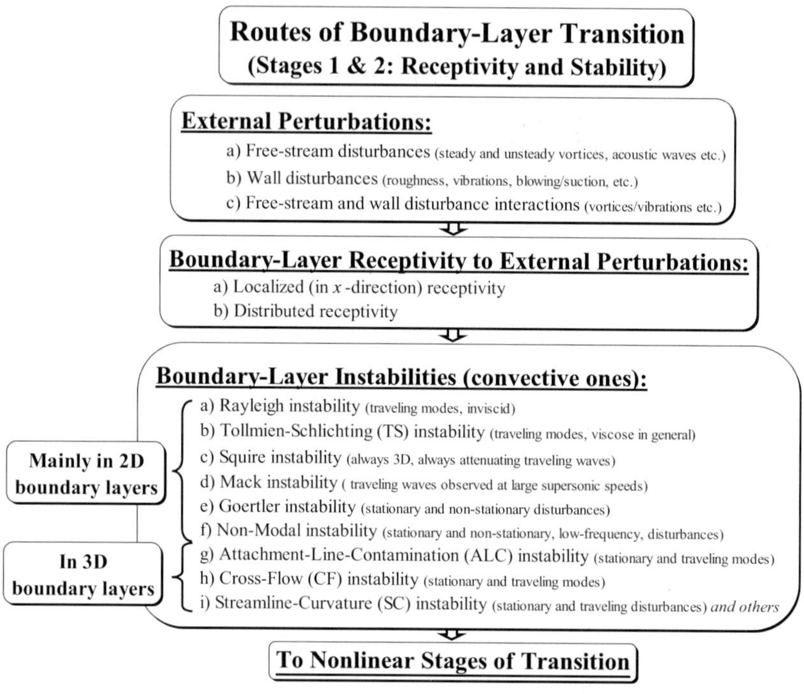

Figure 2. Boundary-layer receptivity and most important convective instabilities.

The receptivity process represents an initial stage of the transition process (see Fig. 2 and reviews in [3 - 5]), which starts, in fact, from the very leading edge, where the free stream just touches with a body. Consequently, almost any laminar boundary layer can be regarded simultaneously as a transitional boundary layer, at least when the Reynolds numbers are not too low.

There are two main kinds of receptivity mechanisms: (i) the localized one and (ii) the distributed one. In the first case, the streamwise scale of the region of excitation of boundary-layer perturbations is small in comparison to that of the region of their subsequent amplification and, hence, the receptivity and the instability problems are separated in space and can be decoupled. In the second case, the receptivity takes place in a long streamwise domain. The boundary-layer perturbations are simultaneously excited (due to the receptivity) and amplified (due to an instability) in the same streamwise regions. This circumstance leads to coupling up the distributed receptivity and the instability problems and makes the receptivity study much more complicated.

There are several receptivity mechanisms, most important for transition, found at present in subsonic boundary layers. *In 2D boundary layers* generation of TS-waves (2D and 3D ones) occurs due to: (i) surface vibrations, (ii) scattering of acoustic waves on surface non-uniformities and vibrations, (iii) resonant scattering of non-stationary free-stream vortices with wall-normal vorticity on distributed surface roughness, and (iv) generation of streaky structures occurring mainly under the influence of steady (wall-normal and streamwise) free-stream vortices. *In 3D boundary layers (with dominating CF-instability)* generation of CF-modes (vortices and waves) occurs due to: (i) surface roughness, (ii) surface vibrations, (iii) streamwise oriented steady free-stream vortices, and (iv) resonant scattering of non-stationary free-stream vortices with streamwise vorticity on distributed surface roughness. At the same time, a large number of other receptivity mechanisms has not been studied yet.

4. VARIETY OF CONVECTIVE INSTABILITIES

There are many types of convective boundary-layer instabilities (Fig. 2), leading to growth of various kinds of perturbations. The most important of them (studied in case of rigid surfaces) are: (*a*) Rayleigh waves, (*b*) Tollmien-Schlichting (TS) waves, (*c*) Squire modes (always decaying waves with wall-normal vorticity), (*d*) Mack waves (observed at large supersonic speeds), (*e*) Görtler vortices, (*f*)streaky structures (optimal disturbances), (*g*) cross-flow (CF) vortices and waves, (*h*) attachment-line-contamination (ALC) instability modes, (*i*) streamline-curvature (SC) instability modes, and some others. Some of them are mainly stationary (like CF-vortices, Görtler vortices, and streaky structures); others are non-stationary (like TS- or CF-waves). Let us characterize these instabilities in more detail.

The instability modes important primarily for *2D boundary layers* are the following (Fig. 2).

The *Rayleigh waves* represent a kind of traveling (2D or 3D) disturbances associated with essentially inviscid instability, which occurs in presence of an inflexion point in the mean velocity profile. This instability is very important in adverse-pressure-gradient (APG) boundary layers (both two- and three-dimensional) and has a physical nature similar to that of Kelvin-Helmholtz instability observed in free-shear layers. The *Tollmien-Schlichting waves* represent a viscous extension of Rayleigh waves. They are very important in a great variety of transition scenarios observed in both 2D and 3D boundary layers, both with and without streamwise pressure gradients. The *Mack waves* (whose other names are second mode or acoustic mode) are observed at large supersonic or at hypersonic speeds and represent a branch different from TS-waves developing in compressible flows (called also the first or vortical mode).

The *Görtler instability modes* represent mainly steady streamwise vortices (but also unsteady ones) associated with inertial Görtler instability appearing due to combined influence of wall curvature $K = \partial^2 Y/\partial x^2$ and wall-normal mean-velocity gradient $\partial U/\partial y$. This instability may occur when the product $K\partial U/\partial y$ is positive and exceeds a certain threshold (which depends on Reynolds number). Görtler vortices represent a convective instability; therefore, they are amplified from certain initial (priming) perturbations and the receptivity aspect of the corresponding transition problem is as important as in cases of instabilities (*a*) to (*d*) discussed above.

The *streaky structures* (*optimal disturbances*) represent a rather exotic kind of boundary-layer instability called *non-modal instability* or mechanism of *transient growth* of perturbations. This instability is related to near-fields of disturbance sources and connected with so-called *lift up effect* and *non-orthogonality of eigen modes* of the associated linear stability problem including so-called *modes of continuous spectrum* and Squire modes. This instability can lead to a rather strong (but always restricted) amplification of steady or quasi-steady elongated streamwise vortices with streamwise wavenubers close to zero, for which disturbance-source near fields have relatively large streamwise extents. Such mechanisms can be very important in cases of enhanced external perturbations (such as free-stream turbulence, surface vibrations, unsteady blowing-suction etc.) because spatial ranges of possible disturbance amplification are always rather restricted (the growth is *transient*) leading to restricted maximum amplification factors as discussed above.

The instability modes important primarily for *3D boundary layers* are the following (Fig. 2).

The *cross-flow vortices and waves* are observed in 3D boundary layers only. They represent modes of CF-instability the physical nature of which is similar to Rayleigh- and TS-instability but occurs on the spanwise component of the mean velocity profile. Similar to Görtler vortices the steady CF-modes represent *counter-rotating* vortices, whose axes, however, are not exactly parallel to the potential flow direction (as in case of Görtler instability) but rather inclined at a small angle of about 2 degrees. In contrast to Görtler instability, the CF-steady modes being *superimposed on the 3D base flow* produce *co-rotating* vortices typically observed in flow visualizations. Traveling CF-modes (the CF-waves) have the same physical nature as that of steady CF-modes and represent just CF-vortices moving downstream, with axes inclined (usually) at larger angles to the flow direction (the corresponding wave-vectors have smaller angles of inclination).

The *attachment-line-contamination instability modes* can be amplified in the 3D boundary layer formed along the attachment line of a swept wing or a swept cylinder when the characteristic Reynolds number $\overline{R}$ based on spatial scale $\delta_l = [\nu_e/(\partial U_e/\partial x)]^{1/2}$ and free-stream spanwise velocity W_∞ reaches a certain threshold value. The most dangerous ALC-perturbations are stationary and they are usually excited by surface roughness (contamination).

The *streamline-curvature instability modes* occur in 3D boundary layers with high values of sweep angle and chordwise pressure gradient when the streamlines are significantly curved in the plane of the spanwise and streamwise coordinates. Similar to the CF-modes these perturbations can be either steady or unsteady and usually appear together with the former.

At linear stages of transition (when the disturbance amplitudes are small enough) the boundary-layer perturbations listed above do not interact with each other, but a conversion of one kind of disturbances into another seems to be possible in some cases when the boundary-layer properties change downstream. In particular, the ALC- and Görtler modes can play the role of initial disturbances for the beginning of the CF-instability in the vicinity of the attachment line, while the CF-waves seem to be able to excite TS-waves and new CF-waves in those regions of swept-wing boundary layers where favorable pressure gradient changes to an adverse one. The associated mechanisms of transformation are not sufficiently studied yet.

5. WEAKLY-NONLINEAR STAGES OF TRANSITION

When (and if) the disturbance amplitudes reach certain threshold values (between 0.1 to 10% depending on the instability type) some nonlinear disturbance interactions become important in the transition process, which enters a nonlinear stage (Fig. 3). Initial nonlinear stage of transition is called usually the *weakly nonlinear* one. At this stage, the boundary layer perturbations still represent traveling or steady waves (instability modes), which do not develop, however, independently (as at linear stages) and start to interact with each other. The interactions can occur either between different spectral modes belonging to the same type of instability (say between TS-waves with different spanwise wavenumbers and frequencies) or between different kinds of instability modes (for instance between CF-vortices and TS-waves). The nonlinearity can lead to either enhancement of the disturbance amplification or to its reduction and, even, to suppression of disturbance growth. The resonant interactions observed at this stage can be very strong, leading to the double-exponential (i.e. exponent-in-exponent) behavior of perturbations and influencing the turbulence onset very substantially. The weakly nonlinear interactions can provide also a significant enrichment of the disturbance frequency-wavenumber spectrum compared to that present at the antecedent linear stage.

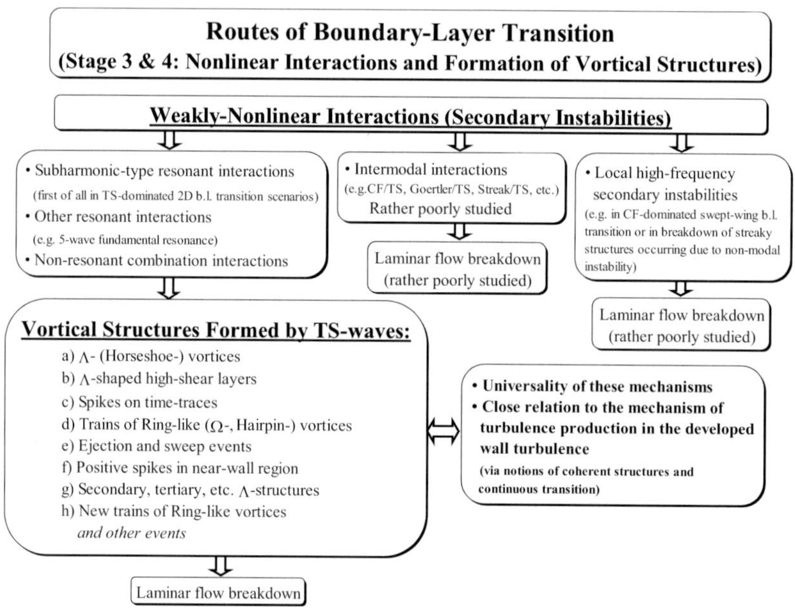

Figure 3. Most important nonlinear mechanisms of boundary-layer breakdown.

A set of other important nonlinear phenomena observed at weakly nonlinear stages of transition is associated with the *secondary instability* (Fig. 3). This notion implies usually a linear instability of the base flow *disturbed* by some primary, finite-amplitude perturbations. The local, inflexional, high-frequency secondary instability represents one of the brightest examples. This instability is able to provide a very rapid growth of boundary-layer perturbations. It can occur when the spatial gradient of either the mean or the instantaneous (low-frequency) velocity of the disturbed boundary layer reaches a certain threshold. In particular, this is observed in 3D boundary layers with dominating CF-instability and in 2D boundary layers at enhanced free-stream turbulence levels. In the former case the role of primary perturbations forming the velocity gradients is played by CF-vortices or (and) CF-waves, and by streaky structures in the latter case.

The subharmonic resonance of TS-waves (at its parametric, quasi-linear stage) can be also regarded as a kind of secondary instability. In this case the mathematical formulation of the problem can be based on a linear Floquet theory (or, alternatively, on weakly nonlinear theories or asymptotic theories). The disturbances amplified most rapidly by the subharmonic kind of the secondary instability are mainly (but not only) the low-frequency ones. In fact, this instability is able to amplify a very broad spectrum of 3D TS-waves with frequencies significantly exceeding those amplified by primary instability. In general, the secondary instabilities can either trigger (nearly immediately) the laminar flow breakdown (as in 3D boundary layers) or just accelerate the turbulence onset (as in the subharmonic resonance case), but they can also suppress the disturbance growth, as in the case of the subharmonic resonance with the 'anti-resonant' phase shifts between the involved instability modes. Some reviews of experimental and theoretical results obtained in this field can be found e.g. in [6, 7].

6. LATE STAGES OF TRANSITION AND COUPLING WITH WALL TURBULENCE

Late nonlinear stages of the boundary layer transition are often called *essentially nonlinear* ones. They are characterized by a transformation of instability modes (traveling and steady waves) into intense concentrated vortices, localized in physical space. This change of the objects under consideration represents one of the most complicated problems for the theoretical description of transition from the weakly nonlinear stage to the essentially nonlinear one. There are several other kinds of instability found at late nonlinear stages but their physics is very much different from that characteristic of the linear and weakly nonlinear stages (see e.g. [6, 8 – 10]).

The most typical phenomena observed at late stages of 2D-boundary-layer transition (initiated by TS-instability) are the following: (a) Λ-vortices, (b) Λ-shaped high-shear layers, (c) spikes on time-traces, (d) trains of ring-like vortices, (e) ejection and sweep events, (f) positive spikes, (g) secondary, tertiary, etc. Λ-structures, (h) new trains of ring-like vortices.

All these structures and events observed at late stages of transition are also found in developed turbulent boundary layers [10]. There is also a strong similarity of their properties. The same is true (at least in some cases) for the turbulence production mechanisms. There is a viewpoint that the turbulent boundary layer can be regarded (in a certain sense) as a continuous laminar-turbulent transition. If this hypothesis is true, the turbulent flow can be also regarded as a kind of transitional flow. Taking into account the receptivity stage of the transition, one may also say that the whole boundary layer over a body (from its leading to trailing edge) can be regarded as a transitional one when the Reynolds number exceeds a certain critical value (which is different for different base flows and environmental perturbations).

ACKNOWLEDGEMENTS

This work is supported by the Russian Foundation for Basic Research (Grants N 03-01-00299 and 03-01-04003).

REFERENCES

1. Reynolds O. On the experimental investigation of the circumstances which determine whether the motion water shall be direct or sinuous, and the law of resistance in parallel channels, *Phil. Trans. Roy. Soc.*, **vol. 174**, pp. 935-982, 1883.
2. Prandtl L. Über Flüssigkeitsbewegung bei sehr kleiner Reibung, *Verhandlg. III. Intern. Math. Kongr.*, Heidelberg, 1904, pp. 484-491.
3. Kachanov Y.S. Three-dimensional receptivity of boundary layers, *Eur. J. Mech., B/Fluids*, **vol. 19, no. 5**, pp. 723-744, 2000.
4. Gaponenko V.R., Ivanov A.V., Kachanov Y.S., Crouch J.D. Swept-wing boundary-layer receptivity to surface non-uniformities, *J. Fluid Mech.*, **vol. 461**, pp. 93-126, 2002.
5. Würz W., Herr S., Wörner A., Rist U., Wagner S., Kachanov Y.S. Three-dimensional acoustic-roughness receptivity of a boundary layer on an airfoil: experiment and direct numerical simulations, *J. Fluid Mech.*, **vol. 478**, pp. 135-163, 2003.
6. Kachanov Y. Physical mechanisms of laminar boundary-layer transition, *Ann. Rev. Fluid Mech.*, **vol. 26**, pp. 411-482, 1994.
7. Borodulin V.I., Kachanov Y.S., Koptsev D.B. Experimental study of resonant interactions of instability waves in self-similar boundary layer with an adverse pressure gradient: I. Tuned resonances, *Journal of Turbulence*, **vol. 3, no. 062**, 2002.
8. Bake S., Fernholz H.H., Kachanov Y.S. Resemblance of K- and N-regimes of boundary-layer transition at late stages, *Eur. J. Mech., B/Fluids*, **vol. 19, no. 1**, pp. 1-22, 2000.
9. Borodulin V.I., Gaponenko V.R., Kachanov Y.S., Meyer D.G.W., Rist U., Lian Q.X., Lee C.B. Late-stage transitional boundary-layer structures. Direct numerical simulation and experiment, *Theo. Comp. Fluid Dynamics*, **vol. 15**, pp. 317-337, 2002.
10. Kachanov Y.S. On a universal mechanism of turbulence production in wall shear flows, *Notes on Numerical Fluid Mechanics and Multidisciplinary Design*, **vol. 86.** *Recent Results in Laminar-Turbulent Transition*, Berlin, Springer, 2003, pp. 1-12.

INSTABILITIES IN BOUNDARY-LAYER FLOWS AND THEIR ROLE IN ENGINEERING‡

J.D. Crouch
Boeing Commercial Airplanes, P.O. Box 3707, Seattle, WA 98124-2207, U.S.A.
jeffrey.d.crouch@boeing.com

Abstract: Two classes of instabilities in boundary-layer flows are considered. The first are instabilities associated with the boundary-layer profiles, leading to a transition from laminar to turbulent flow. The second are instabilities to the global flow field associated with a separated boundary layer, leading to unsteady buffeting. Analyses for these instabilities are used to augment steady-state computations in engineering. The paper describes the basic formulation of these stability problems, and addresses the questions of when the boundary layer should be regarded as turbulent, and when the entire flow field should be regarded as unsteady.

Key words: Buffet Onset, Instability, Transition, N-Factor, Unsteady Flow

1. INTRODUCTION

The prediction of boundary-layer flows is at the heart of any external aerodynamic design or performance estimation. In some applications the boundary layer equations are solved explicitly, along with an inviscid outer-flow solution. However, it is becoming increasingly common to directly solve for the steady-state solution to the Reynolds-Averaged Navier-Stokes (RANS) equations. Although the unsteadiness is known to play critical roles in defining the flow structure, full-scale unsteady calculations are still not generally used in engineering design because of the large computational resources that are required. In some engineering problems, stability theory

‡ In tribute to Dr. W.-H. Jou

G.E.A. Meier and K.R. Sreenivasan (eds.), IUTAM Symposium on One Hundred Years of Boundary Layer Research, 105-114,

can be used to augment the steady-state computations. Instabilities may forecast changes in the flow structure, or be harbingers to the breakdown of the steady-state approach. We consider two classes of instabilities in boundary-layer flows that play important roles in engineering. The first are instabilities associated with the boundary-layer profiles, leading to a transition from laminar to turbulent flow. The second are instabilities to the global flow field associated with a separated boundary layer, leading to unsteady buffeting.

Prandtl established the critical role of the boundary layer in determining the basic performance of airfoils, and in controlling flow separation [1]. He proposed a hierarchical treatment for solving boundary-layer flows – in which the external inviscid flow is first calculated and then used to calculate the viscous boundary-layer flow. This approach is still commonly used for calculating attached flows with negligible separation. In studying the instabilities leading to transition, this hierarchical approach has proven very successful – starting from the original analysis of Tollmein [2], and continuing up to the current state of the art in transition prediction.

Separated flows, however, require a full coupling between the inviscid and viscous flow; these are generally calculated by solving the steady RANS equations. Similarly, the prediction of global flow instabilities leading to unsteady buffet requires that the boundary layer and the inviscid flow be calculated simultaneously. Although this is a departure from the process proposed by Prandtl, the underlying mechanisms (which drive any design or control) are tied to his fundamental description of the flow.

We consider the basic formulation of these two stability problems and address the questions of when the boundary-layer flow should be regarded as turbulent, and when the entire flow field should be regarded as unsteady.

2. STABILITY FORMULATION

Following classical stability theory, the state vector $\mathbf{q}$ is decomposed into a basic state $\mathbf{Q}$ and a perturbation $\mathbf{q}'$, $\mathbf{q}=\mathbf{Q}+\mathbf{q}'$. Substituting into the Navier-Stokes equations, linearizing about $\mathbf{Q}$, and assuming $\mathbf{Q}$ is a solution to the steady Navier-Stokes equations yields

$$L_Q[\mathbf{q}'] = 0. \tag{1}$$

The linear operator L_Q, in combination with boundary conditions, governs the linear perturbations about the basic state $\mathbf{Q}$.

2.1 Boundary-layer instabilities

The formulation of the stability problem for boundary-layer-transition prediction is very clearly aligned with Prandtl's concept of a hierarchal treatment of the boundary layer [1]. In fact, Prandtl was one of the first to consider boundary-layer stability – where he analyzed the stability using piecewise linear profiles [3]. In current methods, the basic flow $\mathbf{Q}$ is typically obtained by solving the steady boundary-layer equations. Even when the basic flow is calculated from the Navier-Stokes equations, the stability problem is still based on the boundary-layer formulation, with boundary conditions requiring that the disturbances vanish at the surface and outside the boundary layer.

Following the original work of Tollmien [2], the streamwise variation of the boundary-layer profiles are neglected in the stability analysis, leading to the Orr-Sommerfeld equation, or its generalization. This leads to the so-called quasi-parallel approximation, in which the x variation in $\mathbf{Q}$ is treated parametrically,

$$\mathbf{q}(x,y,z,t)=\mathbf{Q}(y;x)+\mathbf{q}'(y;x)e^{i\alpha x}e^{i(\beta z-\omega t)}. \qquad (2)$$

The frequency ω and the spanwise wavenumber β are prescribed, and the complex streamwise wavenumber α is calculated as an eigenvalue problem. The state vector is given by $\mathbf{q} = (u,v,w,T,\rho)$ for compressible flows, and by $\mathbf{q} = (u,v,w)$ for incompressible flows. Numerical solution of the Orr-Sommerfeld equation is now routine, using spectral methods or higher-order finite-difference methods. A relatively recent formulation of the stability problem by Bertolotti et al. [4] incorporates more fully the non-parallel effects of the boundary-layer growth. In these so-called parabolized stability equations (PSE), the base flow and the perturbation are treated as weak (but continuous) functions of x.

2.2 Global flow instabilities

To study the onset of unsteadiness associated with boundary-layer separation, the boundary layer can not be decoupled from the external flow. Rather, the entire flow field must be considered in the stability analysis. Here, the basic flow is obtained by solving the steady Navier-Stokes equations, or in the case of a turbulent boundary layer, the RANS equations. Focusing on a two-dimensional mean flow, the state vector is decomposed as

$$\mathbf{q}(x,y,z,t)=\mathbf{Q}(x,y)+\mathbf{q}'(x,y)e^{i\beta z}e^{-i\omega t}. \qquad (3)$$

The spanwise wavenumber β is prescribed, and the complex frequency ω is calculated from an eigenvalue problem. By contrast to the study of boundary-layer instabilities (which date back to Prandtl), this type of study is relatively new. This form of stability analysis was used by Jackson [5] and Zebib [6] to study the onset of laminar-flow vortex shedding on cylinders. More recently, Theofilis [7] has applied this formulation to analyze a number of laminar-flow instabilities. The global instability considered here, which is associated with unsteadiness of a turbulent boundary-layer flow, was first considered by Crouch, Garbaruk, Shur & Strelets [8].

For buffet-onset prediction, the boundary layers are typically turbulent. The state vector for a compressible turbulent flow is given by $\mathbf{q} = (u, v, w, T, \rho, \nu_T)$, where ν_T is the turbulent eddy viscosity. Here, both the mean flow and the eigenmode are two-dimensional functions. The stability equation is discretized using finite differences, on the same grid used for calculating the basic state. This leads to a rather large eigenvalue problem of $O(10^6)$ unknowns, compared to $O(10^3)$ for boundary-layer instabilities. To solve such large problems, the implicitly-restarted Arnoldi method is used [9]. This enables the calculation of a small number of eigensolutions in the neighborhood of some prescribed frequency ω^*.

3. TRANSITION PREDICTION

Transition predictions in engineering are motivated by the dramatic differences between laminar and turbulent boundary layers, impacting the skin friction, the heat transfer, and even the pressure distributions. Prandtl noted the important role of transition in explaining the change in the flow pattern around a sphere as it passes through the critical Reynolds number [10]. In other problems the effects may be less dramatic than going from a laminar to a turbulent separation, but accounting for transition is none-the-less critical to predicting the flow behavior. For example, the existence of laminar flow over the first 20% of the wing chord leads to roughly a 20% reduction in profile drag compared to a fully-turbulent boundary layer at flight Reynolds numbers. In transonic flow, the shock position depends on the boundary-layer thickness in addition to the geometry. At a wind-tunnel Reynolds number of 3.5 million, a 20% run of laminar flow results in a 4 to 5% downstream movement of the shock compared to a fully turbulent condition. This has consequences on the lift as well as the drag. As described by Prandtl, separation occurs when the momentum of the fluid near the wall is too low to overcome a pressure rise imposed by the external flow. The momentum "health" of the boundary layer depends on its upstream history – so a small run of laminar flow can influence the initiation and extent of the separated flow, even when the flow is turbulent at separation.

There are a number of methods used for estimating the location of the laminar-turbulent transition. By far, the most commonly-used method in engineering applications is the so-called e^n method – originally devised by Smith & Gamberoni [11] and Van Ingen [12]. For linear perturbations, a disturbance mode can be characterized by a single amplitude $A(x; \omega,\beta) = A_0 e^{n(x;\omega,\beta)}$, where $n(x;\omega,\beta) = -\int^x \alpha_i(s;\omega,\beta)ds$. The amplification factor is defined as the envelope of all modal growth curves

$$n(x) = \max_{\omega} \max_{\beta}\left(-\int_{x_0}^{x} \alpha_i(s;\omega,\beta)\,ds\right). \tag{4}$$

In the e^n method, transition is assumed to occur when the amplification factor n reaches a critical value N, where N is established by correlation with experiments. For controlled experimental conditions, this provides an effective prediction method. However, when applied away from the correlation conditions the method does not yield consistent results. The primary shortcoming of the basic e^n method is that the receptivity and nonlinear-breakdown physics cannot be adequately accounted for in a single value of n. Most notably, the differences in the receptivity (responsible for the initial amplitude A_0) need to be accounted for in any generalizable prediction method.

To overcome the major shortcomings of the basic e^n method, various amplitude-based methods have been proposed [13]. These methods require a receptivity calculation to determine initial amplitudes, a linear growth phase similar to e^n, and an amplitude criterion to signify transition. The receptivity accounts for influences such as free-stream turbulence, noise, surface waviness, roughness, or localized irregularities. The simplest amplitude criterion predicts that transition occurs when the linear amplitude exceeds a threshold value, $A \geq A_T$. More sophisticated relationships involving multiple modes have also been proposed [14]. The applicability of amplitude methods has been limited by the lack of detailed information about the free-stream and model-surface conditions.

Variable n-factor methods provide an approach that is intermediate to an amplitude method and the basic e^n method. Here the value of N is given as a function of the external conditions, which influence the receptivity or locally change the growth rate. The methods are based on experimental correlation or a combination of correlation and theory. One of the first variable n-factor relationships was proposed by Mack [15] to account for the influence of free-stream turbulence on transition due to Tollmien-Schlichting (TS) waves. The transitional TS-wave n-factor is given by

$$N_{TS} = -8.43 - 2.4\ln(Tu), \tag{5}$$

where Tu is the turbulence intensity. Figure 1a shows the transition Reynolds number predicted from N_{TS} along with experimental data of Schubauer & Skramstad [16] and Dryden [17]. If the turbulence affects the transition through the receptivity, equation (5) implies that the initial amplitude A_0 is proportional to $Tu^{2.4}$, and the threshold amplitude is $A_T \sim 0.02\%$ (which is too small to cause transition). As of now, there is not theoretical support for this relationship. None-the-less, the variable TS-wave n-factor yields generally good results when compared to experiments for values of Tu down to 0.07% and greater than 0.5%. For $Tu < 0.07\%$, the data of Schubauer & Skramstad show a fixed transition Reynolds number which is related to wind-tunnel acoustics. Data from a quieter facility would likely follow the n-factor relationship to lower values of Tu.

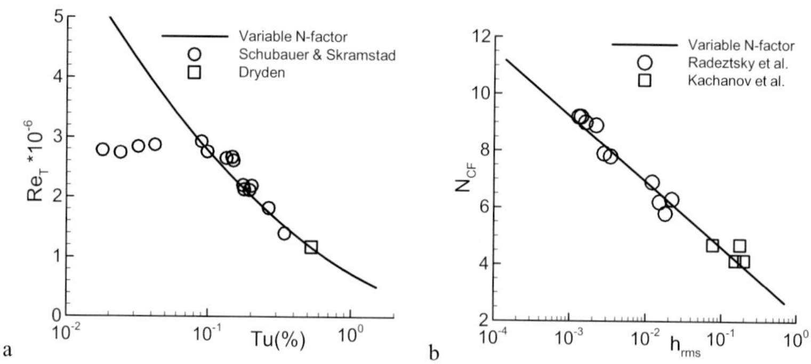

Figure 1. Variable N-factor results: (a) Flat-plate transition Reynolds number as a function of Tu, based on N_{TS}, (b) Swept-wing N_{CF} as a function of surface roughness.

Another important factor affecting TS-wave transition is the presence of step discontinuities in the surface. Large steps can trigger an immediate transition, while smaller steps only enhance the natural growth of TS waves. The small-step effects can be modeled by a change of ΔN_{TS} in the baseline (smooth surface) critical n-factor. Data for ΔN_{TS} has recently been obtained by Wang & Gaster [18] for a zero pressure-gradient boundary layer, and by Crouch, Kosorygin & Ng [19] for favorable and adverse pressure gradients. These results show that step effects can be accounted for by using a ΔN_{TS} for step heights up to 1.5 times the local boundary-layer displacement thickness.

In low-turbulence environments ($Tu<0.1\%$), transition due to cross-flow instability is caused by stationary vortices, which are excited by surface roughness. Receptivity theory shows the cross-flow vortex amplitude varies linearly with the roughness height. Crouch & Ng [20] provide a variable n-factor expression for cross-flow instability transition, based on a uniform receptivity spectrum

$$N_{CF} = 2.3 - \ln(h_{rms} / \delta^{*}). \tag{6}$$

The rms roughness height h_{rms} is normalized by the boundary-layer thickness at the neutral point for the critical mode δ^{*}. Figure 1b shows the variable n-factor expression along with experimental n-factor values from Radeztsky et al. [21] and Kachanov et al. [22]. This data represents two different wind-tunnel facilities, and several different model-roughness configurations. This variable n-factor expression captures the roughness effects over a wide range of nondimensional roughness heights, from below *0.001* to greater than *0.2*, where the n-factor goes from greater than *9* to values smaller than *4*.

4. BUFFET-ONSET PREDICTION

Buffet onset predictions in engineering are motivated by the – typically undesirable – unsteady loads that result from large-scale unsteady separation. The sequence of separation onset, followed by an increased zone of separation and then unsteadiness, is a characteristic of initial buffeting observed on airfoils. Increasing the wing lift, by raising the angle of attack, results in increasingly-adverse pressure gradients toward the trailing edge – eventually leading to flow separation. In transonic flow, the increase in lift is associated with an increase in the shock strength. Some transonic boundary-layer flows can exhibit a shock-induced separation bubble prior to any significant trailing-edge separation. At some point during the increase in wing angle of attack, the flow field becomes globally unsteady. This can lead to an unsteady buffeting of the airplane structure, thus limiting the airplane flight envelope.

Current methods for predicting the occurrence of excess airplane buffeting (a structural response) are based on empirical relationships. However, many features of the airplane response are also observed in experiments on unsteady-buffet onset for airfoil flows. Unsteady CFD has been used to predict this flow-field buffet onset, with some degree of success. However, unsteady CFD calculations are time consuming and expensive, and they are not well suited for identifying the boundary between steady and unsteady flow. In fact, knowledge about the onset conditions for unsteady flow could greatly compliment the application of unsteady CFD.

Global flow instabilities have recently been used to try and predict the unsteady-buffet onset for airfoils based on steady RANS solutions [8]. A simple test case for such an approach is the laminar onset of vortex shedding for a circular cylinder [5,6,8]. The steady flow around a cylinder at a Reynolds number $Re = UD/\nu = 60$ is unstable to a mode with a growth rate of 0.046 and a frequency of $\omega D/U = 0.74$. Coupling this mode (with a finite amplitude) to the steady solution results in an unsteady flow producing the expected Karman vortex street. The onset of vortex shedding is predicted to occur at $Re=48$, in good agreement with earlier works.

The application of the global-instability approach for transonic flow is complicated by the existence of turbulent boundary layers and shock waves. The turbulent boundary layers are accommodated by using a turbulence model, and the shocks are treated by smoothing the steady-flow solution with a very-fine grid around the shock. Figure 2a shows the Mach contours for the flow over an 18%-thick bi-convex airfoil at $M=0.74$. This flow is unstable to a mode with a growth rate of 0.25 and a frequency of $\omega c/U=0.76$. The real part of the u-velocity perturbation for this mode is shown in figure 2b. The perturbation is concentrated around the shock, with lower-level contours in the wake. This frequency is in reasonable agreement with experiments of McDevitt et al. [23]. In the experiments, buffet onset occurred at $M{\sim}0.76$ with increasing Mach number, but persisted down to $M{\sim}0.73$ with decreasing Mach number. The stability theory predicts buffet onset at $M{\sim}0.73$.

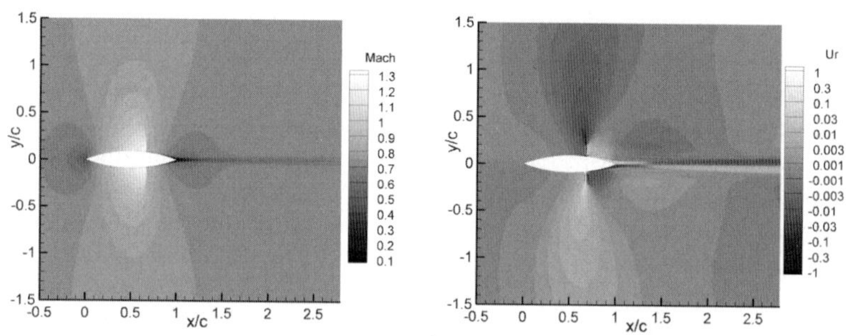

a b

Figure 2. Bi-convex airfoil at $M=0.74$ showing: (a) Steady-state Mach contours, and (b) Perturbation-velocity contours.

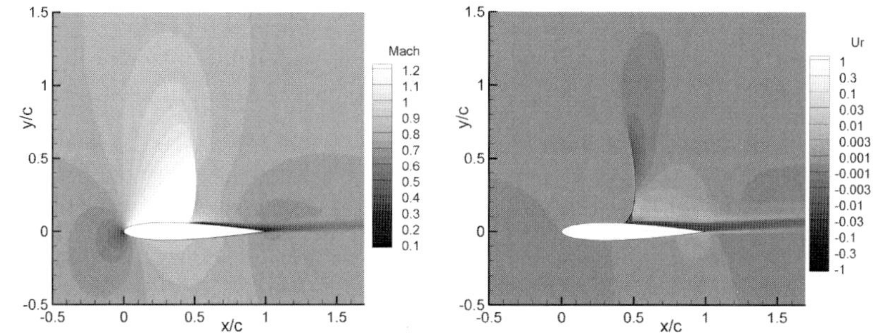

a b

Figure 3. NACA 0012 airfoil at $M=0.76$ showing: (a) Steady-state Mach contours, and (b) Perturbation-velocity contours.

A final example is the application to the NACA 0012 airfoil. Figure 3a shows the Mach contours for the steady flow at $M=0.76$ and $\alpha=3.2^o$. This flow is found to be unstable with a growth rate of 0.05 and a frequency of $\omega c/U=0.25$. The real part of the u-velocity perturbation is given in figure 3b. For this Mach number, the experiments [24] showed a buffet onset at $\alpha\sim3.1^o$, and the stability theory predicts $\alpha\sim3.05^o$. These initial results show the stability theory to provide a good predictor for unsteady-buffet onset.

5. CONCLUSIONS

The formulations for two different stability problems have been presented. Boundary-layer instabilities are based on a formulation that stems from the original works of Prandtl [3] and Tollmien [2]. These types of instabilities are the bases for state-of-the-art transition-prediction methods. Global flow instabilities cannot be analyzed within the hierarchical framework proposed by Prandtl; rather, a full coupling between the boundary layer and the external flow is required. These instabilities are used to predict the onset of unsteady flow. Future methods for transition prediction (short of direct numerical simulation) will continue to exploit the basic boundary-layer formulation of Prandtl. Buffet-onset predictions will bypass the boundary-layer formulation, but the basic mechanisms are integrally tied to the boundary-layer flow.

REFERENCES

1. Prandtl, L. 1905. Verh. Int. Math. Kongr., 3rd. Heidelberg, 1904, p. 484. Transl. 1928. NACA Memo. No. 452.
2. Tollmien, W. Nachr. Ges. Wiss. Göttingen, Math.-phys. Kl., p. 21, 1929.
3. Prandtl, L. "Bemerkungen über die Entstehung der Turbulenz", *ZAMM I*, pp. 431-436, 1921.
4. Bertolotti, F.P., Herbert, Th., Spalart, P.R. "Linear and nonlinear stability of the Blasius boundary layer," *J. Fluid Mech.*, **vol. 242**, pp. 441-474, 1992.
5. Jackson, C.P. "A finite-element study of the onset of vortex shedding in flow past variously shaped bodies," *J. Fluid Mech.*, **vol. 182**, pp. 23-45, 1987.
6. Zebib, A. "Stability of viscous flow past a circular cylinder," *J. Engr. Math.*, **vol. 21**, pp.155-165, 1987.
7. Theofilis, V. "Advances in global linear instability of nonparallel and three-dimensional flows," *Prog. Aero. Sci.*, **vol. 39**, pp. 249-315, 2003.
8. Crouch, J.D., Garbaruk, A., Shur, M., Strelets, M. "Predicting buffet onset from the temporal instability of steady RANS solutions," *Bull. Am. Phys. Soc.*, **vol. 47**, p. 68, 2002.
9. Lehoucq, R.B., Sorensen, D.C., Yang, C. *ARPACK user's guide*, Philadelphia, SIAM, 1998.
10. Prandtl, L. "Der Luftwiderstand von Kugeln," Nachr. Ges. Wiss. Göttingen, Math.-phys. Kl., pp. 177-190, 1914.
11. Smith, A.M.O., Gamberoni, A.H. "Transition, pressure gradient, and stability theory," Douglas Aircraft Co., Rept. ES26388, El Sequndo, 1956.
12. Van Ingen, J.L. "A suggested semi-empirical method for the calculation of the boundary layer transition region," Univ. of Technology, Rept. UTH1-74, Delft, 1956.
13. Crouch, J.D. "Transition prediction and control for airplane applications," AIAA Paper No. 97-1907, 1997.
14. Herbert, Th., Crouch, J.D. "Threshold conditions for breakdown of laminar boundary layers," *Laminar-Turbulent Transition*, IUTAM, Toulouse, France, 1990, pp. 93-101.
15. Mack, L.M. "Transition prediction and linear stability theory," *Laminar-Turbulent Transition*, CP-224, AGARD, 1977, pp. 1/1-22.
16. Schubauer, G.B., Skramstad, H.K. "Laminar boundary-layer oscillations and transition on a flat plate," National Bureau of Standards, Res. Paper 1722, 1947.
17. Dryden, H.L. "Airflow in the boundary layer near a flat plate," NACA Rep. 562, 1936.
18. Wang, Y.X., Gaster, M. "Effect of surface steps on boundary layer transition," *Exp. in Fluids*, 2004, to appear.
19. Crouch, J.D., Kosorygin, V.S., Ng, L.L. "Modeling the effects of two-dimensional steps on transition due to Tollmien-Schlichting waves," *Laminar-Turbulent Transition*, IUTAM, Bangalore, India, 2004, to appear.
20. Crouch, J.D., Ng, L.L. "Variable n-factor method for transition prediction in three-dimensional boundary layers," *AIAA J.*, **vol. 38**, pp. 211-216, 2000.
21. Radeztsky, R.H., Reibert, M.S., Saric, W.S., Takagi, S., "Effect of micron-sized roughness on transition in swept-wing flows," AIAA Paper No. 93-0076, 1993.
22. Kachanov, Y.S., Borodulin, V.I., Koptsev, D.B. "Effect of distributed roughness on swept-wing boundary-layer transition," ITAM Report, unpublished, 1999.
23. McDevitt, J.B., Levy, L.L. Jr., Deiwert, G.S. "Transonic flow about a thick circular-arc airfoil," *AIAA J.*, **vol. 14**, pp. 606-613, 1976.
24. McDevitt, J.B., Okuno, A.F. "Static and dynamic pressure measurements on a NACA 0012 airfoil in the Ames high Reynolds number facility," NASA Tech. Paper 2485, 1985.

IN-FLIGHT INVESTIGATIONS OF TOLLMIEN-SCHLICHTING WAVES

Arne Seitz, Karl-Heinz Horstmann
DLR Institute of Aerodynamics and Flow Technology, Lilienthalplatz 7, 38108 Braunschweig Germany; phone: +49 531 295 2888; fax: +49 531 295 2320; e-mail: arne.seitz@dlr.de

Abstract: Flight tests were performed in order to learn more about the technically important case of naturally occurring Tollmien-Schlichting waves in boundary layer transition. A multi-element hot-film array, placed on the right hand wing of the flying test bed LFU-205, was used to sense skin friction fluctuations provoked by Tollmien-Schlichting disturbances propagating in the laminar boundary layer. Fourier analysis in time and space of the experimental data revealed that a broadband spectrum of 2d and 3d Tollmien-Schlichting waves is involved in the signals, typical for wave packets. Occurrence of these wave packets happens stochastically and it was found that they vary in intensity as well as in their spanwise extent. Examination of signal time histories from sensors distributed in chordwise direction showed that breakdown to turbulence, indicated by spikes in the signal, starts from within the wave packets with the strongest amplitudes.

Key words: boundary layer, laminar, turbulent, natural transition, Tollmien-Schlichting wave, flight test, hot-film, wave packet

1. NOMENCLATURE

A', A'_0	amplitude and initial amplitude of TS-instability
$e = E_0 + e'$	anemometer output voltage, mean value and fluctuation
Ma_∞	free stream Mach number
Re_c	Reynolds number based on chord length
t	time

G.E.A. Meier and K.R. Sreenivasan (eds.), IUTAM Symposium on One Hundred Years of Boundary Layer Research, 115-124,

$U_\infty, p_\infty, T_\infty, \rho_\infty$	free stream velocity, static pressure, static temperature and density
$u', \hat{u}$	disturbance velocity and its amplitude
x/c	dimensionless chordwise coordinate
y, z	coordinates in spanwise direction and normal to the wall
α_r, β_r	wave numbers of TS-wave
μ	viscosity
τ', τ_{w0}	wall shear stress fluctuation and mean wall shear stress
f, ω_r	frequency and angular frequency of TS-wave
ω_i	amplification rate of TS-wave

2. INTRODUCTION

Initiated by Ludwig Prandtl, two of his students, Walter Tollmien and Hermann Schlichting, worked on the problem of boundary layer stability. They essentially solved the problem and today Tollmien-Schlichting waves are well known as primary instabilities in transition of boundary layers from laminar to turbulent flow [**1**–**2**]. But since almost all experimental investigations on this type of instability have been gained in wind-tunnel tests, mainly under conditions of controlled disturbance excitation, very little is known about the technically important case of naturally occurring Tollmien-Schlichting waves in free-flight. Recently, Horstmann and Miley [**3**] measured free stream excited fluctuations of the boundary layer by means of multiple hot-wire probes distributed across an aircraft's wing. The fluctuations were found to be of broad spectral character, typical for wave packets. Nevertheless, measurements indicated also, that these wave packets must be of small spanwise extent, since correlation between signals of neighbouring probes (20mm apart) was poor. Because the rods supporting the hot-wires triggered bypass transition, a detailed investigation of the downstream development of disturbances was impossible.

In order to clarify the situation, further flight tests were performed with the following objectives: (i) determination of frequencies, spanwise wavenumbers and amplitudes of those two and three-dimensional Tollmien-Schlichting waves involved in the naturally exited wave packets and comparison with linear theory, (ii) determination of the maximum amplitudes and spanwise width of wave packets, (iii) observation of the downstream development of disturbances near the onset of turbulence.

For this series of experiments it was decided to use hot-films in order to measure the skin friction fluctuations provoked by Tollmien-Schlichting waves. With hot-films, a very dense distribution of test points on a given area can be realised. Furthermore they will not trigger bypass transition nor will they introduce any significant disturbances into the laminar boundary layer.

3. DETECTION OF TOLLMIEN-SCHLICHTING WAVES WITH HOT-FILM SENSORS

For the present work, the wing boundary layer under consideration is nearly two-dimensional. Therefore, instabilities of the laminar flow will be essentially Tollmien-Schlichting waves, which exhibit in their *u*-disturbance velocity profiles a maximum close to the wall, Fig. 1

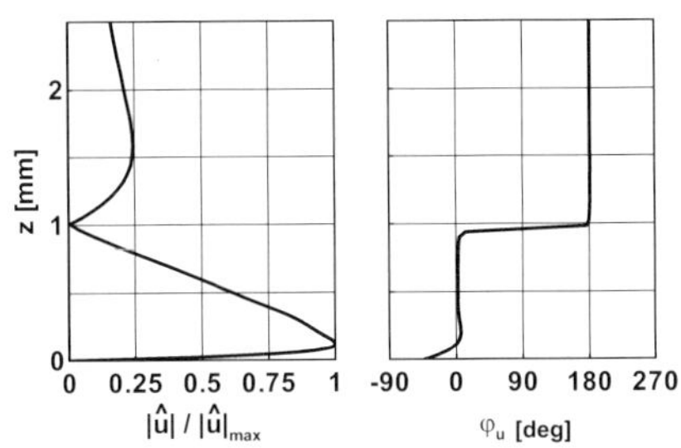

Figure 1: Amplitude and phase distribution of u-disturbance velocity profile.

Here, amplitude and phase function correspond with the wave ansatz from temporal linear stability theory:

$$u' = \hat{u}(z) \cdot \exp(\omega_i t) \cdot \exp(\alpha_r x + \beta_r y - \omega_r t + \varphi_u(z)) \tag{1}$$

Using the definition for the wall shear stress, it may be derived that the velocity disturbance leads to a fluctuating skin friction according to

$$\tau'_w = \mu \left.\frac{d\hat{u}}{dz}\right|_w \cdot \exp(\omega_i t) \cdot \exp(\alpha_r x + \beta_r y - \omega_r t + \varphi_u(z=0)), \tag{2}$$

with $\hat{\tau}_w = \mu \cdot (d\hat{u}/dz)_w$ being the amplitude. It is obvious that Tollmien-Schlichting instabilities leave traces at the wall which contain all the information about the waves like frequency $\omega_r = 2\pi \cdot f$, wave numbers α_r, β_r and amplification ω_i. Here, hot-film sensors are employed to measure the skin friction fluctuations. Originally introduced by Ludwieg [**4**] as a device to determine the mean wall shear stress in turbulent flow, progress in the development of sensors as well as anemometry systems soon allowed also for an investigation of unsteady flow phenomena. The principle of hot-film anemometry is based on convective heat transfer from an electrically heated sensing element. Cooling of the element by the flow depends upon the magnitude of the wall shear stress and, hence, a relation exists between the wall shear stress and the electric voltage needed to keep the sensor at a constant temperature. The instantaneous anemometer output voltage $e = E_0 + e'$ may be split up into a mean value E_0 which is connected to the mean shear stress τ_{w0} and a fluctuating part e', which can be related to τ'_w via a calibration procedure described in [**5**].

4. THE EXPERIMENT

4.1 The multi element hot-film array

A hot-film array with a total of 69 sensors was especially designed to measure the propagation of wave packets formed by Tollmien-Schlichting instabilities. Fig. 2 gives an overview of the layout.

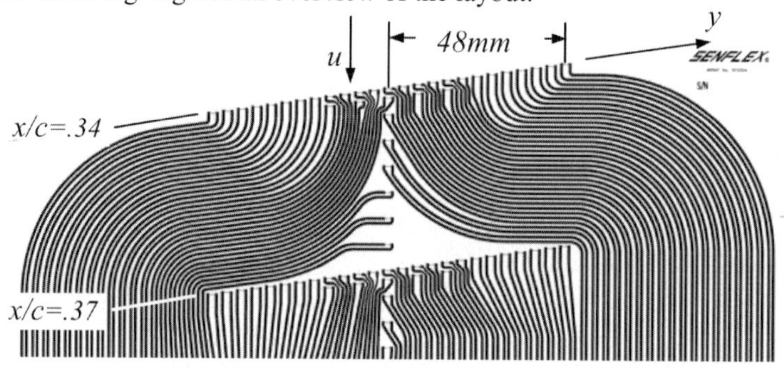

Figure 2: Hot-film array with 69 sensing elements.

Each element consists of a 0.2μm thick, 1.5mm wide and 0.1mm deep nickel film connected to 8μm thick copper coated leads. The 0.1mm thick substrate is of polyimide and was bonded to the wing surface. In order to get an installation without forward or backward facing steps, edges of the substrate foil were filled and sanded. For the same reason, the electric leads were routed away from the surface into the interior of the wing before being soldered to the wiring. The hot-film elements are arranged on the array mainly in two spanwise rows at constant chord positions of $x/c = 34\%$ and $x/c = 37\%$. Each row consists of 25 sensors 4mm apart in spanwise direction.

Both rows are connected by a chordwise column of 7 sensors on the centre line of the array.

4.2 Experimental setup and data acquisition

Fig. 3 shows the flying testbed LFU-205 of DLR. It is a light aircraft equipped with a laminar glove made from composite material on the right hand wing. The airfoil in the glove region was especially designed to allow for a moderate growth of Tollmien-Schlichting waves in a constant pressure region or under a slightly adverse pressure gradient. The 69 element hot-film array is placed in the middle of the glove, which is also equipped with 48 orifices in one chordwise row to measure the actual pressure distribution, Fig. 4. With an infrared thermo vision system placed in the cockpit of the aircraft the location of turbulence onset in the glove region is monitored. Further instrumentation comprises a pitot-tube with total and static pressure ports located on the left wing-tip and a total temperature probe on the lower side of the left hand wing. Data gained from these probes are utilized to compute free stream values of velocity, static pressure, static temperature and density as well as Mach number and Reynolds number.

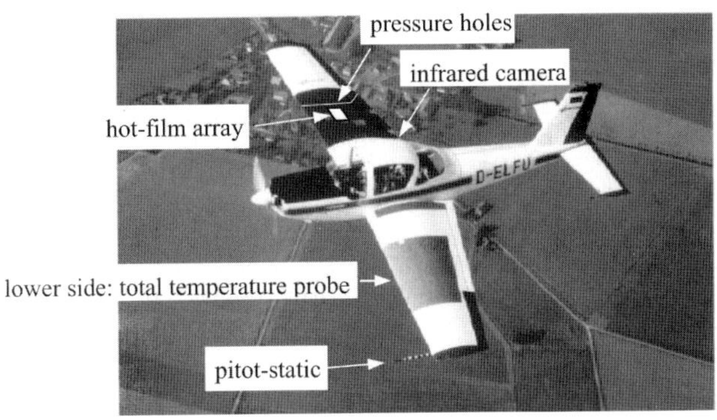

Figure 3: Flying test bed LFU-205 with instrumentation.

All test flights were performed in still air within a stable high pressure region. Although not measured within this series of experiments, previously performed tests with the LFU-205 under similar weather conditions showed that the free stream turbulence level is not higher than $Tu = 0.05\%$, [**6**].

The hot-films were operated in the constant temperature (CTA) mode, with the temperature being held at 150° C. The fluctuating parts of the sensor signals were sampled simultaneously for one second at a rate of

48kHz. Fig. 5 shows as an example a 10ms long stretch cut out from the time histories $\tau'_w(t)$ of eleven sensor elements located around the centre line (y= 0mm) of the array at a chord position of x/c = .37.

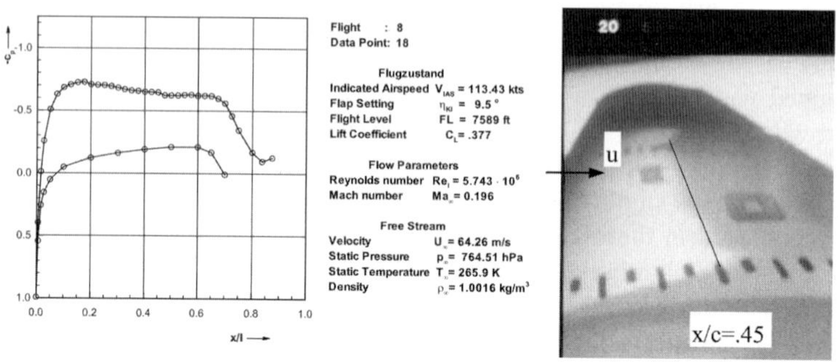

Figure 4: C_P-distribution, free stream values of $U_\infty, p_\infty, T_\infty, \rho_\infty, Ma_\infty, \mathrm{Re}_c$ and infrared image from a typical test point.

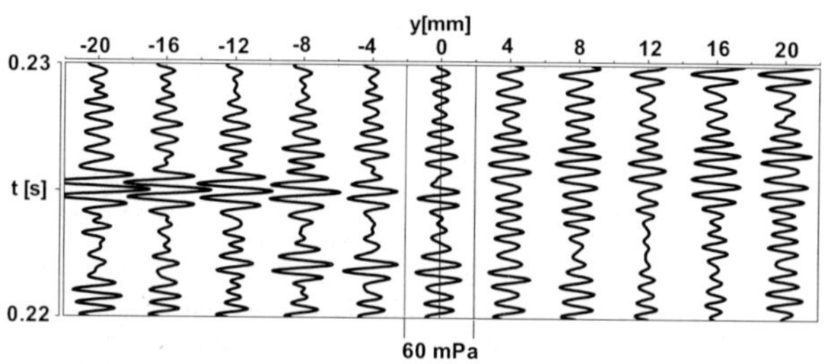

Figure 5: Time histories $\tau'_w(t)$ of signals from eleven hot-film sensors located around the centre line of the array (y = 0mm) at a chord position of x/c = .37.

4.3 Results of Data Analysis

A first visual inspection of the time histories $\tau'_w(t)$ presented in Fig. 5 reveals the principal topology of the disturbance flow. The horizontal locations of the records are proportional to the spanwise positions of the corresponding sensor elements on the wing. Thus, one can clearly discern several coherent structures, which are characterized by a sudden increase of the amplitude over three or four cycles, a pattern that is extended in spanwise direction over several records.

The amplitude modulation found is typical for the occurrence of wave packets. In order to determine the Tollmien-Schlichting components involved in the formation of these wave packets, Fourier decomposition in time and space of the experimental data, which are available at discrete points in the t, y plane for both spanwise rows of hot-films, was performed. From the resulting Fourier coefficients smoothed amplitudes $\hat{\tau}_w = \hat{\tau}_w(f, \beta_r)$ were calculated. Fig. 6 shows as an example the analysis based on the 25 time histories $\tau'_w(t)$ measured at x/c = 37 %. The Tollmien-Schlichting amplitudes $\hat{\tau}_w$ are plotted as contours over a spectrum of frequencies f and spanwise wave numbers β_r. Additionally, the propagation angle ψ of the waves can be read from this chart, with ψ being based on the direction of the boundary layer edge flow.

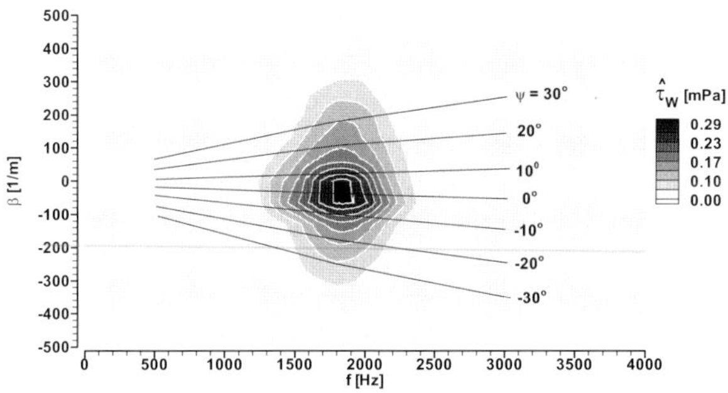

Figure 6: Frequency against wave number spectrum.

The spectrum reveals that Tollmien-Schlichting waves with a frequency of $\approx 1800\,\text{Hz}$ and a propagation direction parallel to the edge flow ($\psi = 0^\circ$) contribute most to the wave packets, but that oblique waves are also involved. Nevertheless, with increasing propagation angle, three dimensional disturbances loose weight quickly. A cross-check of this experimental finding with the results of linear stability theory was made by comparison of the measured amplitude spectrum for the $\psi = 0^\circ$ direction with a computed one. In order to accomplish this, the amplitude ratio A'/A'_0, with A'_0 being the initial amplitude of a Tollmien-Schlichting disturbance at its point of neutral stability, was calculated for waves of different frequency and propagation direction parallel to the edge flow, using a linear stability analysis method. The mean flow was determined on the basis of the measured pressure distribution employing a laminar boundary layer method. Fig. 7 displays the result. Good agreement can be found with respect to the range of disturbance frequencies and also for the peak position of the amplitude distribution. Since the slopes of measured and calculated spectra fit quite well, it can be concluded that initial amplitudes of naturally excited Tollmien-Schlichting waves forming a wave packet are all roughly of the same order of magnitude, as is assumed in linear theory.

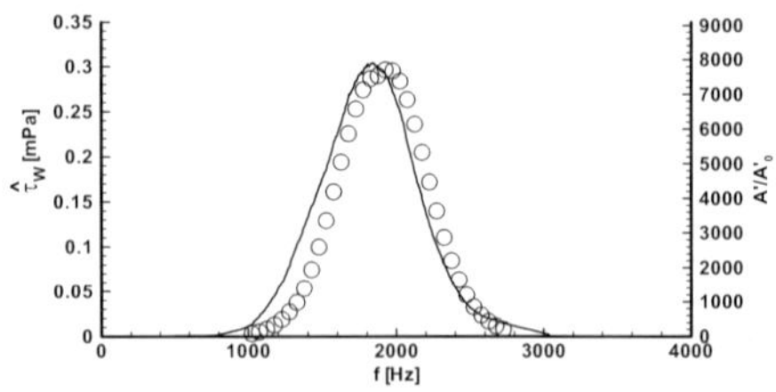

Figure 7: Comparison of measured (line) and calculated (circles) amplitude spectra, $\psi = 0°$.

As mentioned before, the spectra from the Fourier decomposition of the measured data were smoothed. Therefore, an average wave packet was analyzed rather than a particular one. On the other hand, as already could be seen in Fig. 5, wave packets passing a certain position on the wing may differ in their appearance. Thus, further investigations were performed regarding shape and intensity of individual disturbances expressed by their spanwise extent and their maximum amplitudes. For this purpose it is convenient to display the measured data in a slightly different way. For each time history $\tau'_w(t)$ an envelope or amplitude function $\tilde{\tau}_w(t)$ was specified. These functions can subsequently be used to draw a contour plot of the disturbance amplitudes in the t, y plane, which was done for the chord position at $x/c = 37\%$, Fig. 8. As intended, shape and intensity of the disturbed regions can be distinguished clearly. Only a short cut-out of the time history of $\tilde{\tau}_w(t)$ can be seen in this picture, but a visual inspection of the complete one second long data sample uncovers the fact that the wave packets occur stochastically; no periodicity or even repetition of certain patterns was observed. Furthermore, maximum amplitudes as well as spanwise extents of the wave packets vary strongly.

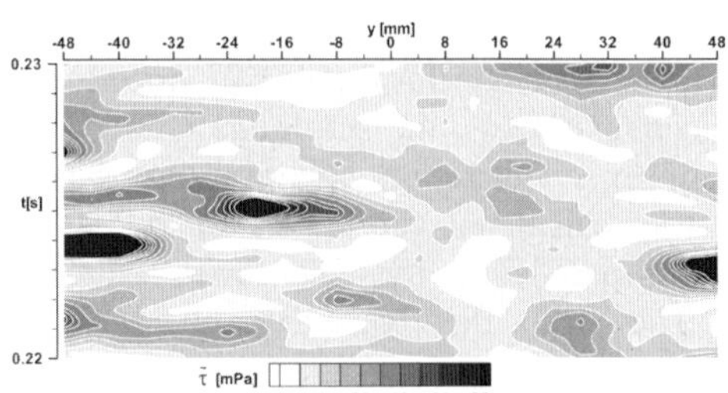

Figure 8: Contour plot of amplitude function.

In order to get a better overview, a simple algorithm was developed that automatically searches for wave packets and records their individual maximum amplitude as well as their width. Based on these data two histograms were compiled, which are presented in Fig. 9. It can be seen, that the average wave packet in this case had a width of 17 mm (disturbance wavelength: ≈13 mm) and maximum amplitude of 19 mPa (mean wall shear stress: 1400 mPa). Nevertheless, disturbances that are wider in span and also more intense in amplitude were found, with the biggest packet being 56 mm wide, while its maximum amplitude was 59 mPa. It should be mentioned, that the packet outlines differed considerably and no specific shape, like for example that of a wave packet generated by a pulse from a point source (Gaster and Grant [7]), could be found.

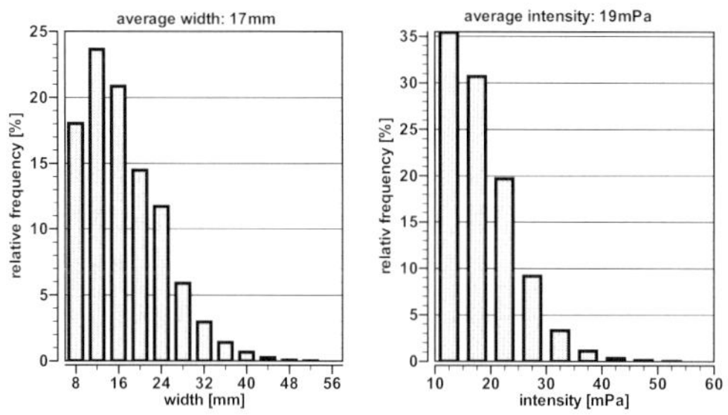

Figure 9: Histograms showing the relative frequency of wave packets with different intensities and width.

The results presented so far refer to a fixed chord position, but the distribution of sensors on the wing allows also for an observation of the downstream development of wave packets. As an example, Fig. 10 shows eight time histories of those sensors located on the chordwise column in the center of the array.

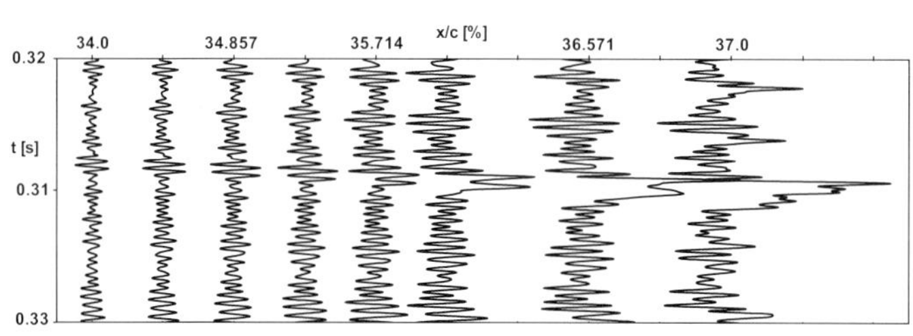

Figure 10: Downstream development of wave packets near breakdown to turbulence.

It can be seen that amplitudes of the wave packets gradually grow downstream. Breakdown to turbulence then starts near to a chord position of $x/c = 37$ % from within those disturbances with the biggest amplitudes.

5. CONCLUSIONS

Flight tests were performed in order to learn more about the technically important case of free-stream excited Tollmien-Schlichting waves. A multi-element hot-film array, placed on the right hand wing of the flying test bed LFU-205, was used to sense the skin friction fluctuations provoked by Tollmien-Schlichting disturbances propagating in the laminar boundary layer.

Fourier analysis in time and space of the experimental data revealed that a broadband spectrum of 2d and 3d Tollmien-Schlichting waves is involved in the signals and was typical for wave packets. The comparison of an experimentally determined amplitude spectrum for a wave propagation direction parallel to the flow at the boundary layer edge with that calculated by means of linear stability theory shows good agreement. Further investigations referred to shape and intensity of the wave packets, which showed that their spanwise extent and maximum amplitudes both vary strongly for the disturbances that passed a fixed chord position. Furthermore it was observed that the occurrence of wave packets happens stochastically; no periodicity or even repetition of certain patterns was found.

Time histories of signals from sensors distributed in the chord direction showed that breakdown to turbulence, indicated by spikes in the signal, starts from within the wave packets with the strongest amplitudes.

REFERENCES

1. Tollmien W. "Über die Entstehung der Turbulenz", *Nachr. d. Ges. d. Wiss. zu Göttingen, Math.-Phys. Klasse,* pp. 21-44, 1929.
2. Schlichting H. "Zur Entstehung der Turbulenz bei der Plattenströmung", *Nachr. d. Ges. d. Wiss. zu Göttingen, Math.-Phys. Klasse*, pp. 192-208, 1933.
3. Horstmann K.-H., Miley S.J. "Data Report of Flight and Wind-Tunnel Investigations of Tollmien-Schlichting Waves on an Aircraft Wing", *DLR Institutsbericht IB 129-91/18*, 1991.
4. Ludwieg H. "Ein Gerät zur Messung der Wandschubspannung turbulenter Grenzschichten", *Ing. Arch. Vol. 17*, 1949.
5. Seitz A., Horstmann K.-H. "Propagation of Tollmien-Schlichting waves in a wing boundary layer", *Notes on numerical fluid mechanics and multidisciplinary design, vol. 86*, 2003.
6. Riedel H., Sitzmann M. "In-flight investigations of atmospheric turbulence", *Aerospace Science and Technology, no. 5*, pp. 301-319, 1998.
7. Gaster M., Grant I. "An experimental investigation of the formation and development of a wave packet in a laminar boundary layer", *Proc. R. Soc., A347*, pp. 253-269, 1975.

THE INFLUENCE OF ROUGHNESS ON BOUNDARY LAYER STABILITY

M. Gaster
Engineering Department, Queen Mary (University of London)
Mile End Road, London E1 4NS.
E-mail m.gaster@qmul.ac.uk

Abstract: A theory is proposed for the influence of *shallow* two-dimensional periodic roughness on the stability of waves in a boundary layer. The approach is based on the linearized perturbation equations for disturbances in a locally parallel flow. The mean flow induced by two-dimensional ripples is modelled by Fourier expansions in the stream direction coupled with a Taylor series in the direction normal to the wall. The required no-slip boundary condition that applies over the ripple surface is transferred, through the Taylor expansion, to the x-axis. It turns out that even for quite small roughness heights it is necessary to use more than the first term of the Taylor expansion to define the boundary condition adequately. The resulting mean boundary layer is then defined by the usual boundary layer equations, but with a negative slip boundary value on the x-axis. The *slip* velocity arises from the nonlinear coupling between the Fourier terms in the definition of the ripple shape and the perturbation velocities. Although the slip velocity is small, it does have a significant effect on the stability of Tollmien-Schlichting waves. Wind tunnel measurements of the effect of two-dimensional ripples on the evolution of Tollmien-Schlichting waves have been made for a range of ripple heights.

Key words: Roughness, stability, experimental measurement.

1. INTRODUCTION

100 years ago, Prandtl published his model of the viscous flow close to a wall and established the "Boundary layer". He showed that at high Reynolds numbers the Navier-Stokes equations for the flow over a body could be decomposed into an outer inviscid potential solution coupled to the boundary through a thin zone close to the wall where the controlling equations had a parabolic structure.

If the boundary surface is smooth, the boundary conditions of zero slip and normal velocity can be applied directly to the parabolic equations. If the surface is rough, however, the local perturbation to the flow is elliptic and occurs in a narrow zone close to the wall. Here we consider a boundary surface composed of periodic two-dimensional roughness ripples. When the

G.E.A. Meier and K.R. Sreenivasan (eds.), IUTAM Symposium on One Hundred Years of Boundary Layer Research, 125-134,

ripples are sufficiently small and shallow, the equation defining the perturbation to the basic boundary layer flow can be linearized. Using the additional approximation that the flow can be treated as approximately parallel; the equations reduce to the Orr-Sommerfeld equation without the time dependent term. These equations have been used to determine the roughness perturbation field to the boundary layer.

If the height of the roughness is small enough it will be shown that the resulting mean boundary layer is indeed unaffected. At larger ripple amplitudes, however, it turns out that the boundary layer is modified though the mean boundary conditions. In particular, the flow is displaced outwards from the surface, although the shape of the velocity profile is unaffected.

This paper reports the results of wind tunnel measurements of the amplification of instability waves propagating through a boundary layer modified by a region of periodic wall roughness. The results are discussed in the light of a simple theoretical model of the disturbed flow.

2. THEORY

The base flow is the flat plate Blasius boundary layer. The surface over which the boundary layer develops has a series of two-dimensional ripples of rectangular waveform of mark-space ratio unity as shown on figure 1. The perturbation to the mean boundary layer flow created by these ripples is defined by the steady Orr-Sommerfeld equation. Solutions are then sought that have the appropriate boundary conditions on the surface of the ripples.

The ripple surface can be defined in terms of the Fourier series:-

$$y^* = \frac{4h}{\pi}\left[\cos(\alpha x) - \frac{1}{3}\cos(3\alpha x) + \ldots\right], \tag{1}$$

where h is the ripple half-height, and α is the wavenumber. The perturbation created by this waveform can also be expressed as the Fourier series:-

$$u(x,y) = u_0(y) + \Re\left[u_1(y)e^{-i\alpha x} + u_2(y)e^{-2i\alpha x} + u_3(y)e^{-3i\alpha x} + \ldots\right], \tag{2}$$

where the terms $u_n(y)$ refer to the solution of the perturbation equations for the streamwise velocity component of that degree. These solutions can be expressed as Taylor series in the normal direction. On substituting the expression for the ripple surface into the flow perturbation equations and applying the no-slip boundary conditions a set of equations for each of the

Fourier elements can be derived. The process has only been carried out up to the 1st harmonic, but clearly more terms could be obtained.

The non-periodic component

$$u_0 + \frac{2h}{\pi} u'_{1r} + \frac{h^2}{2} u''_0 = 0. \tag{3}$$

The real and imaginary components of the fundamental

$$u_{1r} + \frac{4h}{3\pi} u'_{2r} + \frac{h^2}{2} u''_{1r} = -\frac{4}{\pi}\left[\frac{dU}{dy} + hu'_0\right], \tag{4}$$

$$u_{1i} + \frac{8h}{3\pi} u'_{2i} + \frac{h^2}{2} u''_{1i} = 0. \tag{5}$$

The real and imaginary terms of the first harmonic

$$u_{2r} + \frac{4h}{3\pi} u'_{1r} + \frac{h^2}{2} u''_{2r} = 0, \tag{6}$$

$$u_{2i} + \frac{4h}{3\pi} u'_{1i} + \frac{h^2}{2} u''_{2i} = 0. \tag{7}$$

Numerical values for the various derivatives of the Fourier modes on the x-axis were determined by integrating the Orr-Sommerfeld equation. The resulting mode shapes are shown on figure 2 for the parameters of the experiment.

The solution of the above set of equations provided values of the various Fourier components of the perturbation velocity field. In particular, the quantity u_0 defines a mean perturbation to the basic mean velocity profile on the x-axis. The modified mean-flow velocity profile is identical to the original base flow, but with a modified origin, that effectively displaces the profile away from the axis. The Fourier components for the fundamental and first harmonic provide the spatially wavy structure of the flow that decays rapidly away from the x-axis.

The magnitude of u_0 depended on a number of flow parameters, but for a given flow setup was a function of ripple height. It is clear from figure 3 that the value of the boundary slip-velocity increases very rapidly with until the equations fail to provide meaningful solutions above 150 microns.

The wavelength of the unstable Tollmien-Schlichting waves is much longer that the induced spatially periodic ripples and their behaviour is unaffected directly by the spatial perturbations. The growth of periodically excited travelling instability waves is, however, influenced by the modified velocity profile. The neutral and amplification contours for two different boundary *slip* velocities are shown on figures 4. The neutral loop moves to lower critical Reynolds numbers and the upper branch moves outwards to create a larger amplification zone containing increased amplification rates. A trajectory for the particular Fnumber used in the experiments is shown together with the domain over which measurements were made.

A periodic point source was used to generate instability waves in the flow. Calculations of the perturbation flow field for a similar setup was based on the solutions of the usual linearized parallel flow Orr-Sommerfeld equation. The individual modes for specific spanwise and streamwise wavenumbers were evaluated by direct numerical integration from the free stream towards the wall, taking care to remove the parasitic divergent solutions. Inverse transforms provided the velocities in physical space around and downstream of the source. The contours of the streamwise velocity fluctuations are plotted on the $x{\sim}y$ plane through the centre of the wedge on figure 7 & 8 and for the smooth plate and a rough plate.

3. EXPERIMENT

The experiment was set up in the Queen Mary College low-turbulence wind tunnel on the boundary layer of a 1.5 m long flat plate designed so that different roughness panels could be inserted in the plate. The tunnel was of a conventional closed circuit low speed design with a number of removable working sections of 0.9 m square and 2.0 m long. A computer controlled three-axis traverse enabled the hot-wire probe to be positioned to any specified location with good precision. Ribbed surfaces were fabricated on insert panels 300 mm long in the direction of flow.

Artificial periodic excitation was provided by a small speaker on the rear of the plate through a 1 mm dia hole on the centre-line at 374 mm from the leading edge. All the results discussed were obtained at one value of Fnumber and one unit Reynolds number. Environmental parameters such as atmospheric pressure and temperature were periodically measured so that the tunnel speed could be controlled to a value appropriate to the chosen unit Reynolds number to better than 0.1%. By also adjusting the excitation frequency as the velocity changed similarity was maintained. The periodic excitation signal was stored in the buffer of a D/A converter and clocked out to the speaker at the same time as the hot-wire output was sampled. An experiment could be carried out completely under computer control. At each station the recorded signals of the hot-wire anemometer and that feeding the speaker were ensemble averaged and the first three Fourier components stored. The wedge shaped disturbance zone generated by a periodic point source could be explored by taking sections through the region. Here only the sections recorded over the $x{\sim}y$ plane passing through the source are shown.

Only data obtained from surfaces with roughness wavelengths of 6.25-mm will be discussed. As in the earlier work [1] adhesive plastic “Letraset” strips were initially used to create a ribbed surface 126 microns thick. Increased thicknesses were made with other plastic and paper material glued on the plate. The levels of the surfaces were adjusted by the push-pull screws so that the centre of the ribbed surface was aligned to the smooth part of the plate.

4. RESULTS

Figure 5 shows the contours of the fundamental Fourier component downstream of a periodic source measured over the smooth plate. The amplitude levels of the calculated pattern on figure 7 have been adjusted so that the two patterns are similar close to the source. To take some account of boundary layer growth the calculations have been carried out in three sections with appropriate length scales. The contour patterns appear virtually identical to one another close to the source, but still compare reasonably well downstream.

Five insert panels with ribs of various heights were also used. In each case a streamwise traverse through the source was made. Space only allows only two sets of data to be shown in this short paper. A panel with a single thickness of “Letraset” with $h = 65$ microns was tried. The resultant contour plot was virtually identical to that obtained on the smooth panel. Another surface was made by gluing paper strips on top of the “Letraset” to form ridges with $h = 125$ microns. From figure 3, it could have been expected that this would have had a moderate influence on the amplification rates as a slip velocity of around -0.035 was indicated. However, measured growth rates were not as large as might have been expected. Panel 3 was made from thin strips of plastic sheet and had an h of 218 microns, which is outside the range of the predictions shown on figure 3 and was predicted to create a very significant effect. Figure 6 shows that the amplitudes were increased at the last station by a factor of 3 over those measured on the smooth surface. Calculations with a slip velocity of -0.05 are shown on figure 8.

At this stage, there was some disappointment that the “blow-up” with increasing values of h suggested by figure 3 did not materialize. Two more panels were therefore prepared with even larger values of h of 255 and 500 microns. The measured disturbance patterns were virtually identical to those obtained with the previous panel of 218 microns.

5. DISCUSSION

It is clear from the few results shown that roughness influences the growth of instability waves. The theoretical model, only developed up to *h* squared terms, failed to provide accurate predictions of the amount of slip velocity induced and hence of the growth of periodic disturbances. The failure of the model solution was almost certainly because insufficient terms of the expansion were used, rather than the basic principle of the mechanism proposed. The approach is only applicable to small ripples as used in [1]. The theoretical model appropriate to shallow ripples assumes that the streamlines flow along the pattern of the ripples, but when the ripple is large the flow may separate. It would appear that once this regime has been reached further increase in ripple height has virtually no effect on the flow and hence on stability. The maximum increase in 'N' factor induced by 300 mm of rippled surface was about only one.

Fine roughness induces mean disturbances that decay rapidly with distance from the surface and the elliptic structure only has to be catered for very close to the boundary.

The roughness explored here was not only two-dimensional but also periodic. Real roughness is random and three-dimensional and both these factors are vitally important in modifying the base flow.

The linear theoretical calculation to predict the perturbation field from a point source provided reasonable agreement with the measurements but clearly needs to be modified to take at least some account of the boundary layer development with downstream distance.

6. CONCLUSIONS

The growth of linear disturbances from a periodic point source was measured on both a smooth and a ribbed flat plate.

Small ripples have very little influence on wave growth, but larger ones increased the amplification.

The increase in amplification was linked to the apparent negative slip velocity imposed on the mean flow by the nonlinear coupling between the ripple shape and the steady perturbation induced by the ripple.

Above some threshold, ripple height any increase in height had no further effect on the flow.

REFERENCE

[1] Gaster M. "The Influence of Roughness on Boundary Layer Transition", *Symposium on Advances in Fluid Mechanics,* Bangalore, India July 2003. pp.176-187.

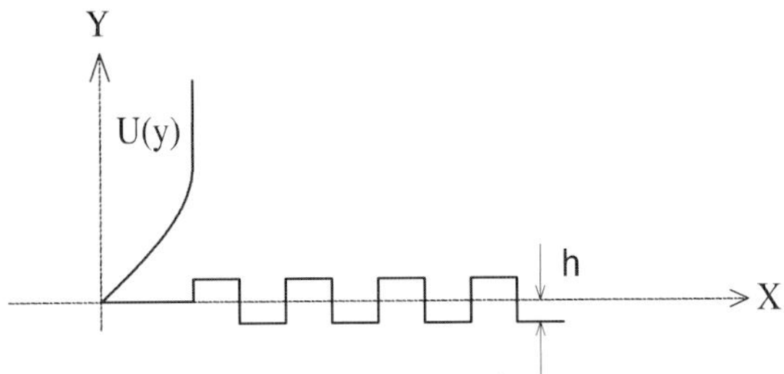

Figure 1 Model of Ribbed Surface

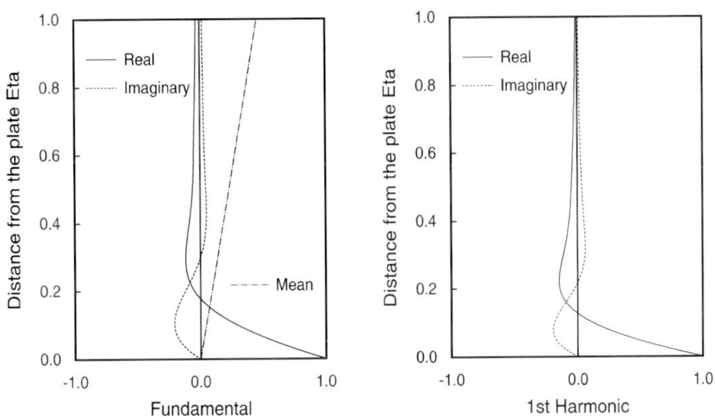

Figure 2 Mean Perturbation modes

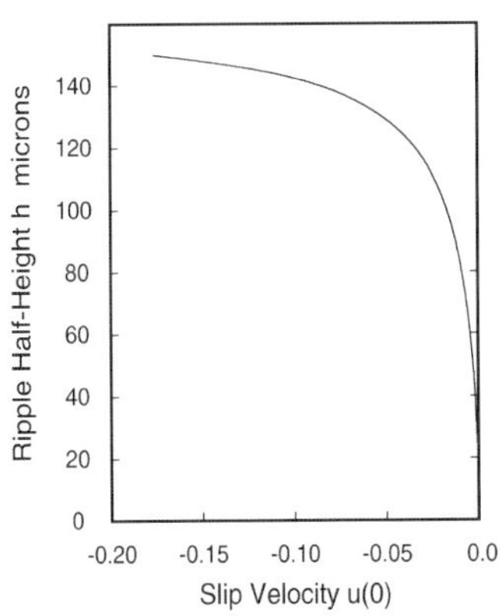

Figure 3 Slip Velocity versus Ripple height

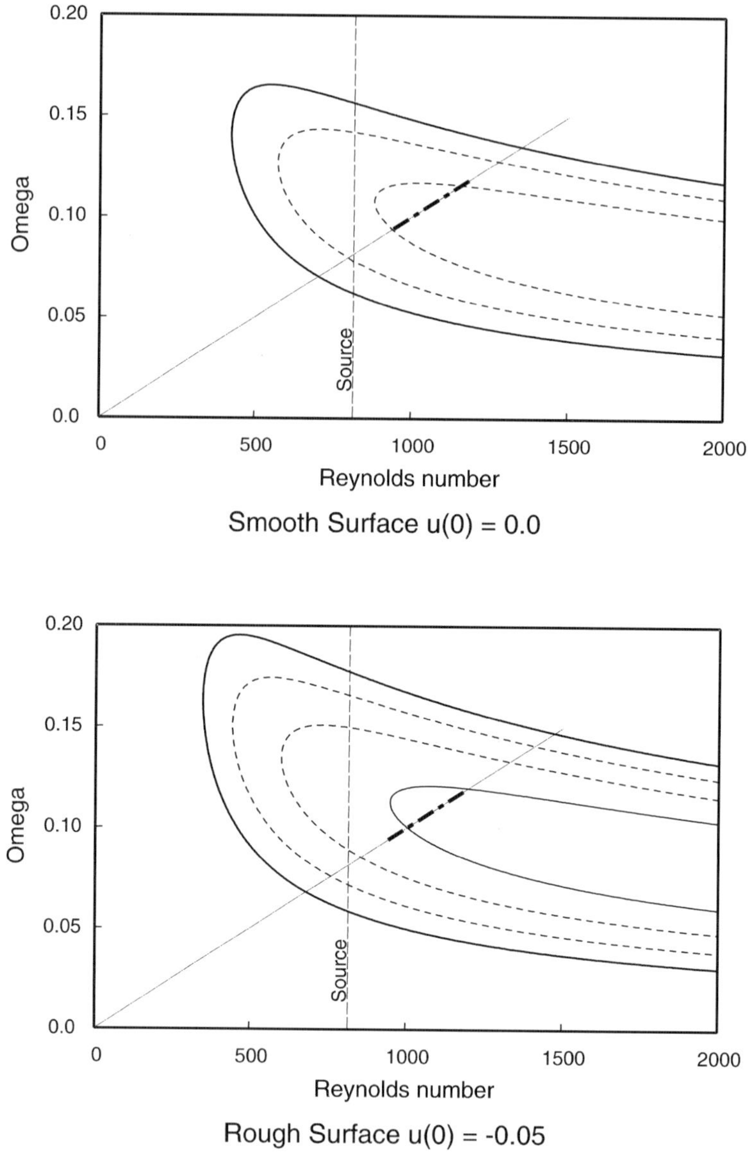

Figure 4 Neutral Loops for different Slip Velocities

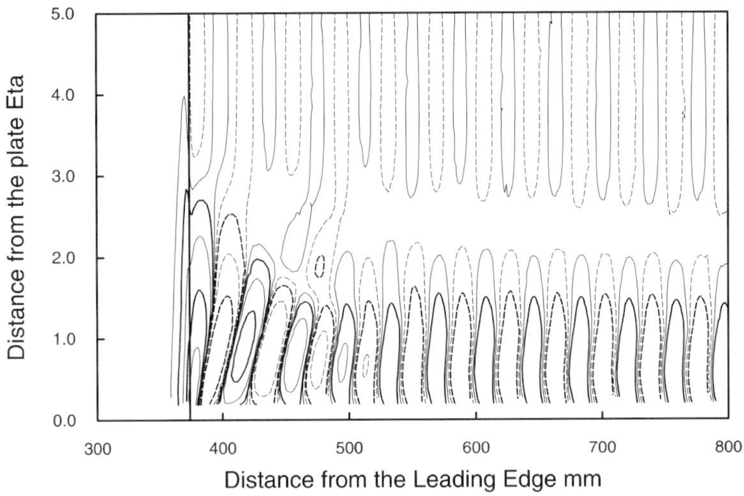

Smooth Plate

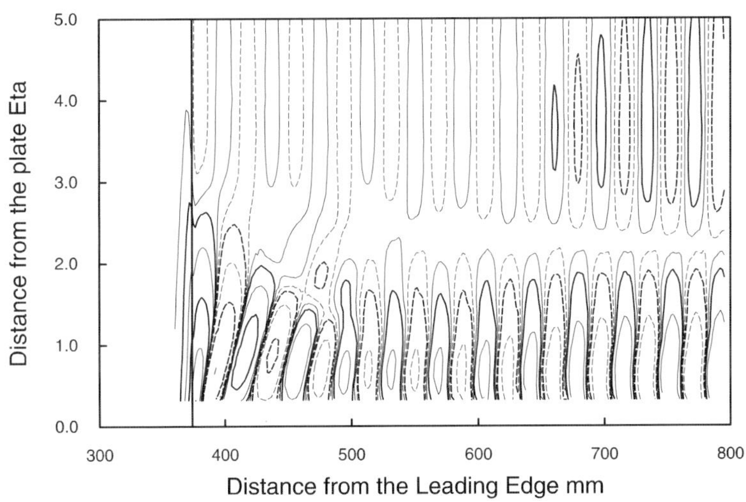

Rough Panel with h = 218 microns

Figures 5 & 6 Experimetal Measurements

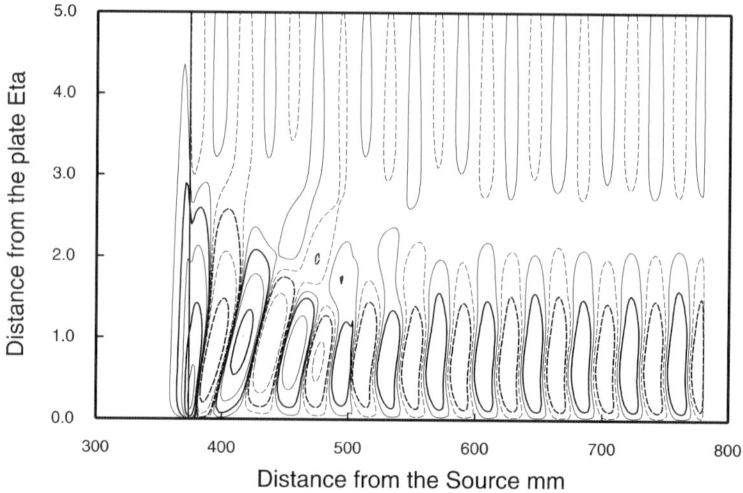

Smooth Plate

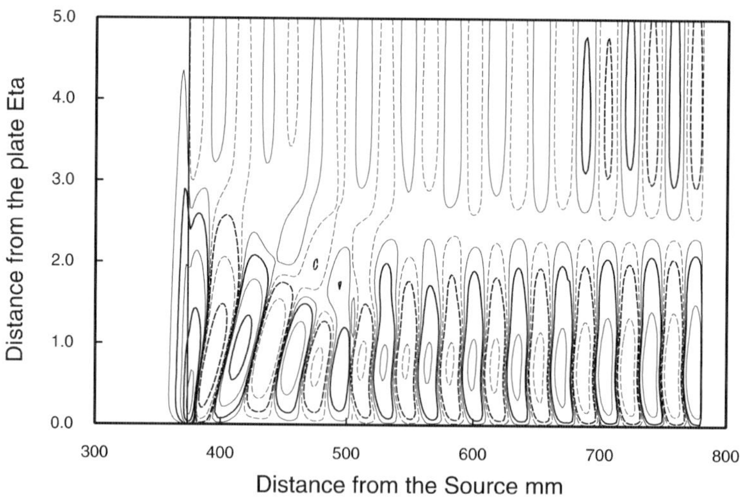

Slip velocity -0.05

Figures 7 & 8 Theoretical Predictions

BOUNDARY-LAYER INSTABILITY IN TRANSONIC RANGE OF VELOCITIES, WITH EMPHASIS ON UPSTREAM ADVANCING WAVE PACKETS

Oleg S. Ryzhov and Elena V. Bogdanova-Ryzhova
[1]*Department of Mechanical and Aeronautical Engineering, University of California Davis, Davis, CA 95616, USA*

[2]*GE Global Research, Fluid Mechanics Laboratory, Schenectady, NY 12305, USA*

Abstract: A generalized triple-deck approach is developed for high subsonic and low supersonic velocities, assuming the Reynolds number Re to be large. A long-standing paradox about the boundary-layer stability predicted by the classical version of the triple deck for $M_\infty > 1$ is resolved by discovering a new unstable eigenmode that emerges at some critical value of a transonic similarity parameter. The frequency and both wavenumbers of unsteady spiral Görtler vortices on a concave surface are shown to asymptotically obey in scaled variables the same dispersion relation as incompressible disturbances of a similar type. The famous boundary-layer equations introduced by Prandtl [**1**] a century ago underlie the extended theory.

Key words: Boundary layer, instability, wave packets, transonic flows, supersonic flows.

1. GOVERNING EQUATIONS

Within the framework of the conventional triple-deck approach, the streamwise and spanwise reference lengths, l_x and l_z, are of equal order in magnitude of a small parameter $\varepsilon = \mathrm{Re}^{-1/8}$ provided that the Mach number M_∞ at the upper edge of the boundary layer does not approach 1. However an additional transformation of all non-dimensional quantities includes $\delta = 1 - M_\infty^2$ tending to zero with $M_\infty \to 1$. The choice of equal l_x and l_z leads to meaningless results in an attempt to simplify the Navier-Stokes equations

G.E.A. Meier and K.R. Sreenivasan (eds.), IUTAM Symposium on One Hundred Years of Boundary Layer Research, 135-144,

in the transonic range of velocities. Starting from this observation, let l_z increase to $O\left(\varepsilon^3|\delta|^{-7/8}\right)$ and l_x preserve its triple-deck scaling $O\left(\varepsilon^3|\delta|^{-3/8}\right)$ on the assumption that $|\delta| = \varepsilon^{8/9}$. Then the perturbed velocity potential φ_1 in the outer inviscid sublayer satisfies an equation

$$\frac{\partial^2 \varphi_1}{\partial t \partial x} + K_\infty \frac{\partial^2 \varphi_1}{\partial x^2} - \left(\frac{\partial^2 \varphi_1}{\partial y_1^2} + \frac{\partial^2 \varphi_1}{\partial z^2} \right) = 0 \tag{1}$$

cast in the canonical form. The transonic similarity parameter [**2**]

$$K_\infty = \frac{M_\infty^2 - 1}{\varepsilon^{8/9} \left(2M_\infty^2\right)^{8/9} C^{1/9} \tau_w^{2/9} \left(T_w^* / T_\infty^*\right)^{4/9}} \tag{2}$$

depends on the normalized skin friction τ_w, the Chapman constant C and the ratio T_w^*/T_∞^* of the wall temperature T_w^* to the ambient temperature T_∞^*.

The curvature effects become crucial in the main deck when studying the periodic structure of spiral-type Görtler vortices on a concave cylindrical wall with generators aligned with the spanwise direction. At the bottom $y_2 = 0$ of this sublayer the excess pressure p is related to the potential function φ_1, the surface curvature κ and the instantaneous displacement thickness $-A(t,x,z)$ through the interaction law

$$p = -\frac{\partial \varphi_1(t,x,0,z)}{\partial x} + |\delta|^{7/2} K_\kappa A \tag{3}$$

The second similarity parameter

$$K_\kappa = \kappa \left(2M_\infty^2\right)^{1/9} C^{7/18} \tau_w^{-11/9} \left(T_w^* / T_\infty^*\right)^{14/9} \tag{4}$$

grows linearly with the surface curvature κ.

The celebrated boundary-layer equations by Prandtl [**1**] control the disturbance pattern in the lower viscous sublayer. They are

$$\frac{\partial u}{\partial x} + \frac{\partial v}{\partial y} + \frac{\partial w}{\partial z} = 0,$$

$$\frac{\partial u}{\partial t} + u\frac{\partial u}{\partial x} + v\frac{\partial u}{\partial y} + w\frac{\partial u}{\partial z} = -\frac{\partial p}{\partial x} + \frac{\partial^2 u}{\partial y^2}, \tag{5a,b}$$

$$\frac{\partial w}{\partial t}+u\frac{\partial w}{\partial x}+v\frac{\partial w}{\partial y}+w\frac{\partial w}{\partial z}=-|\delta|K_z\frac{\partial p}{\partial z}+\frac{\partial^2 w}{\partial y^2} \tag{5c}$$

where $p = p(t, x, z)$ prescribed by the interaction law (3) does not vary across the thin near-wall region. However, contrary to the original postulate put forward by Prandtl [**1**], the pressure variations are not given in advance, rather they are to be determined simultaneously with the velocity field. The third similarity parameter

$$K_z = \left(2M_\infty^2\right)^{8/9} C^{1/9} \tau_w^{2/9} \left(T_w^* / T_\infty^*\right)^{4/9} \tag{6}$$

underlies the pulsation distribution in the spanwise direction.

The outer-flow equation (1) has to be integrated along with the system of boundary-layer equations (5a-c) coupled together by the interaction law (3), the limit condition

$$u - y \to A \quad \text{as} \quad y \to \infty \tag{7}$$

at the outer reaches of the near-wall sublayer and the no-slip conditions $u = v = w = 0$ at a smooth surface $y = 0$. The formulation of the problem in eigen-values is complete.

2. DISPERSION RELATION

The Tollmien-Schlichting wave in a boundary layer on a concave surface is fixed by the frequency ω, streamwise wavenumber k and spanwise wavenumber m. With the interaction law allowing for the curvature effects, the dispersion relation connecting these three quantities takes on the form

$$\Phi(\Omega) = Q\left(\omega, k, m; K_\infty, |\delta|^{7/2} K_\kappa, |\delta| K_z\right), \quad \Omega = i^{-2/3}\omega k^{-2/3}, \tag{8a,b}$$

Here the left-hand side

$$\Phi(\Omega) = \frac{d\,\mathrm{Ai}(\Omega)}{dY}\left[I(\Omega)\right]^{-1}, \quad I(\Omega) = \int_\Omega^\infty \mathrm{Ai}(Y)\,dY \tag{9a,b}$$

involves the first derivative and the improper integral of the Airy function $\mathrm{Ai}(\Omega)$ depending on the auxiliary variable Ω. The right-hand side

$$Q = i^{1/3}\frac{k^2 + |\delta| K_z m^2}{k^{5/3}}\left(\frac{k^2}{\lambda} + |\delta|^{7/2} K_\kappa\right), \quad \lambda = D^{1/2},$$
$$D = i\omega k - K_\infty k^2 + m^2 \qquad (10a\text{-}c)$$

is a function of the frequency and both wavenumbers entering λ that designates a branch of $(i\omega k - K_\infty k^2 + m^2)^{1/2}$ with positive real part to ensure the decay of disturbances at infinity $y_1 \to \infty$.

3. SUPERSONIC BOUNDARY LAYERS

It is a long-standing paradox that conventional triple-deck scheme predicts stability of the supersonic boundary layer on a flat plate against normal Tollmien-Schlichting waves (oblique eigenmodes can be unstable depending on the obliqueness angle). To resolve the paradox we apply the extended version of the theory equating with zero both the surface curvature κ and the spanwise wavenumber m. Then the dispersion relation

$$\Phi(\Omega) = (ik)^{1/3} k^2 \lambda^{-1}, \quad \lambda = D^{1/2}, \quad D = i\omega k - K_\infty k^2 \qquad (11a\text{-}c)$$

ensuing from (10a-c) contains only the transonic similarity parameter K_∞. The dispersion curves in the auxiliary complex Ω-plane prove to be most suitable for tracing stability properties with K_∞ increasing. For subsonic velocities of the oncoming stream and $K_\infty < 0$ the changes in the shape and position of the dispersion curves are insignificant. A passage through the critical values $M_\infty = 1$ and $K_\infty = 0$ in which case $\lambda = \sqrt{i\omega k}$ goes smoothly. However a new feature arises if we deal with a supersonic boundary layer and $K_\infty > 0$. Figure 1 serves to clarify the matter assuming $k > 0$ for the sake of definiteness. An additional root $\hat{\Omega}(k; K_\infty)$ is seen to originate at a point $\Omega = \Omega^{**} = \Omega_r^{**} + \Omega_i^{**} i = -0.738 - 0.426i$ located on the ray $\arg(\Omega) = -5\pi/6$ separating the domains of stable and unstable oscillations. For K_∞ as low as 1.560, a loop made up in the Ω-plane by this root is tiny. With K_∞ increasing the loop stretches very fast towards the first dispersion curve $\Omega_1(k; K_\infty)$ which deforms in such a way as to meet $\hat{\Omega}(k; K_\infty)$ at the point $\Omega = \Omega_{exch} = \Omega_{exch,r} + \Omega_{exch,i} i = -1.012 - 0.384i$ in the domain of stable disturbances. The curve $\Omega_1(k; K_\infty)$ springing up from the point Ω_{d1} gives rise to the boundary-layer instability as long as $K_\infty < K_{\infty,exch}$, where $1.858 < K_{\infty,exch} < 1.859$. The two dispersion curves $\hat{\Omega}(k; K_\infty)$ and $\Omega_1(k; K_\infty)$ collide when the similarity parameter reaches a value $K_{\infty,exch}$. With greater values of K_∞, the first dispersion curve

$\Omega_1\ k;K_\infty$ terminates in Ω^{**} and therefore induces only stable oscillations. In consequence of the unusual mode exchange the new dispersion curve $\hat{\Omega}(k;K_\infty)$ goes to infinity upon crossing the ray $\arg(\Omega) = -5\pi/6$ at the neutral point Ω_* and thus becomes responsible for the generation of exponentially enhancing disturbances in the range $K_\infty > K_{\infty,exch}$.

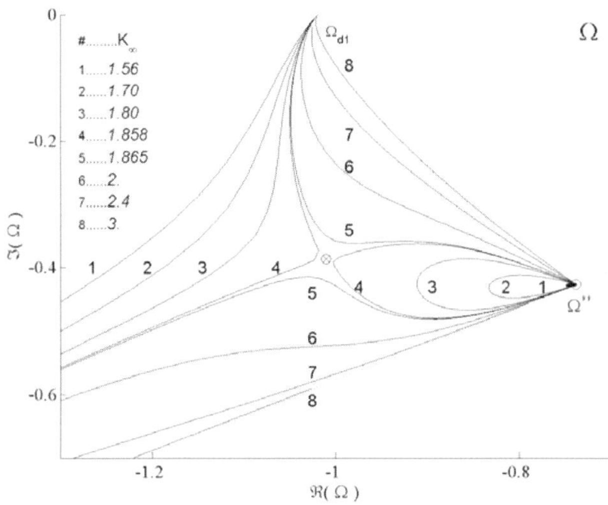

Figure 1. Auxiliary complex Ω-plane.

Further analysis relies on the computation of the dispersion curves in the plane of complex frequencies. As might be inferred from the above results, the first root $\omega_1(k;K_\infty)$ of the dispersion relation (11a-c) is associated with exponentially amplifying oscillations provided that $K_\infty < K_{\infty,exch}$. The formulation for the subsonic boundary layer is retrieved with $K_\infty \to -\infty$. To the contrary, unstable oscillations are induced by $\hat{\omega}(k;K_\infty)$ if $K_\infty > K_{\infty,exch}$. Figure 2 illustrates the mode exchange at $K_\infty = K_{\infty,exch}$. In the limit as $K_\infty \to \infty$, the neutral oscillations are specified by the frequency $\omega_*(K_\infty)$ and wavenumber $k_*(K_\infty)$ coming from [**2**]

$$\omega_*(K_\infty) \sim \frac{\omega_*^{(l)3}}{k_*^{(l)2}} \frac{1}{K_\infty^2} \left\{ 1 - 2\left[\frac{\omega_*^{(l)}}{k_*^{(l)}}\right]^8 \frac{1}{K_\infty^9} + \dots \right\},$$
$$k_*(K_\infty) \sim \frac{\omega_*^{(l)3}}{k_*^{(l)2}} \frac{1}{K_\infty^3} \left\{ 1 - 3\left[\frac{\omega_*^{(l)}}{k_*^{(l)}}\right]^8 \frac{1}{K_\infty^9} + \dots \right\} \qquad (12a,b)$$

where $\omega_*^{(l)} = 2.2968$ and $k_*^{(l)} = 1.0005$ are the neutral lower-branch frequency and wavenumber characteristic of the incompressible Blasius boundary layer. As a matter of principle, (12a,b) settle the question of how the triple-deck theory should be extended to provide a passage through the threshold value $M_\infty = 1$ and then cover a range of moderately supersonic velocities. It should be emphasized that the inclusion of the mixed time-space derivative $\partial^2 \varphi_1 / \partial t \partial x$ is the cornerstone of the generalized scheme.

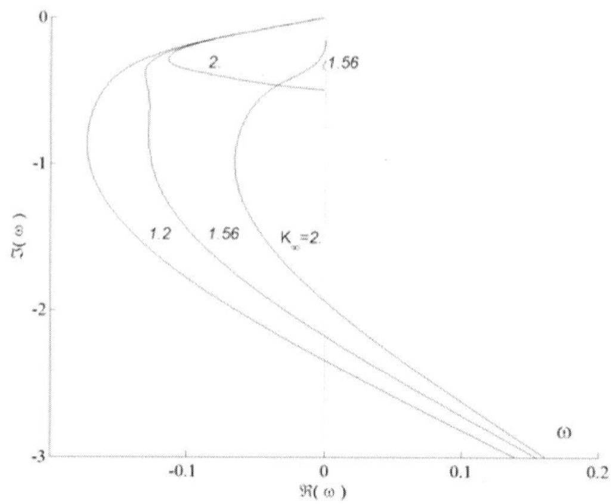

Figure 2. Plane of complex frequencies at supersonic velocities.

4. INSTABILITY OF GÖRTLER-VORTEX EIGENMODES

The surface curvature brings about new effects by creating centrifugal forces which, in turn, give rise to the streamwise Görtler vortical structures periodically spaced in the spanwise direction. One of the families of Görtler vortices endowed with the triple-deck disturbance field is known to be absolutely unstable in the streamwise direction at high subsonic Mach numbers [3]. This property may provoke earlier transition or, conversely, be exploited to delay transition by exciting less amplifying secondary modes. The transonic regime has to be investigated to get insight into a possibility for vortical instabilities to occur at moderately supersonic Mach numbers.

The dispersion relation with the right-hand side (10a-c) offers the clues to studying the disturbance pattern on a curved surface. All three similarity parameters K_∞, K_κ and K_z become operative in this case. By definition, both terms in brackets entering (10a) should be regarded of equal order in magnitude when dealing with the spiral-type vortices. This requirement leads to the renormalization

$$\omega = |\delta|^{5/7}\left(\frac{K_\kappa}{K_z}\right)^{2/7}\omega_G, \quad k = |\delta|^{15/14}\left(\frac{K_\kappa}{K_z}\right)^{3/7} k_G, \qquad (13a\text{-}c)$$
$$m = |\delta|^{-19/14} K_\kappa^{-1/7} K_z^{-6/7} m_G$$

of the frequency and both wavenumbers as applied to a side band appearing in the Tollmien-Schlichting wave spectrum. Related to new variables, (10a-c) can be cast as

$$\Phi(\Omega_G) = i^{1/3}\frac{|\delta|^{27/7} K_\kappa^{8/7} K_z^{-1/7} k_G^2 + m_G^2}{k_G^{5/3}}\left(\frac{k_G^2}{\lambda_G}+1\right), \quad \lambda_G = D_G^{1/2}, \qquad (14a\text{-}c)$$
$$D_G = i|\delta|^{9/2} K_\kappa K_z \omega_G k_G - |\delta|^{34/7} K_\infty K_\kappa^{8/7} K_z^{6/7} k_G^2 + m_G^2$$

with $\Omega_G = \Omega$ being invariant under the affine transformation (13a-c). In the limit $|\delta| \to 0$ that is associated with the Görtler frequencies and wavenumbers proper, the dispersion relation reduces to its incompressible counterpart [**4**]

$$\Phi(\Omega_G) = i^{1/3}\frac{m_G^2}{k_G^{5/3}}\left(\frac{k_G^2}{|m_G|}+1\right) \qquad (15)$$

independent of whether the oncoming stream is subsonic or supersonic. However the Cauchy problem turns out to be ill-posed in the framework of the simplified approach owing to ignoring the contribution from the Tollmien-Schlichting eigenmodes. The full dispersion relation in either of the forms (10a-c) or (14a-c) provides a mathematical description of the wave /vortex interaction, the vortical low-frequency side band making, in appropriately normalized variables, an integral part of the spectrum for all Mach number regimes, including $M_\infty = 0$.

Figure 3 presents the first dispersion curves $\omega_{G1}(k_G)$ determined by (15) in the complex ω_G-plane for small and moderate values of k_G and $m_{G0} = 0.4; 0.5; 0.6$ (only the lower half-plane is shown, the upper half-plane is its mirror image). These parts of the dispersion curves are responsible for the Görtler vortical eigenmodes of prime concern here. The contribution to the disturbance system from the Tollmien-Schlichting wave eigenmodes comes about from larger values of the streamwise wavenumber that obey the full dispersion relation (14a-c). In the limit as $k_G \to \pm 0$ it follows from (15) that

$$\Im[\omega_{G1}(k_G)] \to -\frac{m_{G0}^2}{k_G} \to \mp\infty, \quad \Re[\omega_{G1}(k_G)] \to \frac{\sqrt{2}}{2}\frac{|k_G|^{3/2}}{|m_{G0}|} > 0. \quad (16a,b)$$

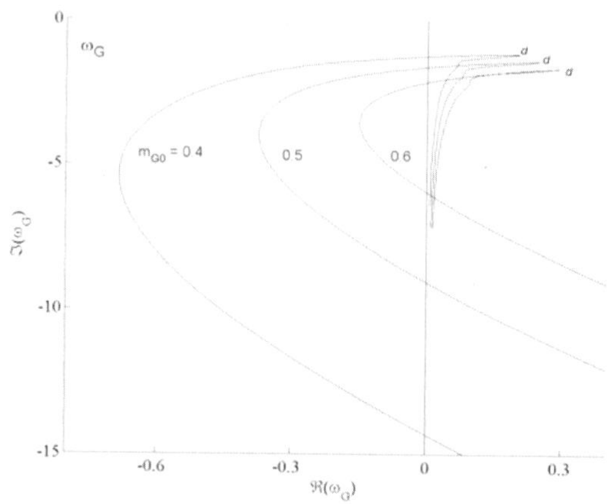

Figure 3. Plane of complex frequencies for Görtler vortices.

The same result holds in the context of (14a-c). The most important inference to be drawn from this asymptotics is that the first dispersion curve falls apart into two separate branches located in the upper and lower half-planes of the complex ω_G-plane. In turn, each branch consists of two lobes connected through a loop located not far from the origin. The formation of the loop is a distinctive feature intrinsic to the vortical eigenmodes. With loosely spaced vortices illustrated by $m_{G0} = 0.4; 0.5; 0.6$ in Figure 3, it is just the loop which gives rise to the left-hand lobes approximately described by (15). The dispersion curves for tightly spaced vortices with larger values of the spanwise wavenumber do not involve such a loop, nevertheless the left-hand lobes are still present in the forked shapes of their upper and lower branches and terminate in (16a,b) as $k_G \to \pm 0$.

It is common knowledge that the derivative $d\Im(\omega_{G1})/dk_G < 0$ at each point of the right-hand lobes that trigger the downstream moving wave packets at the heart of convective instability. To the contrary, $d\Im(\omega_{G1})/dk_G > 0$ along the left-hand lobes thereby inducing modulated disturbances in front of a perturbing agency. The negative sign of $d\Im(\omega_{G1})/dk_G$ changes to the positive sign when passing through a point d on the dispersion-curve loops in Figure 3 where $\Re(\omega_{G1})$ attains its local maximum. Accordingly, we may anticipate that any z-periodic external source mounted on a concave surface is capable of emitting in the pulse mode the wave packets advancing against the oncoming stream. Hence the boundary layer has come to be known as absolutely unstable in the streamwise direction [3]. Whatever the Mach number, the modulated disturbances are driven upstream by the vortical eigenmodes maintained by centrifugal forces. They do not arise in the Blasius boundary layer on a flat plate.

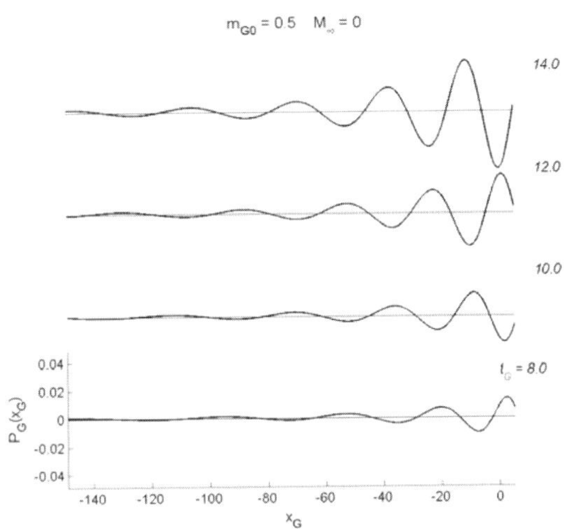

Figure 4. Wave packet moving upstream of vibrating ribbon.

The global maximum $\max_g[\Re(\omega_{G1})]$ of $\Re(\omega_{G1})$ on the right-hand lobes of the first dispersion curve (not seen in Figure 3) controls highly modulated wave packets with short-scaled oscillation cycles that sweep in the region downstream of a perturbing agency. An example of such large-sized disturbance at the bottom of convective instability can be found in [3] for a high subsonic Mach number. The local maximum $\max_d[\Re(\omega_{G1})]$ of $\Re(\omega_{G1})$ at the point d takes place with rather small values of the streamwise wavenumber, $\max_d[\Re(\omega_{G1})]$ being far less compared to $\max_g[\Re(\omega_{G1})]$ for any m_{G0}. It follows herefrom that the lower-amplitude disturbances responsible for absolute instability in the streamwise direction should be composed of long-scaled oscillation cycles and not severely modulated. Even though these disturbances build up fairly slowly their amplitude grows according to an exponential law. With time increasing, they enter an essentially nonlinear stage changing the flow pattern ahead of the exciting source.

Figure 4 computed with $m_{G0} = 0.5$ demonstrates the pulsation system of Görtler vortices penetrating upstream in line with the above general conclusions from the wave physics. The wavelength $\lambda_G \approx 24$ of a pulsation cycle is consistent with a value $k_G = k_{Gd} \approx 0.27$ at the point d where $\Re(\omega_{G1})$ attains its local maximum.

5. CONCLUSIONS

The boundary-layer concept put forward by Prandtl [**1**] a century ago takes new life in the modern stability theory. However, the pressure cannot be given in advance from a solution for the outer inviscid problem. Rather, it is to be evaluated simultaneously with the velocity field.

The time-dependence of disturbances in the potential flow region appears to be a key element in resolving a long-standing triple-deck paradox that the boundary layer does not experience the Tollmien-Schlichting instability at supersonic velocities. A mode exchange is shown to result in a new unstable eigenmode coming into being at a certain value of the transonic similarity parameter.

The Görtler vortical eigenmodes underlie absolute instability in the streamwise direction in sub-, trans- and supersonic regimes. They form a spectral side band with the appropriately normalized frequency and both wavenumbers governed by the dispersion relation for incompressible flow. The wave packets composed of Görtler vortices penetrate the boundary layer ahead of a perturbing source.

REFERENCES

1. Prandtl L., “Über Flüssigkeitsbewegung bei sehr kleinen Reibung”, In: *Verhandl. III, Intern. Math. Kongr. Heidelberg*, pp. 484-491, Leipzig, Teubner, 1905.
2. Ryzhov O.S., “An asymptotic approach to separation and stability problems of transonic boundary layer”, In: *Transonic Aerodynamics, Problems in Asymptotic Theory (Ed. L.P. Cook)*, Philadelphia, SIAM, 1993.
3. Ryzhov O.S., Bogdanova-Ryzhova E.V. Boundary layer instabilities in transonic range of velocities. In: *Proc. IUTAM Symposium Transsonicum IV (Ed. H. Sobeieszky)*, Dordrecht, Kluwer Academic, 2002.
4. Ruban A.I. “Propagation of wavepackets in the boundary layer on a curved surface.” *Izv. Akad Nauk SSSR, Mekh. Zhidk. Gaza*, **No 2,** pp. 59-68, 1990 (in Russian; English translation: *Fluid Dyn.*, **25**, pp. 218-221, 1990).

LAMINAR-TURBULENT-LAMINAR TRANSITION CYCLES

RODDAM NARASIMHA
Jawaharlal Nehru Centre for Advanced Scientific Research, Jakkur, Bangalore 560 064, India; Ph: 91 80 2208 2789; Fax: 91 80 2360 0865; e-mail: roddam@caos.iisc.ernet.in

Abstract: Recent experimental work in the leading edge region of a 60°-swept wing at an angle of attack of 18° shows three distinct minima within the first 10% of the chord in the rms value of the signal from surface hot film gauges and in an intermittency coefficient. Qualitatively similar but weaker variations are seen at other high angles of attack and on other wings tested in the experiments. Based on an analysis of the variation of pressure gradient, curvature and Reynolds number in these flows, it is proposed that the most direct interpretation of the observations is that there is a sequence of transitions in the flow, from laminar to turbulent, reverting towards laminar, back again towards turbulent etc. In the case mentioned six such transitions, not all of them complete, can be identified. Such 'transition cycles' may account for the problems encountered in scaling maximum lift coefficient from tunnel to flight Reynolds number.

Key words: Transition, relaminarization, transition cycles, swept wings, maximum lift coefficient

1. INTRODUCTION

In the hundred years since Prandtl's extraordinary paper on the boundary layer idea was published, a major and continuing concern has been the transition from laminar to turbulent flow. Prandtl himself pursued an enduring effort to understand transition, convinced that flow instability must be the major reason for it. It is not so well known that he also studied flows in which stable stratification tends to suppress turbulence [1], as in work done with Reichardt [2] on the boundary layer developing below a hot surface.

In the 1960s and 70s relaminarizing flows were studied extensively [3]. In the last ten years there has been growing interest in the flow near the leading edge of a swept wing at relatively high angles of attack, in an effort to understand the complex scale effects that have been observed on such wings, especially with respect to maximum lift coefficient [4]. A recently completed experimental study [5] provides valuable data for further anlaysis.

G.E.A. Meier and K.R. Sreenivasan (eds.), IUTAM Symposium on One Hundred Years of Boundary Layer Research, 145-154,

2. THE EXPERIMENTS

The experiments were carried out in the 1.5 m × 1.5 m low-speed wind tunnel at the National Aerospace Laboratories, Bangalore [5]. The tunnel has speed control better than 1%, and the freestream turbulence is less than 0.1%. The test configuration is shown in Figure 1. Figure 2 shows the three airfoil sections chosen for the study, with key geometric details in Table 1. Wing A is an NLR 7301 airfoil-flap geometry swept at 45°. Wing B is the same airfoil modified near the leading edge with a larger nose radius and swept at 60° to promote attachment line transition even at moderate angles of attack. Finally, Wing C is an airfoil with a circular nose of large radius (50 mm) swept at 60°, chosen to provide a different history of favourable pressure gradients, including a twin peak in the acceleration parameter K_s (see notation list at end). Each wing model had an end plate to avoid tip effects and a span of 2.5 times the airfoil chord to provide conditions approximating infinite sweep. The inboard section of the wing was mounted on a base plate with a 2D wedge in front so that advantage could be taken of the tunnel wall turbulent boundary layer flowing over the wing root to promote attachment line transition.

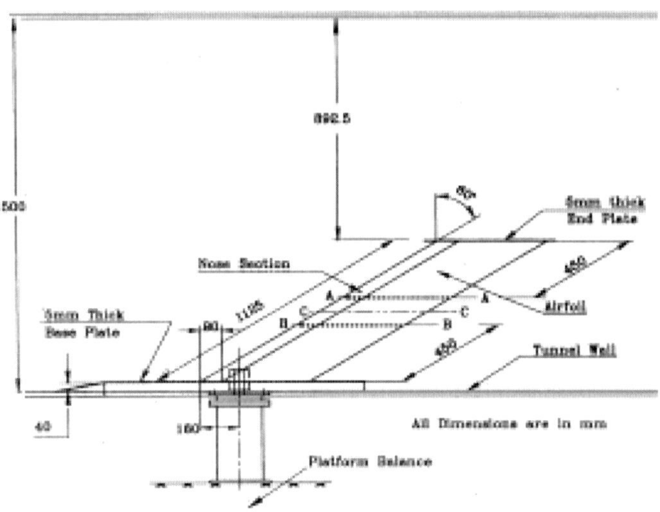

Figure 1: Schematic of the experimental arrangement in the tunnel, wing C (Table 1)

Model surface pressure distributions were measured along sections AA and BB (Figure 1) using low-pressure SETRA transducers integrated with 48-port scanivalves. Single element surface hot-films (Micro Measurements Co, USA, 50 ohms resistance) were employed as primary diagnostics for inferring the state of the boundary layer. Typically 10-12 gauges were bonded to the surface in the leading edge zone (within the first 15% of the chord on the pressure and suction surfaces); the bonding was done with great care, ensuring the films were flush with the surface. The active film was nearly normal to the local surface flow direction, determined as indicated below. The films were staggered in the spanwise direction between sections AA and BB of Figure 1, so as to minimize thermal or mechanical wake effects. In any case sensitivity to flow direction was not high over the range 6-18° in model incidence. The hot-films were used with a DISA 55M constant temperature anemometer system at a fixed overheat ratio of 1.20. The gauge signal was sampled at 2kHz for about 30 seconds. The other diagnostic was surface flow visualization using titanium di-oxide in oil to detect surface streamlines, the attachment line, laminar-separation bubbles, etc. All tests were carried out at a nominal freestream velocity of 48 m/s, yielding a chord Reynolds number of about 1.3×10^6.

From the extensive programme of tests carried out, we analyse data from three selected experiments [5], referred to respectively as C-18, A-17 and B-12, the code indicating the wing section and angle of attack in degrees (see Table 1).

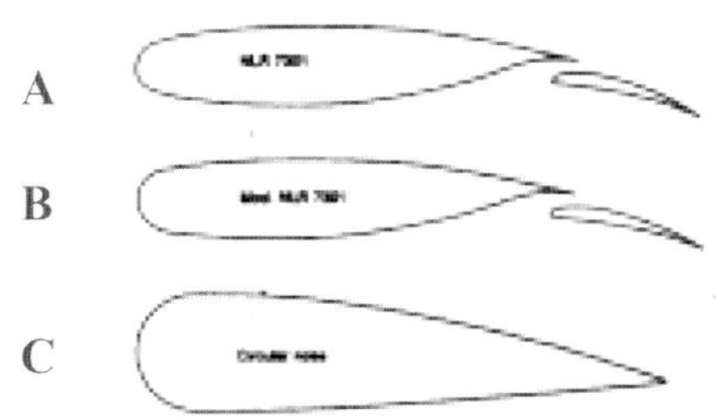

Figure 2: Airfoil sections of the three wings (A, B, C) tested; sections parallel to free stream

Table 1: Experiments analysed

Code	Model	Airfoil chord, m	Extended chord with flap, m	Sweep, deg.	Angle of attack, deg
A-17	NLR 7301	0.450	0.54	45	17
B-12	Mod. NLR 7301	0.4275	0.512	60	12
C-18	Circular Nose	0.40	-	60	18

3. RESULTS

We begin by analysing the C-18 flow, because here the results are most dramatic. Figure 3 shows the time traces of signals (e') from film gauges at several stations in the leading edge region. The attachment line is in the neighbourhood of Station C, and there is a short separation bubble between Stations N and O. These locations are confirmed by surface oil flow visualizations. Going along the wing surface from the attachment line on the pressure side we see a rapid increase in fluctuation levels indicating that the boundary layer, which must have been close to a laminar state at the attachment line, undergoes transition to turbulence with increasing Reynolds number downstream. This is one of the relatively rare instances in the series of experiments where the attachment line boundary layer did not seem to be fully turbulent. The reason for this is that the Reynolds number $\overline{R}$, which is often used as the criterion for indicating the state of the attachment line boundary layer, was generally low, staying below the critical value of about 250 up to $\alpha = 12^{\circ}$ (see Figure 3, [5]). Beyond this incidence there is a slight increase, and although at 18° the value is close to 400, the film gauge signals indicate that the flow is still far from being fully turbulent.

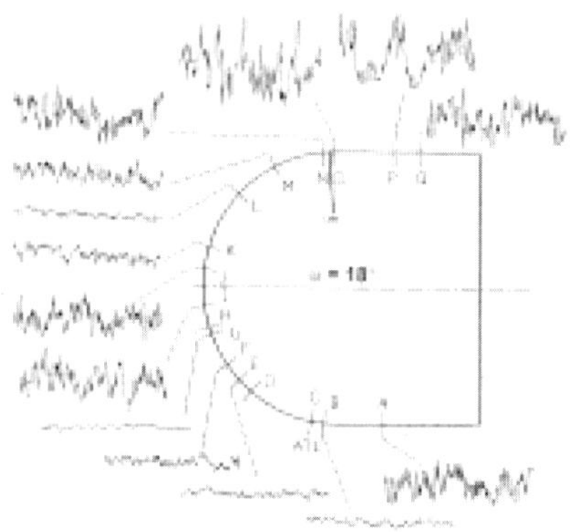

Figure 3: Hot-film traces at different stations (A to Q) around leading edge region, flow C-18

The variations in the signals as we move towards the suction surface are the more intriguing. First of all they show a mild increase from Station B/C towards Station F, indicating a trend towards turbulent flow. There is however a sharp decrease at G, followed by a spectacular increase in signals from stations I and J. Then follows one more decrease to low levels at Station L, and a relatively long increase through N to very high values at gauges N-Q following immediately after the separation bubble centred around NO.

The intensity of these signals in quantitative terms is shown in Figure 4 in the form of spectra. These are consistent with the trends displayed in Figure 3, but they further show that the changes at the low wave number end can some times be two orders of magnitude or more (note that the stations represented in Figure 4 are A, F, I, K, N and Q, and do not include G and L where signal levels are very low).

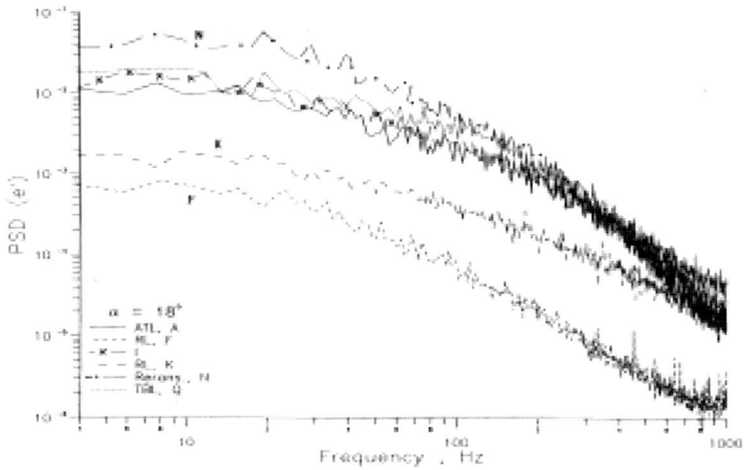

Figure 4: Power spectral density of hot-film signals from different stations, flow C-18

Figure 5 collects data on the spatial variation of a variety of parameters on flow C-18, all plotted against the non-dimensional arc length *s/c*, with origin at the geometric leading edge. It will be seen that the curvature of the surface is highest around *s/c* = 0, and that the pressure gradient parameter K_s has two peaks, both higher than the critical value of 3×10^{-6} often stated as necessary for a turbulent boundary layer to relaminarize. (These twin peaks in K_s can be seen as corresponding to the usual lower- and upper-surface suction peaks on the airfoil; this is possible because, at $\alpha = 18^{\circ}$, the attachment line is downstream of the high curvature regions around the leading edge that result in those suction peaks.) The traces in Figure 3 are seen to be generally consistent with the rms signal intensities labelled e_{RMS}. Starting from the attachment line there is an indication of slight increase in e_{RMS} on both the pressure and the suction sides of the wing. A weak maximum is seen at $s/c \simeq -0.04$, a stronger one around $s/c = 0$ to $+0.01$, a minimum around $s/c = 0.04$, and the highest maximum at $s/c \simeq 0.18$.

Many of these differences are highlighted in the curve labelled γ, which represents an intermittency coefficient. This coefficient was computed using an algorithm written for transition studies [6] and widely used in our group; no change has been made in the algorithm or in the implementing software. In the present flows a clear alternation between laminar and turbulent epochs, as in classical transitional flow, is not easily seen. But the value of γ given by the code can still provide a measure of the closeness to fully turbulent state in the flow. Keeping this in mind we can see from the bottom panel of Figure 5 that γ displays three maxima between $s/c = -0.05$ and $s/c \simeq 0.2$; even more clearly it exhibits three well-defined minima respectively at $s/c = -0.05$, -0.015 and $+0.05$. We see here that, even within the short space of a few percent points in the chord near the leading edge, the intermittency shows strong variations.

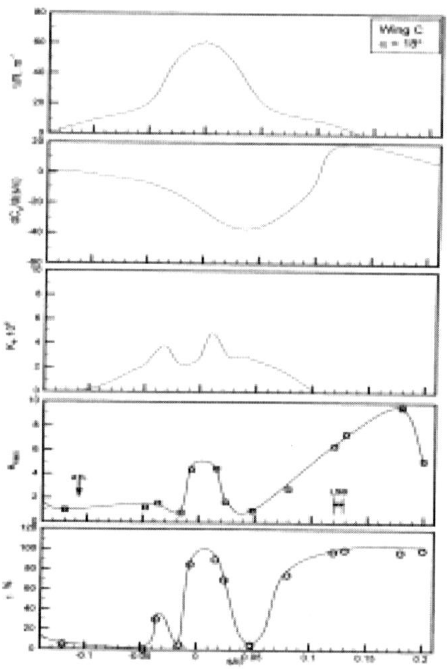

Figure 5: Variation of different parameters along arc length in flow C-18. From the top: curvature of surface R^{-1}, non-dimensional surface pressure gradient, non-dimensional pressure gradient parameter, r.m.s. value of hot-film signal e' and intermittency coefficient γ

From this curve, reading it together with the traces shown in Figure 3, it is natural to interpret the flow as indicating a total of six transitions. Starting from the attachment line, there is first of all a transition towards turbulent flow on the pressure side of the wing ($s/c < -0.1$). Similarly there is also a transition towards turbulent flow around $s/c = -0.04$ to -0.05. That is, a low-γ, nearly laminar attachment line boundary layer is transitioning towards turbulence on either side of itself. The intermittency exceeds 30% at around -0.04, but dips immediately thereafter to nearly 0 at -0.015, indicating that even before the transition is far from complete there is a suppression of turbulence in a relaminarizing regime. This is immediately followed by a substantial increase in γ which now reaches a value of almost 1 at $s/c \simeq 0$ and drops again thereafter as it relaminarizes at $s/c = 0.05$. Beyond this point there is once again another transition to a turbulent state around $s/c = 0.12$, which in terms of intermittency is virtually complete. This is also where a laminar separation bubble appears in the flow; because of the highly unstable nature of such a bubble e_{RMS} keeps increasing till $s/c \simeq 0.18$, although the intermittency itself has now virtually saturated at 1. It is interesting that beyond $s/c = 0.18$ e_{RMS} drops although the intermittency does not. This is a sure indication that the specially high levels of fluctuation associated with the separation bubble are now dropping to values characteristic of the classical attached turbulent boundary layer.

It is therefore clear that in swept wing flows of the kind that are being studied here sharp changes in the character of the flow can occur in a small fraction of the chord near the leading edge, and it is plausible to associate these changes with transitions, either complete or incomplete, between the laminar and turbulent states in either direction.

The indication of such 'cycles' of transition, as I would like to call them, is strongest in flow C-18, but they are seen in other experiments in the series as well, although usually in milder form. Figure 6 shows flow A-17 in which both e_{RMS} and γ behave somewhat similarly, even though they are not as striking. In flow B-12 (Figure 7), which is subjected to much milder pressure gradients (the critical value $K_s = 3 \times 10^{-6}$ is barely crossed for about 2% of the chord along the arc on the upper surface), the changes in e_{RMS} and γ are very small. Nevertheless it is extraordinary that the undulations seen in the intermittency curve here are in the same direction as in the more dramatic case of flow C-18. It would of course be difficult to take the variations seen in flow B-12 as indications of transitions, as the magnitude of the undulations is so small; the value of γ in each of these weak cycles changes as little as from 0.8 to 0.9, down to 0.6, up to 1 and so on. But with the hind sight afforded by C-18 one can see how exquisitely sensitive the flow is to small changes in pressure gradient (and possibly curvature).

4. DISCUSSION

We propose here a plausible explanation for the remarkable variations seen in γ and other flow parameters in Figure 5.

We can start from the attachment line, where the flow, if not laminar, is still far from being fully turbulent, with a γ of less than 0.05. Moving on the pressure side towards the trailing edge ($s < 0$) the local Reynolds number will increase as the flow accelerates slowly, and a classical transition follows. The same thing happens on the suction side as well, and γ increases to 0.3. However the flow now encounters the first ('lower-surface') suction peak, and the corresponding favourable pressure gradient crosses the critical value $K_s = 3 \times 10^{-6}$ at $s/c \simeq -0.05$, pushing the flow towards relaminarization and a sharp drop in γ.

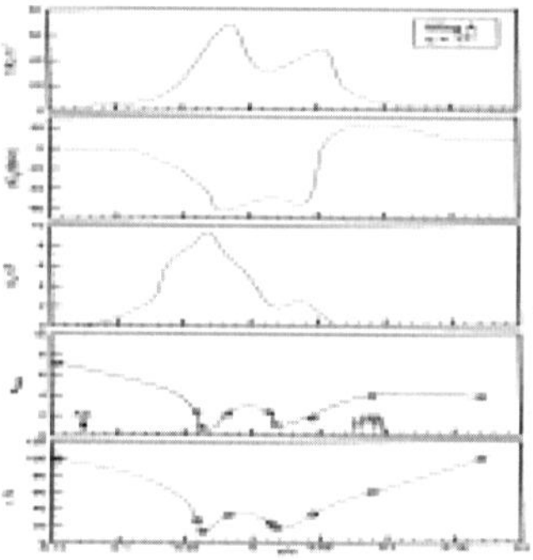

Figure 6: Variation of different flow parameters in A-17; same format as Figure 5

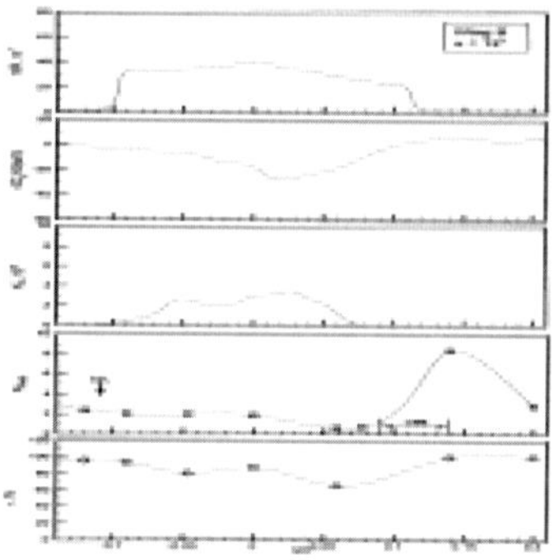

Figure 7: Variation of different flow parameters in B-12; same format as Figure 5

Immediately thereafter, however, K_s drops, and a quick retransition occurs; this can be extremely rapid [3], consistently with data in Figure 5. The flow now encounters the second ('upper-surface') peak of nearly 5×10^{-6} in K_s, and moves towards a second relaminarization, reaching $\gamma < 0.05$ at $s/c \sim +0.05$. A drop in K_s follows, and the flow goes through a second forward transition to turbulence. The presence of a short separation bubble at $s/c \simeq 0.12$-0.13 raises turbulent fluctuations to very high levels, but as the flow reattaches downstream of the bubble turbulent intensities drop to values characteristic of a normal turbulent boundary layer.

A summary of the six transitions as set out in the above picture is presented in Figure 8.

One question that needs further anlaysis concerns the role of surface curvature. Now surface curvature has two effects on the flow: what we may call a 'static' effect, which determines the pressure distribution, and a 'dynamic' effect, which influences the structure of the turbulence. The explanation given above depends basically on the static effect, suggesting that the dynamic effect is not decisive. An analysis of turbulence intensities in 2D flows [7] suggests, however, that the dynamic effect on turbulent fluctuations is not negligible, and the stabilizing effect of convex curvature (often described through an equivalent Richardson number) reinforces the relaminarizing effect of a favourable pressure gradient. In flow C-18 as well as the others such a reinforcing role is probably in operation, and helps to ensure rapid relaminarization, but to determine this in more quantitative terms demands turbulence measurements across the boundary layer.

A similar question arises about the role of three-dimensionality, which is a strong factor in transition on swept wings. Once again more

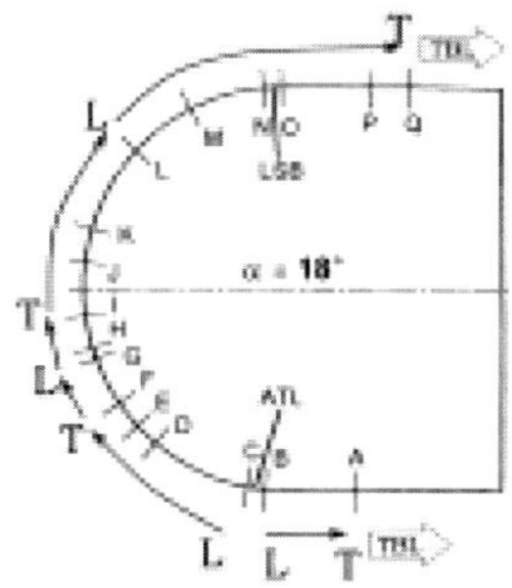

Figure 8: Transition cycle (L→T→L. . .) in flow C-18; L, T stand for laminar and turbulent respectively; TBL = turbulent boundary layer

detailed spanwise measurements would be required to settle this question. A third question is whether the present observations are triggered by some obstacle upstream. This is extremely unlikely for several reasons. First of all the staggered arrangement of the hot-film sensors, in a region where the flow corresponds well to that over an infinitely swept wing, ensures that they are not sampling flow over a single curve, and that the wake effect, if present, is very weak. Secondly very similar effects have been detected on other models and at different α; flow C-18 is only the most dramatic. For example, the signal from the same sensor in C-16 is very different from that in C-18, although the qualitative nature of the cycles is similar. For all these reasons there is strong evidence that the phenomena reported here constitute a genuine flow effect.

5. CONCLUSIONS

The main conclusions of the present work are that (i) within something like 10 per cent of the chord near the leading edge region, several (possibly incomplete) transition cycles between laminar-like and turbulent-like flows can occur: in one case we show there are as many as six transitions; (ii) qualitatively, these are largely understandable in terms of 2D mechanisms involving strong favourable pressure gradients, which, at high angles of attack, are bimodal for a fluid particle travelling over to the suction side from the attachment line; (iii) based on experiments in 2D flows, it appears plausible that the dynamic effect of curvature assists in more rapid relaminarization, but does not otherwise play a decisive role; (iv) even at angles of attack as low as 12°, and (consequently) weaker favourable pressure gradients, the signatures of (largely incomplete) transition cycles can still be detected, in the form of weak undulations in the intermittency coefficient.

ACKNOWLEDGEMENTS

I am most grateful to Dr P R Viswanath and Dr R Mukund for extensive discussions on the experimental data as well as the suggested mechanisms. The data were acquired on a project supported by Boeing through Dr Jeff Crouch, whose comments on the work have always been full of insight. The present analysis is carried out as part of a programme supported by a grant from DRDO, for which special thanks are due to Dr V Siddhartha.

NOTATION

c	airfoil chord
C_p	surface pressure coefficient based on free-stream conditions
e'	signal from surface hot-film gauge
e_{RMS}	r.m.s. value of e'
K_s	streamwise acceleration parameter = $(\nu/Q^2_e)\,(dU_e/ds)\cos^2\psi_e$
Q_e	component of free-stream velocity along (outer flow) streamline
R	local radius of curvature of airfoil surface
$\bar{R}$	$W_\infty(\nu\,U'_e)^{-0.5}$
s	arc length along surface in direction of freestream
U_e	velocity at edge of boundary layer along direction normal to leading edge
W_∞	component of free-stream velocity along wing leading edge
α	angle of attack
γ	intermittency coefficient
ν	kinematic viscosity
ψ_e	angle between outer streamline and the normal to the leading edge

REFERENCES

1. Prandtl L "Einfluss stabilisierender Krafte auf die Turbulenz" (1930). In: *Gesasmmelte Abhandlungen* 2:778, Springer-Verlag, Berlin, 1961.
2. Prandtl L, Reichardt H "Einfluss von Wärmeschichtung auf die Eigenschaften einer turbulenten Strŏmung". *Dtsch. Forschung* 21:110-121, 1934.
3. Narasimha R, Sreenivasan K R "Relaminarization of Fluid Flows" *Adv. in Appl. Mech.* 19:221-309, 1979.
4. Van Dam C P, Vijgen P M H W, Yip L P, Potter R C "Leading-Edge Transition and Relaminarization Phenomena on a Subsonic High-Lift System" *AIAA Paper* 93-3140, 1993.
5. Viswanath P R, Mukund R, Narasimha R, Crouch J D "Relaminarization on Swept Leading Edges under High-Lift Conditions" *AIAA Paper* 2004-0099, 2004.
6. Jahanmiri M, Rudra Kumar S, Prabhu A "A Method for Generating the Turbulent Intermittency Function" *Dept. Aero. Engg., Ind. Inst. Science, Bangalore*, Report 91 FM 13, 1991.
7. Mukund R "Relaminarization in a Short Acceleration Zone on a Convex Surface" *PhD Thesis, Dept. Aero. Engg., Ind. Inst. Science, Bangalore*, 2002.

A CENTURY OF ACTIVE CONTROL OF BOUNDARY LAYER SEPARATION: A PERSONAL VIEW

Israel J. Wygnanski
AME Department, The University of Arizona 1130 N. Mountain Avenue, Tucson, AZ 85721
email: wygy@ame.arizona.edu

Abstract: Active control of separation involves a multitude of parameters that are closely coupled and make the complete understanding of the flow difficult if not impossible. Since clear technological advantages are derived from its utilization we should endeavor to define these parameters and assess their relative significance. The present paper is an attempt in this direction.

Key Words: Boundary Layer Separation, Circulation Control, Active Flow Control

1. SEPARATION & CIRCULATION CONTROL: A BRIEF HISTORICAL BACKGROUND

We came to celebrate and honor Professor Prandtl whose epoch making paper gave a new impetus to fluid mechanics by predicting drag at high Reynolds numbers. The thousands of reports that deal with the behavior of the boundary layer under different circumstances are a clear testament to his success. Prandtl also carried out the first experiments on separation and circulation control by introducing suction from the rear portion of a circular cylinder. He carried out these and many other experiments in order to verify his theory[1], but in the process he revealed the complex phenomenon of flow separation. Goldstein[2] suggested that separation was Prandtl's main goal when he wrote his 1904 article. The quest for high lift that is hindered by separation led to the invention of slotted wings by Hendley Page and Lachmann (passive boundary layer control) and to tangential blowing by Baumann[3]. Suction and blowing became the two main stays of active

G.E.A. Meier and K.R. Sreenivasan (eds.), IUTAM Symposium on One Hundred Years of Boundary Layer Research, 155-165,

control of the boundary layer. The former did so by removing the slow vortical fluid through a slot, thus accelerating the boundary layer flow upstream, while the latter added momentum to the boundary layer.

An extensive investigation of slot suction was made by Schrenk[4] on thick airfoils that were otherwise plagued by early separation. Steady suction was historically characterized by a mass flow coefficient, C_Q, since its primary purpose was the removal of low momentum fluid from the boundary layer of a given velocity. Excessive suction could also provide "Circulation Control" (CC) that implies an increase of lift above and beyond the expected value generated by incidence and camber. The use of conformal mapping correctly predicted the lift enhancement and drag penalty that are both proportional to the sink strength generated by the suction[5]. Slot suction for the purpose of lift enhancement (CC) did not withstand the test of time because of the associated drag increase and the large ducts that were required to remove the low pressure, external fluid. At that time the thickness of airfoils diminished rapidly due to our quest to increase the speed and they could not accommodate large internal ducting. Nevertheless, surface suction and multiple slot suction is still considered to be useful for drag reduction by delaying transition to turbulence.

The availability of highly compressed air on jet-propelled airplanes made the boundary layer by blowing attractive. Experiments on airfoils having variable slot widths[6] indicated that narrower slots were more effective than wide ones and led to the replacement of C_Q by the momentum flux coefficient, C_μ, that collapsed the data and became the most significant parameter used for separation and circulation control[7]. The lift increment attained by preventing separation is also proportional to C_μ when the latter is small, since the size of the separated flow region is reduced. For an input of momentum that exceeds a critical value $(C_\mu)_{cr}$, the lift increment became smaller, $\Delta C_L \propto \sqrt{C_\mu}$ due to what was termed as a "jet-flap" effect. The critical value of C_μ, separating the two regimes, varies between $2\% < (C_\mu)_{cr} < 5\%$ depending on the specific application. When blowing is applied from the shoulder of a highly deflected flap on a symmetrical airfoil $(C_\mu)_{cr}$ is mostly dependent on flap deflection provided the incidence, $\alpha = 0^\circ$. At $C_\mu = (C_\mu)_{cr}$ the pressure distribution over the airfoil approximates the distribution predicted by ideal flow analysis.

The models used to compute the lift increment associated with CC are mostly inviscid. They replace the jet leaving the trailing edge of a wing by a curved streamline[8-11] or a vortex sheet. The angle at which the jet is inclined to the chord-line is determined by the flap deflection, or in the absence of a sharp trailing edge, by a small cusp that establishes the Kutta condition. In some instances the problem was linearized[8,9] by assuming that the incidence and the jet deflection are small. In all the above theoretical models, the mixing of the jet with the ambient flow was neglected, predicting that the entire jet momentum is recovered as thrust regardless of the jet's initial inclination angle relative to the oncoming stream[12]. This is approximately valid up to a flap deflection of 60° for which most of C_μ is recovered as thrust provided C_μ is large. In reality the jet entrains fluid from its surroundings and that entrainment is well represented by placing a suitable distribution of sinks along its path[13,14]. When a jet flows over the upper surface of an airfoil, this sink distribution adds to lift an increment $\Delta C_L \propto \sqrt{C_\mu}$ similarly to the jet-flap effect. Whenever the jet is not required to overcome separation, the model predicts the lift increment properly (fig. 1b). The streamlines shown in fig. 1a suggest that a significant fraction of C_μ is spent on overcoming the leading edge separation at $\alpha \neq 0°$ and enclosing a large bubble extending up to the blowing slot. Thus, steady blowing has a strong upstream effect. When the jet is emitted from the trailing edge of a

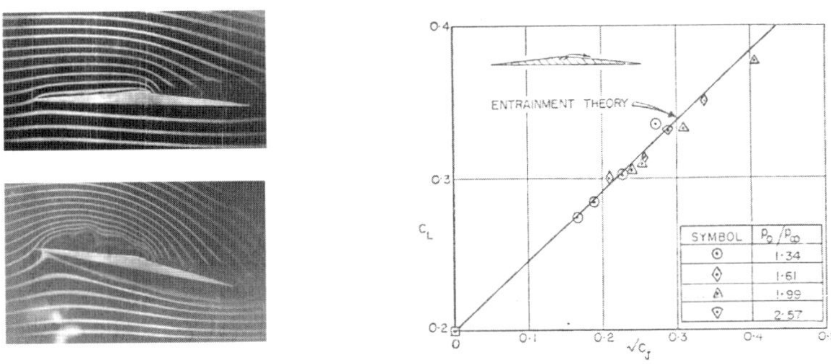

Figure 1. Blowing from the mid chord of a 5% thick triangular airfoil. (a) Flow visualization at two angles of incidence; (b) A comparison between measured and predicted results at α=0°.

bluff body, the entrainment that takes place on both sides of the jet may eliminate separation, but it also contributes to form drag[14] due to the low pressure generated on the aft portion of the body.

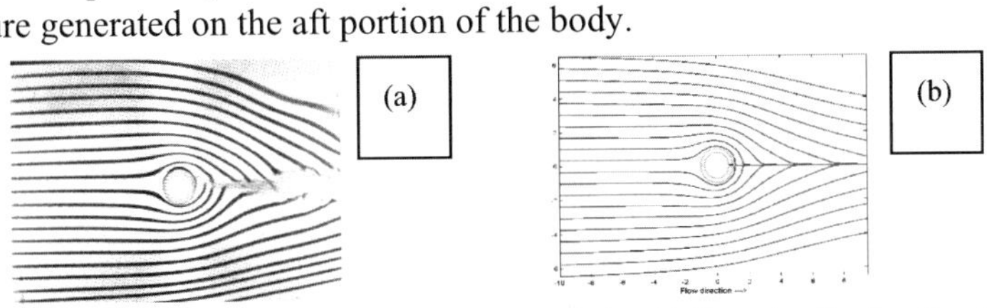

Figure 2. A jet emerging from the rear of a circular cylinder: (a) flow visualization $C_\mu = 0.37$ (b) entrainment model $C_\mu = 0.24$.

The streamlines shown in fig. 2a were observed in a smoke tunnel, and the ones on the right were calculated using the inviscid representation of jet entrainment. For the two patterns to match, a disparity in C_μ is required ($\Delta C_\mu = 0.13$) to overcome losses resulting from separation. In spite of their simplicity, these models establish the boundaries between separation and circulation control, as well as the parameters that govern the flow in both regimes. Although many potential applications of circulation control were considered, none of them reached mass production.

The introduction of periodic excitation through a slot reduced the required level of $<c_\mu>$ by an order of magnitude and even further blurred the criteria that used to separate between boundary layer control and circulation control. Although the amplitude of the input oscillations may be defined in terms of a time dependent $<c_\mu>$, the manner in which the flow responds to this input is less clear than it was for steady blowing. A comparison with steady suction is only possible if C_Q would be replaced by $-C_\mu$. Complications arise from the addition of periodicity (in both time and in space) because effective excitation may trigger a suite of instabilities dormant in the flow that increase in strength and generate the vortices that periodically scour the wing surface and transport the necessary momentum to change the state of the flow.

The present paper attempts to classify the leading parameters affecting flow separation and its control.

2. SIMPLE MODELS REPRESENTING THE EFFECTS OF BLOWING, SUCTION & PERIODIC EXCITATION

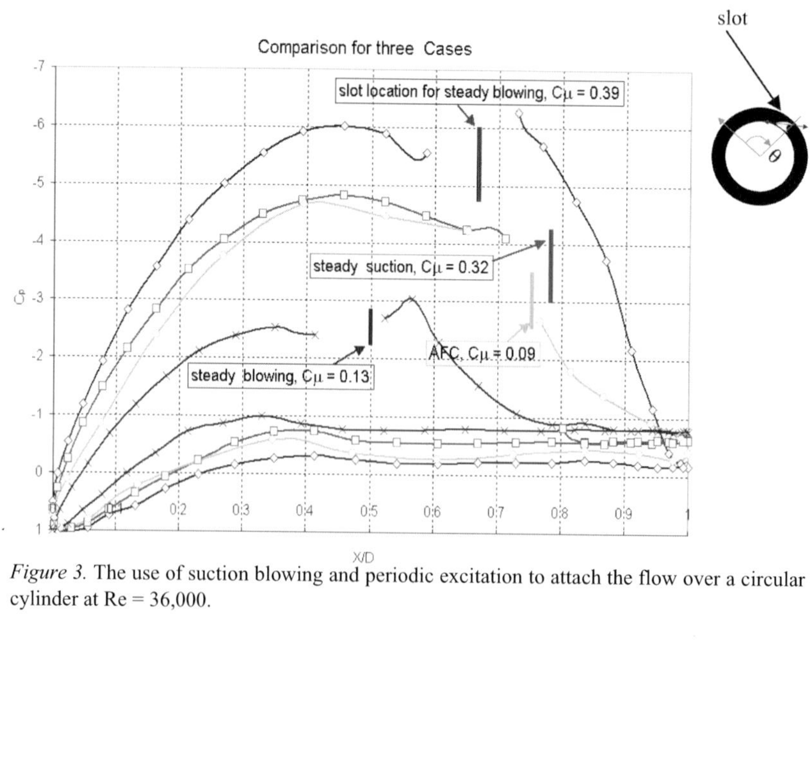

Figure 3. The use of suction blowing and periodic excitation to attach the flow over a circular cylinder at Re = 36,000.

The efficacy of blowing, suction, or periodic excitation in maintaining attached flow over the upper surface of a circular cylinder is compared in fig. 3 where the pressure distribution is plotted in "airfoil like coordinates." The same tangential slot was used in exercising all three methods of flow control. Since the optimum location of the slot depends on the method and on the momentum input, the cylinder was rotated until the best results were achieved for the control method used. For steady blowing at $C_\mu = 13\%$, $\theta = 90^\circ$ (x/D = 0.5) while for much stronger $C_\mu = 39\%$, the best slot location moved to the rear (i.e. to $\theta = 110^\circ$). The $C_\mu = 39\%$ when ejected normally to the surface at the trailing edge was also able to keep the flow attached over the entire cylinder (see fig. 2) but without generating lift. Steady, tangential blowing attaches the flow downstream of the slot and changes the Kutta condition on the upper surface. The upstream effect of jet entrainment is coupled to control of circulation, because it reduces the minimum pressure upstream of the slot depending on the C_μ input. The optimum location of steady suction at $C_\mu = -32\%$ (note that C_μ is used rather then C_Q to characterize the suction) is located farther downstream than for comparable steady blowing and it corresponds to x/D = 0.78 (or $\theta = 125^\circ$). The upstream effect of the steady suction is only slightly inferior to steady blowing; however, the flow downstream of the slot is entirely separated resulting in an inferior enhancement of lift. Periodic excitation at $<c_\mu> = 9\%$ emanating from a slot located at $\theta = 120^\circ$ generated almost identical upstream effect as the much larger suction, but it also attached the flow downstream of the actuator. Thus, steady blowing and periodic excitation attach the flow downstream of the slot while changing the pressure gradient upstream as well, while slot suction fails to provide the required control of separation downstream.

The ideal flow models representing suction and blowing in circulation control (CC) experiments are different. Suction is modeled by a single sink coinciding with the location of the slot while blowing is modeled by a distributed line-sink whose strength represents the entrainment into a turbulent jet calculated from continuity on the basis of experimental data. The sink strength per unit length of the jet's trajectory[13] is proportional to $\sqrt{(C_\mu/x)}$, clearly weakening with increasing distance from the nozzle. The measured and computed pressure distributions on the surface of the cylinder agree quite well with one another, provided the sink strength representing the suction in the model is approximately 1/3 of the experimental C_μ value used. The upstream influence of the suction extends to the leading stagnation point, while the flow downstream of the suction slot is almost stagnant at base pressure dictated by the separation from the lower surface. Strong blowing affects the pressure distribution both upstream and downstream of the slot. The sinks representing its path accelerate the flow upstream (fig. 1), while the jet helps the flow to overcome large adverse pressure gradients downstream of the slot.

In an attempt to expand this model to periodic excitation emanating from a slot the following hypotheses were made:

(i) Each cycle of the periodic excitation is divided equally into suction and blowing parts whose strength varies sinusoidally with time.

(ii) The suction portion of the cycle is represented by a point sink located at the slot.

(iii) During the blowing portion of the cycle, sinks of variable strength (amplitude) emerge from the slot at a constant rate and are advected downstream at a direction slightly diverging from the surface (due to the divergence of the shear flow in which they are embedded) with a velocity that is proportional to the free stream velocity. (This aspect is similar to the simulation of a jet start-up problem).

(iv) The sinks generated during blowing also undergo spatial amplification or decay according to inviscid jet stability criteria. Since the width of the jet increases with downstream distance, the amplification criteria are defined by a local dimensionless frequency ($f Y_2/U$ where U is the free stream velocity and Y_2 is the characteristic width of the flow) that increases in the direction of streaming even for a constant frequency of excitation.

(v) When the next suction cycle starts, the finite array of sinks generated during blowing proceeds downstream and spatially amplifies or decays according to the stability criteria established until a new blowing cycle starts again.

(vi) After a number of cycles (depending on the excitation frequency), a steady state is achieved because the initial sinks departed the computational domain and the flow simply oscillates at the excitation frequency.

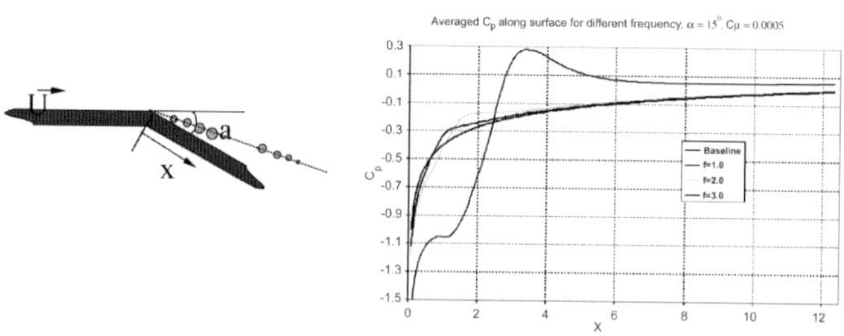

Figure 4. Pressure distribution on a generic flap deflected at 15° with model actuation emanating from the flap shoulder at $<c_\mu> = 0.05\%$.

This model was tried first on a simplified version of the "generic flap" because this flow was investigated extensively in recent years [15,16]. In this case only the distance from the flap shoulder (i.e. the computational domain) matters. Starting with a pressure distribution that is commensurate with the turning angle of the flow, the pressure recovery over the computational domain increased depending on the initial amplitude of the excitation and on its frequency (fig. 4). It appears that the best pressure recovery occurred at a dimensionless frequency of unity. The flap deflection and the level of $<c_\mu>$ selected correspond to experiments carried out in [15 & 16], obviously the pressure distribution of the baseline flow that is represented by an inviscid model is incorrect.

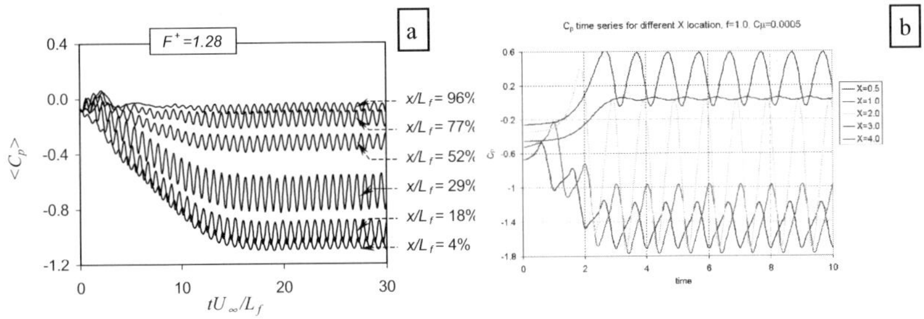

Figure 5. Phase locked pressure distribution on the flap during reattachment: (a) measured on a generic flap during reattachment, (b) modeled from the start of the simulation.

The instantaneous pressures at a given location on the surface varied in sinusoidal manner with time while the amplitude of this variation depends on the distance from the corner (fig. 5). The amplitudes necessary to change the pressure recovery are of the same order as those used experimentally. It took a number of periods from the initiation of excitation until the new, mean pressure distribution was established. This time scale is in agreement with the recent measurements of Darabi[16]. One may improve on the model by changing the trajectory of the sinks so that the pressure gradient normal to the streamline on which these sinks are located will balance the centrifugal force across it and represent the occurrence of a bubble or a dynamic stall vortex. This model contains empirical input such as spatial amplification of the sinks, and it will not stand up to proper large eddy simulations and probably not even to time dependent RANS. Its usefulness is in its simplicity because it provides quick qualitative answers to complex physical phenomena, and it can be easily modified to include amplitude modulations or actuation in burst mode.

3. PERIODIC EXCITATION IN CONTROL OF SEPARATION

We have seen that on the circular cylinder (fig. 3) and presumably on all other bluff bodies, there is no clear demarcation between separation and circulation control. In fact steady suction seems to control circulation (i.e. yield lower C_p levels upstream) without prior control of separation because the flow downstream of the slot remains separated, in spite of the strong circulation generated. In this case, as in many other observations, periodic excitation is more effective than either steady blowing or suction. It emerges as a new tool for controlling separation[17]. The old definition of circulation control as being: "an artificial increase in circulation over that which could be expected from incidence and camber in unseparated flow"[18] is inappropriate, since in most applications on airfoils the pressure distribution tends to but does not exceed the values predicted by inviscid solutions[19].

To the uninitiated, active control of separation of a two dimensional boundary layer by periodic excitation involves the introduction of harmonic (single frequency) oscillations emanating from a point source (single line) at a given amplitude. The amplitude of the periodic velocity perturbation $<u_j>$ should be commensurate with the velocity existing at the edge of the boundary layer. However, since the latter varies with location, $<u_j>$ is compared with the free stream velocity forming a dimensionless ratio of $(<u_j>/U_\infty)$. The question arises whether this velocity ratio uniquely represents the control authority over separation, as it is claimed by Nagib [20], or should it involve the addition of oscillatory momentum, $<j> = \rho t<u_j>^2$, where t represents the width of the slot, as it was shown for separation control by steady blowing [7]. There is some evidence to suggest that the slot width does not play a role in this problem (in moving from the laboratory to flight test on the XV-15, the slot width was not scaled), but what is the length scale to be considered in the definition of $<c_\mu>$? Clearly the entire chord of a flapped airfoil (i.e. an airfoil with a deflected trailing edge flap) is not involved with separation control over the flap, although it may represent circulation control over the entire airfoil. Another complication that does not occur when either steady blowing or suction are involved, stems from the use of instabilities to magnify the desired input. Mean flow can be distorted when $(<u_j>/U_\infty)^2$ becomes finite (assuming a single frequency infinitesimal input in otherwise two dimensional flow). Thus, one relies on spatial amplification to make $(<u_j>/U_\infty)^2 > [(<u_j>/U_\infty)^2]_{crit}$ and this depends on the input amplitude, $<u_j>$, its frequency, f, and the shape of the mean velocity profile at the location of the actuator. It brings forth the strong dependence of the actuator location on the success of separation control by periodic excitation. Clearly, a separated boundary layer or a boundary layer that is on the verge of separation amplifies a select range of frequencies by its inflected velocity profile, while actuation emanating far upstream of separation may decay. This is also the reason why actuation downstream of a separation bubble is much less effective than upstream and why strong free stream turbulence reduces the effectiveness of the excitation.

What is the best choice of the frequency of the excitation? By being interested in the control of separation only, the length of the separated flow provides the clue and it may correspond to the length of a deflected flap, L_f, provided that one is interested in increasing its effectiveness. In this case the reduced frequency, $F^+ = (fL_f/U_c)$ where U_c is the phase velocity of the harmonic input represents the ratio between the length of the flap and the wave-length of the input. It is often assumed that $U_c \propto U_\infty$, although the proportionality constant is close to ½. Often however, circulation is being controlled while the flow over the flap remains separated and that involves the entire length of the airfoil [19,20], complicating further the selection of the length in the definition of F^+. This brings into question the selection of the right frequency on "hump" experiments where the rear part of an airfoil is represented by a hump on the wind tunnel wall of "infinite" length. The separated flow over a finite chord airfoil entrains fluid that passed over its lower loft, but the separated flow over a hump can only entrain fluid from the bubble's interior. The curvature of the separating streamline bounding the bubble that establishes the pressure within the bubble also represents the entrainment capacity of the large eddies in the mixing layer that bounds the bubble. This brings forth another important variable in the control of separation: the initial state of the flow. The prevention of separation requires a different range of parameters than the enforcement of reattachment [15,16].

We had assumed that the flow is two dimensional and incompressible but it need not be. Concave curvature or slotted flaps in two-dimensional mean flows may create meandering eddies resulting from centrifugal instability that also affect separation and the best strategy for its control. Concentrated vortical flows (e.g. flow over a delta wing at incidence) may respond differently to periodic excitation as may be the case for three dimensional flows on yawed bluff bodies or swept back wings. The coupling between the actuator location and a shock is another parameter about which we know very little. Finally, harmonic excitation in two dimensions is relatively "simple" to analyze, but it may not be the most effective to utilize. It may be eventually replaced by more complex modes in space and in time (e.g. burst mode emanating from finite orifices) that couple better to a preexisting flow.

It took 26 years between the introduction of blowing for the control of separation by Baumann and the introduction of C_μ by Ph. Poisson-Quinton in 1947. Active control of separation by periodic excitation superimposed on suction or blowing is much more complex, and the definition of the hierarchy of the affecting parameters will require a major effort. Nevertheless, we may start taking advantage of the concept while we continue to search for its understanding. In closing a quotation from Albert Einstein is appropriate: *"One reason why mathematics enjoys special esteem, above all other sciences, is that its laws are absolutely certain and indisputable, while those of other sciences are to some extent debatable and in constant danger of being overthrown by newly discovered facts."* The parameters governing active flow control are a case in point.

ACKNOWLEDGEMENT

The author wishes to acknowledge the help of G. Han in formulating and computing the inviscid model representing periodic excitation.

REFERENCES

[1] A. Betz, Boundary Layer Control in Germany, *Boundary layer and flow control, its principles and application,* G.V. Lachmann (ED), Pergamon Press NY., 1961.

[2] S. Goldstein, Fluid Mechanics in the first half of this century, Annual Reviews of Fluid Mechanics 1, 1-28, 1969.

[3] A. Baumann, Tragflugel fur Flugzeugemit Luftraustrittsoffnungen in der Ausenhaut, Deutsches Reichs Patent 400806, 1921.

[4] Schrenk, O., Versuche an einem Absaugflugel, ZFM 22, p. 259, 1931.

[5] J. Legras, Contribution Theorique a L'effet de puit C.R. Acad. Sc. Paris, 1947.

[6] Schwier W., Auftriebsanderung Durch Einem auf der Flugeldruckseite Ausgeblasenen Luftstrahl UM31912, 1944.

[7] Ph. Poisson-Quinton & L. Lepage French Research on Control of Boundary Layer and Circulation in: *Boundary layer and flow control, its principles and application.* (Lachmann, G.V.-editor), Pergamon Press, New York, 1961.

[8] Spence, D.A. The Lift Coefficient of a Jet-Flapped Wing. *Proc. Roy. Soc. A* Vol. 238, p. 46, 1956.

[9] Legendre, R. Influence de l'Emission d'un Jet au bord de Fuite d'un Profil sur l'Ecoulement autour de ce Profil. *Comptes Rendus, Académie des Sciences,* Paris, 1956.

[10] Woods, L.C. Some Contributions to Jet-Flap Theory and to the Theory of Source Flow from Aerofoils. A.R.C. Current Paper 388, 1958.

[11] Malavard, L. Sur une Théorie Lineaire du Soufflage au bord de Fuite d'un Profil d'Aile. *Comptes Rendus, Académie des Sciences,* Paris, 1956.

[12] Davidson, I.M., The jet flap *J. Royal Aero. Soc.* Vol. 60, pp. 25, 1956.

[13] Wygnanski, I. & Newman, B. G. The effect of jet entrainment on lift and moment for a thin airfoil with blowing. *Aeronautical Quarterly*, Vol. XV, p. 122, 1964.

[14] Wygnanski, I., "The Effect of Jet Entrainment on Loss of Thrust on a Two-Dimensional Jet-Flap Aerofoil," *Aeronautical Quarterly* **17**, pp. 31-51, 1966.

[15] Nishri B. & Wygnanski I. (1998) Effects of periodic excitation on turbulent flow separation from a flap. *AIAA Journal* **38** (4).

[16] Darabi A. On the mechanism of forced flow reattachment. PhD Thesis, Tel Aviv University, 2000 Also *J. Fluid Mechanics* 510, 105, 2004.

[17] D. Greenblatt & I. Wygnanski, The control of flow separation by periodic excitation, *Progress in Aerospace Sciences* 36, 487, 2000.

[18] Williams, J., British Research on Boundary Layer Control for High Lift. *Boundary layer and flow control, its principles and application.* (Lachmann, G.V.-editor), Pergamon Press, New York, 1961.

[19] I. Wygnanski,The Variables Affecting the Control of Separation by Periodic Excitation. AIAA paper 2004-2505, 2004.

[20] H. Nagib & J. Kiedaisch, A new look at scaling parameters for evaluating effectiveness of active control of separation, AIAA 2003-0056, 2003.

BOUNDARY LAYER SEPARATION CONTROL BY MANIPULATION OF SHEAR LAYER REATTACHMENT

P R Viswanath
Experimental Aerodynamics Division, National Aerospace Laboratories
Bangalore - 560 017, India, E-mail : Vish@ead.cmmacs.ernet.in

Abstract : A brief review of recent experimental research in our group on separation control is presented. Examples in which a significant modification of a turbulent separated flow is achieved by direct manipulation of shear layer reattachment are discussed. It is suggested that such a strategy could be attractive and beneficial.

Key words : Separation control, tangential blowing

1. INTRODUCTION

The problem of boundary layer separation and control has attracted considerable attention over several decades [1-3], both because of the fundamental flow physics and technological applications. Some of the essential ideas related to boundary layer separation and the need to prevent the same from occurring have been addressed by Prof. Prandtl[1].

Methods of separation control, traditionally, have involved energization of the boundary layer upstream of the separation point (e.g. suction or tangential blowing at the wall) so that the modified boundary layer can negotiate the adverse pressure gradients without separating. Since a separated flow involves important elements like shear layer development, recirculating (or dead air) zone and shear layer reattachment (or wake closure), ideas for separation control, in general, could involve manipulation of the above flow elements. The use of suction downstream of separation in controlling a diffuser flow was demonstrated by Prof. Prandtl and group [1]; other examples include base bleed for base drag reduction [4] and use of a splitter plate for suppressing vortex shedding behind a bluff body [5,6].

In this paper, we address the issue of control or a significant modification of a turbulent separated flow essentially by a "direct manipulation of the

G.E.A. Meier and K.R. Sreenivasan (eds.), IUTAM Symposium on One Hundred Years of Boundary Layer Research, 167-176,

shear layer reattachment", which is a key element in the dynamics of separated flows [7]. We demonstrate the above idea through three different examples from recent experimental research in our laboratory. The use of tangential blowing downstream of separation, but inside the bubble (or reversed flow), for suppressing turbulent separated flows at low speeds, with two different boundary conditions at reattachment will be illustrated [8,9]. In the third example, we show the effectiveness of a passive flow control device, applied locally in the region of shear layer reattachment, for reducing the intensity of surface pressure fluctuations in a transonic, turbulent separated flow [10]; the device utilized involves a porous surface with a cavity beneath.

2. BROAD FEATURES OF SEPARATED FLOWS

Turbulent boundary layer separation generally occurs (Fig. 1) due to sustained adverse pressure gradients (or at a sharp corner as in the case of back step or base flow). The separated shear layer develops nominally under zero pressure gradient conditions initially and the velocity along the dividing streamline increases ; the shear layer entrains fluid both from the reversed flow zone as well as from the outer inviscid flow (external to the shear layer). This is followed by the reattachment of the shear layer onto a downstream wall (or recompression associated with the confluence of top and bottom shear layers); the reattachment (or recompression) pressure rise and the region over which it occurs may be expected to depend primarily on the shear layer characteristics and the boundary conditions at reattachment. The mass entrained by the shear layer from the reversed flow is returned at reattachment which forms a recirculating bubble. The pressure field in a separated flow is significantly influenced by the shear layer reattachment process and therefore it is a key element in the dynamics of separated flows. Our knowledge of the reattachment process has not improved much beyond the early work in search of reattachment criteria or models and many fundamental issues still remain unanswered.

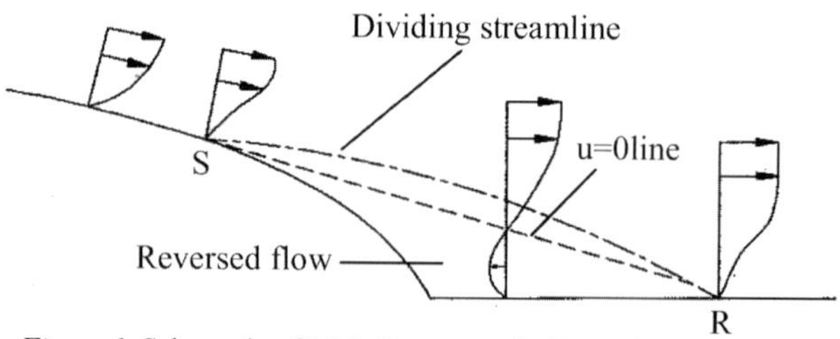

Figure 1. Schematic of Main Features of a 2D Turbulent Separated Flow

Any flow control technique applied either ahead of separation or downstream of it in the bubble will have a certain influence, directly or indirectly, on the development of the shear layer and therefore on the reattachment process and the resulting pressure field. In contrast, one could attempt to manipulate or interfere with the reattachment process somewhat directly in order to realize a certain control or modification of a separated flow; this idea has received very little attention in literature. Since the reattachment process provides the closure condition to a separated flow, its manipulation could offer an effective means of control. We briefly review recent experimental research involving direct manipulation of shear layer reattachment for turbulent separation control and demonstrate the benefits.

3. RESULTS AND DISCUSSION

3.1 Separation Control by Tangential Blowing Downstream of Separation, but Inside Bubble

The usefulness of steady tangential blowing through a narrow slot ahead of separation is well documented in literature [2,3]. In this context, it is useful to distinguish between blowing upstream of separation (U-type) and blowing downstream of separation (D-type) but within the bubble. The effectiveness of D-type blowing has been investigated on three, different configurations involving turbulent separated flows and some of the salient details of the experiments are summarized in Table 1.

Table 1: Experimental Details

Geometry	M_∞	slot ht mm	U_j/U_∞	Remarks
2D ramp (Ref.11)	2.50	1.08	0.87	Reattachment on solid surface
Axisymmetric (Ref. 8)	Low speed U_∞= 20 m/s	2.5	0.75, 1.25, 1.55	,,
2D Airfoil-like body (Ref. 9)	Low speed U_∞= 25 m/s	1.5	0.75, 1.0	Shear layer closure in wake

Viswanath et al [11] discussed briefly some of the difficulties associated with U-type blowing in supersonic flow and provided the first assessment of D-type blowing in a ramp-induced turbulent separated flow at a Mach 2.50; based on measurements of surface pressures and limited pitot pressures in the separated zone, they showed that D-type blowing was more effective than U-type.

In a subsequent paper recently, Viswanath et al [8] demonstrated through flow-field measurements in an axisymmetric separated flow at low speeds that tangential injection or blowing inside the bubble can be an effective means of separation control. Fig. 2 shows increased static pressure recovery due to blowing at all three values of jet velocity (U_j). In particular, the pressure plateau region associated with separation is eliminated suggesting suppression of wall flow reversal even for jet velocity ratio (U_j / U_∞) = 0.75. The mean velocity profiles show the benefits of blowing in the interaction zone ; the reversed flow in the separated zone (x = -18mm, 25mm) is eliminated at both values of U_j (=1.25 and 1.55U_∞). The increased mean velocity all across the layer suggests efficient mixing of the injected jet with the surrounding flow. They [8] also showed the changes and the complex nature of turbulent shear stress and kinetic energy profiles arising out of blowing.

Having seen the success of D-type blowing in two different flows and speed regimes with shear layer reattachment onto a solid surface, Viswanath and Madhavan [9] recently extended their work on D-type blowing to a trailing-edge separated on an (elongated) airfoil-like body at low speeds; in this case the shear layers merge and close in the near-wake. They studied two blowing slot locations (both downstream of separation) and the results of surface pressure distributions and mean velocity profiles for the optimum case of slot location (closer to separation point) are presented in Fig. 3; the results are shown for the rear part of the model in order to observe the details. Blowing results in the elimination of the plateau region in surface pressure suggesting wall flow reversal is suppressed; significant pressure recovery all the way upto the trailing-edge may be seen at both values of U_j. The mean velocity profiles reveal elimination of both wall and wake flow reversals (at x=875 and 890mm) at both values of U_j leading to attached flow at the trailing-edge; this results in increased circulation around the model and a significant lift enhancement.

In summary, the effectiveness of tangential blowing downstream of separation, but within the bubble as a means of separation control has been well demonstrated with shear layer reattachment onto a solid wall as well as shear layer closure in the wake. The major flow mechanisms with this type of control include: a) elimination of wall flow reversal due to the interaction of the injected jet (having higher momentum) with reversed-flow (low momentum) boundary layer which results in surface pressure recovery and b) jet entrainment of the reversed flow in the bubble, a strong factor promoting increased mixing near the wall. In essence, the reattachment or wake closure is removed or eliminated, as a direct influence of the injected jet. The blowing requirements for the D-type are small and comparable to U-type and details can be found in Ref. 9.

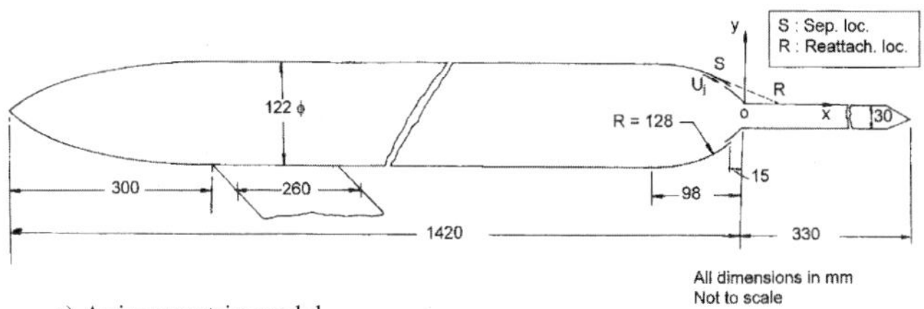

a) Axisymmetric model

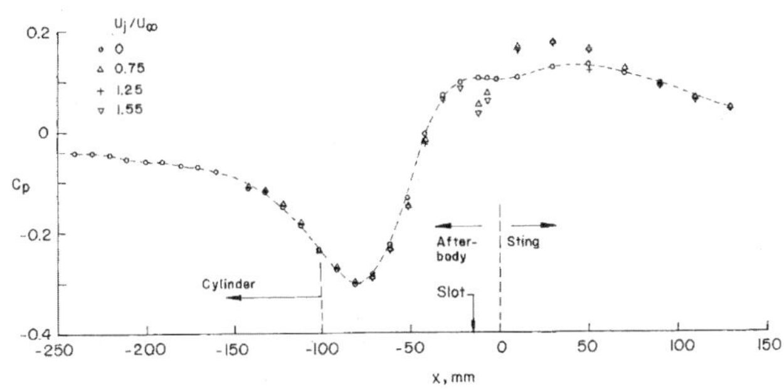

b) Effect of blowing on surface pressure distributions

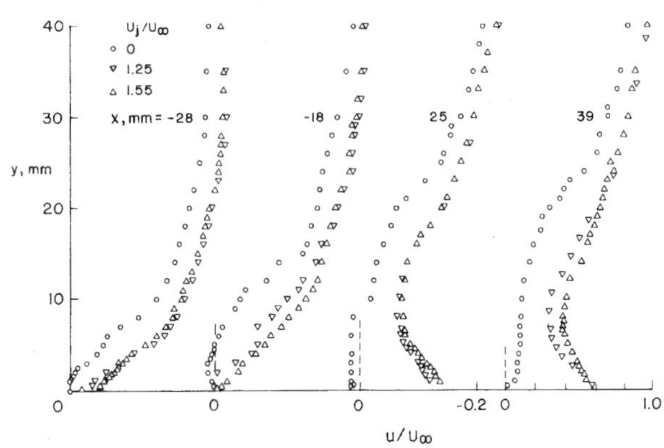

c) Effect of blowing on mean velocity profiles in separated region

Figure 2. Separation Control by Tangential Blowing Inside Bubble

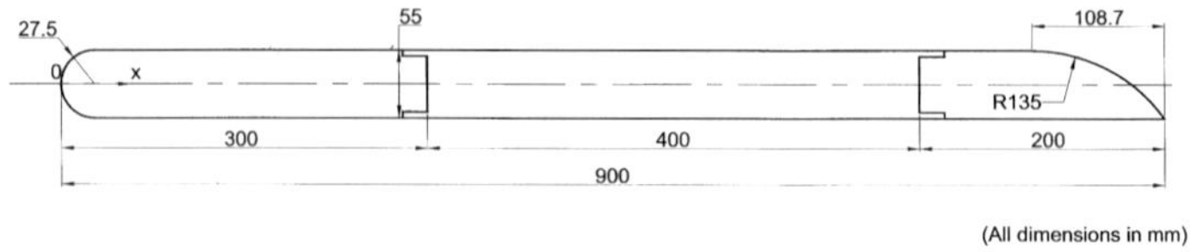

a) Geometric details of flat plate-contoured aft-section model

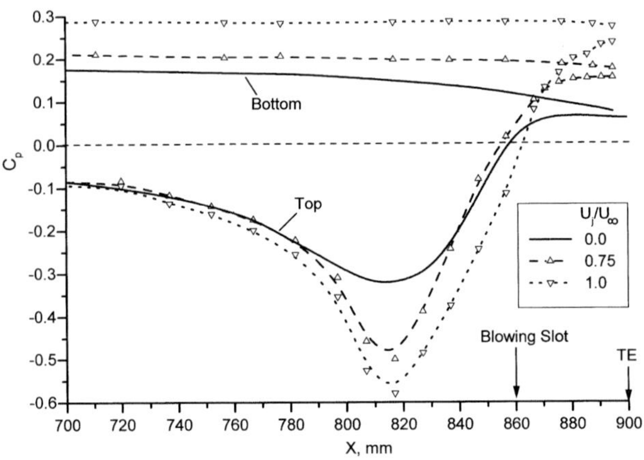

b) Effect of blowing on model static pressure distributions

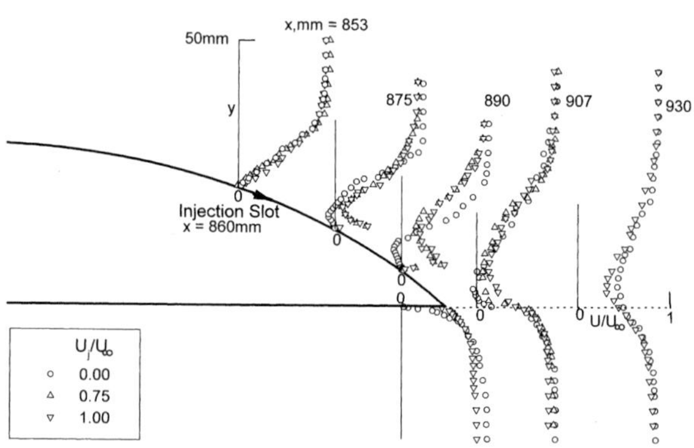

c) Effect of blowing on mean velocity profiles in the separated zone

Figure 3. Control of Trailing-edge Separated Flow by Tangential Blowing Inside Bubble

3.2 Passive Control of Surface Pressure Fluctuations in Transonic Separated flows

The concept of passive control involving a porous surface and a cavity underneath located in the region of shock-boundary layer interaction has been investigated for wave drag reduction on airfoils at transonic speeds [12]. The pressure rise across the shock wave results in flow through cavity from downstream to upstream of the shock wave: this is equivalent to a combination of suction downstream and blowing upstream of the shock, in turn increasing the communication across the shock wave. In addition to providing wave drag reduction, the above passive control methodology has been shown to reduce surface pressure fluctuations due to shock-boundary layer interaction as well.

Rajan Kumar and Viswanath [10] recently adapted the above passive control concept for reducing the surface pressure fluctuations in reattach-

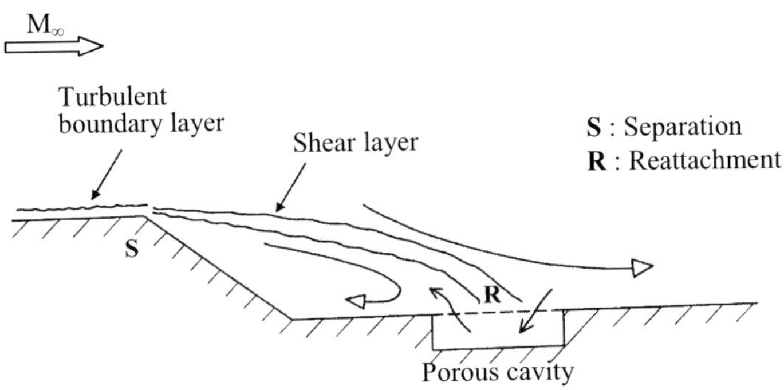

Figure 4. Schematic of Passive Control Concept for Reattaching Flows

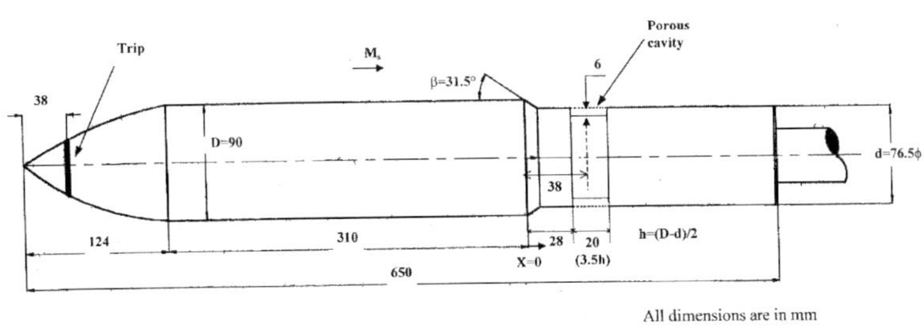

Figure 5. Geometric Details of a Generic Axisymmetric Body

ing flows (Fig. 4); the control was applied locally in the reattachment zone. These experiments were made on a generic axisymmetric body (Fig. 5) in the Mach number range of 0.8 to 1.20.The cavity employed had a porosity of 15% and a depth of 6mm and unsteady

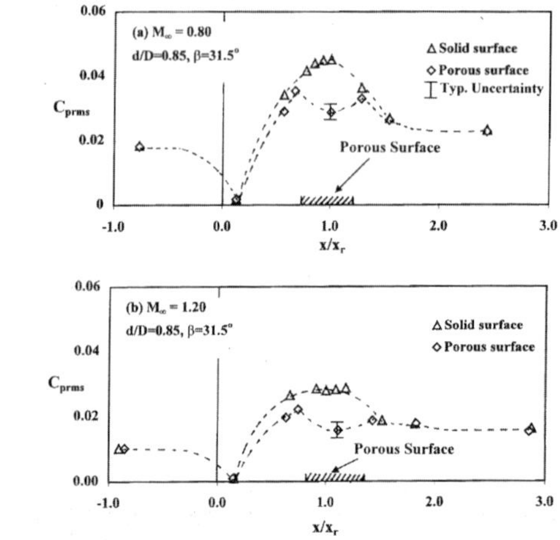

a) Effect of passive control on rms pressure fluctuations

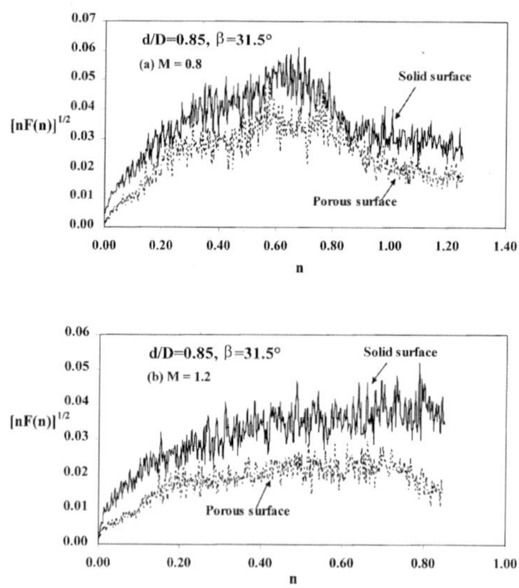

b) Spectra of pressure fluctuations in the vicinity of reattachment

Figure 6. Effect on Passive Control on Surface Pressure Fluctuations in Reattaching Flows

pressure measurements were made using a number Kulite transducers located in the separated zone; more experiment details are available in Ref. 10.

The effectiveness of the passive control in lowering the r.m.s. value of pressure fluctuations and on the spectral characteristics at a Mach number of 0.80 and 1.20 are displayed in Fig. 6. Passive control results in reducing the peak pressure fluctuations in the reattachment zone by as much as 35% with some lower reduction in the separated flow region as well. Typical spectra in the vicinity of reattachment shows that the energy is appreciably reduced over a wide range of frequencies with a larger effect at low frequencies. Certain small changes in the mean surface pressures in the separated zone were observed as well as due to the application of passive control [10].

The effectiveness of passive control in reducing the surface pressure fluctuations is a direct result of a change in boundary condition at shear layer reattachment, that is, from a solid wall to a porous surface with a cavity beneath it. The control device results in suction downstream and blowing upstream and therefore some changes in the shear layer entrainment are to be expected; further, the spectral characteristics suggest that large scale (or low frequency) unsteadiness including possibly turbulence in the shear layer is inhibited appreciably by the porous surface.

4. CONCLUSIONS

Examples from recent experimental research in our laboratory are reviewed to show the effectiveness of direct or nearly direct manipulation of shear layer reattachment for turbulent separation control. It is suggested that such a methodology could provide an alternate strategy for separation control/management.

ACKNOWLEDGEMENTS

The author is grateful to his colleagues Drs. K T Madhavan, G Ramesh and Mr. Rajan Kumar who have contributed to separation control research over the years at NAL.

REFERENCES

1. Prandtl L. *Essentials of Fluid Dynamics*, London, Blackie & Son Ltd, 1952.
2. Lachmann GV. *Boundary Layer and Flow Control*, Vol.1 and 2, Oxford, Pergamon Press, 1961.

3. Chang PK. *Control of Flow Separation*, Hemisphere Publishing Corporation, Washington, 1976.
4. Bearman PW. AGARD-CP-4, Part II, 480-507, 1966.
5. Roshko A. NACA TN 3169, 1954.
6. Bearman PW. "Investigation of flow behind two-dimensional model with a blunt trailing edge and fitted with splitter plates", *J. Fluid Mech*, V21, 241-255, 1965.
7. Roshko A. "A look at our present understanding of separated flows", AGARD Symposium on Separated Flows, Brussels, 1966.
8. Viswanath PR, Ramesh G, Madhavan, KT. "Separation control by tangential blowing inside the bubble", *Expt. Fluids*, V29, 96-102, 2000.
9. Viswanath PR, Madhavan KT. "Control of trailing-edge separation by tangential blowing inside bubble: A novel approach", AIAA Paper 2003-223, 2003.
10. Rajan Kumar, Viswanath PR. "Passive control of surface pressure fluctuations in reattaching flows", *Aeronautical Journal,* V106, 669-674, 2002.
11. Viswanath PR, Sankaran L, Sagdeo PM, Narasimha R, Prabhu A. "Injection slot location for boundary layer control in shock-induced separation", *J. Aircraft,* V20, 726-732, 1983.
12. Raghunathan S, "Passive control of shock-boundary layer interaction", *Prog. Aero. Sci*, V25, 271-296, 1988.

STABILITY, TRANSITION, AND CONTROL OF THREE-DIMENSIONAL BOUNDARY LAYERS ON SWEPT WINGS

William Saric and Helen Reed
Arizona State University, Tempe, AZ. 85287-6106 USA, phone (480) 965-2822, fax (480) 965-1384, email: saric@asu.edu

Abstract: A review of the most recent contributions to the understanding of the stability and transition of swept-wing boundary layers in very-low-disturbance environments is given. Laminar flow control is discussed with respect to different flight regimes.

Key words: Swept wing, 3-D boundary layers, stability, transition, control

1. INTRODUCTION

The topic of swept-wing boundary layers is of particular interest to this symposium since the flow over a swept wing creates a three-dimensional boundary layer whose description was first elucidated by Prandtl [1] in 1945. Prandtl's contribution was important because he formulated the 3-D boundary layer equations in this case, emphasized Pohlhausen's method as an approximate solution, and showed that the crossflow velocity profile was inflectional. With the onset of transonic and supersonic flight after World War II, Prandtl's paper was as timely and important as many of his other original contributions. This pioneering work spawned a number of papers that eventually resulted in the more well-known works by Cooke [2, 3]. Thus, the Falkner-Skan-Cooke similarity solutions were the model for approximating 3-D boundary layers for 45 years. Although the complete 3-D boundary-layer solutions been common place for over a decade, the archaic similarity solutions still occasionally appear.

With Prandtl's demonstration that the crossflow profile was inflectional, it was apparent the stability characteristics of swept-wing boundary layers were going to be different than the streamwise

G.E.A. Meier and K.R. Sreenivasan (eds.), IUTAM Symposium on One Hundred Years of Boundary Layer Research, 177-188,

instability in the form of Tollmien-Schlichting (T-S) waves. This was borne out during Pfenninger's X-21 laminarization program in the early 60's [4]. Thus, this paper addresses the special features of 3-D boundary layers that make transition control different from the 2-D case.

Whereas stability calculations, transition correlations, and control methods are well-understood for 2-D boundary layers (Reed et al [5]; Saric et al [6] the same cannot be said for swept-wing flows. The recent review by Saric et al [7] illustrates the difficulties. Essentially, 2-D boundary layers are subject to T-S instabilities that can be passively controlled by accelerating flow as far back as possible. Thus, maximum thickness is moved aft and the pressure-recovery region is minimized. However, accelerating the flow increases the crossflow and thus makes the situation more unstable for crossflow waves. The inflectional profile also has a stability response that includes both traveling and stationary waves.

The presence of stationary waves in the form of co-rotating streamwise vortices is a strong modulator of the mean flow. Although the initial growth rates may follow linear theory, the changes in the meanflow necessitates a nonlinear analysis. Whereas the traveling-wave passband is more unstable than the stationary waves, stationary waves have a particular sensitivity to surface roughness and weak freestream turbulence not found in traveling wave instabilities [7]. Thus, stationary waves are the dominant mode in low-disturbance environments.

Transition control (or Laminar Flow Control - LFC) has been of particular importance in modern transports because of the resulting reduction in turbulent skin-friction drag. The idea of LFC is to control disturbances while they are still governed by linear stability and the details of the breakdown process are not always important. This scheme is effective principally in low-disturbance environments. For engineering applications, passive control (i.e. static manipulation of the boundaries or flow field) seems more feasible than active feedback control through wave cancellation.

The paper will review recent progress in understanding control of the transition process in 3-D boundary layers in both subsonic and supersonic flow applications. A passive control scheme using periodic discrete roughness elements is discussed that offers the possibility of LFC without suction systems. The validity of this scheme at low Reynolds numbers was demonstrated experimentally and with nonlinear PSE and DNS [8-10]. Higher Reynolds number examples will be given from subsonic and supersonic wind-tunnel experiments and flight-test experiments.

The general principle that governs all flow-control implementation is that the flow-control strategy must be incorporated within the conceptual and preliminary design phases, and not just the final design of the aircraft. This optimizes the effects of control consistent with the design constraints of the aircraft. While it is generally accepted that it is not practical to laminarize the fuselage or the engines, there are possibilities to achieve some laminar flow on the wings and empennage and it is in these areas that the discussion is applicable. In a low-disturbance environment such as flight, transition to turbulence within the boundary layer occurs as a result of flow instabilities or attachment-line contamination. There are five basic instability mechanisms that can contribute to transition on a wing. These are described first before proceeding with the means for achieving LFC.

Curvature Induced Instabilities – Görtler Vortices: Regions of concave curvature can give rise to a basic instability first recognized by Rayleigh. For boundary layers on concave surfaces in open systems, Görtler vortices appear. These are stationary, streamwise oriented, counter-rotating vortices that nonlinearly modify the mean flow and lead to secondary instabilities and turbulence [11]. This mechanism is not too important because it can be easily controlled by the appropriate profile design. Usually concave curvature, if used, is aft on the pressure side of the airfoil. If critical Görtler numbers are reached an informal suggestion by Pfenninger that locally alternates concave and convex curvature was shown to work by [11] because convex curvature is more stabilizing than the concave curvature is destabilizing.

Attachment-Line Contamination and Instability: The combination of leading-edge radius and sweep gives rise to attachment-line contamination and instability [4, 12] but can be controlled in a straightforward manner. To avoid wing-root disturbances from propagating along the attachment line and feeding into and tripping the boundary layer, it is necessary to keep the attachment-line momentum-thickness Reynolds number, $Re_\theta < 100$ [13]. The flow along the attachment line can also undergo a shear-layer instability but this occurs at a higher value of Re_θ. For most applications, practical control is achieved by simply limiting the nose radius for a specific sweep angle.

Receptivity and Streamwise Instabilities – Tollmien-Schlichting Waves: Streamwise instabilities related to the Tollmien-Schlichting mechanism result in the appearance of T-S waves. This was, for years, the prototype instability that dominated transition to turbulence thinking for much of the 20th century and is the order-one problem for LFC because it is important for both swept and unswept wings under all flight conditions [5, 14-17]. The instability waves are slow growing and their behavior is governed by the Orr-Sommerfeld equation (OSE) with some modification for nonparallel effects. However, T-S waves are very sensitive to freestream conditions and should really be governed by an initial-value problem (parabolized stability equations) rather than and eigenvalue problem (OSE). The initial amplitudes of the waves come from the freestream and the transfer function is called receptivity. A recent review [6] discusses the progress in this area. Receptivity issues will always confuse the conversion of wind-tunnel data to flight. This is especially true of supersonic tunnels. The T-S mechanism is especially sensitive to freestream sound and to 2-D roughness (steps and gaps) so care must be exercised in these cases.

T-S waves typically occur in the mid-chord region and transition induced by T-S can be reasonably correlated with linear stability theory [5]. There is no dearth of important papers on the actual breakdown process itself but these are of little interest in LFC. The basic idea of LFC for T-S waves is to control linear disturbances. Once the nonlinear and 3-D effects are present, only heroic efforts can re-laminarize the boundary layer.

Since linear theory of T-S waves, with an accountability for non-parallel effects in the form of parabolized stability equations (PSE), is a reliable tool [18], one can now concentrate on control with serious attention paid to freestream disturbances. Successful control strategies for T-S waves are discussed briefly in Section 2.

Receptivity and Crossflow Instabilities: The combination of sweep and pressure gradient create a flow within the boundary layer that is perpendicular to the inviscid streamlines. This velocity profile is inflectional and undergoes an instability at low x-Reynolds numbers. The instability takes the form of co-rotating vortices aligned approximately in the inviscid flow direction. Both traveling and stationary waves can be present – the former excited by freestream turbulence and the later by micron-sized surface roughness.

Transition to turbulence in crossflow-dominated, swept-wing boundary layers has received considerable attention over the past decade or so. The reason is the obvious engineering benefit that would result from enabling laminar flow over most of the wing. The difficulty faced in confronting this problem has been the strongly nonlinear nature of the crossflow instability. Linear methods have provided almost no useful results in predicting transition and therefore tremendous effort has been given to understanding the nonlinear aspects of the phenomenon. The recent reviews of swept wing stability have been given by Arnal [17], Bippes [19, 20], Crouch [21], Haynes and Reed [9], Herbert [16], Kachanov [22], Reshotko [15], and Saric et al [7, 8, 23]. The relevant review for this paper is Saric and Reed [24].

The primary instability region is now very well understood and excellent agreement between NPSE computations [9], DNS [10], and experiments [8] has been achieved. The quality of the agreement suggests that all the features important for the primary instability, including curvature and details of the nonlinear effects, are adequately modeled and other crossflow-dominated configurations can be computed with some confidence.

The location at which the saturated vortices produced by this instability break down and lead to turbulence is not nearly as well documented. What is observed in stationary-wave-dominated transition experiments is that, at some point aft of where the vortices saturate, breakdown to turbulence occurs very rapidly along a jagged front. This suggests that the final stage of transition occurs over a very short streamwise distance and is the result of a secondary instability described experimentally [13, 25] and computationally [26-28]. Recent work [10, 29] has put the secondary instability on firm ground. The net result of these efforts is a very complete understanding of the primary crossflow instability, including details of the nonlinear saturation of the dominant stationary mode and the growth of harmonics. An important consequence is that a means of transition suppression has been developed [8] that exploits the nature of the nonlinearities.

Detailed physical receptivity mechanisms for crossflow have not been investigated experimentally. However, recent work has provided a parametric understanding of receptivity. Surface roughness is an important crossflow receptivity mechanism [24, 30]. The Deyhle and Bippes [31] experiments established that for low levels of freestream turbulence, the transition process is dominated by stationary crossflow waves, while at high disturbance levels, traveling waves dominate because of the larger amplitude unsteady initial conditions. However, the stationary modes may be the most important practical case because of the low freestream turbulence observed in flight situations.

Transient Growth: Transient growth is fundamentally different from T-S wave growth because it results from an inviscid rather than viscous mechanism and produces algebraic rather than exponential disturbance growth. This is due to the fact that the Orr-Sommerfeld equation is not self-adjoint and thus the eigenvectors, although linearly independent, are not orthogonal. Thus, the sum of two stable modes governed by the OSE-Squires system can experience algebraic growth. Disturbances with algebraic growth eventually decay exponentially, but prior to this decay they are capable of undergoing very significant growth upstream of first branch of neutral curve. Therefore, since the growth factors can be large and since the mechanism is active upstream of T-S waves, transient growth may play a significant role in the transition process under certain conditions. See Reshotko [33] for an up-to-date review on the subject that has relevance to practical systems.

It is generally accepted that transient growth plays an important role in the transition process when surface roughness and freestream turbulence are present [34]. One assumes that LFC surfaces will not have significant roughness and that the flight environment will have low turbulence levels. This assumption will have to be re-addressed of course if flight data for LFC systems show otherwise.

2. LAMINAR FLOW CONTROL

Since linear-stability behavior can be easily calculated, passive control schemes are usually designed based on LST. However, since the initial conditions (receptivity) are not generally known, only relative comparisons between control schemes are possible and must be between systems with similar environmental conditions.

Experience with LST shows that a fuller boundary layer (more negative $\partial^2 U/\partial y^2$) usually results in lower disturbance growth. Thus, wall-normal suction, accelerating pressure gradient, wall cooling in air, or wall heating in water are possibilities as control for LFC [14].

Natural Laminar Flow (NLF) control involves the tailoring of the pressure gradient region to control disturbances and delay transition far back along the chord of a 2-D airfoil. The combination of suction in the leading-edge region and NLF (in the mid-chord region) is then termed Hybrid Laminar Flow Control (HLFC). See Joslin [34] for a review.

A new technique for swept wings is proposed [7, 8, 24, 24] called Swept Wing Laminar Flow Control (SWLFC), that capitalizes on the nonlinear nature of crossflow and suggests a passive control scheme. Whereas LFC, NLF, and HLFC discourage the growth of crossflow and control T-S, SWLFC naturally stabilizes T-S and promotes the growth of selected stationary crossflow vortices to delay transition.

Suction and Thermal Control: The paper by Joslin [34] reviews the state of the art of LFC from the 1930's through the 1990's and as such is recommended as a complement to this paper. He discusses studies of the benefits of LFC for subsonic and supersonic aircraft, early studies related to manufacturing tolerances and insect contamination, and results of concept studies in wind tunnels and in flight. Reported are flight-test results from major United States and European programs. These programs show that laminar flow can be maintained under operational conditions, at high Reynolds numbers, and at both subsonic and supersonic conditions. Joslin then points out that opportunities will now be innovations in addressing the need for added systems, uncertainty in maintenance requirements, long-term structural integrity, and long-term operational and reliability characteristics.

For 3-D boundary layers, LST shows that cooling the wall has a stabilizing effect in air. However, the effect is very small because crossflow instability is a dynamic instability due to the presence of an inflection point away from the wall in the crossflow-velocity mean-flow profile. The properties of the inflection point are hardly affected by the wall conditions. For the same reason, larger values of suction are required for stabilizing 3-D boundary layers than with the 2-D case. Because a pressure gradient is the cause for instability on a 3-D wing, NLF is unfeasible to control crossflow.

Passive Swept-Wing Laminar Flow Control – SWLFC: The crossflow instability has been the primary Chimera holding back LFC. Favorable pressure gradients used to stabilize streamwise instabilities destabilize crossflow. In low-disturbance environments such as flight or special low-turbulence wind tunnels, the dominant crossflow instability is a stationary co-rotating vortex structure. Contrary to streamwise instabilities, this instability is not sensitive to 2-D roughness or freestream sound. Extensive tests with freestream sound and roughness in the ASU experiments have verified this behavior. On the other hand, also contrary to streamwise instabilities, crossflow is highly sensitive to 3-D roughness and freestream vorticity. In high-Mach-number flow, therefore, the more conventional blowdown tunnels with sound the prevailing disturbance can be used for crossflow tests on wings whose leading edges are swept beyond the Mach angle and one need not use a "quiet" supersonic tunnel. Of course, if one were to study streamwise instabilities or blunt-body effects, one must use a quiet facility. Also, sound passing through a bow shock will produce vorticity, so that in cases where the leading edge is unswept or swept ahead of the Mach angle, a quiet facility is also suggested.

Experiments with periodic discrete roughness elements (DRE) on a swept wing demonstrated a scheme for laminarization. Two important observations concerning the DRE results of Saric et al [8] are: (1) unstable waves occur only at integer multiples of the primary disturbance wavenumber; (2) no subharmonic disturbances are destabilized. Spacing apart the roughness elements with wavenumber $k = 2\pi / \lambda$, excites harmonic disturbances with spanwise wavenumbers of 2*k*, 3*k*, ... n*k* (corresponding to $\lambda / 2, \lambda / 3, ... \lambda / \mathrm{n}$) but does not produce any unstable waves with "intermediate" wavelengths or with wavelengths greater than λ. Thus, excitation of subcritical wavelengths cause these waves to grow and prevent more unstable waves from growing. The subcritical waves eventually decay leaving nothing [8, 23, 24]. In a DNS solution, Wassermann and Kloker [10] have shown the same stabilization due to subcritical forcing using the same independent approach regarding the calculation of the basic state. Saric and co-workers also showed that holes and glow discharge work equally as well as bumps, and that bump shape is not important - just the spacing.

Airfoil Design Criteria for SWLFC: According to [35], the main ideas to consider during the design of the airfoil are to encourage crossflow, eliminate streamwise and attachment-line instabilities, and allow shorter wavelengths to grow sufficiently, early enough for control of the most unstable wavelength. The initial part of the design procedure is to have an accelerated flow that is nearly subcritical to T-S waves. When considering natural or passive LFC under flight Reynolds numbers of 15 million or more, it is injudicious to work at the margins of T-S instability. The present design philosophy is to minimize streamwise instabilities and concentrate on meanflow modifications to reduce the growth of crossflow waves.

To implement distributed roughness (or holes or glow discharge) for laminar flow control, one recognizes that in the flight environment, stationary crossflow is the dominant instability. One first identifies the most unstable stationary crossflow wavelength, λ_{crit}. Linear stability theory accurately predicts this critical wavelength and the location at which it first becomes unstable (neutral point). Then one studies stationary crossflow of shorter, *subcritical* wavelengths, λ_{sub}. These are the waves we will force by roughness for control. Therefore it is necessary that these waves grow strongly earlier than the critical wave, but then decay downstream after *O*(40%) chord. The observation is that the C_p distribution can be so designed that waves of about half the wavelength of the most unstable wave will grow sufficiently and then decay, thus changing the basic state and not allowing the most unstable wave to take hold keeping their respective N-factors low.

Therefore, an airfoil conducive to laminar flow control by distributed roughness must feature uniformly accelerated flow so that T-S waves are stable. With wing sweep, this favorable pressure gradient will be very unstable to crossflow. The associated C_p distribution must allow shorter-wavelength disturbances to grow sufficiently in the leading-edge region to nonlinearly modify the basic state and inhibit the growth of the longer-wavelength, most-unstable disturbance. This delays transition.

Traveling crossflow waves are more unstable than stationary crossflow waves according to linear stability theory. However, everything is forced at the shorter wavelength, λ_{sub}, and these traveling-wave growth rates are much lower and should not lead to transition. Traveling crossflow is not an issue with the distributed roughness.

The control proposed is to be applied at the leading edge within the first 2-5% chord which is the most effective position. Downstream of this location, it has not effect [30]. However, the roughness height is restricted to not be high enough to locally trip the boundary layer to turbulence i.e. $Re_k < 150$. Here $Re_k = O(1)$.

3. LFC IN DIFFERENT FLIGHT REGIMES

Lowspeed – General Aviation and UAVs: For unswept wings, one can achieve passive LFC by designing an airfoil with a pressure minimum at say 70% chord and achieve laminar flow over the entire accelerated flow region. Crossflow and attachment-line problems do not exist and it is easy to keep T-S waves subcritical on the pressure side and sub-transitional (not marginal) on the suction side. Some care has to be taken with surface roughness – in particular 2-D steps or gaps, but in general, the analysis tools and technology are available. LFC increases loiter time for UAVs and reduces fuel costs for general aviation.

Subsonic – SensorCraft Aircraft and Business Jets: For a SensorCraft type vehicle, time "on station" is an important consideration and LFC can make a contribution if it is simple and passive. Assume a prototype flight envelope of: Mach 0.6 – 0.7; altitude 60 – 70 kft; chord Reynolds number 6 – 7 million; sweep angle 25 – 35 deg. Under these conditions, one expects both T-S and crossflow. Because of the modest sweep and Reynolds numbers, the T-S and crossflow both can be controllable by design to give significant laminar flow runs. The design is not trivial since it must be based on integrated stability calculations and flowfield calculations within the C_L and C_M constraints. The use of passive SWLFC (periodic discrete roughness elements or DRE) for crossflow control will extend laminar flow beyond the natural case and could be limited by the pressure-recovery region.

Although it is usually recommended not going to flight unless there is nothing left to do in a wind tunnel, this is an example of a tailor-made solvable problem. The main issue is that wind-tunnel unit Reynolds numbers and turbulence levels are higher than flight and crossflow is very sensitive to turbulence.

Transonic Transports: This flight envelope has Mach 0.8; altitude 30 – 40 kft; chord Reynolds number up to 25 million; sweep angles up to 35 deg. This is a more difficult problem because both crossflow and T-S are strongly amplified. The review of Joslin [34] indicates that the technology, if not ready, could be made ready. One would have thought that the projected 25% savings in fuel would have kept this work active. However, wind-tunnel experiments are extraordinarily difficult in the Mach number range and flight tests are expensive and limited.

Supersonic Transports: During FY02 to FY05, DARPA supported the Quiet Supersonic Platform (QSP) program which, besides the shaped sonic boom studies, had a strong element of LFC. This flight envelope had Mach 2.4; altitude 40 – 50 kft; chord Reynolds number up to 50 million; and typically sweep angles beyond 67 deg. (subsonic leading edge). LFC tests were conducted in a small wind tunnel with limited success [35] and in flight on the NASA-DFRC F-15B also with limited success. Tests in the NASA-LaRC 4 x 4 UPWT at $M = 2.17$ were not successful because of leading-edge separation bubbles.

Wind tunnels may be out of the question unless small models are used in huge tunnels. The large sweep angles to achieve a subsonic leading edge make the attachment-line Re_θ marginal at high unit Reynolds numbers. On the other hand, if a supersonic leading edge is used to reduce Re_θ, the bow shock will create vorticity from the sound field which will acerbate the crossflow problem.

ACKNOWLEDGEMENTS

This work was sponsored (in part) by the Air Force Office of Scientific Research, USAF under grant number F49620-97-1-0520 and the DARPA QSP program under Grant MDA972-01-2-0001.

REFERENCES

1. Prandtl L. Über Reibungsschichten bei dreidimensionalen Strömungen. Festschrift zum 60. Geburtstage von A. Betz, Göttingen S. 134-141, 1945. (Ludwig Prandtl Gesammelte Abhandlungen, 2, 679-686, Springer-Verlag 1961).

2. Cooke JC. The boundary layer of a class of infinite yawed cylinders. *Proc. Camb. Phil. Soc.* **46**, 645-648, 1950.
3. Cooke JC. Pohlhausen's method for three-dimensional boundary layers. *Aeronaut. Quart.* **3**, 51-60, 1951.
4. Pfenninger W. Laminar flow control. *AGARD Rep. No. 654* (Special course on drag reduction), VKI, Rhode-St.-Genese, Belg, 1977.
5. Reed HL, Saric WS, Arnal, D Linear Stability Theory Applied to Boundary Layers. *Ann. Rev. Fluid Mech.* **28**, 389-428, 1996.
6. Saric WS, Reed HL, Kerschen EJ. Boundary-Layer Receptivity to Freestream Disturbances. *Ann. Rev. Fluid Mech.* **34**, 291-319, 2002.
7. Saric WS, Reed HL, White EB. Stability and Transition of 3-D Boundary Layers. *Ann. Rev. Fluid Mech.* **35**, 413-440, 2003.
8. Saric WS, Carillo RB, Reibert MS. Nonlinear stability and transition in 3-D boundary layers. *Meccanica* **33**, 469-487, 1998.
9. Haynes T, Reed HL. Simulation of swept-wing vortices using nonlinear parabolized stability equations. *J. Fluid Mech.* **405**, 325-349, 2000.
10. Wassermann P, Kloker M. Mechanisms and control of crossflow-vortex induced transition in a 3-D boundary layer. *J. Fluid Mech.* **456**, 49-84, 2002.
11. Saric WS. Görtler vortices. *Ann. Rev. Fluid Mech.* **26**, 379-409, 1994.
12. Poll DIA. Some observations of the transition process on the windward face of a long yawed cylinder. *J. Fluid Mech.* **150**, 329-56, 1985.
13. Reed HL, Saric WS. Stability of three-dimensional boundary layers. *Ann. Rev. Fluid Mech.* **21**, 235-84, 1989.
14. Reshotko E. 1984. Laminar Flow Control. *In Special Course on Stability and Transition of Laminar Flows, AGARD Report 709.*
15. Reshotko E. Progress, accomplishments and issues in transition research. *AIAA Pap. No. 1997-1815.*
16. Herbert T. Transition prediction and control for airplane applications. *AIAA Pap. No. 1997-1908.*
17. Arnal D. Laminar-turbulent transition: Research and applications in France. *AIAA Pap. No. 1997-1905.*
18. Herbert, T. Parabolized stability equations. *Ann. Rev. Fluid Mech.* **29**, 245-83, 1997.
19. Bippes H. Environmental conditions and transition prediction in 3-D boundary layers. *AIAA Pap. No. 1997-1906.*

20. Bippes H. 1999. Basic experiments on transition in three-dimensional boundary layers dominated by crossflow instability. *Prog. Aero. Sci.* **35**, 363-412.
21. Crouch JD. Transition prediction and control for airplane applications. *AIAA Pap. No. 1997-1907.*
22. Kachanov YS. Experimental studies of three-dimensional instability of boundary layer. *AIAA Pap. No. 1996-1976.*
23. Saric WS, Carrillo RB, Reibert MS. Leading-edge roughness as a transition control mechanism. *AIAA Pap. No. 1998-0781.*
24. Saric WS, Reed HL. Crossflow Instabilities – Theory and Technology. *AIAA Paper No. 2003-0771.*
25. Kohama Y, Saric WS, Hoos JA. A high-frequency, secondary instability of crossflow vortices that leads to transition. In *Proc. Royal Aero. Soc. Conf. on Boundary-Layer Transition and Control,* 1991.
26. Malik MR, LF, Chang CL. Crossflow disturbances in three-dimensional boundary layers: Nonlinear development, wave interaction and secondary instability. *J. Fluid Mech.* **268**, 1-36, 1994.
27. Malik MR, Li F, Choudhari MM, Chang CL. Secondary instability of crossflow vortices and swept-wing boundary layer transition. *J. Fluid Mech.* **399**, 85-115, 1999.
28. Janke E, Balakumar P. On the secondary instability of three-dimensional boundary layers. *Theoret. Comput. Fluid Dyn.* **14**, 167-94, 2000.
29. White EB, Saric WS. Secondary instability of crossflow vortices. In press *J. Fluid Mech.* 2004.
30. Radeztsky RH Jr, Reibert MS, Saric WS. Effect of isolated micron-sized roughness on transition in swept-wing flows. *AIAA J.* **37**, 1371-7, 1999.
31. Deyhle H, Bippes H. Disturbance growth in an unstable three-dimensional boundary layer and its dependence on initial conditions. *J. Fluid Mech.* **316**, 73-113, 1996.
32. Reshotko E. Transient Growth – A factor in by-pass transition. *Phys. Fluids* **13**, 1067, 2001.
33. White EB, Reshotko E. Roughness induced transient growth in a flat-plate boundary layer. *AIAA Paper No. 2002-0138.*
34. Joslin RD. Aircraft laminar flow control. *Ann. Rev. Fluid Mech.* **30**: 1-29, 1998.
35. Saric WS, Reed HL. Supersonic laminar flow control on swept wings using distributed roughness. *AIAA Pap. No. 2002-0147.*

TRANSITION TO TURBULENCE IN 3-D BOUNDARY LAYERS ON A ROTATING DISK

– Triad resonance

Thomas C. Corke and Eric H. Matlis

University of Notre Dame, Aerospace and Mechanical Engineering, Center for Flow Physics and Control, Notre Dame, IN 46556, USA. email: tcorke@nd.edu

Abstract: This work is an experimental study of the resonant interaction between traveling cross-flow modes and low-mode-number stationary modes in transition to turbulence of the three-dimensional boundary layer flow over a rotating disk. A distributed array of ink dots are placed on the disk surface to enhance a narrow band of azimuthal and radial wave numbers of both stationary and traveling modes. The size of the dots is small ($d = 1.6$ mm, $h = 0.06$ mm, $h(\omega/\nu)^{1/2} = 0.16$) so that the disturbance they produce is linear. Two hot-wires are used to perform spatial correlations measurements giving the wavenumber vector. The time series were sorted in terms of the separate contributions of the traveling and stationary modes. Cross-bicoherence was used to identify triad phase locking. The presence of wave number matching which indicates a triad resonance was then verified.

Key words: Cross-flow mode transition, rotating disk flow, triad wavenumber resonance

1 Introduction

The boundary layer flow over a rotating disk in a quiescent fluid has frequently been used as a canonical three-dimensional flow which exemplifies the cross-flow instability. In this flow, the instability appears as outward-spiraling waves.

In our work, we use the flow over a rotating disk as a means to study mechanisms for transition to turbulence originating from the cross-flow instability. The emphasis is on documenting the development of traveling modes, and determining their role in the transition process with an eye towards nonlinear interactions with stationary modes. In an effort to better control the initial conditions, and owing to the sensitivity of this instability to surface roughness, we use an earlier technique [1, 2] and apply an array of roughness "dots" to the surface of the disk. The azimuthal number and spiral pattern of dots is chosen based on the azimuthal wave number that is most likely to occur naturally (that is, be most amplified based on linear theory). The scale (height and diameter) of the dots is sufficient to put

G.E.A. Meier and K.R. Sreenivasan (eds.), IUTAM Symposium on One Hundred Years of Boundary Layer Research, 189-198,

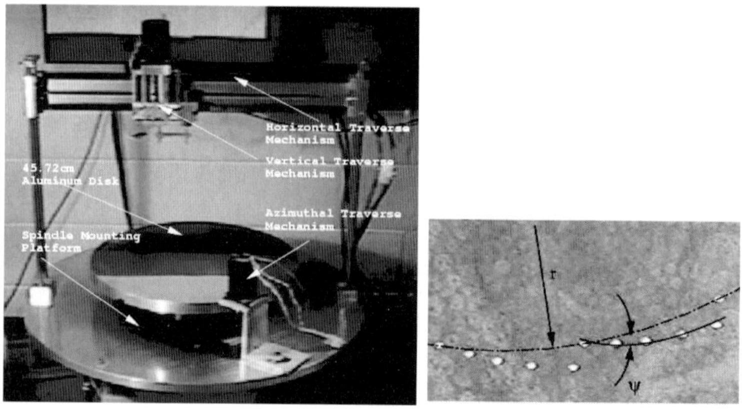

Figure 1. Photographs of rotating disk setup (left) and sample $n = 19$ spiral dot arrangement (right) used in experiment.

energy into a narrow band of modes, but not so large that it changes the basic flow. The choice of the different dot conditions allows us to isolate the respective contributions of stationary and non-stationary components, and their interaction, that leads to transition to turbulence.

Using this technique, Corke & Knasiak [1, 2] documented phase locking between pairs of traveling cross-flow waves and low mode number stationary modes as being an important mechanism in transition to turbulence. With frequency phase locking being the first requirement of a triad resonance, this paper documents the second requirement which is wavenumber matching between the interacting modes.

2 Experimental Setup and Procedure

The rotating disk facility consisted of a polished aluminum disk mounted on an air-bearing with an integrated dc-motor. The dc-motor was controlled by a dedicated digital controller which took feedback from an optical encoder. The encoder produces ttl-pulses every 0.5° of rotation. The whole system leads to minimum vibration of the disk and a constant rotation speed to within 0.003%. The measurement surface consisted of a 3.175 cm thick, 45.72 cm diameter aluminum disk. The disk was ground and diamond lapped to be flat and parallel to 0.0038 mm, and polished to a 2 micron finish. A photograph of the rotating disk setup is shown in Figure 1.

A motorized traversing mechanism was mounted above the surface of the disk. It allowed two directions of motion: radial and wall-normal, with an accuracy of 0.0025 and 0.00025 mm respectively. Motion was controlled through software with a digital data acquisition and control computer.

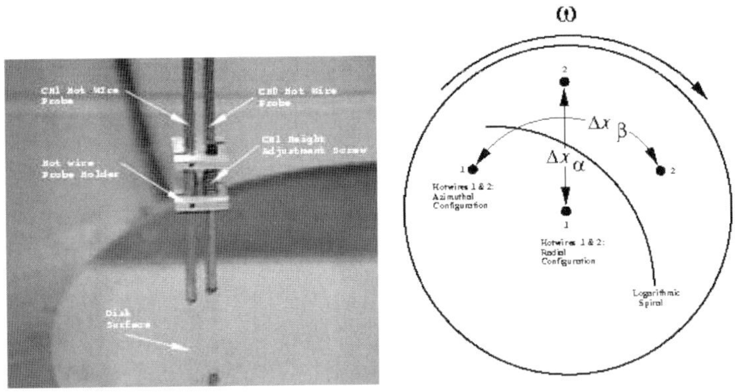

Figure 2. Photograph of dual hot-wire probe arrangement (left) and schematic of sensor positions used in measuring wave number vector.

The surface roughness consisted of small ink dots which were applied in a pattern on the surface of the disk. To do this, an inking pen was mounted on the traversing mechanism using a custom designed holder. In order to place dots at different azimuthal positions, a friction wheel driven by a stepper motor was mounted to contact the outer edge of the disk. The motor can be seen at the bottom right part of the photograph of the disk setup in Figure 1. During the experiment, the friction wheel was removed.

The rotational placement of the dots used the optical pickup on the disk position feedback for gross positioning. Fine positioning used the stepper motor calibration. The maximum resolution in dot placement was determined to be 0.053°. The location of the dots was always referenced to the same rotational position on the disk. This position also corresponded to the first point in a velocity data acquisition time series.

Two hot-wire sensors were used to acquire velocity time series simultaneously at two points in space. A photograph of these over the disk surface is shown in Figure 2. They are referred to as sensors 1 and 2. A schematic showing the arrangement of the two sensors for determining the separate wave numbers, α and β is shown in the right part of Figure 2. The sensors were always oriented to be most sensitive to the azimuthal component of velocity.

2.1 Disk Surface Conditions

Following Corke & Knasiak [1, 2], controlled disturbances were introduced by placing a distributed array of ink dots on the disk surface. The objective was to enhance a narrow band of azimuthal and radial wave numbers of cross-flow modes without disturbing the mean flow. Two cases were examined. The first was 19 spiral patterns of dots at a fixed radius around the disk, where $n = 19$ was most amplified. The spiral pattern was designed to match the most amplified spiral wave angle of $\psi = 11.2^\circ$. A photograph of the spiral dot pattern is shown in the right part of Figure 1.

Table 1. Disk Surface Periodic Roughness Conditions

Dot Pattern	r/r_c
$n = 19$ spiral	$0.955 \leq r/r_{c_I} \leq 1.005$
$n = 27$ dot	$r/r_{c_I} = 1.212$

Table 2. Experimental Conditions and Critical Radii

Parameter	Value
ω	104.7 s^{-1}
ν	1.56 x $10^{-5} m^2/s$
R_{c_I}	285 (theory)
r_{c_I}	11.11 cm
$R_{c_{II}}$	69 [3]
$r_{c_{II}}$	2.66 cm
R_{c_A}	510 [4]
r_{c_A}	19.86 cm

The other case involved placing 27 evenly spaced dots in the azimuthal direction at a fixed radius where $n = 25$ was most amplified. This was intended to produce a condition that favored a less amplified stationary mode and subsequently a different azimuthal mode number, β, and spiral wave angle, ψ. The two dot cases are summarized in Table 1.

2.2 Experimental Procedure

The disk was operated at a fixed rotation speed, $\omega = 104.7$ s^{-1}. The subsequent critical radii were then defined. These are listed in Table 2.

The velocity measurements consisted of digitally sampling contiguous voltage time series points proportional to the azimuthal velocity (U_θ), for the two sensors simultaneously, at discrete points in the wall-normal (z) direction for different radial (r) locations. The sensor voltages were converted to velocity through a calibration relation that was determined before making measurements.

The sampling rate was sufficient to resolve frequencies corresponding to stationary cross-flow modes with azimuthal mode numbers (n) up to 75. The contiguous records were 1024 points in length which corresponded to 6.82 revolutions of the disk. A total of 32 records were sampled at each z-position. The absolute height of the sensor at the closest position above the wall was set using a cathetometer with a resolution of 0.002 mm.

One of our principle objectives was to separate the velocity fluctuations in the time series into contributions from stationary and traveling instability modes. To do this, we first constructed the rotation-averaged time series. This corresponded to averaging together the velocity-series of successive rotations of the disk. Each disk rotation was marked by an analog pulse that was simultaneously digitized with the hot-wire voltages. The acquisition frequency of 2500 Hz, and disk rotation rate of 16.66 Hz gave 150

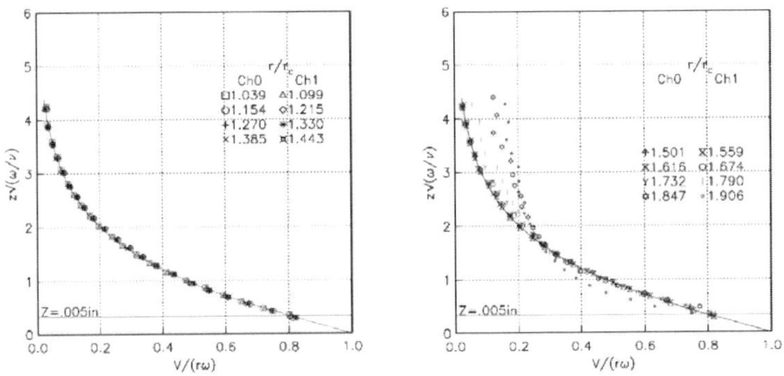

Figure 3. Mean velocity profiles at "linear" (left) and "nonlinear" (right) radii for 19-spiral case.

points/rotation. With 1024 points in 32 records this gave 192 averages. To obtain the part of the time series fluctuations due to traveling modes, the rotation-averaged series was subtracted from each 150-point segment of the total time series. This process was performed for each spatial point sampled.

3 Results

Because of space limitations, the results will only focus on the 19 spiral dot case. Complete results are available from Matlis [5]. Mean velocity profiles which document the basic flow are shown in Figure 3. These have been normalized by the similarity forms and compared to the analytic solution for flow over an infinite rotating disk. Shown for reference as the horizontal line at the bottom of the plot is the height of the dots in similarity units. The left plot corresponds to the "linear range" of radial locations. The right plot to the "nonlinear range", where the mean profile deviates from the laminar distribution. The "linear range" results show excellent agreement with the theoretical profile. The mean velocity profiles only begin to deviate when the amplitude of the cross-flow modes reach sufficient level to distort the mean flow. This is typical of both cases.

3.1 Dispersion Relation - Stationary Modes

The velocity time series corresponding to the stationary cross-flow modes was extracted from the total hot-wire time series by forming ensemble averages that were triggered on a single disk location. From these, a spatial correlation analysis was conducted using the cross-correlation between the two hot-wire sensors to determine the two-dimensional wavenumber vector at each frequency. This wave vector κ is a function of the radial and azimuthal wave numbers α and β, respectively. Dispersion curves for β are presented in Figure 4.

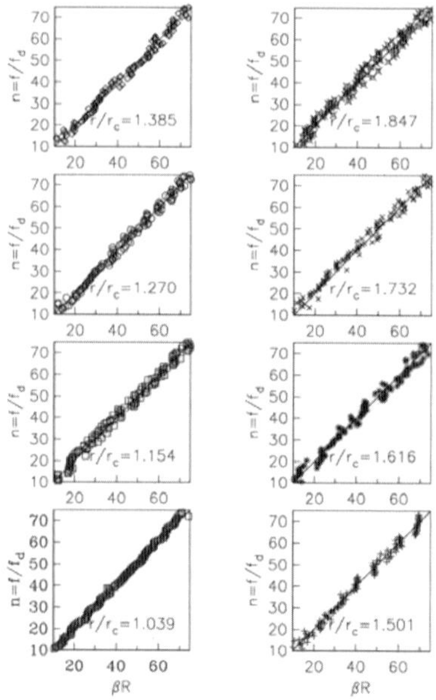

Figure 4. Stationary mode dispersion curves for β. 19-spiral case.

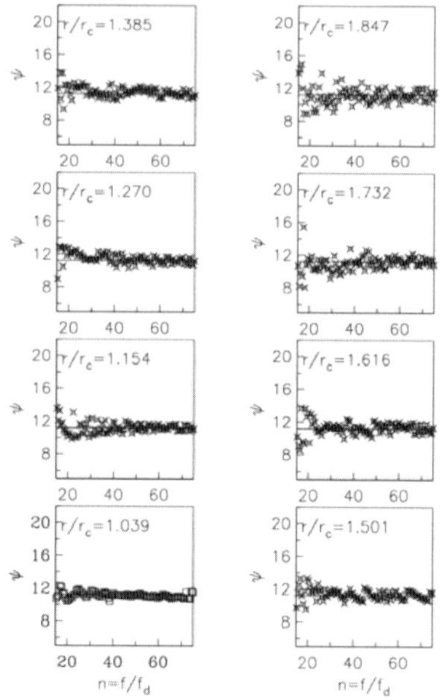

Figure 5. Stationary mode spiral wave angles, ψ. 19-spiral case.

The stationary dispersion curves were used effectively to confirm the accuracy of the wave number measurements. Because the rotation ensemble-averaged flow is stationary in an a priori manner, its phase speed is exactly that of the disk. This enabled the stationary flow to be used as a control by plotting the dispersion curves for β. The results verify the expected dispersion relation, $n = \beta R$.

Dispersion curves were also determined for the radial mode number α. These were used along with those for β, to determine the spiral angle ψ of the instability wave vector, $\psi = \tan^{-1}(\beta/\alpha)$. These are presented in Figure 5. This verifies that the stationary modes in this case follow the theoretically preferred angle of 11.2°, shown by the solid line.

3.2 Dispersion Relation - Traveling Modes

The traveling-mode time series were obtained by subtracting from the total velocity time series, the time series that was stationary with the disk rotation. The dispersion curves for β for the traveling modes are shown in Figure 6. In the linear region, the values are very close to falling on the 45° line, indicating wave speeds that are close to the disk speed. However, in the nonlinear region, starting at $r/r_c = 1.501$, we observe a change in the slope of the

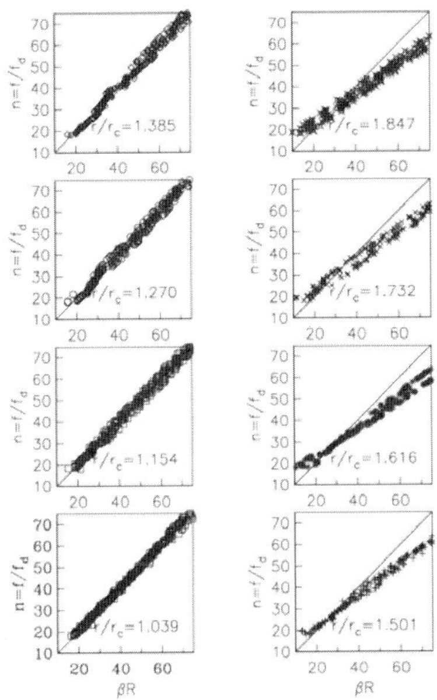

Figure 6. Traveling mode dispersion curves for β. 19-spiral case.

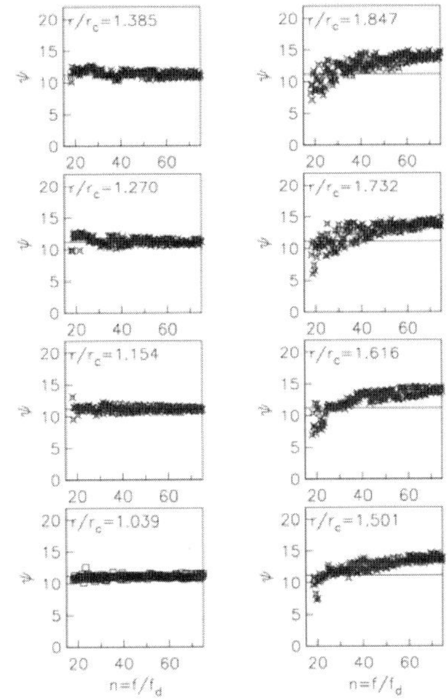

Figure 7. Traveling mode spiral wave angles. 19-spiral case

dispersion curve whereby wavenumbers, $\beta R \leq 25$ are traveling significantly faster than the disk speed, and $\beta R > 25$ are traveling significantly slower. The radial wavenumber dispersion for the traveling modes were found to vary linearly with Reynolds number so that the change in β resulted in a change in the wave angles of the traveling modes in the nonlinear region. This is shown in Figure 7. In the linear region, the spiral wave angle of the traveling modes is the same as the stationary modes. However, in the nonlinear region, lower frequencies, $f/f_d \leq 25$, have lower wave angles than the stationary modes, and higher frequencies, $f/f_d > 25$, have wave angles larger than the stationary modes. Higher wave angles signify that these waves are spiraling outward faster than the stationary modes.

3.3 Triad Wavenumber Resonance

Corke and Knasiak [1, 2] had documented the growth of a low wavenumber ($n = 3$) stationary mode during the later stages of transition to turbulence on the disk, and speculated that it was due to a triad resonance with a pair of traveling modes. For this, they demonstrated a triple phase locking, which is the first condition of proof of a triad resonance. They were unable to check the second condition, which is a wavenumber matching between the three modes, namely $k_1 \pm k_2 = k_3$, where $k_i = (\alpha_i^2 + \beta_i^2)^{1/2}$.

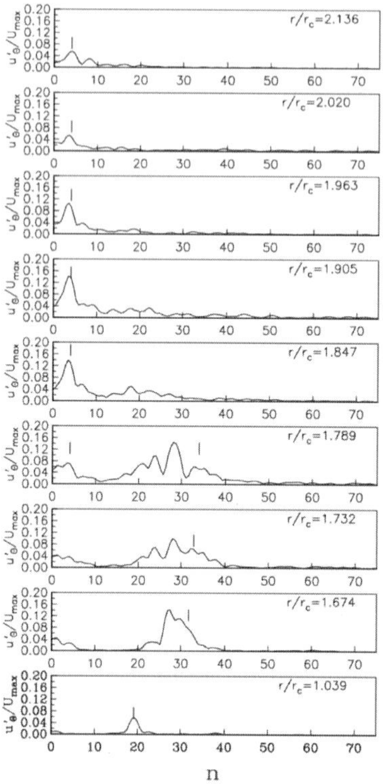

Figure 8. Spectra of velocity fluctuations of stationary modes at different radii. 19-spiral case.

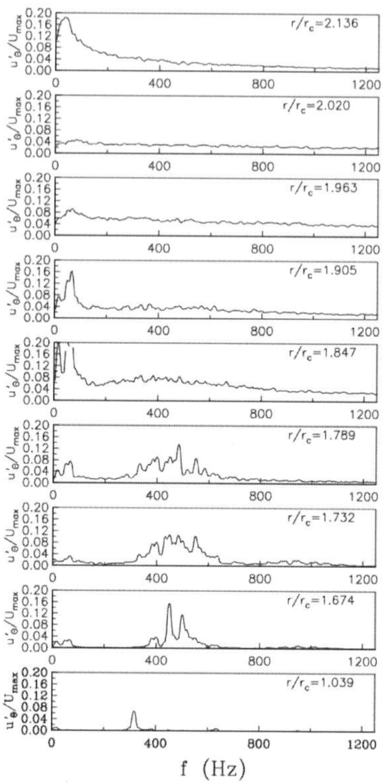

Figure 9. Spectra of velocity fluctuations of traveling modes at different radii. 19-spiral case.

In the present results, for the same dot condition, spectra of velocity fluctuations of the stationary modes shown in Figure 8, reveal a similar growth of a low wavenumber stationary mode, although it now appears to be closer to $n = 4$. The spectra of velocity fluctuations of traveling modes at the same radii are shown in Figure 9. Cross-bicoherence (CBC) analysis has been performed on the time series used in generating these spectra. These revealed a triple phase locking between two traveling modes whose difference frequency equaled that of the $n = 4$ mode. For example two of the involved traveling modes were at $f_1 = 510$Hz and $f_2 = 443$Hz so that $f_1 - f_2 = 67$Hz which when divided by the disk frequency (16.66 Hz) gives $n = 4$.

Table 3 was constructed to show that the same wavenumber matching existed between the three modes. This involves all of the triad interactions at $r/r_c = 1.616$ that were revealed by the CBC. The absolute uncertainty in the wavenumber vector is largest at the lower frequencies, and therefore affects the low wavenumber stationary mode the most. For the traveling

Table 3. Triad frequency and wavenumber matching for difference interaction leading to stationary $n = 4$ mode, $r/r_c = 1.616$, $f_d = 16.66$Hz. 19-spiral case.

Traveling			Stationary	Traveling			Stationary
f_2	f_1	$(f_2 - f_1)/f_d$	n	k_2	k_1	$k_2 - k_1$	k_3
517.3	449.0	4.10	4.10	22.73	19.61	3.12	2.65±0.35
522.3	454.0	4.10	4.10	22.95	19.84	3.12	2.65±0.35
527.1	458.8	4.10	4.10	23.17	20.06	3.12	2.65±0.35
532.0	463.6	4.10	4.10	23.40	20.28	3.12	2.65±0.35
537.0	468.5	4.10	4.10	23.62	20.50	3.12	2.65±0.35
541.8	473.5	4.10	4.10	23.84	20.73	3.12	2.65±0.35
546.6	478.3	4.10	4.10	24.05	20.95	3.10	2.65±0.35
551.5	483.1	4.10	4.10	24.27	21.17	3.10	2.65±0.35
556.4	488.1	4.10	4.10	24.50	21.39	3.10	2.65±0.35
561.3	493.0	4.10	4.10	24.72	21.62	3.10	2.65±0.35
566.1	497.8	4.10	4.10	24.94	21.84	3.10	2.65±0.35
571.1	502.8	4.10	4.10	25.17	22.06	3.10	2.65±0.35
575.9	507.6	4.10	4.10	25.39	22.28	3.10	2.65±0.35
580.8	512.5	4.10	4.10	25.61	22.51	3.10	2.65±0.35

Table 4. Triad frequency and wavenumber matching for summing interaction leading to stationary $n = 58$ mode, $r/r_c = 1.616$, $f_d = 16.66$Hz. 19-spiral case.

Traveling			Stationary	Traveling			Stationary
f_2	f_1	$(f_2 + f_1)/f_d$	n	k_2	k_1	$k_2 + k_1$	k_3
517.3	449.0	58.0	57.5	22.73	19.61	42.34	42.16±0.35
522.3	454.0	58.6	57.5	22.95	19.84	42.79	42.16±0.35
527.1	458.8	59.2	57.5	23.17	20.06	43.23	42.16±0.35

modes, being at higher frequencies, the wavenumber vectors for the difference interaction are accurate to the second decimal place shown. The two right-most columns show that within the experimental uncertainty, the triad condition is satisfied. Such a triad wavenumber matching condition only existed for frequencies that indicated a significant CBC, and these only existed in the nonlinear instability development region of the flow.

Some of the frequencies listed in Table 3 were also found to have a high CBC in a summing interaction with a stationary $n = 58$ mode. The documentation of the wavenumber matching in this case is shown in Table 4. This illustrates how the triad resonance affects a broad range of wavenumbers that will lead to rapid spectral broadening associated with turbulence transition.

4 Summary

The results revealed conclusive evidence of a triad resonance between pairs of traveling cross-flow modes and low mode number stationary modes. This was found to be the dominant mechanism for spectral broadening in transition to turbulence in this flow, and may account for large wavelength "jagged" transition fronts observed in flow visualization on swept wings were transition to turbulence develops from a cross-flow instability.

References

[1] Corke, T. & Knasiak, K. 1994. Cross-flow instability with periodic distributed roughness. *Transition, Turbulence & Combustion, Vol I*, Kluwer Acad. Pub., 43.

[2] Corke, T. & Knasiak, K. 1998. Stationary-traveling cross-flow mode interactions on a rotating disk. *J. Fluid Mech.*, **355**, pp. 285-315.

[3] Faller, A.J., and Kaylor, R.E., "Investigation of Stability and Transition in Rotating Boundary Layers," Dynamics of Fluids and Plasmas, pp. 309-329, Academic Press, 1966.

[4] Lingwood, 1995. An experimental study of absolute instability of the rotating-disk boundary layer flow. *J. Fluid Mech.*, **314**, pp. 373-405.

[5] Matlis, E. H. 1997. Wavenumber analysis and resonance of stationary and traveling cross-flow modes on a rotating disk. M.S. Thesis, Ill. Inst. Tech.

CONTROL AND IDENTIFICATION OF TURBULENT BOUNDARY LAYER SEPARATION

Inspired by Prandtl's centennial contributions

Avi Seifert* and LaTunia Pack Melton**

**School of Mechanical Engineering, Faculty of Engineering*
Tel-Aviv University, Tel-Aviv 69978, ISRAEL (seifert@eng.tau.ac.il)

*** Flow Modeling and Control Branch, NASA Langley Research Center*
Hampton, VA 23861, USA (l.p.melton@larc.nasa.gov)

Abstract: Effective delay of turbulent boundary layer separation could be achieved via closed-loop control. Constructing such a system requires that sensor data be processed, real-time, and fed into the controller to determine the output. Current methods for detection of turbulent boundary layer separation are lacking the capability of localized, fast and reliable identification of the boundary layer state. A method is proposed for short-time FFT processing of time series, measured by hot-film sensors, with the purpose of identifying the alternation of the balance between small and large scales as the boundary layer separates, favoring the large scales. The method has been validated by comparison to other criteria of separation detection and over a range of baseline and controlled flow conditions on a "simplified" high-lift system, incorporating active flow control.

Key words: Boundary layer, turbulent, separation, sensors, hot-films, closed-loop, active flow control, actuators, high-lift.

1. INTRODUCTION

The identification of turbulent boundary layer separation and its control are of great practical importance [**1-3**]. Most applications require closed-loop control to perform efficiently, and should use knowledge of the boundary layer "state" (i.e. attached or separated and intermediate

G.E.A. Meier and K.R. Sreenivasan (eds.), IUTAM Symposium on One Hundred Years of Boundary Layer Research, 199-220,

states) over the separation prone region. Moreover, the knowledge as to how close the boundary layer is to separating should be updated at the highest possible rate to provide feed-back to the controller. Several flow features can be used as indicators of the "state" including, vanishing skin friction [**1**], local minimum of the wall pressure fluctuations [**4-5**] and vanishing pressure gradient [**6**]. Measurement techniques often used include; cross-spectral information from neighboring hot-film sensors [**7-12**], information from wall mounted miniature LDV system, and floating MEMS devices [**3**].

An appealing technique is the surface mounted hot-film, due to its simplicity, availability and low cost. Transition detection using the standard deviation alone or with either the skewness or the flatness of un-calibrated hot-film data is well established [**7-13**], as is the identification of stagnation points and laminar separation bubble using cross sensor information for identifying the "phase reversal" feature [**8-13**].

In a turbulent boundary layer, the phase reversal technique cannot be used to determine where separation occurs. Several researchers [**11,13**] have presented data using the correlation coefficient of adjacent hot-film sensors for detecting separation. None of the above-mentioned methods were successful at determining the turbulent boundary layer separation location using the present data. Furthermore, existing methods rely on cross sensor information transfer, slowing down the loop. Therefore, the search for a new turbulent boundary layer separation detection criterion is warranted.

The subsequent sections of the paper will describe the experimental setup, the proposed criterion, the algorithm developed, and the application of the method to several data sets acquired on different configurations and at different flow conditions.

2. EXPERIMENTAL SET-UP

2.1 The wind tunnel

Tests were conducted in the Basic Aerodynamic Research Tunnel (BART), at NASA Langley Research Center. This is a low-speed open-circuit wind tunnel, with a 0.71 m high by 1.02 m wide by 3.05 m long test section. The maximum speed of the tunnel is 60 m/s ($Re/m = 0.345 \times 10^6$). The wind tunnel is well instrumented and allows optical access for Particle Image Velocimetry (PIV) measurements.

2.2 Simplified High-Lift Model

The simplified high-lift version of the NASA EET (Energy Efficient Transport) airfoil [**14**] was designed in a modular manner so that zero-net-mass-flux actuators could replace solid regions in the model near the leading and trailing edge flap shoulders (Fig. 1a). The 406.4 mm chord model has a 15% chord leading edge flap (LEF) that can be deflected from 0° to -30° and a 25% chord trailing edge flap (TEF) that can be deflected from 0° to 60°. Angle of attack settings for the airfoil, the TEF and the LEF were all automated and computer controlled. The model has 78 streamwise static pressure taps located at mid-span. In addition to the static pressure taps, there are nine unsteady pressure transducers on the model surface (Fig. 1b) and at least one unsteady pressure transducer in each actuator cavity for monitoring the pressure fluctuations produced by the actuator and correlating the wind tunnel experiment with the bench-top actuator calibration tests. The data presented herein has not been corrected for wind tunnel wall interference.

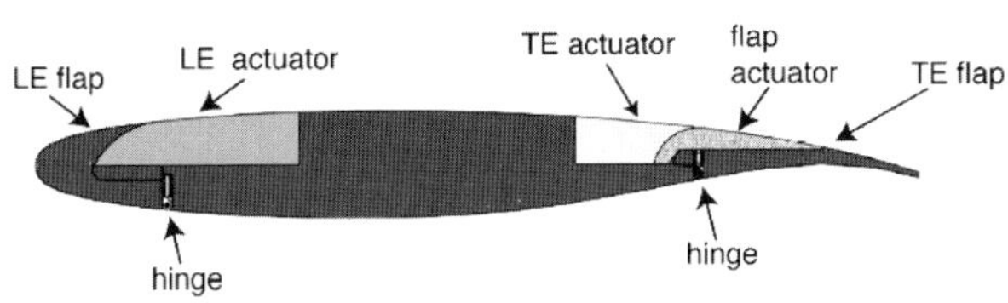

Fig. 1a EET airfoil with actuators regions represented by lighter gray-level.

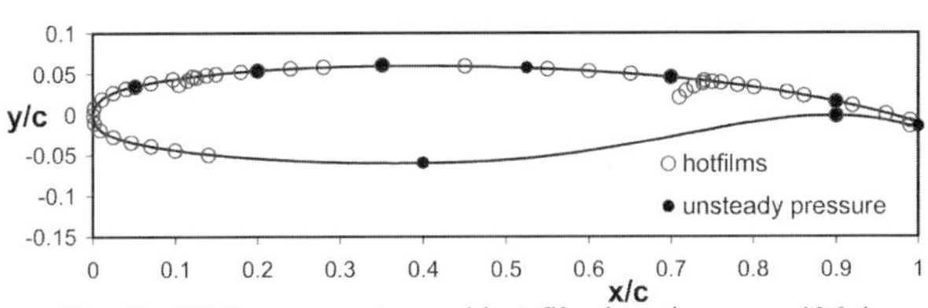

Fig. 1b EET pressure tap and hot-film locations, c = 406.4mm.

2.3 Hot-film Arrays

To aid in determining the locations of transition and separation, 48 hot-films were installed on the model. On each element (LEF, TEF, and main) of the model there are 16 hot-film sensors approximately 50.8 mm to the right of the model centerline (and therefore from the pressure sensors, Fig. 1b). A 16-channel constant temperature anemometer, operated at an overheat ratio of 1.2, coupled with a switch matrix was used to operate the three groups of 16 hot-films. The hot-film data was low-pass filtered at 10 kHz and then digitized by a 16-bit high-speed analog-to-digital converter at 25.6 kHz. The hot-film sensors, 0.4 mm long each, were etched onto a polyimide sheet and then bonded to the model. The polyimide sheet covered one half of the span of the model. A 0.1 mm step exists at the juncture between the polyimide sheet and the model. Body filler was used to fair the step. The frequency response of the hot-films was evaluated to be flat up to 5-7 kHz.

2.4 Actuators

An internal piezoelectric fluidic actuator was used in each of the actuator regions shown in Figure 1a. The actuator slots were inclined downstream at an angle of about 30 deg to the surface. The flap actuator has four alternative excitation slots. Only one slot was active during each experiment. The slot located at $x/c = 0.757$ with the TE flap undeflected is discussed in this paper. This slot is 0.635 mm wide and has 19 segments that are 0.051 m long. The remaining slots were sealed using either a water-soluble filler or 0.051 mm thick adhesive Kapton tape.

2.5 PIV Set-Up

Digital PIV was used to measure the instantaneous flow fields, phase synchronized with the actuators' cycle. The PIV system includes two 1.3 k by 1 k cameras, equipped with 105 mm Macro lens, installed side-by-side. The field of view from each camera was overlapped to cover a larger region above the airfoil. The magnification of the imaging system was about 9:1 with the measurement plane about 70 mm wide. The interrogation area at each grid point had a 24x24 pixel resolution. This corresponds to about 1.5 mm square at the measurement plane. A maximum overlap of 50% between adjacent interrogation regions was used. Smoke (average size of 1 micron), introduced upstream of the contraction, by a commercial smoke generator, was used for seeding. Dual Nd-Yag lasers were used to illuminate a light sheet, placed about 50 mm off the model centerline. The laser pulse separation was set at 55 microsecond to cover a free stream velocity of about 10 m/s. Time averaged data acquired using approximately 100 image pairs for each flow condition are presented.

2.6 Experimental Uncertainty

The α's presented are accurate to within $\pm 0.03°$. The LEF and TEF deflection angles are accurate to within $\pm 0.25°$, $C\mu$ (excitation momentum coefficient, [22]) is accurate to within 20%, the chord Re is accurate to within 3%. PIV velocity uncertainty for the current test condition is about 3%.

3. ALGORITHM

If one examines the wall normal distribution of the turbulent eddies in an attached turbulent boundary layer [**15**], one finds that the smaller eddies are active close to the wall while the large eddies are active across the entire boundary layer. As the turbulent boundary layer approaches separation, due to an adverse pressure gradient, the level of the turbulent activity increases and the relative magnitude between the small and large scales, as interpreted by the spectra of the near wall velocity [**5**], the wall shear-stress or the wall pressure fluctuations [**4-5, 16-18**], is altered such that the lower frequencies-larger eddies contain an increasingly larger fraction of the turbulent kinetic energy. This process, however, is not steady [**16**], but intermittent. *Our physical interpretation of these findings is that at incipient detachment, the small but still energy containing eddies will stop interacting with the wall in an intermittent manner.* The largest scales would still interact with the wall even when the flow is mostly separated, affecting the attached flow periods, and therefore, also reinstating the high frequency content of the measured wall mounted sensor signal. Hence, a flow separation criterion was developed, that is based on the intermittent disappearance of the high frequency content from the spectra of a hot-film sensor signal. This criterion does not require cross sensor information exchange and could hopefully be used also in complex 3D flows.

The method is constructed of the following steps:

1. Short time FFT (STFFT) of the hot-film voltage time history,
2. High-pass filtering of the FFT coefficients, above a predetermined cross-over frequency (to be defined below),
3. Summation of the high frequency coefficients to generate the "high-frequency standard deviation" (HFSDV) of the signal,
4. Time averaging of all the HFSDV's calculated in step 3,
5. Normalizing the time dependent HFSDV's by their mean,
6. Calculating the standard deviation of the normalized HFSDV.

The crossover frequency (f_c) was selected based on interrogation of a large number of data sets to be $f_c/Ue \approx 5$, where Ue is the velocity at the boundary layer edge, calculated from the pressure distribution. The result of the above procedure is independent of the flow conditions due to the auto normalization. Time windows of 40ms were currently used, at the current low speed conditions, but it can be significantly shorter especially at higher speeds. A threshold level of 0.225 was found to be adequate for identifying baseline as well as controlled separation over a wide range of configurations and Reynolds numbers.

The addition of large or small-scale coherent structures [19-23], for the purpose of controlling the turbulent boundary layer separation, complicates the separation identification procedure, by adding coherent motion to the complex separating boundary layer spectra. However, a separation of scales exists between the small-scale motion, which the current criterion searches for its disappearance, and the excitation imposed large-scale structures. The method also performed well when high frequency excitation was introduced, due to the high dissipation rate of these scales.

4. RESULTS

4.1 Baseline and controlled flow

Due to the limited scope of the current paper, only one flow condition will be described in detail. Fig. 2 describes the pressure distributions of the baseline and controlled flows. The model is at $\alpha = 6^{o}$, the leading and trailing edge flaps are deflected -25° and 20°, respectively and the chord Reynolds number, Re_c, is $0.24x10^6$.

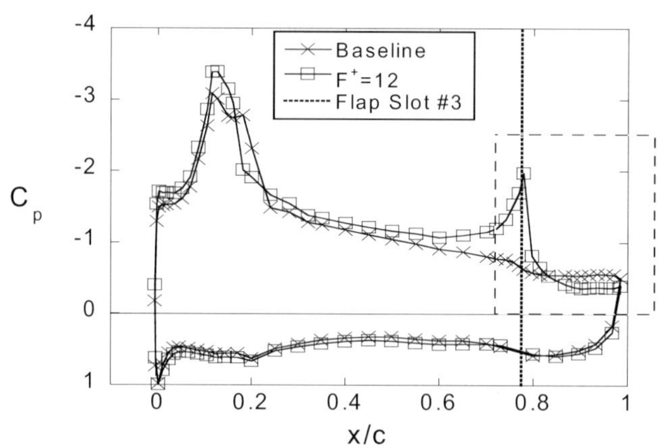

Fig. 2 Pressure distributions of baseline and controlled flows [22]. Dashed box indicates PIV interrogation region of Figures 3.

The baseline flow separates at x/c≈0.8, i.e. slightly downstream of the deflected flap shoulder, based on the vanishing pressure gradient criterion. When high frequency excitation at rather high momentum coefficient, $C\mu \approx 1.0\%$, was applied from the flap actuator slot, located at x/c=0.77 (δ_f=20°), the flow partly reattached to the flap. This actuation resulted in delayed separation and increased circulation around the entire airfoil. The lift coefficient was increased in this case from 1.65 to 1.82 while the form-drag coefficient increased from 0.074 to 0.078.

4.2 Application of the algorithm

The mean wall pressures (Cp) could only provide an indirect indication of the location of the mean separation location. A more direct indication of the mean separation location can be obtained from analyzing the PIV measured streamwise velocities presented in Figures 3a and 3b, for the baseline and controlled flows, respectively. Note the seam between the two partly overlapping images acquired simultaneously by the two PIV cameras. One should take into account the wall normal averaging of the PIV technique, due to the interrogation cell size. The PIV data indicates, using the zeroing of the near wall velocity [1] as guidance, that the baseline flow separates at x/c = 0.78. According to the same criterion, separation is delayed to x/c = 0.87, using the control input.

Applying the separation detection criterion based on the hot-film signals and using the suggested threshold level of 0.225, as shown in Fig. 4, indicates that the baseline flow separates at x/c =0.78 and that separation is delayed to x/c =0.87 in the controlled flow. The agreement between the Cp, the PIV data and the new separation detection criterion is satisfactory, validating the criterion. The criterion was validated also at different Re_c and LEF and TEF deflections and excitation locations [**23**].

Sensitivity tests for the sampling rate, STFT window width, low pass filter setting to simulate a lower band-width sensor, and threshold levels were conducted and resulted in low sensitivity.

5. CONCLUSIONS

A new turbulent boundary layer separation detection method has been proposed, formulated and validated using experimental data obtained over the NASA-Tel Aviv University "simplified high-lift system".

The method makes use of the changing balance between small and large-scale turbulent structures, favoring the large scale, as the boundary layer separates. Raw surface mounted hot-film sensor data was acquired and analyzed. The method and the separation detection criterion were validated using mean pressures and PIV data over a range of Reynolds numbers, geometries, and excitation locations.

Further study is required in order to enable the use of the new method and criterion as part of a closed-loop active separation control system, with the aim of efficient-distributed control of turbulent boundary layer separation.

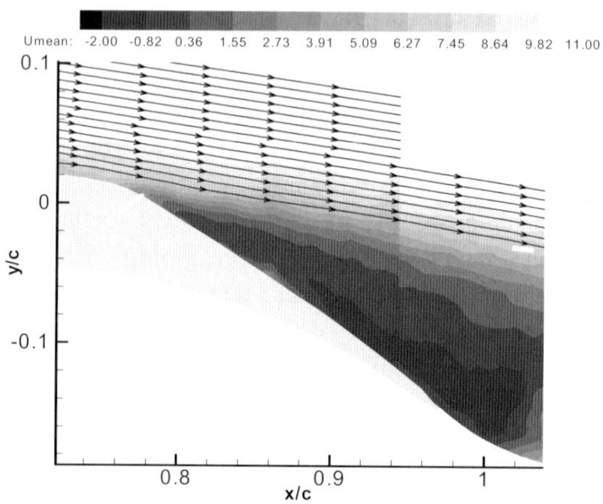

Fig. 3a PIV data of the flow above the baseline deflected flap (dashed region in Fig. 2).

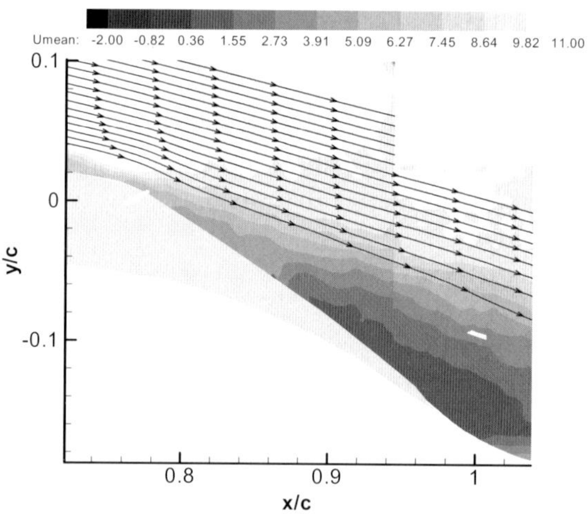

Fig. 3b PIV data of the flow above the controlled deflected flap (dashed region in Fig. 2).

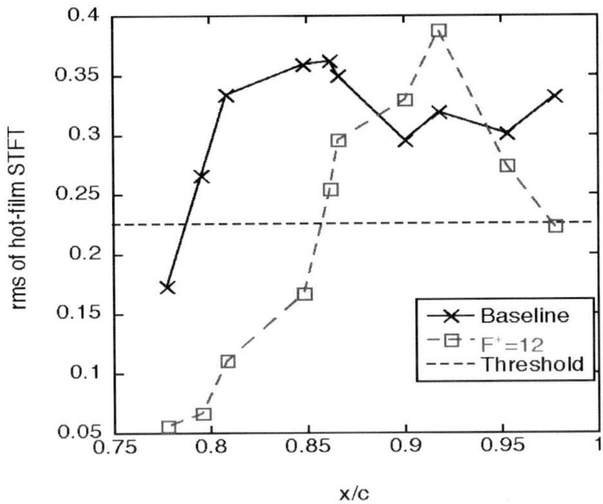

Fig. 4 Separation location based on a threshold level of 0.225 for the flow conditions of Figures 1-3.

ACKNOWLEDGEMENT

The authors would like to express their gratitude to the following for substantial support in the reported research: I. Fono (TAU), C.S. Yao, W. L .Sellers, M. J. Walsh, A. E. Washburn, L. N. Jenkins, R. D. White, G.C. Hilton, J. Mau, L. M. Hartzheim, S. O. Palmer, R. D. Lewis, and A. R. McGowan.

REFERENCES

1. Prandtl, L. (1904, Translation: Motion of Fluids with Very Little Viscosity, NACA TM 452, 1928).
2. Chang, P.K.: *Control of Flow Separation: Energy Conservation, Operation Efficiency, and Safety*, Hemisphere Publishing Corporation, 1976.
3. Gad-El-Hak, M.: *Flow Control, Passive, Active and Reactive Flow Management*. Cambridge University Press, August, 2000.
4. Na, Y. and Moin, P.: Direct Numerical Simulation of a Separated Turbulent Boundary Layer. *J. of Fluid Mech.*, Vol 374, pp. 379-405, 1998.
5. Simpson, R.L., Ghodbane, M., McGrath, B.E.: Surface Pressure Fluctuations in a Separating Turbulent Boundary Layer, *J. Fluid Mech.*, 177, pp. 167-186, 1987.

6. Allan, B.G., Jer-Juang, N., Raney, D., Seifert, A. Pack, L.G., Brown, D.E.: Closed-loop separation control using oscillatory flow excitation, *ICASE Report* 2000-32,2000.
7. Bertelrud, A.: Transition on a Three-Element High Lift Configuration at High Reynolds Numbers, *AIAA paper* 98-0703, January 1998.
8. Nakayama, A., Stack, J.P., Lin, J.C., and Valarezo, W.O.: Surface Hot-Film Technique for Measurements of Transition, Separation, and Reattachment Points, *AIAA paper* 93-2978, July 1993.
9. Stack, J.P., Mangalam, S.M., and Kalburgi, V.: The Phase Reversal Phenomenon at Flow Separation and Reattachment, *AIAA paper* 88-0408, 1988.
10. Hausmann, F., Schröder, W. and Limberg, W.: Development of a Multi-Sensor Hot-film Measuring Technique for Transition Detection in Cruise Flight, *AIAA paper* 2002-0534.
11. Meijering, A. and Schröder, W.: Experimental Analysis of Separated and Transonic Airfoil Flow, *AIAA paper* 2001-2987. June 2001.
12. Lee, T. and Basu, S.: Nointrusive Measurements of the Boundary Layer Developing on a Single and Two Circular Cylinders, *Exp. in Fluids*, Vol. 23, pp. 187-192 Springer-Verlag, 1997
13. Krause, E., Abstiens, R., Fuhling, S., Vetlutsky, V.N.: Boundary-layer investigations on a model of the ELAC 1 configuration at high Reynolds numbers in the DNW, *Eur. J. Mech. B-Fluids*, 19, pp. 745-764, 2000.
14. Lin, J.C. and Dominik, C.J.: Parameteric Investigation of a High-Lift Airfoil at High Reynolds Numbers, *AIAA J.*, Vol. 34, No. 4, 1997, pp. 485-491.
15. Klebanoff, P.S.: Characteristics of Turbulence in Boundary Layer with Zero Pressure Gradient. *NACA Report* 1247, 1955.
16. Simpson, R. L.: Turbulent Boundary Layer Separation, *Ann. Rev. Fluid Mech.*, pp. 205-234, 1989.
17. Na, Y. and P. Moin: The Structure of Wall-Pressure Fluctuations in Turbulent Boundary Layers with Adverse Pressure Gradient and Separation, *J. Fluid Mech.*, Vol. 377, pp. 347-373, 1998.
18. Simpson, R.L., Chew, Y.-T., and Shivaprasad, B. G.: The Structure of a Separating Turbulent Boundary Layer. Part 2. Higher-Order Turbulence Results. *J. Fluid Mech.*, 113, pp. 53-73, 1981.
19. Seifert, A. and Pack, L.G.: Oscillatory Control of Separation at High Reynolds Numbers. *AIAA J.*, Vol. 37, No. 9, 1999 pp. 1062-1071.
20. Seifert, A. and Pack, L.G.: Active Flow Separation Control on Wall-Mounted Hump at High Reynolds Numbers. *AIAA J.*, Vol. 40, No. 7, July 2002, pp. 1363-1372.
21. Pack, L.G., Schaeffler, N.W., Yao, C.S., and Seifert, A.: Active Control of Separation from the Slat Shoulder of a Supercritical Airfoil. *AIAA paper* 02-3156, June 2002. Submitted to *AIAA J. of Aircraft*, April 2004.
22. Pack, L.G., Yao, C.S., and Seifert, A.: Active Control of Separation from the Flap Shoulder of a Supercritical Airfoil. *AIAA paper* 03-4005, June 2003. Submitted to *AIAA J.*, July 2004.
23. Pack Melton, L., Yao, C.S., and Seifert, A.: Application of Excitation from Multiple Locations on a Simplified High-Lift System, *AIAA Paper* 04-2324, June 2004.

THE NEAR-WALL STRUCTURES OF THE TURBULENT BOUNDARY LAYER

Javier Jiménez[1] and Genta Kawahara[2]
[1] *ETSI Aeronáuticos, Universidad Politécnica, 28040 Madrid, Spain and Centre for Turbulence Research, Stanford University.*
jimenez@torroja.dmt.upm.es
[2] *Aeronautics and Astronautics, Kyoto University, Kyoto 606-8501, Japan*
gkawahara@kuaero.kyoto-u.ac.jp

Abstract: Models for the viscous and buffer layers over smooth walls are reviewed. Much of the friction coefficient in wall-bounded flows depends on this near-wall region. It is shown that there is a family of numerically-exact nonlinear structures which account for about half of the energy production and dissipation in the wall layer. The other half can be modelled in terms of their unsteady bursting. Many of the best-known characteristics of the wall layer, such as the dimensions of the dominant structures, are well predicted by these models.

Key words: Wall turbulence, nonlinear systems.

1. INTRODUCTION

In the same way that Prandtl showed that the global flow over obstacles can depend on the behaviour of very thin boundary layers, the flow within the turbulent boundary layer itself is in large part controlled by a thin region near the wall. This talk reviews the current theories about the flow in the immediate vicinity of smooth walls. Its modern study began experimentally [1, 2] in the 1970's, and got a strong impulse with the advent of direct numerical simulations [3] in the late 1980's and 1990's. The numerical emphasis is partly a personal bias of the author, but it is not altogether arbitrary. The near-wall region is well suited to simulation and difficult to explore experimentally. In §2 we outline the models that have been proposed for this

G.E.A. Meier and K.R. Sreenivasan (eds.), IUTAM Symposium on One Hundred Years of Boundary Layer Research, 209–220,

part of the flow, including how recent work on equilibrium solutions is related to turbulence. In §4 we discuss time-dependent bursting.

2. THE STRUCTURE OF NEAR-WALL TURBULENCE

Wall-bounded turbulence over smooth walls can be described in terms of two sets of scaling parameters [4]. Viscosity is important near the wall, and the length and velocity scales in that region are constructed with the kinematic viscosity and with the friction velocity u . Magnitudes expressed in these 'wall units' are denoted by $^+$ superscripts. The near-wall layer extends at most to $y^+ = 150$ [5] and, since y^+ can be interpreted as a Reynolds number, its relatively low values make turbulence near smooth walls a good candidate for simple models.

Far from the wall the velocity also scales with u , but the length scale is the flow thickness h. Between those two regions there is an intermediate solution with no length scale, where the mean velocity is given approximately by

$$U^+ = \quad {}^{-1} \log y^+ + A. \tag{1}$$

The Kármán constant, ≈ 0.4, is an essentially universal property of turbulence in the logarithmic layer, but the intercept constant, $A \approx 5$ for smooth walls, is a boundary condition that depends on the viscous inner region.

That inner layer is extremely important for the flow as a whole. The ratio between the inner and outer length scales is the friction Reynolds number, h^+, which may range up to $h^+ = 5 \times 10^5$ for large water pipes. In that limit the near-wall layer is only about 3×10^{-4} times the pipe radius, but it follows from (1) that 40% of the velocity drop takes place below $y^+ = 50$. Turbulence is characterized by the expulsion of the energy dissipation away from the large energy-containing eddies. In wall-bounded flows that separation occurs not only in scale space, but also in the shape of the mean velocity profile. The singularities are expelled both from the large scales, and from the centre of the flow towards the wall. Because of this singular nature, the near-wall layer is not only important for the rest of the flow, but it is also essentially independent from it [6].

If the equation for the total kinetic energy is written for the layer below some wall distance y, it yields an approximate expression for the intercept constant A. This has practical implications, because the centreline velocity $U_c^+ = (2/c_f)^{1/2}$ determines the friction coefficient c_f. If we estimate U_c^+ by extending (1) to $y = h$, the component due to the logarithm depends only on the Reynolds number, and the drag can only be manipulated by changing A.

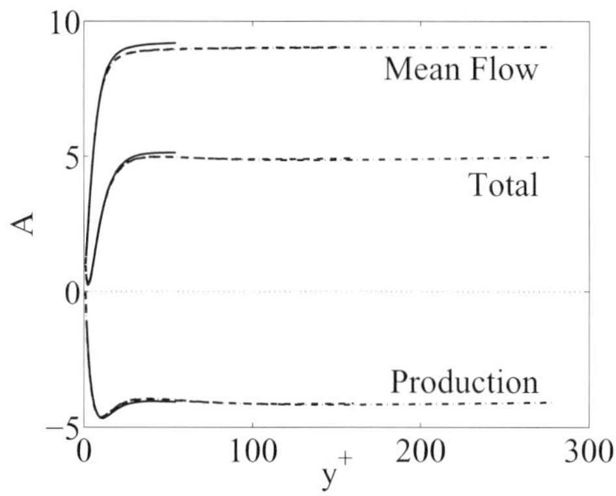

Figure 1. The terms in the expression (4) for the intercept constant, as explained in the text. Data are from computational channels below $y/h = 0.3$ [7]. ———, $Re = 180$; - - - -, $Re = 550$; —·—, $Re = 950$.

Using wall units, the energy balance can be written as

$$U^+ - \int_0^{y^+} y^+ (\ _yU)^+ \,\mathrm{d}y^+/h^+ = \int_0^{y^+} [\ ^+ + (\ _yU)^{+2}]\,\mathrm{d}y^+, \quad (2)$$

where

$$^+ = \left[1 - y^+/h^+ - (\ _yU)^+\right](\ _yU)^+, \quad (3)$$

is the turbulent energy production. In the logarithmic layer, $(\ _yU)^+$ can be approximated by $^{-1}/y$ in the integral in the left-hand side of (2), with an error which is only $O(1/h^+)$. Equation (2) can then be manipulated into

$$A = \int_0^{y^+} \ ^+ \,\mathrm{d}y^+ + \ ^{-1}(y^+/h^+ - \log y^+) + \int_0^{y^+} (\ _yU)^{+2} \,\mathrm{d}y^+. \quad (4)$$

Equation (4) makes the dependence of A on the viscous layer explicit. Its right-hand side can be separated into two parts. The first integral is the turbulent energy production, and the logarithmic and linear terms that follow it are the production that would result from the velocity profile (1). In the logarithmic layer both contributions approximately cancel, and their sum tends to a constant. The second integral is the viscous dissipation coming of the mean profile, which resides almost exclusively in the viscous sublayer. The behaviours of the two contributions, and of the resulting value for the intercept, are shown in figure 1, which shows that A is determined below $y^+ \approx 40$.

3. MODELS FOR THE SUBLAYER

Because of this global influence, the region below $y^+ \approx 100$ has been intensively studied. It is dominated by coherent streaks of the streamwise velocity and by quasi-streamwise vortices. The former are an irregular array of long $(x^+ \approx 1000)$ sinuous alternating streamwise jets superimposed on the mean shear, with an average spanwise separation [8] of the order of $z^+ \approx 100$. The vortices are slightly tilted away from the wall [9], and stay in the near-wall region for $x^+ \approx 200$. Several vortices are associated with each streak [10], with a longitudinal spacing of the order of $x^+ \approx 400$. Most of them merge into disorganized vorticity after leaving the wall neighbourhood [11].

It was proposed very early that streaks and vortices form a regeneration cycle in which the vortices are the results of an instability of the streaks [12], and the streaks are caused by the advection of the mean velocity gradient by the vortices [1, 13]. It is for example known that disturbing the streaks inhibits the formation of the vortices [6]. The manipulation is only effective if the flow is perturbed between $y^+ \approx 10$ and $y^+ \approx 60$, suggesting that it is between those levels that the streaks are involved in the vortex-generation process. There is a substantial body of numerical [14, 15, 16] and analytic [17, 18] work on the linear instability of model streaks. It shows that they are unstable to sinuous perturbations associated with inflection points of the perturbed velocity profile, whose eigenfunctions correspond well with the shape and location of the observed vortices. This type of models imply a time-dependent cycle in which streaks and vortices are created, grow, generate each other, and eventually decay. Additional references can be found in [6].

A slightly different point of view is that the regeneration cycle is organized around a nonlinear travelling wave, a fixed point in some phase space, which represents a nonuniform streak. This is not too different from the previous model, which essentially assumes that the undisturbed streak is a fixed point, and that the cycle is an approximation to an orbit along its unstable manifold. The new models however consider fixed points which are nontrivially perturbed streaks, and therefore separates the dynamics of turbulence from those of transition.

Nonlinear equilibrium solutions of the three-dimensional Navier–Stokes equations, with the right characteristics, have been obtained numerically in the past few years for plane Couette flow [19, 20], plane Poiseuille flow [21, 22, 20], and autonomous wall flows [23]. All those solutions contain a wavy low-velocity streak flanked by a pair of staggered quasi-streamwise vortices of alternating signs [24, 18], closely resembling the spatially-coherent objects educed from the

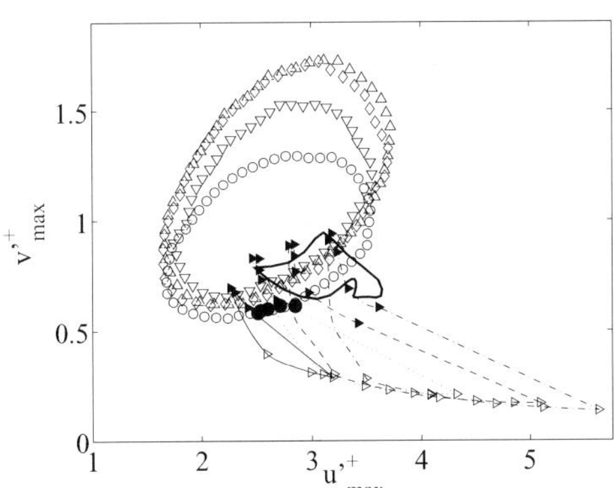

Figure 2. Comparison of some exact solutions with the near-wall turbulent structures, in terms of the maxima of the u' and v' r.m.s. profiles taken over boxes of size $b_x^+ \times b_z^+ \times y^+ = 380 \times 110 \times 50$. ▷ , Nagata's solutions for Couette flow [30]. Solid symbols are 'upper branch' solutions, and open ones are 'lower branch'. ● , autonomous permanent waves [23]. The solid loop is an exact limit cycle in plane Couette flow [25]. Other open symbols are probability isocontours from large-box Poiseuille flows [7]: △ , $h^+ = 1880$; ◊ , 950; ▽ , 550; ○ , 180. They contain 90% of the p.d.f.

near-wall region of true turbulence. Their mean and fluctuation intensity profiles are reminiscent of the experimental ones [23, 20], and the same is true of other properties. For example, the range of spanwise wavelengths in which the nonlinear solutions exist is always in the neighbourhood of the observed spacing of the streaks in the sublayer, $z^+ \approx 100$.

In those cases in which the stability of these solutions has been investigated, they have been found to be unstable saddles in phase space at the Reynolds numbers at which turbulence is observed. They are not therefore expected to exist as such in real turbulence, but the flow could spend a substantial fraction of its lifetime in their neighbourhood. Exact limit cycles and heteroclinic orbits based on these fixed points have been found numerically [25, 26], and reduced dynamical models of the near-wall region have been formulated in terms of low-dimensional projections of such solutions [27, 28, 29].

Two questions remain open: whether all the exact solutions that have been published for wall-bounded flows are related to each other and to near-wall turbulence, and whether real turbulence is best described in terms of steady structures or of unsteady events.

The first question is addressed in figure 2. The earliest and best-known nontrivial steady solutions of a wall-bounded Navier-Stokes shear flow are those in [19], which were recently extended in [30] to a wider range of parameters. They can be classified into 'upper' and 'lower' branches in terms of their mean wall shears. It is shown in [30] that most of the known wall-bounded solutions by other authors can also be classified into one or the other branch. The 'lower' solutions have strong and essentially straight streaks and weak vortices. 'Upper' solutions have weaker sinuous streaks flanked by stronger vortices. They consequently have weaker root-mean-square streamwise-velocity fluctuations u', and stronger wall-normal ones v' than those in the lower branch. Their velocity fluctuation profiles agree well with those of real turbulence.

The relative strength of both types of fluctuations for a particular solution can be characterized by the maximum values of its profiles of r.m.s. u' and v'. Different solutions can then be compared among themselves, and with fully-turbulent flows, by means of those two numbers. The r.m.s. profiles of the exact solutions, which are computed over periodic domains of size $L_x^+ \times L_z^+ \approx 400 \times 100$ parallel to the wall, cannot however be compared directly with the fluctuation profiles compiled from full experiments or from computations, which typically have domains of the order of $L_x^+ \times L_z^+ \approx 10,000 \times 5,000$.

To allow the comparison in figure 2, each wall of the large computational boxes was divided into 'minimal' sub-boxes with the same wall-parallel dimensions as the computational boxes of the exact solutions, $b_x^+ \times b_z^+ \approx 380 \times 110$, and the statistics were compiled over them. Each sub-box was characterized by its maximum r.m.s. intensities below $y^+ = 50$, and the values for different sub-boxes were summarized as a joint probability density function of the two quantities, compiled over the different sub-boxes and over time. Each flow was not therefore characterized by a single point, but by the probability distribution of the possible states of the sub-boxes.

The results of the figure suggest that only the 'upper-branch' exact solutions are representative of real turbulence. They also show that the correspondence is reasonably good, but that there are fluctuations in the near-wall region of real turbulent flows which are substantially stronger than those of the exact solutions.

4. BURSTING VERSUS STEADY SOLUTIONS

The next question is whether those stronger fluctuations can best be described by a different kind of steady solutions or by unsteady ones. The unsteady models discussed in section 2 follow the original interpretation of the visualizations of the sublayer [1], which was that the streaks regenerate through intermittent 'bursting'. That interpretation has sometimes been dismissed as a visualization artifact, and even the original authors acknowledged that their visualizations could be consistent with advecting permanent objects [31]. Bursting became associated with the ejections observed by stationary velocity probes, specially after numerical simulations showed that the velocity streaks were long-lived. The events identified in the analysis of single-point data were associated to the passing of quasi-streamwise vortices, intermittent in space but not necessarily in time [11]. This explanation bypassed the question of whether the observed temporally-intermittent sublayer events were artifacts or really existed.

The difficulty of following for long times individual structures in fully turbulent flows complicates the distinction between permanent

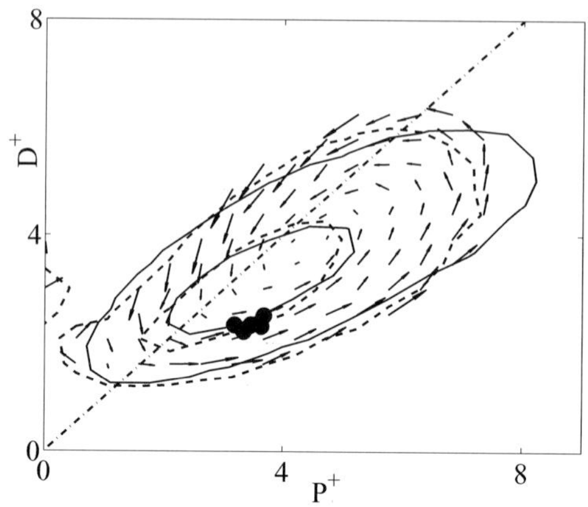

Figure 3. Joint probability density functions of the turbulent energy production and dissipation below $y^+ = 35$. $h^+ \approx 180$. The diagonal line is energy equilibrium, $P = D$. The arrows are explained in the text. ----, minimal Poiseuille flow, $L_x^+ \times L_z^+ = 450 \times 125$; ———, full channel, analysed over similar sub-boxes. The isolines contain 40% and 90% of the data. ●, equilibrium solutions from [23].

structures and time-dependent processes with a long period, but intermittent breakdown of near-wall turbulence is observed in minimal-flow numerical simulations [10] in which spatial intermittency is not an issue. In these simulations the wall-parallel periodic dimensions of the computational box are small enough to produce a periodic array of identical essentially-single structures [10], and the analysis is simplified because those structures can easily be followed in time. An example is given in figure 3, where the evolution of the flow near the wall in a minimal channel is represented in terms of the production P and of the dissipation D, integrated below $^+ = 50$. Each instantaneous state of the minimal flow is represented by a point in the (P, D) plane, and the joint p.d.f. in figure 3 is compiled as the system evolves in time. The arrows in the figure represent the mean evolution velocity in parameter space.

The flow describes a cycle in the (P, D) plane, during which the fluctuations accumulate energy when $P > D$, grow, cross into the dissipative part of the plane where $D > P$, and finally decay. The period of the cycle is of the order of $T^+ \approx 400$ in the high-Reynolds number limit [30], during which the structures advect about $x^+ = 5000$. The steady upper-branch equilibrium waves described above are in this representation production-dominated in the near-wall layer and relatively quiescent. The Couette limit cycle which was included in figure 2, but which is not plotted here for clarity, is a miniature version of the energy cycle in figure 3, with whom it shares many characteristics [30].

Figure 3 shows that the minimal flow 'bursts' in the sense of the original unsteady descriptions in reference [1]. The same temporal information is not accessible for full flows, because of the problem of identifying individual structures, but a joint p.d.f. of P and D can be compiled for them over minimal sub-boxes. Such a p.d.f. is included in figure 3, and it is similar enough to that of the minimal case to strongly suggest that the full flow is also bursting.

Sublayer bursting is important for the flow as a whole. We saw at the beginning of this paper that a large part of the friction coefficient is associated with the energy balance below $y^+ = 50$. It turns out that, if we define 'bursts' from figure 3 as points where $P^+ \gtrsim 4$, they account for roughly half of that balance.

Note that these bursts are not ejections associated with the passing of quiescent vortices. The sub-boxes that we have used to analyse the flow are large enough to always contain a full vortex pair, and in particular they are large enough for their mean wall-normal velocity to be always very close to zero. The scatter in figure 3 is due to differences between quiescent and excited full sets of structures. The mean profiles of most quantities change according to their location in the bursting cycle [30]. The general effect of bursting is to move the active structures away from the wall.

5. CONCLUSIONS

We have reviewed the present understanding of the dynamics of turbulent flows near smooth walls. This is a subject that, like most others in turbulence, is not completely closed, but which has evolved in the last two decades from empirical observations to relatively coherent theoretical models. It is also one of the first cases in turbulence, perhaps together with the structure of small-scale vorticity in isotropic turbulence, in which the key technique responsible for cracking the problem has been the numerical simulation of the flow. The reason is that the Reynolds numbers of the important structures are low, and therefore accessible to computation, while experiments are difficult. For example the spanwise Reynolds number of the streaks is only of the order of $z^+ = 100$, which is less than a millimetre in most experiments, but we have seen that it is well predicted by the range of parameters in which the associated equilibrium solutions exist.

The thinness of the layer in which the dynamics takes place makes the flow very sensitive to small perturbations at the wall. Roughness elements of the order of a few wall units, microns in a large pipe, completely destroy the delicate cycle that has been described here, and can increase the friction coefficient by a factor of two [32]. Conversely it only takes a concentration of polymers of a few parts per million in the near-wall region [33] to decrease the drag by 40%. The same can be said of the control strategies based on the manipulation of the near-wall structures [34, 35]. It has often been questioned whether such strategies, which have mostly been developed in low-Reynolds-number numerical simulations, would lose effectiveness at higher Reynolds numbers. The analysis in section 2 shows that they will, because the dissipation in the logarithmic layer cannot be avoided, but that the degradation with h^+ is only logarithmic.

The preparation of this paper was supported in part by the CICYT grant DPI2003–03434. I am indebted to J.C. del Álamo, G. Kawahara and M.P. Simens for providing most of the data used in the figures.

REFERENCES

1. Kim, H.T., Kline, S.J. and Reynolds, W.C., The production of turbulence near a smooth wall in a turbulent boundary layers, *J. Fluid Mech.* **50**, 133–160 (1971)
2. Morrison, W.R.B., Bullock, K.J. & Kronauer, R.E., Experimental evidence of waves in the sublayer. *J. Fluid Mech.* **47**, 639–656 (1971)
3. Kim, J., Moin, P. & Moser, R., Turbulence statistics in fully developed channel flow at low Reynolds number. *J. Fluid Mech.* **177**, 133–166 (1987)
4. Tennekes, H. and Lumley, J.L., *A first course in turbulence*, chapter 8. MIT Press (1972)
5. Österlund, J.M., Johansson, A.V., Nagib, H.M. & Hites, A note on the overlap region in turbulent boundary layers, *Phys. Fluids* **12**, 1–4 (2000)

6. Jiménez, J. & Pinelli, A., The autonomous cycle of near wall turbulence, *J. Fluid Mech.* **389**, 335–359 (1999)
7. del Álamo, J.C., Jiménez, J., Zandonade, P. & Moser, R.D., Scaling of the energy spectra of turbulent channels, *J. Fluid Mech.* **500**, 135–144 (2004)
8. Smith, C.R. & Metzler, S.P., The characteristics of low speed streaks in the near wall region of a turbulent boundary layer. *J. Fluid Mech.* **129**, 27–54 (1983)
9. Jeong, J., Hussain,F., Schoppa,W. & Kim,J., Coherent structures near the wall in a turbulent channel flow. *J. Fluid Mech.* **332**, 185–214 (1997)
10. Jiménez, J. & Moin, P., The minimal flow unit in near wall turbulence. *J. Fluid Mech.* **225**, 221–240 (1991)
11. Robinson, S.K., Coherent motions in the turbulent boundary layer. *Ann. Rev. Fluid Mech.* **23**, 601–639 (1991)
12. Swearingen, J.D. & Blackwelder, R.F., The growth and breakdown of streamwise vortices in the presence of a wall. *J. Fluid Mech.* **182**, 255–290 (1987)
13. Bakewell, H.P. & Lumley, J.L., Viscous sublayer and adjacent wall region in turbulent pipe flow. *Phys. Fluids* **10**, 1880–1889 (1967)
14. Hamilton, J.M., Kim, J. & Waleffe, F., Regeneration mechanisms of near-wall turbulence structures. *J. Fluid Mech.* **287**, 317–348 (1995)
15. Waleffe, F., On a self-sustaining process in shear flows. *Phys. Fluids* **9**, 883–900 (1997)
16. Schoppa, W. & Hussain, F., Coherent structure generation in near-wall turbulence. *J. Fluid Mech.* **453**, 57–108 (2002)
17. Reddy, S.C., Schmid, P.J., Baggett, J.S. & Henningson, D.S., On stability of streamwise streaks and transition thresholds in plane channel flows. *J. Fluid Mech.* **365**, 269–303 (1998)
18. Kawahara, G., Jiménez, J., Uhlmann, M. & Pinelli, A., Linear instability of a corrugated vortex sheet – a model for streak instability, *J. Fluid Mech.* **483** 315–342 (2003)
19. Nagata, M., Three-dimensional finite-amplitude solutions in plane Couette flow: bifurcation from infinity, *J. Fluid Mech.* **217**, 519–527 (1990)
20. Waleffe, F., Homotopy of exact coherent structures in plane shear flows, *Phys. Fluids* **15**, 1517–1534 (2003)
21. Toh, S. & Itano, T., On the regeneration mechanism of turbulence in the channel flow, *Proc. Iutam Symp. on Geometry and Statistics of Turbulence*. (eds. T. Kambe, T. Nakano and T. Muiyauchi), Kluwer. 305–310 (2001)
22. Waleffe, F., Exact coherent structures in channel flow, *J. Fluid Mech.* **435**, 93-102 (2001)
23. Jiménez, J. & Simens, M.P., Low-dimensional dynamics in a turbulent wall flow, *J. Fluid Mech.* **435**, 81–91 (2001)
24. Waleffe, F., Three-dimensional coherent states in plane shear flows. *Phys. Rev. Letters* **81**, 4140–4143 (1998)
25. Kawahara, G. & Kida, S., Periodic motion embedded in plane Couette turbulence: regeneration cycle and burst, *J. Fluid Mech.* **449**, 291–300 (2001)
26. Toh, S. & Itano, T., A periodic-like solution in channel flow, *J. Fluid Mech.* **481**, 67–76 (2003)

27. Aubry, N., Holmes, P., Lumley, J.L. & Stone, E., The dynamics of coherent structures in the wall region of a turbulent boundary layer, *J. Fluid Mech.* **192**, 115–173 (1988)
28. Sirovich, L. & Zhou, X., Dynamical model of wall-bounded turbulence. *Phys. Rev. Lett.* **72**, 340–343 (1994)
29. Waleffe, F., On a self-sustaining process in shear flows, *Phys. Fluids* **9**, 883–900 (1997)
30. Jiménez, J., Kawahara, G., Simens, M.P., Nagata, M. & Shiba, M., Characterization of near-wall turbulence in terms of equilibrium and 'bursting' solutions, submitted to *Phys. Fluids*.
31. Offen, G.R. & Kline, S.J., A proposed model for the bursting process in turbulent boundary layers, *J. Fluid Mech.* **70**, 209–228 (1975)
32. Jiménez, J., Turbulent flows over rough walls, *Ann. Rev. Fluid Mech.* **36**, 173–196 (2004)
33. McComb, W.D. *The physics of fluid turbulence*. Oxford U. Press (1990)
34. Choi, H., Moin, P. and Kim, J., Active turbulence control and drag reduction in wall-bounded flows. *J. Fluid Mech.* **262**. 75–110 (1994)
35. Jiménez, J., On the structure and control of near wall turbulence, *Phys. Fluids* **6** 944–953 (1994)

TURBULENCE IN SUPERSONIC AND HYPERSONIC BOUNDARY LAYERS

Alexander J. Smits and M. Pino Martin
Department of Mechanical and Aerospace Engineering, Princeton University, Princeton NJ 08544, USA., Tel: 609-258-5117, Fax: 609-258-6123, Email: asmits@princeton.edu

Abstract: A summary is given of the behavior of turbulent boundary layers in supersonic and hypersonic flow where the effects of compressibility have a direct influence on the turbulence. Experimental and DNS results are presented and compared.

Key words: Turbulence, supersonic, hypersonic, shocks, shocklets, boundary layer.

1. INTRODUCTION

In 1951, van Driest published a seminal paper entitled Turbulent boundary layer in compressible fluids [1] that founded the study of turbulent boundary layers at high speed. Four years later, he was present at the Braunschweig meeting on Fifty years of boundary layer research [2], and reported on the status of that fledgling field. The fifty years were, of course, counted from the date Prandtl delivered his famous lecture on flows with very small friction at the Third Mathematical Congress in Heidelberg in 1904, thereby establishing for the first time the concept of the boundary layer [3]. The occasion of the present symposium, another fifty years on from the Braunschweig meeting, offers an excellent opportunity to consider where we currently stand, and what prospects there are for future progress in understanding the behavior of turbulent boundary layers in supersonic and hypersonic flow, that is, where the effects of compressibility have a direct influence on the turbulence.

G.E.A. Meier and K.R. Sreenivasan (eds.), IUTAM Symposium on One Hundred Years of Boundary Layer Research, 221-230,

The most important parameter in the description of incompressible turbulent boundary layer behavior is undoubtedly the Reynolds number. For compressible flows, the Mach number becomes a further scaling parameter. Within the boundary layer, the flow is supersonic in the outer layer and subsonic near the wall, although the sonic line is located very close to the wall at high Mach number. A temperature gradient develops across the boundary layer due to the conversion of kinetic energy to heat as the flow velocity decreases. In fact, the static-temperature variation can be very large even in an adiabatic flow, resulting in a low-density, high-viscosity region near the wall. In turn, this leads to a skewed mass-flux profile, a thicker boundary layer, and a region in which viscous effects are somewhat more important than at an equivalent Reynolds number in subsonic flow.

The temperature variations across the layer also cause the fluid properties to vary. For example, for an air flow with a freestream Mach number of 3 on an adiabatic wall, the density varies across the boundary layer by a factor of about 5, while the kinematic viscosity varies by a factor of about 17 [4].

Intuitively, one would expect to see significant dynamical differences between subsonic and supersonic boundary layers. However, it appears that many of the apparent differences can be explained by accounting for the fluid-property variations that accompany the temperature variation. This suggests a rather passive role for the density differences in these flows, most clearly expressed by Morkovin's hypothesis, which states that the dynamics of a compressible boundary layer are expected to follow the incompressible behavior closely, as long as the Mach number associated with the fluctuations remains small. We interpret this to mean that the fluctuating Mach number, M', must remain small, where M' is the rms perturbation of the instantaneous Mach number from its mean value, taking into account the variations in velocity and sound speed with time. If M' approaches unity at any point, we expect direct compressibility effects such as local shocklets and pressure fluctuations to become important. If we take $M' = 0.3$ as the point where compressibility effects could start to become important for the turbulence behavior, Smits & Dussauge [4] estimated that for zero-pressure-gradient adiabatic boundary layers at moderately high Reynolds numbers this point would be reached with a freestream Mach number of about 4 or 5 (see Figure 1a).

In fact, we find from experiments performed by Baumgartner *et al.* [5] that compressibility effects on turbulence in zero pressure gradient flows are weak, even at $M = 8$. The only significant effect seems to be on the integral length scale, which decreases significantly with Mach number (see Figures 1b and 2). All other statistical measures, such as the Reynolds stresses and the higher order moments, do not show any obvious influence of Mach number, as long as they are scaled (when appropriate) by the local density rather than the density at a fixed point [4]. This scaling is called the

Morkovin scaling. Other quantities, such as the intermittency, also do not appear to be a function of Mach number, despite earlier evidence to the contrary (for example, compare the intermittency data in [5] with that in [6]).

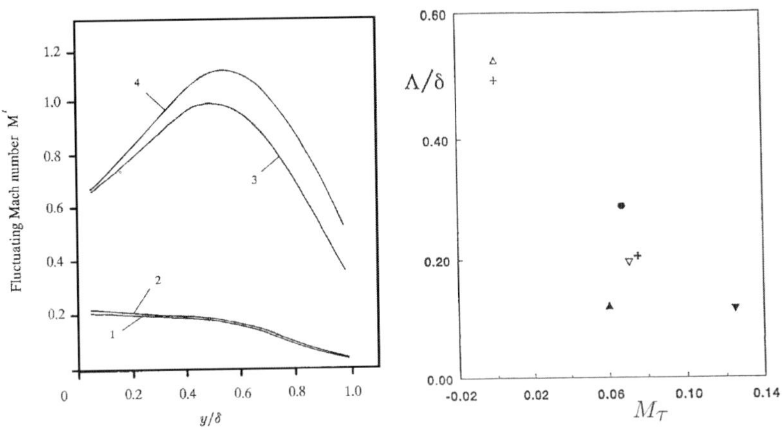

Figure 1. (a): Fluctuating Mach number distributions (estimated). Flow 1: M = 2.32; Flow 2: M = 2.87; Flow 3: M = 7.2; Flow 4: M = 9.4. (b) Integral scale as a function of friction Mach number in boundary layer flow (by experiment). Figures from [4], where original references are given.

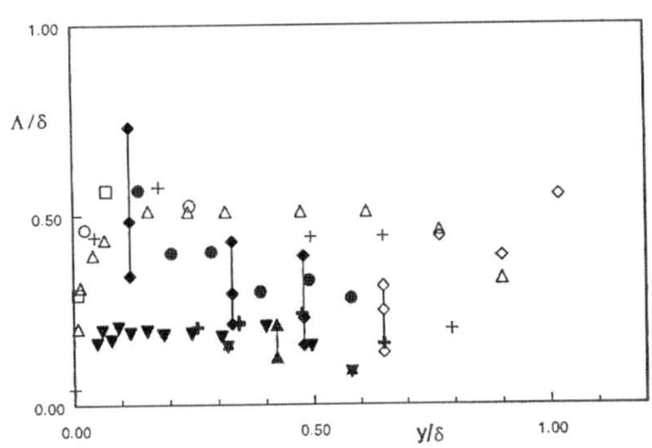

Figure 2. Integral scale as a function of freestream Mach number in boundary layer flow (by experiment). Open symbols: subsonic flow. Closed symbols: supersonic and hypersonic flow. Figure from [4], where original references are given.

Nevertheless, the role of intense Mach number fluctuations is interesting from a fundamental point of view, as well as perhaps playing a part in decreasing the integral length scale as the Mach number increases. In this respect, shocklets (local shocks occurring within the boundary layer due to supersonic relative motions) may be important. Although shocklets have been visualized in hypersonic boundary layers [5], only Direct Numerical Simulations (DNS) can be used to examine their structure and their possible role in rescaling turbulent motions. Many other open questions exist, regarding, for example, the level of the pressure fluctuations within the boundary layer, the velocity-temperature correlation, the accuracy of Reynolds analogies, the kinetic energy budget, and, most intriguingly, the structure of the coherent motions. Most of these quantities cannot be measured accurately, or even at all, and DNS, appropriately validated, may be crucial in making further progress.

Direct numerical simulations of boundary layers in supersonic flow have only become available recently. The first results were reported in 2000 by Adams [7], who studied the flow over a compression ramp at M = 3 and a Reynolds number based on momentum thickness $Re_q = 1685$. The impact of Adam's work was somewhat limited by the low value of his Reynolds number, in that there were no experiments available for comparison with his results at that time. More recently, Martin [8-9] reported DNS of hypersonic turbulent boundary layers at higher Reynolds numbers, allowing comparison with data taken at IMST in France [11-14]. Having established the plausibility of her computations, Martin studied the behavior of the boundary layer as the Mach number varied from 3 to 8 while keeping the Reynolds number in wall units approximately constant. The purpose of the present contribution is to discuss these results as well as recent experimental data, and consider how they may change our understanding of high-speed turbulent boundary layers.

2. COMPARISON OF DNS AND EXPERIMENT

Martin's computational data are available for the conditions given in Table 1 [9]. Typical values for grid resolution and domain size are $L_x/\delta = 7.9$, $L_y/\delta = 2.0$, $L_z/\delta = 15.4$, $\Delta x^+ = 7.6$, $\Delta y^+ = 2.8$, $N_x = 384$, $N_y = 256$, $N_z = 110$ (Case M4). Here x, y and z are the streamwise, wall-normal, and spanwise directions, respectively. The wall was isothermal, and the effects of varying the wall temperature were assessed. Although the results were presented for a time-developing layer, and Martin showed that only small differences existed between that temporal and spatially-developing layers. Martin also demonstrated very good agreement with experimental data for Case M2, establishing a high degree of confidence in the entire data set.

The DNS over this Mach number range (3 to 8) support many observations gleaned from experiment. For example, the mean velocity profiles transformed according to van Driest collapse with the usual scaling using inner and outer variables. Also, the Reynolds stresses collapse using

Morkovin's scaling at about the same level of accuracy as seen in experiment, the intermittency profile shows little influence of Mach number (although there is an unexplained peak near the wall that exceeds a value of one, see Figure 3a), and the temperature/velocity correlation R_{uT} is almost independent of Mach number and constant at about 0.7 for most of the layer (Figure 3b), although that is a little lower than the generally accepted experimental value of 0.8.

Case	M_δ	Re_θ	δ^+
M2	2.32	4452	745
M3	2.98	2390	325
M4	3.98	3944	368
M5	4.97	6225	382
M6	5.95	8433	396
M7	6.95	10160	414
M8	7.95	13060	430

Table 1. Boundary layer parameters for DNS computations Cases M2 to M8 [9].

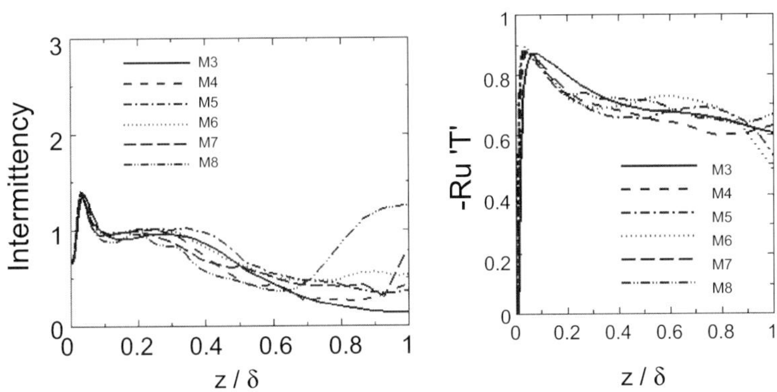

Figure 3. DNS results [9]. Flow conditions given in Table 1. Here, z is the wall normal distance. (a) Intermittency determined by 3/flatness; (b) Temperature/velocity correlation.

For the fluctuating pressure levels, with the wall temperature set approximately equal to its adiabatic value, the fluctuations increased from about 2% in the freestream to 4 or 5% near the wall, with only a weak Mach number dependence (Figure 4a). Experiments in a M = 1.8 flow showed a similar increase but only from about 0.3% in the freestream to 1% at the wall [4]. It seems the DNS results are too high, which may be the result of insufficient averaging. However, the computations reveal that at Mach 5 decreasing the wall temperature by a factor three from its adiabatic value increases the level of the pressure fluctuations by about 75%, which is a new result if confirmed in subsequent work.

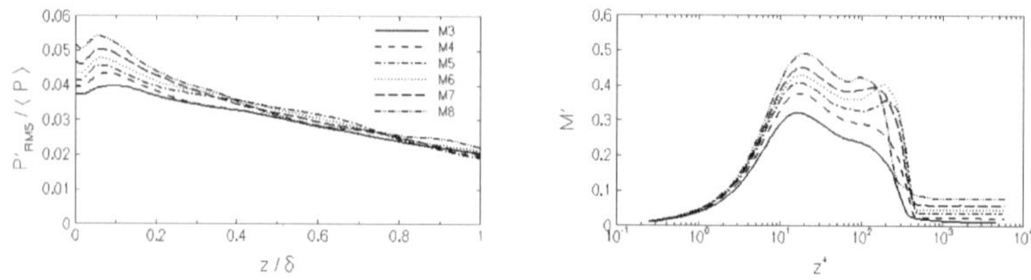

Figure 4. DNS results from Martin [9]. Flow conditions given in Table 1 Here, z is the wall normal distance. (a) Pressure fluctuation; (b) Fluctuating Mach number ($z^+ = zu_\tau / \nu$).

Interestingly, the fluctuating Mach numbers do not indicate the high values expected from the estimates shown in Figure 1a. For example, the DNS results indicate a maximum value of about 0.5 at Mach 8, compared to the estimated value of 1.0 at the same Mach number (Figure 4b). There is apparently little sensitivity with wall temperature, a result that was expected from previous work, but the factor of two difference between the estimate made by [4] and the DNS results is potentially very important, and may help to explain why the Mach number effects detected by experiment [5] are generally rather small.

Despite the small values of M , shocklets were observed in the DNS. Figure 5a shows instantaneous contours of pressure in a Mach 4 turbulent boundary layer from Martin [8]. Shocklets were also observed by Baumgartner *et al.* [5] in Filtered Rayleigh Scattering (FRS) images taken in a Mach 8 boundary layer. In the case of DNS, the presence of shocklets is found by the large magnitude of the correlation between gradients of the divergence (from positive to negative) with large pressure gradients (of opposite sign). Verifying that the Rankine-Hugoniot conditions are met along the instantaneous streamline further corroborates the presence of shocklets. The window in the figure shows the location of a shocklet, which is a small-length, small-time scale shock. Further study of the role of shocklets in the turbulence dynamics are in progress at Princeton.

With respect to the integral length scale, Figure 5b shows the DNS results, demonstrating, as expected, a decrease in size as the Mach number increases. As the freestream Mach number increases from 3 to 8, the value decreases about 50%, which is in good accord with the experiment (Figure 2), given the rather scattered nature of the data.

Under the conditions of negligible fluctuations in total enthalpy, the one-dimensional energy equation gives:

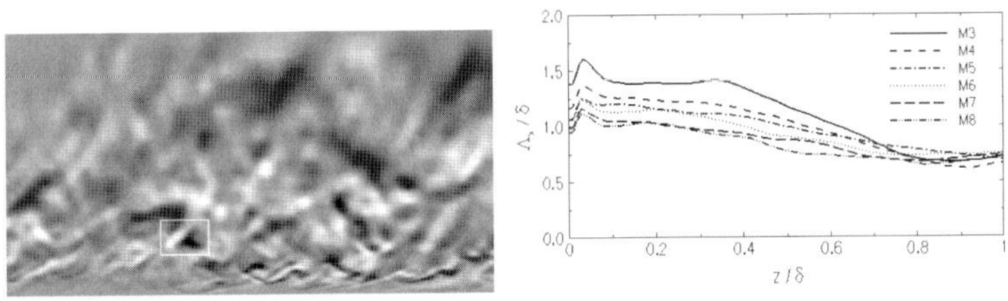

Figure 5. (a) Divergence of velocity for a Mach 4 turbulent boundary layer in a streamwise plane. DNS results from [8], flow is from left to right. (b) Integral length scale computed from DNS data [9]. Here, z is the wall normal distance.

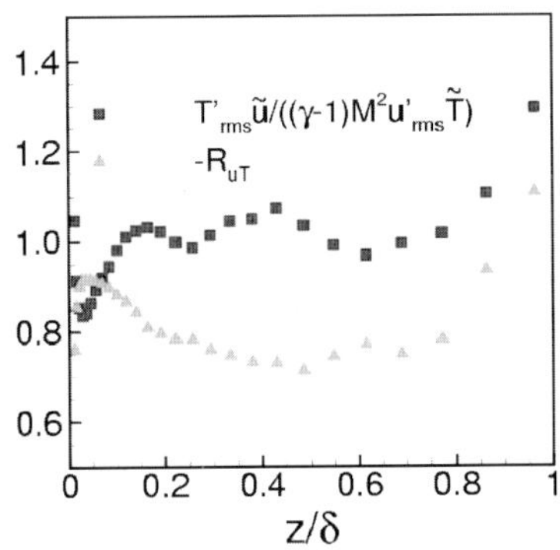

Figure 6. Test of the Strong Reynolds Analogy for a Mach 3 turbulent boundary layer, $Re_\theta = 2{,}400$ ($\delta^+ = 400$); z is the wall normal distance. DNS results, from Wu & Martin [10].

$$\frac{\sqrt{\overline{T'^2}}}{\overline{T}} = (\gamma - 1)M^2 \frac{\sqrt{\overline{u'^2}}}{\overline{U}} \tag{1}$$

and

$$R_{uT} = -\frac{\overline{u'T'}}{\sqrt{\overline{u'^2}}\sqrt{\overline{T'^2}}} = -1 \tag{2}$$

These relations are often called the Strong Reynolds Analogy (SRA), and they are commonly used in experiment to relate the temperature and velocity fluctuations where only one of the two quantities is known [4]. The DNS results for the temperature velocity correlation R_{uT} as a function of Mach number were shown earlier in Figure 3b. For a Mach 3 boundary layer with $Re_q = 2400$, results from additional DNS confirm the experimental observation that the correlation level varies between 0.7 and 0.8 over most of the layer (Figure 6). This value is smaller than that given by Equation 2, but somewhat surprisingly the comparison with Equation 1 is almost perfect, except for a small region near the wall where viscous effects are undoubtedly important. The two observations taken together suggest a phase difference between the velocity and temperature signals, as first discussed by Smith & Smits [15]. Further analysis of the DNS may provide a more definitive answer to this suggestion.

One of the most direct comparisons between DNS and experimental data can be made visually. In Figure 7, we show snapshots in a streamwise plane, taken from two nominally identical flows, one experimental and one DNS. The qualitative similarities are obvious. More detailed analysis of the experimental images will provide quantitative data on streamwise length scales, structure angles, and intermittency. A preliminary result on the intermittency profile is given in Figure 8.

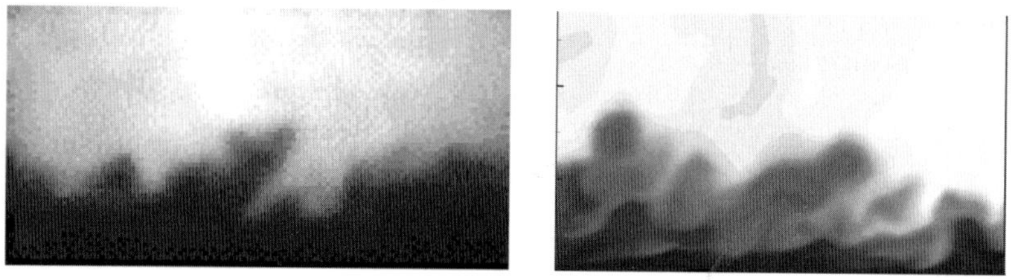

Figure 7. Mach 3 boundary layer, flow is from left to right. (a) Experimental FRS data, $Re_\theta = 2{,}397$. Bookey & Wyckham, private communication. (b) Density contours computed from DNS data, $Re_\theta = 2{,}400$ ($\delta^+ = 400$). Martin, private communication.

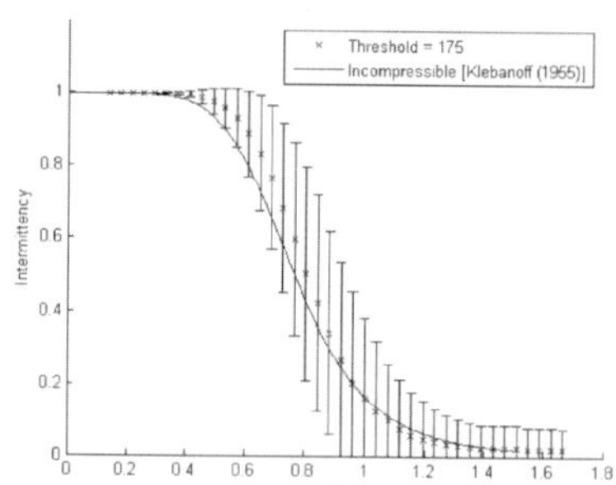

Figure 8. Intermittency profile from experimental FRS data, same boundary layer as in Figure 7. Bookey, private communication.

3. CONCLUDING REMARKS

DNS of turbulent boundary layers at supersonic and hypersonic speeds have only started to appear in the literature within the last four years.

Current results are still limited to relatively low Reynolds numbers, but the Reynolds numbers are already within the range where direct comparisons with experiment are possible. The data from the group headed by Dussauge at IUSTI in Marseille (formerly IMST) has been particularly valuable in this regard. Current efforts at Princeton to obtain experimental data under the same flow conditions as the DNS by Martin are helping to further increase our confidence in the computations.

Despite the limitations on Reynolds number, the detailed information on turbulence statistics and turbulence structure provided by DNS is already enriching our understanding of the behavior of turbulence in supersonic and hypersonic flows. Further analysis of the computations may well lead us to understand more fully the role that compressibility plays in turbulent boundary layers. For example, the indications that fluctuating Mach numbers in high Mach number flows are considerably lower than formerly believed leads to two possible conclusions. First, the analysis used to form the estimates given in Figure 1a may not have been correct. These estimates made use of Morkovin s hypothesis and the Strong Reynolds Analogy in a very simple way, and although the DNS results generally support these scalings, it appears that the sum of the parts does not add up to the effects observed. Second, we may also conclude that the direct effects of compressibility on wall-bounded flows are even smaller than formerly believed, implying also that shocklets do not have a strong influence on the dynamics of turbulence even at Mach 8. Nevertheless, we see, for example, a rapid change in integral length scale with increasing Mach number in experiments and in DNS. The underlying cause of this phenomenon is still to be found, and we expect to make dramatic progress in the next few years as more DNS data become available.

ACKNOWLEDGEMENTS

The support of AFOSR under Grants F-49620-02-1-0124 and F-49620-02-1-0361, directed by Drs. John Schmisseur and Tom Beutner, and NSF CAREER Award # CTS-0238390 is gratefully acknowledged.

REFERENCES

1. van Driest, E. R. (1951) Turbulent boundary layer in compressible fluids. J. Aeronautical Sciences, **18**:145-160.
2. van Driest, E. R. (1955) The turbulent boundary layer with variable PRANDTL number. In Fifty Years of Boundary-Layer Research, Braunschweig, 257-271.
3. Prandtl, L. (1904) ber Fl ssigkeitsbewegung bei sehr kleiner Reibung. Int. Math.-Kongr., Heidelberg, 3rd, pp. 484-491. Leipzig: Teubner, 1905.

4. Smits, A. J. and Dussauge, J. P. Turbulent Shear Layers in Supersonic Flow, AIP Press/Springer Verlag, 1996, 357 pages. Second Edition, Springer Verlag, 2004.
5. Baumgartner, M. L., Erbland, P. J., Etz, M. R., Yalin, A., Muzas, B., Smits, A. J., Lempert, W. and Miles, R. B. (1997) Structure of a Mach 8 Turbulent Boundary Layer, AIAA paper #97-0765, 35th AIAA Aerospace Sciences Meeting, Reno, Nevada.
6. Robinson, S. R. (1986) Space-time correlation measurements in a compressible turbulent boundary layer. AIAA Paper 86-1130.
7. Adams, N. Direct simulation of the turbulent boundary layer along a compression ramp at $M = 3$ and $Re_q = 1685$. J. Fluid Mech., **420**:47-83, 2000.
8. Martin, M.P., Preliminary DNS/LES database of hypersonic turbulent boundary layers, AIAA Paper No. 2003-3726, 2003.
9. Martin, M.P., DNS of hypersonic turbulent boundary layers, AIAA Paper No. 2004-2337, 2004.
10. Wu, M. and Martin, M. P., Direct numerical simulation of two shockwave/turbulent boundary layer interactions at Mach 2.9 and Retheta = 2400, Submitted to AIAA Journal. Also AIAA Paper No. 2004-2145, 2004.
11. Debi ve, J. F., Gouin, H. and Gaviglio, J. Momentum and temperature fluxes in a shock wave-turbulence interaction. Proc. ICHM/IUTAM Symp. Structure of Turbulence and Heat and Mass Transfer, Dubrovnik, 1981.
12. Debi ve, J. F. Etude d'une interaction turbulence/onde de choc. Th se de Doctorat d'Etat, Universit d'Aix-Marseille II, 1983.
13. El na, M., Lacharme, J. P. and Gaviglio, J. Comparison of hot-wire and laser Doppler anemometry methods in supersonic turbulent boundary layers. In Dybb, A. and Pfund, P. A. (eds), International Symposium on Laser Anemometry, ASME, 1985.
14. El na, M. and Lacharme, J. P. Experimental study of a supersonic turbulent boundary layer using a laser Doppler anemometer. J. M canique Th orique et Appliqu, 7:175-190, 1988.
15. Smith, D. R. and Smits, A. J. Simultaneous measurement of velocity and temperature fluctuations in the boundary layer of a supersonic flow. Experimental Thermal and Fluid Science, **7**:221-229, 1993.

THE ROLE OF SKIN-FRICTION MEASUREMENTS IN BOUNDARY LAYERS WITH VARIABLE PRESSURE GRADIENTS

Hans-Hermann Fernholz
Hermann-Föttinger-Institut für Strömungsmechanik
Technische Universität Berlin, Müller-Breslau-Str. 8, D-10623 Berlin, Germany
e-mail: fernholz@pi.tu-berlin.de

Abstract: A survey of skin-friction measurements is presented in combination with mean velocity measurements in incompressible two-dimensional turbulent boundary layers with variable pressure gradients in the streamwise direction. The most important techniques over a span of 100 years will be discussed and compared if possible under the same flow conditions.

Key words: Skin-friction, variable pressure-gradient boundary layers.

1. INTRODUCTION

In his inaugural lecture on the bounday layer in 1904 L. Prandtl [1] drew attention to the considerable effect of friction between a moving fluid and a wall at rest which is caused by a small value of the viscosity and a large gradient of the velocity normal to the wall. Skin friction accounts for the drag in pipes, for a large proportion of the drag of ships and for most of the profile drag of airplanes at subsonic speeds.

Since in the great majority of cases the Reynolds number is high, boundary layers which will be considered here, are turbulent and for reasons of simplification, incompressible two-dimensional or axisymmetric. The wall is aerodynamically smooth. The pressure gradient in the streamwise direction may be adverse (APG) and lead to separation, or favourable (FPG) and may cause relaminarization.

Since we celebrate the first centenary of boundary-layer theory the role of skin friction will be described in three periods over time. The most important measuring techniques will be discussed and compared under the same flow conditions, in order to show their range of validity.

G.E.A. Meier and K.R. Sreenivasan (eds.), IUTAM Symposium on One Hundred Years of Boundary Layer Research, 231-240,

2. THE PERIOD BETWEEN 1911 AND 1946

During this period skin friction in turbulent boundary layers was determined by means of the momentum integral equation, the velocity gradient normal to the wall $(\partial \bar{u}/\partial y)_{y=O}$ or by the use of the Stanton tube.

In a first investigation Stanton [2], established two, now familiar facts. First he demonstrated the validity of Newton's law for skin-friction in turbulent pipe flow since he found that near the wall the "mechanical viscosity" was equal to the viscosity μ. Secondly he showed that the no-slip condition at the wall was valid also for turbulent flows. This lead to the assumption of a "laminar" sublayer next to a wall [3]. In order to measure the mean velocity very close to the wall of a pipe, Stanton used miniature flattened Pitot tubes (height 0.33mm). G.I Taylor [4] measured velocity profiles in a slightly accelerated turbulent boundary layer along a plate, evaluated the momentum equation for a boundary layer and thus calculated the skin-friction.

Stanton [3] meanwhile continued his line of research to determine $(\partial \bar{u}/\partial y)_{y=O}$ but soon found that this problem could not be solved, even when he used still smaller Pitot tubes, since the exact velocity given by the Poiseuille-flow solution, was not in agreement with the measured one if the wall distance was below 0.4mm. In order to remain as close to the wall as possible, he designed a half Pitot probe which, however, had to be calibrated in a Poiseuille-flow to provide the skin friction. The now so called Stanton tube was situated in the viscous sublayer, and had a rather wide range of validity: the flow could be laminar, transitional or turbulent and have zero or variable pressure gradient.

The possibilities of the Stanton tube were exploited for the first time by Fage & Falkner [5] who designed a Stanton tube which could be inserted at various ports along the centerline of a large Joukowski aerofoil (chord=1m, width=2m). Figure 1 shows the distribution of the skin-friction coefficient $c_f/2$ and the pressure coefficient $c_p/2$ on the suction side. The first maximum value of c_f near the nose was associated with laminar flow and the second with turbulent flow in the boundary layer. The measured velocity profiles suggest that transition took place between the two maxima. It should also be noted that beyond the second maximum, skin friction fell steadily with the distance from the nose without, however, reaching zero on any part of the surface.

Fage & Falkner compared the results of three methods to determine the friction drag (at α = -0.18°) and found differences of less than 8%. They used the measured skin friction from the Stanton tube, estimated the surface friction from G. I. Taylor`s form of the momentum equation for the boundary layer, and subtracted the form drag (pressure drag) from the total drag estimated from the total-head loss in the wake.

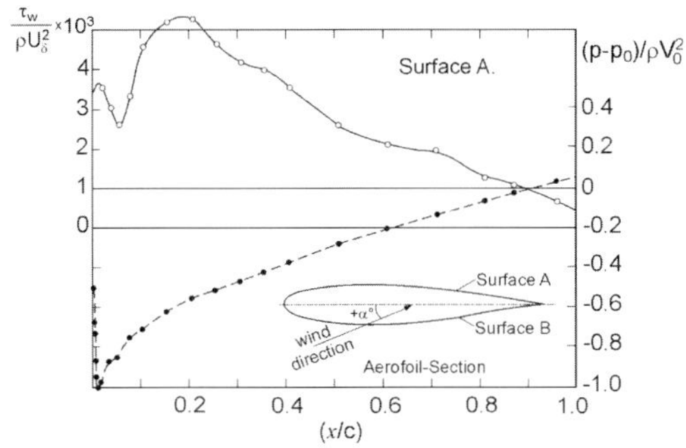

Figure 1. Distribution of skin-friction(Stanton tube) and pressure coefficient in the boundary layer along a symmetrical aerofoil of the Joukowski type at α=5.82° and U=60ft/s. From [5].

Fage & Falkner [6] also applied the Stanton-tube technique to measure skin friction on the surface of a circular cylinder. They confirmed the separation point of the laminar boundary layer at a circumferential location $\varphi \approx 82°$ (fig. 2), already determined by Hiemenz [7] from flow visualization. In a second test at a higher Reynolds number skin-friction was measured through the transition region, showing a second smaller peak in the turbulent boundary layer downstream, followed by turbulent separation. If, at the same Reynolds number, the free-stream turbulence was increased, the separation point moved further downstream.

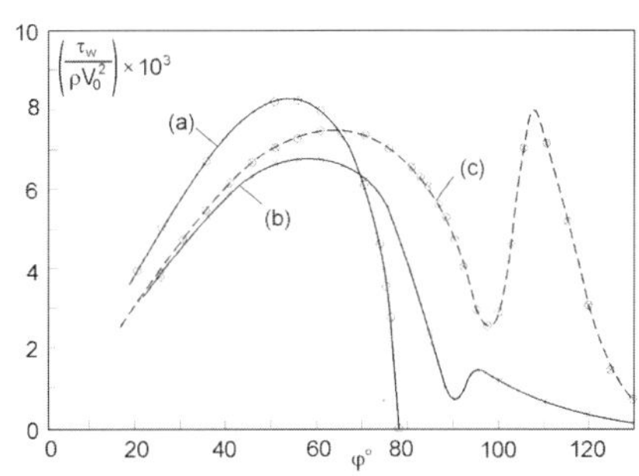

Figure 2. Skin friction distribution on a circular cylinder (Stanton tube) at two Reynolds numbers: (a) Re $=1.06\text{x}10^5$ laminar, (b) Re $=1.66\text{x}10^5$ turbulent, (c) Re $=1.68\text{x}10^5$ (with upstream grid). From Fage & Falkner [6].

In 1931 Gruschwitz [8], a doctoral student of Prandtl's, published measurements in turbulent boundary layers with a strong APG. The first investigation was performed on the suction side of a Göttingen 387 aerofoil at angles of incidence α up to 15°. For α = 12° separation was diagnosed just upstream of the trailing edge. The second rig was the test section of a wind tunnel with a plane test wall and an opposite wall whose contour could be varied in order to set up variable pressure distributions in streamwise direction (fig. 3a). Separation was estimated by inspection of the measured velocity profiles (fig. 3b).

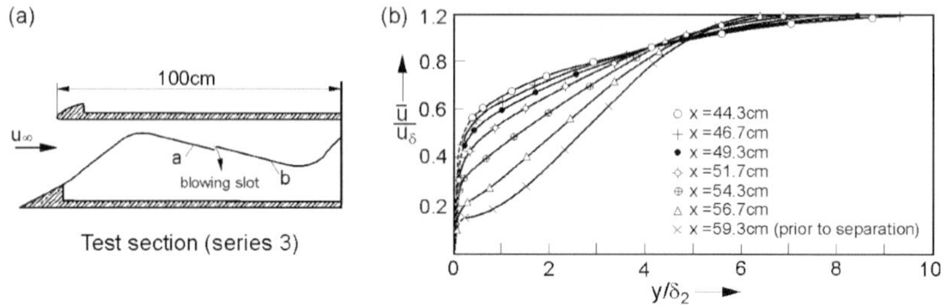

Figure 3. Test section and mean velocity profiles in an APG turbulent boundary layer with separation. From Gruschwitz [8].

The calculation of c_f from the momentum equation resulted in rising values of the skin friction in the APG. The reasons for this trend remained unclear although the author remarked that c_f should be zero at separation. Later experiments by Wieghardt [9] and Tillmann [10], where the skin-friction was evaluated by means of a version of the momentum equation given by Betz, confirmed this unusual behaviour of c_f. Here it appears appropriate to quote Bradshaw & Gregory [11]: "The history of the use of the von Kármán momentum integral equation for skin-friction determination is not a happy one. Slight departures from two-dimensional or axisymmetric flow may produce highly inaccurate estimates of skin friction, particularly in an APG where the skin friction coefficient is the small difference of two large terms in the equation".

3. THE PERIOD BETWEEN 1949 AND 1976

In 1949 Ludwieg [12] designed a heated-element skin-friction gauge using the analogy between skin friction and heat transfer. Ludwieg and Tillmann [13] performed measurements in three turbulent boundary layers, two with mild adverse and one with a favourable pressure gradient. Here skin friction was shown to decrease in an APG and to lie slightly above ZPG values in the FPG case (fig. 4). For this range of streamwise pressure gradients they derived their well known semi-empirical relationship

$$c_f = f(\mathrm{Re}_{\delta_2}, H_{12}).$$

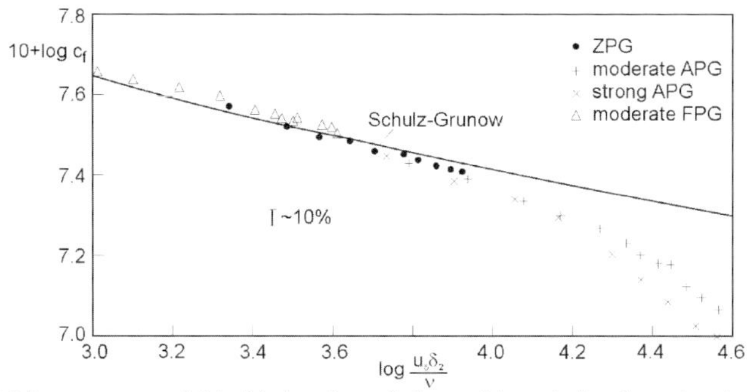

Figure 4. Measurements of skin friction (heated element) in turbulent boundary layers with 4 pressure distributions. From Ludwieg & Tillmann[13].

At the 1968 "Stanford Olympics" there were only five test cases of boundary layers where skin friction had been actually measured: the three cases by Ludwieg & Tillmann [13] and two mild APG cases 2500 and 3300 where Bradshaw [14] had measured skin friction by means of a Preston tube.

Preston's method [15] of measuring turbulent skin friction makes use of a circular Pitot tube resting on the wall and depends on the assumption of a near-wall region of flow similarity common to fully developed pipe flow and turbulent boundary layers. Instead of the mean velocity at a point close to the wall Preston used the pressure difference Δp recorded by a Pitot tube with diameter d on the wall and the local static pressure and obtained from dimensional analysis the relationship

$$\frac{\tau_w d^2}{4\rho v^2} = h(\frac{\Delta p_p d^2}{4\rho v^2})$$

which can be determined from experiments in fully developed pipe flow (see [16] and [17] for the calibration method and curve). Patel [17] found that in severe favourable and adverse pressure gradients the Preston tube overestimated the skin friction, the reason being that the logarithmic law of the wall no longer held under severe APG conditions (see also [18] and [19]).

Mc Allister et al. [20] compared Preston tube measurements with direct force measurements and found that the Preston tube readings were slightly lower ($\approx 4\%$) which is consistent with the results of Brown & Joubert [18].

Head and Rechenberg [16] compared measurements using Preston tubes, Stanton tubes and a surface fence in fully developed pipe flow and in a turbulent boundary layer. They found excellent agreement between the three measuring techniques if the data were taken at exactly the same location in the flow.

The surface fence was first described by Konstantinov & Dragnysh [21]. The measured quantity is the pressure difference upstream and downstream of a razor-blade fence protruding from the wall and situated normal to the flow direction in the viscous sublayer. This device needs no additional static pressure tapping, gives a reading almost double that of a Stanton tube of the same height, is independent of the logarithmic law, i.e. also more independent of disturbances in the turbulence structure [22], and can measure the magnitude and the sign of the skin-friction vector. These latter two properties were used to measure the skin friction in front of a step (fig. 5a+b) in the spanwise [23] and streamwise directions [24]. These two investigations were a major but practically unnoticed step forward in the investigation of separated boundary layers.

Measurements in a Stratford type flow [25], i.e. with nominally zero wall shear stress, showed an important limitation of the surface fence in that it cannot measure the skin-friction at the separation point itself but only in its vicinity.

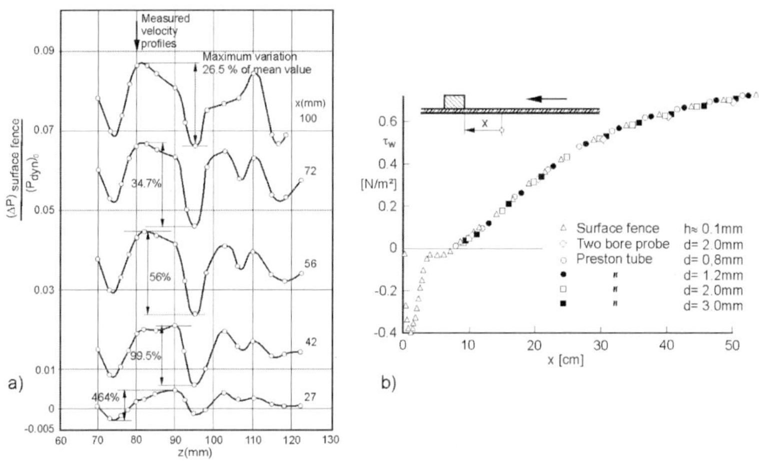

Figure 5. Skin friction distributions in front of a two-dimensional obstacle. a) Spanwise distribution from Fernholz [23] b) Streamwise distribution from Rechenberg [24].

4. THE PERIOD BETWEEN 1976 AND 2004

During this more recent period five "laboratory" measuring-techniques (fig 6) for skin friction have been developed and compared with each other in independent investigations (e.g. Gasser et al. [26], Fernholz et al., [27]).

(a) Direct measurement methods: Floating element balance for large pressure gradients and oilfilm interferometry.

(b) Measuring techniques for mean and fluctuating skin friction: the wall hot-wire probe, the wall pulsed-wire probe, and the MEMS surface friction gauge.

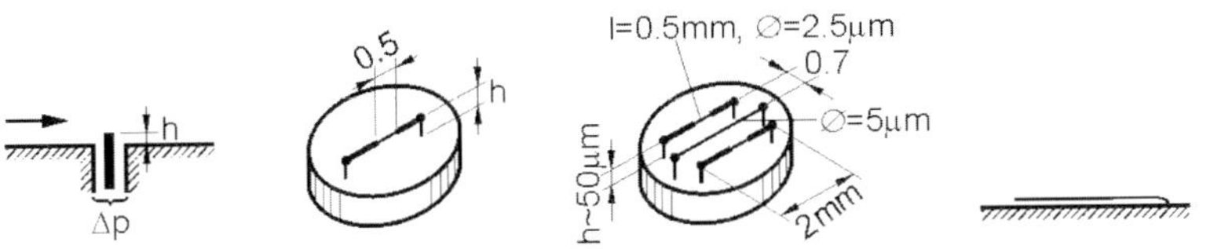

Figure 6. Skin friction measuring techniques.

In his excellent survey Winter [28] discussed the then only direct skin-friction measuring device, the floating element balance, and its problems when used in a pressure gradient. These problems were overcome by Frei & Thomann [29] and Hirt et al. [30] who sealed the gap between the floating element and the wall with a liquid held in place by surface tension. Hirt & Thomann [19] compared floating element and Preston tube data in very strong adverse pressure gradients and found that the Preston tube calibration leads to considerable errors (larger than 10%) for boundary layers of this type.

Oilfilm interferometry determines the skin friction from the movement of interference fringes of a thin oil film as realized by Tanner & Blows [31]. It has a high spatial resolution and is capable of measuring mean values of skin friction in forward and reverse flow. It does not require any assumptions to be made concerning the flow field, is easy to apply and needs little instrumentation. The full theory and details of the application of this technique have been described for example by Monson [32] and Janke [33].

Reichard in 1945 (private communication by H. Eckelmann) suggested the use of a hot wire in wall proximity to measure the skin friction. This technique was first applied by Bradshaw and Gregory [11]. For details of the calibration of wall hot wires see Wagner [34] and Fernholz et al. [27]

The wall pulsed-wire probe was first used by Ginder & Bradbury [35] and described in detail by Castro et al. [36] and by Dengel et al. [37].

Comparisons of the two direct methods with wall pulsed-wire and surface fence were performed in a weak reverse-flow region with low negative skin-friction and in a strong reverse-flow region with high negative skin friction [38].

Gasser et al. [26] performed measurements in several APG distributions in the Zürich boundary-layer wind tunnel using the floating element balance and the Berlin pulsed-wire and surface-fence probes. Dengel & Fernholz [39] and Driver [40] measured the wall shear stress distributions near and in weak separation regions using different measuring techniques.

Comparisons of surface-fence, wall pulsed-wire and oilfilm-anemometry measurements of skin-friction are shown in strong reverse-flow regions. The first separation region was generated by a normal flat plate with a long splitter plate (fig. 7a) whereby the oncoming flow separates at the sharp edge of the normal plate and reattaches on the splitter plate [37].

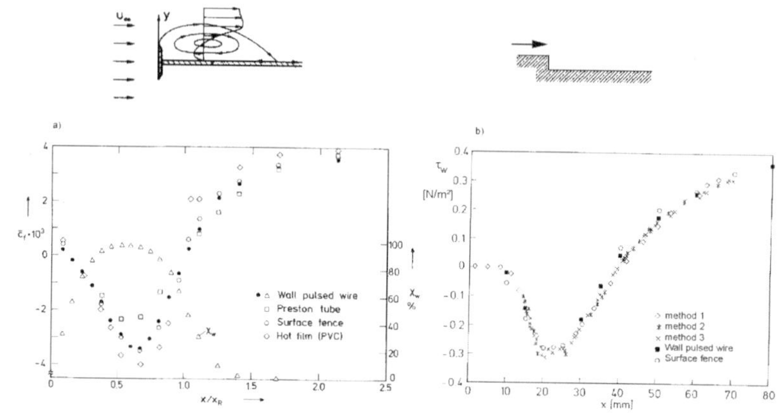

Figure 7. Comparison of four techniques for skin friction measurements in a wall-bounded turbulent shear-layer with a strong reverse-flow region. a) Normal plate with a splitter plate, from Dengel et al. [37], b) Backward facing step, from Janke [33].

The second strong reverse-flow region (fig. 7b) is downstream of a straight backward-facing step [33], and here the wall pulsed-wire and the surface-fence data were compared with those of the oilfilm interferometry (three evaluation methods).

Finally figure 8 shows a comparison between measurements with a micro-fence sensor and a wall pulsed-wire gauge downstream of a fence [41].

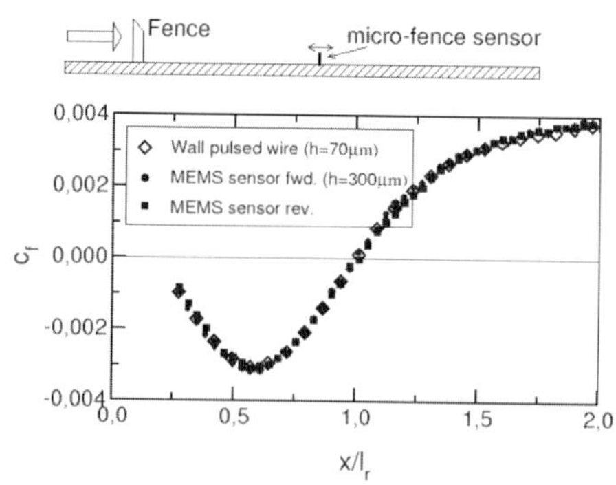

Figure 8. Comparison of skin friction measurements by a wall pulsed wire and a MEMS surface fence in a strong reverse-flow region (data from Schober et al. [41]).

Here the micro fence is made from Silicon (MEMS) and the deflections due to aerodynamic forces on the micro fence are measured by four integrated piezo resistors. This device can also determine the fluctuating wall shear stress in reverse-flow regions.

In strong FPG boundary layers three techniques were compared (Fernholz & Warnack [42]) and the data from oilfilm interferometry, wall hot-wire, and surface fence agree remarkably well except in the relaminarization region where the surface fence shows values which are too high.

REFERENCES

1. Prandtl L. "Über Flüssigkeitsbewegung bei sehr kleiner Reibung", *Verh. des III. Int. Mathematiker-Kongresses*, Heidelberg, 1904, Teubner Verlag Leipzig, pp. 484-451, 1905.
2. Stanton TE. "The mechanical viscosity of fluids" *Roy. Soc. Proc. A* **85**, pp. 366-376, 1911.
3. Stanton TE, Marschall D (Miss), Bryant CN (Mrs.). "On the conditions at the boundary of a fluid in turbulent motion", *Proc. Roy. Soc. A* **97**, pp. 413-434, 1920.
4. Taylor GI "Skin friction on a flat surface", *Rep & Memo ACA No 604 (Scientific papers* **Vol.2**, pp. 102-112), 1918.
5. Fage A, Falkner VM. "An experimental determination of the intensity of friction on the surface of an aerofoil" *Proc. Roy. Soc London A* **129**, pp.378-410 (and R+M 1315),1930.
6. Fage A, Falkner VM. "Further experiments on the flow around a circular cylinder", *Rep & Memo 1369 ARC*, 1931.
7. Hiemenz K. "Die Grenzschicht an einem in den gleichförmigen Flüssigkeitsstrom eingetauchten geraden Kreiszylinder", *Dingler Polytechn. J.*, **326**, p. 321, 1911.
8. Gruschwitz E. "Die turbulente Reibungsschicht in ebener Strömung bei Druckabfall und Druckanstieg" *Ingenieur Archiv.* **H11**, pp. 321-346, 1931.
9. Wieghardt K. "Über die Wandschubspannung in turbulentern Reibungsschichten bei veränderlichem Außendruck", *Aerodyn. Versuchsanstalt 43*, **P 03**, 1943.
10. Tillmann W. "Untersuchungen über Besonderheiten bei turbulenten Reibungsschichten an Platten" *UM 6627* (1945) und *Völkenrode Report VG 34-45T* (1946).
11. Bradshaw P, Gregory N. "The determination of local turbulent skin friction from observations in the viscous sub-layer", *ARC London R+M 3202*, 1959.
12. Ludwieg H. "Ein Gerät zur Messung der Wandschubspannung turbulenter Reibungsschichten" *Ing.-Arch.*, **17**, pp. 207-218, 1949.
13. Ludwieg H., Tillmann W. "Untersuchungen über die Wandschubspannung in turbulenten Reibungsschichten", *Ing.-Arch.*, **17**, pp. 288-298, 1949.
14. Bradshaw P. "The turbulence structure of equilibrium boundary layers", *NPL Ae Rep. 1184*, 1966.
15. Preston JH. "The determination of turbulent skin friction by means of Pitot tubes", *J. Roy. Aero. Soc,* **58**, pp. 109-121, 1954.
16. Head MR, Rechenberg I. "The Preston tube as a means of measuring skin friction", *J.Fluid Mech.*, **14**, pp. 1-17, 1962.
17. Patel VC. "Calibration of the Preston tube and limitations on its use in pressure gradients", *J. Fluid Mech.*, **23**, pp. 185-208, 1965.
18. Brown KC, Joubert PN. "Measurements of skin friction in turbulent boundary layers with adverse pressure gradients", *J. Fluid Mech.*, **35**, pp. 737-757, 1967.
19. Hirt F, Thomann H. "Measurements of wall shear stress in turbulent boundary layers subject to strong pressure gradients", *J. Fluid Mech.*, **171**, pp. 547-562, 1986.
20. McAllister JE, Pierce FJ, Tennant MH. "Preston tube calibrations and direct force floating element measurements in a two-dimensional turbulent boundary layer", *Trans. ASME J. Fluids Eng.*, **104**, pp. 156-161, 1982.

21. Konstantinov NI, Dragnysh GL. "The measurement of friction stress on a surface", English translation, *DSIR RTS 1499*, 1960.
22. Kiske S, Vasanta Ram V., Pfarr K. "The effect of turbulence structure on the Preston-tube method of measuring wall shear stress", *Aeronaut. Quartely*, **31**, pp. 354-367, 1981.
23. Fernholz HH. "Three-dimensional disturbances in a two-dimensional incompressible turbulent boundary layer", *ARC R+M 3368*, London, 1964.
24. Rechenberg I. "Turbulente Grenzschichten und turbulente Oberflächenreibung", *Int. Bericht,* Hermann-Föttinger Institut für Strömungstechnik, TU Berlin, 1964.
25. Fernholz HH. "Experimentelle Untersuchung einer inkompressiblen turbulenten Grenzschicht mit Wandreibung nahe null an einem längs angeströmten Kreiszylinder", *Z. Flugwiss.,* **16**, pp. 401-406, 1968.
26. Gasser D., Thomann H., Dengel P. "Comparison of four methods to measure the wall shear stress in a turbulent boundary layer with separation", *Exp. Fluids*, **15**, 27-32, 1993.
27. Fernholz HH, Janke G, Schober M, Wagner PM & Warnack D. "New developments and applications of skin-friction measuring techniques", *Meas. Technol.*, **7**, 1396-1409, 1996.
28. Winter KG. "An outline of the techniques available for the measurement of skin friction in turbulent boundary layers", *Progr. Aerospace Sci.,* **18**, pp. 1-57, 1979.
29. Frei D., Thomann H. "Direct measurements of skin friction in a turbulent boundary layer with a strong adverse pressure gradient", *J. Fluid Mech.,* **101**, pp. 79-95, 1980.
30. Hirt F, Zurfluh UE, Thomann H. "Skin-friction balances for strong pressure gradients", *Exp. Fluids,* **4**, pp. 296-300, 1986.
31. Tanner L, Blows L. "A study on the motion of oil films on surfaces in air flow, with application to the measurement of skin friction" *J. Phys. E: Sci. Instrum.*, **9**,194-202, 1976.
32. Monson DJ. "A nonintrusive Laser interferometer method for the measurement of skin friction", *Exp. Fluids,* **1**, pp. 15-22, 1983.
33. Janke G. "Über die Grundlagen und einige Anwendungen der Ölfilm-Interferometrie zur Messung von Wandreibungsfeldern in Luftströmungen", *Dissertation,* Technische Universität Berlin, 1993.
34. Wagner PM. "Kohärente Strukturen der Turbulenz im wandnahen Bereich von Ablösegebieten", *Dissertation*, Technische Universität Berlin, 1995.
35. Ginder RB, Bradbury LJ. "Preliminary investigation for skin friction measurements in highly turbulent flows", *ARC Report*, 1973.
36. Castro IP, Dianat M, Bradbury LJS. "The pulsed-wire skin-friction measurement technique" *Proc. 5th Symp. On Turbulence and Shear Flows,* 1987.
37. Dengel P, Fernholz HH, Hess M. "Skin-friction measurements in two-and three-dimensional highly turbulent flows with separation", *Advances in Turbulence,* **40**, ed. G. Compte-Bellot and J. Mathieu (Berlin: Springer), pp. 470-479, 1987.
38. Fernholz HH. "Near-wall phenomena in turbulent separated flows", *Acta Mechanica,* **4**, pp. 57-67, 1994.
39. Dengel P, Fernholz, HH. "An experimental investigation of an incompressible turbulent boundary layer in the vincinity of separation", *J. Fluid Mech.*, **212**, pp. 615-636, 1990.
40. Driver DM. "Reynolds shear stress measurements in a separated boundary layer flow", *AIAA-Paper 91-1787*, 1991.
41. Schober M, Obermeier E., Pirskawetz S, Fernholz HH. "A MEMS skin-friction sensor for time resolved measurements in separated flows", Exp. Fluids, **36**, pp. 593-599, 2004.
42. Fernholz HH, Warnack D. "The effects of a favourable pressure gradient and of the Reyolds number on an incompressible axisymmetric turbulent boundary layer. Part 1 The turbulent boundary layer", *J. Fluid Mech.*, **359**, pp. 329-356, 1998.

THE MEAN VELOCITY DISTRIBUTION NEAR THE PEAK OF THE REYNOLDS SHEAR STRESS, EXTENDING ALSO TO THE BUFFER REGION

K.R. Sreenivasan and A. Bershadskii
International Centre for Theoretical Physics, Strada Costiera 11, 34014 Trieste, Italy; krs@ictp.trieste.it

Abstract: An expression is derived for the mean velocity distribution in pipe and channel flows near the position of the maximum Reynolds shear stress, y_m. This expression agrees well with measurements in a significant region on both sides of y_m, extending to the buffer region on the one hand and almost all the way to the centerline of the flow on the other.

Key words: mean velocity, peak Reynolds shear stress, buffer region.

1. INTRODUCTION

Close to the surface in wall-bounded flows such as pipes, channels and boundary layers, the mean velocity varies linearly with wall-normal distance [1]. Further away from the surface, the traditional understanding has been that the variation is logarithmic [2]. Recent work on this same issue [3] has proposed power law variation as more appropriate (though the suggestion of an empirical power-law fit goes back to Prandtl and his students). Even further out in the flow, the so-called wake function [4] is thought to codify experimental data.

This article does not elucidate the work of Refs. [1-4] directly. We merely use a simple tool to construct an explicit expression for the distribution of the mean velocity near the position of maximum Reynolds shear stress, y_m. The key idea is the logarithmic power series expansion around y_m. Such an expansion is usually suitable when there is some long-range interaction in

G.E.A. Meier and K.R. Sreenivasan (eds.), IUTAM Symposium on One Hundred Years of Boundary Layer Research, 241-246,

the problem, as seems to be true for turbulent flows in general—wall-bounded flows in particular. The validity of the expression so derived extends, on one side of y_m, to the so-called buffer region that exists between the end of the linear region and the beginning of the log/power region. The buffer region is of great importance because of its participation in the turbulence generation mechanism, yet there exist no explicit expressions for the velocity distribution in this region—especially those based on broadly applicable physical principles. On the other side of y_m, the expression seems to be valid almost all the way to the centerline of the flow.

2. ANALYSIS

Let us start from the exact equation valid for pipes and channels

$$-\langle uv\rangle^+ = -dU^+/dy^+ + (1 - y^+/R^+), \tag{1}$$

in which we have used the standard notation: u and v are velocity fluctuations in streamwise and wall-normal directions x and y respectively, U(y) is the mean velocity in the direction x, R is the pipe radius or the channel half-height, and the suffix + indicates normalization by wall variables u_τ and ν, which represent, respectively, the friction velocity and the fluid viscosity. Elementary consideration show that the turbulent stress term $-\langle uv\rangle^+$ increases cubically with y very close to the wall; it changes rapidly into a different form that has not been studied carefully so far before attaining a maximum value in the flow; it subsequently drops off to zero as the flow centreline is approached further outwards [5]. The position of the maximum in the Reynolds shear stress, y_m, is empirically known [6] to obey

$$y_m^+ \approx 1.87\, R_\tau^{1/2}, \tag{2}$$

where the Reynolds number $R_\tau \equiv u_\tau R/\nu$. This fit has been proposed by others as well [7]. Though the multiplicative constant is slightly different in each work, this ambiguity merely reflects the uncertainty associated with the identification of y_m from measured data and is not fundamental. It should be stressed that the distribution of $-\langle uv\rangle^+$ has been obtained by numerically differentiating the measured mean velocity distribution and using (1), and so is not dependent on the inaccuracies that usually plague the Reynolds shear stress measurements.

Let us expand $-\langle uv\rangle^+$ around y_m^+. We had undertaken this exercise already in [6] but had not appreciated the importance of expanding $-\langle uv\rangle^+$

in terms of the logarithm of the distance from y_m^+. It now seems to us that this is the appropriate expansion to make, considering that the Reynolds shear stress varies slowly in the region around its peak. In general, expansions in logarithmic variables are appropriate whenever long-range effects are present, as is the case in wall-bounded flows. We may then write

$$-\langle uv\rangle^+ = c[1-\alpha_1\{\ln(y^+/y_m^+)\}^2 + \ldots + \alpha_n\{\ln(y^+/y_m^+)\}^n + \ldots]. \quad (3)$$

Here, the unknown constants $\alpha_1 \ldots \alpha_n$ are thought to be independent of the Reynolds number, at least when it is high enough. The fit works very well for all Reynolds numbers shown in figure 1, roughly for $y^+ \geq 10$. This region more or less borders the buffer region.

Substituting (3) in (1), and retaining only the first two terms in the expansion (3), we obtain

$$U^+ = \text{const} + y^+ g(y^+) - (y^{+2}/2R^+), \quad (4)$$

where

$$g(y^+) = a_0 + a_1\,[\ln\,(y^+/y_1)]^2, \quad (5)$$

$$\text{with } a_0 = 1 - c + c\alpha_1,\ a_1 = c\alpha_1,\ y_1 = ey_m^+. \quad (6)$$

The expression (4) is technically not expected to be valid all the way to the wall (see figure 1), but we can be somewhat rough and impose the no slip condition $U^+ = 0$ at $y^+ = 0$ to obtain

$$U^+ = y^+[g(y^+) - y^+/2d^+], \quad (7)$$

where $d^+ = 2R^+$. In order to compare the last equation directly with experimental data, it is useful to rewrite it in the form

$$U^+/y^+ + y^+/d^+ = g(y^+) = a_0 + a_1\,[\ln\,(y^+/y_1)]^2. \quad (8)$$

If the present considerations are valid, the left hand side of (8) must show a parabolic variation with respect to y^+ in logarithmic coordinates.

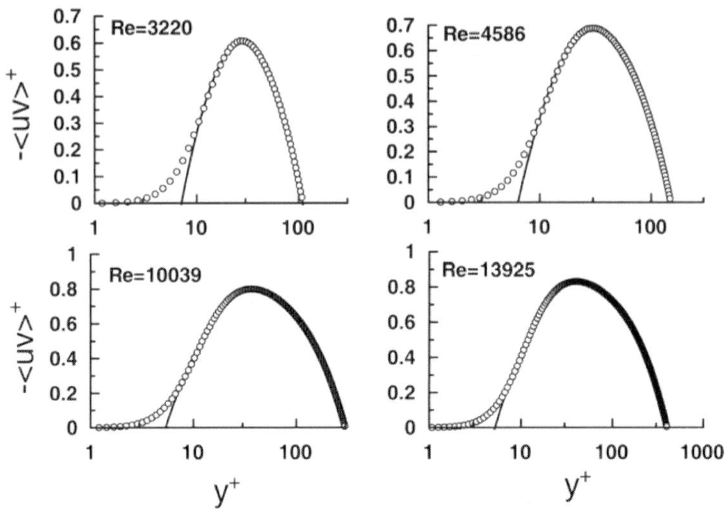

Figure 1. Plots of the Reynolds shear stress from the direct numerical simulations of a channel flow [8], for four different Reynolds numbers, Re, based on the bulk mean velocity and the width of the channel. The data have been fitted by the two term expansion of (3). The fit is very good for $y^+ > 10$.

3. COMPARISON WITH MEASUREMENTS

We show in figures 2 and 3 the recent Princeton data [9] for two Reynolds numbers. The solid parabolas are drawn in order to compare the data with equation (8) in the semi-logarithmical scales. The agreement with the data is excellent almost all the way to y^+ of the order 10 towards the wall, and to y^+ of the order 1000 or more outwards—in fact, almost all the way to the centerline.

4. CONCLUSIONS

In the traditional picture, the Reynolds shear stress attains a constant value of unity, this being the fundamental factor leading to the logarithmic law. That one can identify a maximum value from the distribution of $-\langle uv\rangle^+$ is a reflection that the expected constancy does not obtain at least up to the Reynolds number for which (2) holds. It may be that the relation that holds

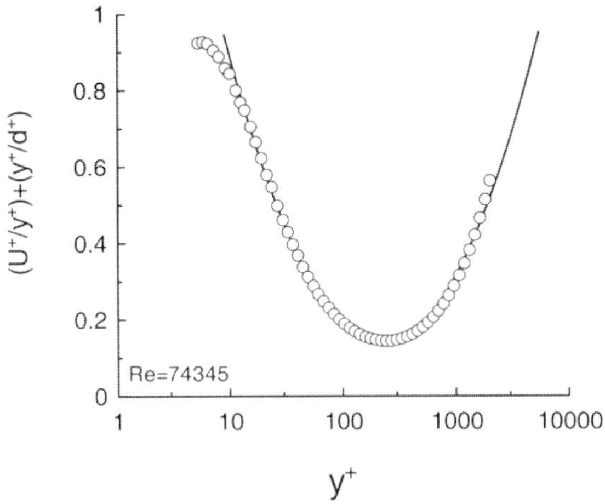

Figure 2. A plot of $U^+/y^+ + y^+/d^+$ against y^+ in semi-log scales (circles), for Re = 74,345, where Re is based on the mean velocity and the pipe diameter. The data are from [9]. The solid parabola indicates correspondence to equation (8).

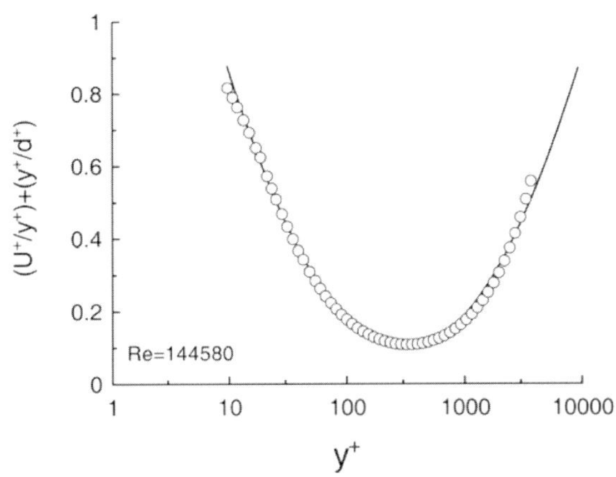

Figure 3. As in figure 2, but for Re = 144,580. Again, the data are from [9].

for "low" Reynolds numbers—in which case the present considerations hold only that range of Reynolds numbers. This possibility is equivalent to the scenario in which y_m^+ remains unchanged beyond a certain Reynolds number. On the other hand, if a maximum can indeed be identified at all Reynolds numbers, this feature has to be taken into account in some way. Such considerations were the subject of [6].

We ourselves view equation (8) as a good fit to the mean velocity data in the buffer region, possibly much further outwards. Whether the proposal is fundamental depends on the status of the logarithmic expansion (3) that we have used. At present, it is hard to resolve the question satisfactorily.

The analysis is strictly valid for only pipe and channel flows (because of (1)), but we expect that it would be valid for constant-pressure boundary layers as well.

REFERENCES

1. Laufer, J. "The Structure of Turbulence in Fully Developed Pipe Flow", *NACA Report* **1174**, 1954.
2. Prandtl, L. *Essentials of Fluid Mechanics,* N.Y. Hafner Publishing Co., New York, 1952.
3. Barenblatt, G.I. "Scaling Laws for Fully Developed Turbulent Shear Flows", *J. Fluid Mech.,* **248**, pp. 513-520, 1993.
4. Coles, D.E., Hirst E.A. *Computation of Turbulent Boundary Layers: 1968 AFOSR-IFP-Stanford Conference Proceedings,* Thermosciences Division, Stanford University, 1969.
5. Sreenivasan, K.R. "A Unified View of the Origin and Morphology of the Turbulent Boundary Layer Structure", in *Turbulence Management and Relaminarisation* (eds. H.W. Liepmann and R. Narasimha), pp. 37-61, Springer-Verlag, 1984.
6. Sreenivasan, K.R., Sahay, A. "The Persistence of Viscous Effects in the Overlap Region, and the Mean Velocity in Turbulent Pipe and Channel Flows", in *Self-Sustaining Mechanism of Wall Turbulence* (ed. R.L. Panton), pp. 253-271, Computational Mechanics Publications, Southhampton and Boston, 1997.
7. Long, R.R, Chen, T.C. "Experimental Evidence for the Existence of Mesolayer in Turbulent Systems", *J. Fluid Mech.*, **105**, pp. 19-59, 1981; Sreenivasan, K.R. "The Turbulent Boundary Layer", in *Frontiers of Experimental Fluid Mechanics* (ed. M. Gad-el-Hak), pp. 159-209, Springer-Verlag, Berlin, 1989.
8. K. Iwamoto, Y. Suzuki and N. Kasagai, "Fully Developed Two-Dimensional Channel Flow", *Int. J. Heat and Fluid Flow,* **23**, 678-, 2002 (Data Code Number CH12_PG.WL6, University of Tokyo).
9. McKeon, B.J., Li, J., Jiang, W., Morrison, J.F., Smits, A.J. "Further Observations on the Mean Velocity Distribution in Fully Developed Pipe Flow", *J. Fluid Mech.* **501**, pp. 135-147, 2004.

TURBULENCE MODELLING FOR BOUNDARY-LAYER CALCULATIONS

Wolfgang Rodi
Institute for Hydromechanics, University of Karlsruhe, Kaiserstr. 12, 76128 Karlsruhe, Germany, e-mail: Rodi@uka.de

Abstract: The paper gives a review of the historical development of turbulence models as used in boundary layer calculations, starting with Prandtl's mixing-length model and ranging to the models currently in use. The interrelations between the main models that have evolved over the years are provided as well as an assessment of their capabilities.

Key words: boundary layers, turbulence models, historical development

1. INTRODUCTION

The concept of the boundary layer, i.e. a thin layer where viscous effects are important in an otherwise inviscid outer flow as introduced by Prandtl in 1904 is of great consequence for flow calculations. Indeed, boundary-layer calculations are an important component of most flow calculations in engineering but also in meteorology. Boundary layers develop on external but also internal surfaces and are of practical importance with respect to their influence on the outer flow, the tendency of the flow to separate from boundaries, the friction, the heat and mass transfer and the turbulence and noise generation in the flow. Boundary-layer calculations can be carried out on their own with the outer flow prescribed, but also coupled with inviscid calculations for the outer flow, or, as has become common recently, be part of the solution of the Navier-Stokes equations for the entire flow field. The geometry of boundary layers is usually fairly simple, but complex flow phenomena occur such as transition and multi-scale turbulent motions with sweeps and ejections near the wall, and the flow may be subject to influences like pressure gradients which may cause separation, free-stream

G.E.A. Meier and K.R. Sreenivasan (eds.), IUTAM Symposium on One Hundred Years of Boundary Layer Research, 247-256,

turbulence, curvature, cross flow, suction/blowing, compressibility, forcing, etc. The details of the turbulent motion can nowadays be calculated by Direct Numerical Simulation (DNS) and to a lesser extent also by Large-Eddy Simulation (LES), but DNS and also well-resolved LES of the near-wall flow are restricted to fairly low Reynolds numbers; further, such calculations are very expensive. Hence, most practical calculations are still carried out by solving the Reynolds-Averaged-Navier-Stokes (RANS) equations in which the effect of all turbulent motions is represented by the turbulent (or Reynolds) stresses. These need to be determined by a turbulence model. In boundary layers it is mainly the turbulent shear stress that needs to be determined and this is hence the main task of a turbulence model in boundary-layer calculations. In this paper, a brief review of the historical development of turbulence models and their use in boundary-layer calculations is given, showing the links between the main models that have evolved over the years and providing an assessment of their capabilities. Attention is restricted to fully turbulent boundary layers as transition modelling is a subject area of its own, and also to 2D incompressible boundary layers along solid walls.

2. MIXING-LENGTH MODELS

Already in 1877, in analogy to the viscous stresses, Boussinsq proposed to relate the turbulent stresses to the mean velocity gradients via a turbulent or eddy viscosity. However, this is not a complete model because the eddy viscosity is not a fluid property but depends on the state of turbulence and hence must also be determined by the turbulence model. The first real turbulence model is the mixing-length hypothesis introduced by Prandtl in 1925 [1] and discussed in more detail in his 1926 papers [2,3]. Through dimensional analysis, Prandtl found the kinematic eddy viscosity ν_t to be proportional to the product of a velocity scale V and a length scale L of the turbulent motion:

$$\nu_t \propto VL \tag{1}$$

Considering thin shear flow he assumed as velocity scale the lateral fluctuations v′and as length scale the diameter of the turbulent fluid elements which he called fluid lumps. The name mixing length appeared first in his 1926 papers where he also describes that this is the length of the path of elements before they loose their identity by mixing with the environment, and he also mentions that it plays a similar role as the mean free path in the molecular diffusion in gases. He then argues that the average longitudinal fluctuation u' is equal to the velocity difference $\ell\left|\partial U/\partial y\right|$ which an element displaced by ℓ has in a layer with velocity gradient $\partial U/\partial y$. Next he assumes that the lateral fluctuations v′ are of the same order as the longitudinal ones so that the velocity scale in (1) is $\ell\left|\partial U/\partial y\right|$. With ℓ as length scale, this then leads directly to the following relations for the eddy viscosity and the turbulent shear stress τ in thin shear layers:

$$\nu_t = \ell^2 \left|\frac{\partial U}{\partial y}\right| \Rightarrow \tau = \ell^2 \left|\frac{\partial U}{\partial y}\right| \frac{\partial U}{\partial y} \tag{2}$$

which is Prandtl's mixing-length model. Prandtl further states that ℓ should be proportional to the flow width in free shear flows and to the wall distance in near-wall flows. In a paper published in 1933 [4] he puts $\ell = \kappa y$, where κ is a constant known as von Karman constant. Assuming further the shear stress τ to be constant he obtains the well-known log law for the velocity distribution. This interrelationship has later often been rederived and used. In the same paper, Prandtl also discusses the "similarity hypothesis" of von Karman and finds that this also yields the log law and hence in the log-law layer is entirely consistant with his mixing-length hypothesis.

The mixing-length hypothesis was later criticized because of the analogy drawn between the momentum transport by lumps of fluid and by molecules. Bradshaw [5] thought that Prandtl got this idea by watching surface photos of flows showing 2D structures, because this was all that could be visualized in these days. However, Prandtl made this remark about the analogy more in passing and clearly expressed the notion that the lumps do not keep their identity when mixing with surrounding fluid. Further, his basic idea that the length scale of turbulence is related to the diameter of the elements (now called eddies) and the velocity scale to the velocity gradient appears quite reasonable in many situations. As can be seen from the turbulent kinetic-energy equation (4) introduced later, turbulence is produced by velocity gradients, and in fact when local equilibrium between production P and dissipation ε prevails, a reduced form of (4) together with the eddy-viscosity relation (5) yields directly the mixing-length formula (2). This clearly shows the usefulness of the concept for certain flows, but also its limitations, namely that the model is suitable for near-equilibrium boundary layers but not for rapidly varying flows in which transport and history effects of turbulence are important and the state of turbulence is not determined directly by the local velocity gradient.

Some versions of the mixing-length model which have been used most in practical calculations are now discussed briefly. Very near the wall where viscous effects are important and the velocity deviates from the log law, the linear $\ell = \kappa y$ distribution yields too high ν_t so that this needs to be damped. This is done usually by introducing a multiplier due to van Driest

$$\ell = \kappa y\left[1 - e^{-y^+/A^+}\right], \qquad with\ y^+ = \left(\tau_w/\rho\right)^{1/2} y/\nu \tag{3}$$

where $A^+ = 26$ and damping is achieved through y^+ via viscous effects. Later it was found that the damping is not predominantly viscous but due to wall blocking of the normal fluctuations, but viscous damping is used in most models, also in the more elaborate ones discussed later. In some of the early boundary-layer methods [6,7], ramp functions were used for ℓ with ℓ from (3) in the inner region (A^+ can depend on various parameters) and $\ell \propto$ boundary-layer thickness δ in the outer region.

A popular model still employed sometimes is due to Cebeci and Smith [8] which actually uses the mixing-length hypothesis only in the inner layer (with van Driest damping including a pressure-gradient effect) while in the outer layer the model relates ν_t directly to a velocity and length scale characteristic for this layer, which is the velocity U_e at the edge of the boundary layer and the displacement thickness δ^*, a proposal already introduced by Clauser in 1956. In the outer layer ν_t is further modified by multiplying it with an intermittency function of y/δ introduced in 1956 by Klebanoff. The matching of inner and outer relations usually occurs in the log layer. Because U_e and δ^* are difficult to determine when separation occurs, even if only locally, Baldwin and Lomax [9] proposed a different model for the outer layer in which they base the velocity- and length-scale determination not on the velocity distribution but on vorticity. Basically the velocity scale is taken as the maximum of wall distance times velocity gradient and the length scale as the distance from the wall where this occurs. Different scale relations are used for wake regions. This model has been used extensively in aerospace applications and has proven to be fairly robust and successful except in rapidly varying flows or when the separation regions become large.

In order to account for non-equilibrium effects, Johnson and King [10] introduced a velocity scale based on the maximum shear stress across the boundary layer and solve an ordinary differential equation for the streamwise development of this. The treatment of the inner layer is similar as in the Cebeci - Smith and Baldwin - Lomax models but with the velocity scale replaced as mentioned; the non-equilibrium feature comes in mainly through the outer-layer model whose core element is the transport equation for the maximum shear stress. As only an ordinary differential equation is solved for this, the model is sometimes referred as ½-equation model.

3. ONE-EQUATION MODELS

Prandtl realized the limited universality of his mixing-length approach and wanted to develop a model that determines the intensity (and hence velocity scale) of turbulence through a more universally valid method. In his 1945 paper [11] he defines the strength of turbulence through its kinetic energy $k = 1/2\left(u'^2 + v'^2 + w'^2\right)$ and derives a transport equation for k, but not from the exact equation for k as was done later by other authors, but in a rational way based mainly on dimensional reasoning. He put the rate of change of k equal to terms due to various mechanisms such as production, dissipation (which he called slackening of turbulence) and diffusion. As length scale he retained his mixing-length ℓ and obtained the following equation

$$\frac{Dk}{Dt} = \underbrace{\nu_t\left(\frac{\partial U}{\partial y}\right)^2}_{P} - \underbrace{c_D\frac{k^{3/2}}{\ell}}_{\varepsilon} + \underbrace{\frac{\partial}{\partial y}\left(\frac{\nu_t}{\sigma_k}\frac{\partial k}{\partial y}\right)}_{Diffusion} \tag{4}$$

ν_t follows from (1) with $V = k^{1/2}$ as velocity scale and ℓ as length scale:

$$\nu_t = \ell k^{1/2} \tag{5}$$

Prandtl calls (4) and (5) his new formula system for developed turbulence. This contains basically 2 constants which were in the paper determined from channel flow measurements. In 1942 Kolmogorov [12] had already proposed a similar system and this is why (5) is known as the Kolmogorov-Prandtl relation. However he had in addition introduced a length scale equation while Prandtl left the determination of ℓ open. Later, versions of Prandtl's model were used by various authors in calculations for the 1968 Stanford Conference on Computation of Turbulent Boundary Layers [13]. Again somewhat later, versions were developed [14,15] in which the length scales ℓ appearing in the eddy-viscosity relation (5) and in the dissipation term of (4) are the same only in the log layer $(= \kappa y)$ but are damped differently closer to the wall with a van Driest relationship. These models were quite successful in calculations of boundary layer, including those with adverse pressure gradient (APG), but because they require an empirical length-scale prescription they were not used widely as complete models but played a role later as near-wall submodels in so-called 2-layer models in which the outer region is calculated by a 2-equation model employing a length-scale equation [16].

Bradshaw et al [17] proposed a model that does not use the eddy-viscosity concept; rather they converted the k-equation into an equation for the shear stress τ by assuming $\tau = \rho a_1 k$, where a_1 is a constant (= 0.3). This assumption is not universally valid but good for boundary layers over a wide range. In their contribution to [13], the authors provided a plot of $\tau / \rho k$ versus production/dissipation of k based on average values observed from experiments in some thin shear layers and this plot (reproduced in Fig. 1) indicates that the parameter $\tau/\rho k$ changes mildly for such layers including free-shear flows and boundary layers, but of course goes to zero in isotropic grid turbulence. At the 1968 Stanford conference there was some controversial discussion on where the Kolmogorov-Prandtl relation (5) would be on this plot - with $\varepsilon = k^{3/2} / \ell$ it yields that $\tau/\rho k$ varies proportional to $(P/\varepsilon)^{1/2}$, i.e. has actually the same trend as and is not too far from Bradshaw et al's curve (see Fig. 1). Later, this relationship will play a role in the Menter SST model. Bradshaw et al's k-equation is somewhat different from (4) as an eddy-viscosity cannot appear. Rather, production P is shear stress times velocity gradient as in the exact k-equation and the diffusion flux is related to a convection velocity and a function of y/δ. The model was clearly one of the better ones at the 1968 Stanford conference [13], but did never become very popular, partly because the assumption $\tau = \rho a_1 k$ is not general enough and is in particular not applicable to flows with velocity maxima as they occur in channels, pipes, jets, wakes.

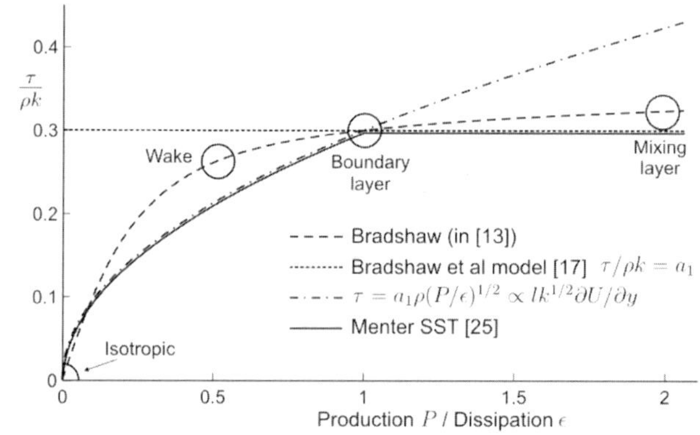

Figure 1. Relation τ/ρk versus P/ε for various models

There is another type of one-equation model which is not based on the k-equation but determines ν_t directly from an empirical transport equation for this quantity. Nee and Kovasnay performed calculations with such a model for the 1968 Stanford conference, but the model was not used further. The idea of a ν_t-equation was taken up again in 1990 by Baldwin and Barth who derived such an equation from the k-ε 2-equation model. Spalart and Allmaras [18] found this derivation too restrictive and devised an entirely empirical ν_t-equation. They proceeded step by step, first considering free shear flows and devising the source term and diffusion terms, but no destruction term; such a term was then introduced for the near-wall region, bringing in the wall distance and a function f_w which is unity in the log layer and decreases in the outer region. The near wall version of the model was first introduced and calibrated for the log- and outer region and then finally adjusted to account for viscous effects very near the wall. Further modifications were introduced for transitional boundary layers. The model does not need a length-scale specification other then the wall distance in near-wall flows. It may not be so suitable for massively separated flows but has been found to give good results for flows with small separation regions and for APG boundary layers. It has therefore become a popular model in aerospace and turbomachinery applications and is used extensively there. As it is based on the concept of a scalar eddy-viscosity, it is not so suitable for complex turbulent flows with large anisotropy effects.

4. 2-EQUATION MODELS AND BEYOND

Because of the difficulties with a general prescription of the length-scale distribution necessary in older one-equation models, there was a move to 2-equation models in which also the length-scale L, in addition to the velocity scale $k^{1/2}$, is determined from a model transport equation. These are the simplest "complete" models that need no prescription of quantities inside the calculation domain but only a specification of boundary conditions. The dependent variable of the length-scale determining equation can be any combination of k and L. In the late 60's and early 70's, some investigations have been carried out at Imperial College, London, with Rotta's [19] kL-equation, but this requires an extra term near walls to conform with the log law. Hence, in the early 70's there was soon a switch over to the $k-\varepsilon$ model which solves an equation for the dissipation rate $\varepsilon \propto k^{3/2}/L$ for determining the length scale L. By choosing the constants appropriately, the model can be made consistent with the log law. The ε-equation, which must be considered largely empirical, is similar to the k-equation (4), with ε replacing k in the rate of change and diffusion terms and the source terms P and ε multiplied by ε / k and each by an empirical constant [20]. In the 70's also the k-ω model basically introduced already by Kolmogorov was revived mainly by Wilcox [21,22]. This solves an equation for $\omega \propto \varepsilon / k \propto k^{1/2} / L$ which may be considered as frequency of the energy containing fluctuations. The form of the ω-equation is similar to that of the ε-equation. For many years, the k-ε model was the dominant one, also at the 1981 Stanford Conference on Complex Turbulent Flows [23], and it was implemented in virtually all general-purpose CFD codes. In practical applications, it is mainly the standard high-Reynolds-number version that is used, together with wall functions [20]. In such calculations the viscous sublayer is not resolved but basically the velocity, k and ε at the first grid point placed in the log layer are related to the friction velocity U_τ (more refined versions are sometimes used [20]). Wall functions lean heavily on the assumption of a logarithmic velocity distribution and the validity of local equilibrium of turbulence (P=ε) at the first grid point. As these assumptions are not generally valid, alternative approaches have been introduced such as low-Reynolds number versions for resolving the near-wall layer or the coupling with a one-equation model in this region (two-layer approach [16]). In the low-Re versions of the kε- model, some of the constants are replaced by functions of ν_t / ν or y^+, bringing in viscous effects. These mainly damp ν_t à la van Driest, but as was discussed already, the physics of the damping is really different. A large variety of different low-Re models have been proposed; Patel et al [24] have reviewed the performance of the ones available in the early 80's, but after that many new ones have appeared, often making use of knowledge from DNS data.

The k-ω model does not need any damping functions - with simply the viscosity appearing in the diffusion terms it can be applied right to the wall, but it can also be used in conjunction with wall functions. All k-ε models were found to perform rather poorly in APG-boundary layers, producing too high wall friction and late separation or not yielding separation at all [25]. This was traced in [26] to the ε-equation which causes an excessive increase of the length scale L in APG flows which are in fact better predicted by one-equation models that prescribe L and also by the mixing-length hypothesis. In [26] it was found that this weakness is due to the fact that the constants were fixed such that the model is consistent with the log-law under zero-pressure gradient conditions. The k-ω model was found to perform consistently better for APG flows and since it does not require any special low-Re damping functions, it was found to be more suitable for simulating the inner region of boundary layers. On the other hand, the model has a very strong sensitivity to the (unknown) free-steam value of ω that needs to be specified at the boundary layer edge and for this reason it is difficult to apply to free shear flows. The k-ε model does not have this problem. This has led Menter [25] to propose a combined model in which the k-ω model is used in the inner near-wall region and the k-ε model (ε-equation converted into an ω-equation) in the outer region near the edge. The transition is controlled by a blending function. This is Menter's base line model - he went one step further and introduced his shear stress-transport (SST) model, based on the observation that the eddy viscosity relation (5) ($\nu_t = k/\omega$ in the k-ω model) yielding the relation $\tau/\rho a_1 k = (P/\varepsilon)^{1/2}$ overpredicts the shear stress in APG flows where the ratio $P/$ can be significantly larger than one. For such situations, Bradshaw's [17] simple assumption $\tau = \rho a_1 k$ yields a better behaviour and hence Menter proposes that this is used when $P/\varepsilon > 1$ (see Figure 1). This leads to the modified eddy-viscosity relation $\nu_t = a_1 k / \max(a_1\omega; \partial U/\partial y)$ which improves further the model predictions in APG flows. This Menter SST model is now widely used and has also been incorporated into many commercial CFD codes.

Another model that has recently become popular and does not involve near-wall damping functions is the V2F model of Durbin [27]. He introduces the lateral fluctuations v'^2 as velocity scale in the eddy-viscosity relation, adds an equation for these and brings in the non-local wall-blocking effects on v'^2 through an elliptic relaxation procedure. The model has not been widely tested for 2D boundary layers but has been applied with considerable success to more complex 3D flows involving also separation.

Already in the early 70's, models not using the eddy-viscosity concept but solving model equations for the individual Reynolds stresses (second moment closures) have been proposed and were tested in the meantime for a number of boundary layers. They are strong on accounting for transport effects and are clearly superior under special, more complex circumstances, but most of them use the ε-equation and therefore suffer from its weaknesses. Also, the models are more complex and numerically demanding and hence not much used in practical calculations. Intermediate between these and the 2-equation models are algebraic stress models and non-linear eddy-viscosity models which can account for anisotropy of turbulence and are hence superior to the simpler models in complex situations, but not really for the 2D boundary layers considered here.

5. CONCLUSIONS

With his ideas on the mixing-length hypothesis and the turbulent kinetic energy equation, Prandtl had a large impact on turbulence model development for many decades. Versions and derivatives of his mixing-length model have been used for many boundary layer calculations with considerable success. A wide variety of models has evolved since his early proposals, and judging from comparison studies in [25,28], the Menter SST and then the Spalart-Allmaras model appear to give the best overall performance in boundary-layer calculations. They are both eddy-viscosity models, and it seems that models at this level will continue to be used in practice for some time. LES will be the future, but even then the turbulence models discussed here will play an important role in the LES/RANS coupling methods that become popular.

REFERENCES

1. Prandtl L., *ZAMM*, **5**, pp. 136-139, 1925
2. Prandtl L., *Hydraulische Probleme*, VDI Verlag Berlin, pp. 1-13, 1926
3. Prandtl L., *Verhandlungen des II. Int. Kongresses für Technische Mechanik 1926*, Zürich: Füßli, pp. 62-75, 1927
4. Prandtl L., *Zeitschrift des VDI*, **77**, pp. 105-144, 1933
5. Bradshaw P., *Nature*, **249**, No. 5453, pp. 135-136, 1974
6. Patankar S.V., Spalding D.B., *Heat and Mass Transfer in Boundary Layers*, 2nd ed. Intertext, London, 1970
7. Crawford M. E., Kays W.M., Stanford Univ., Dept. Mech. Eng., Rept. HMT-23, 1975
8. Cebeci T., Smith A.M.O., *Analysis of Turbulent Boundary Layers,* Academic Press, New York, 1974
9. Baldwin B.S., Lomax H., AIAA Paper 78-257, 1978
10. Johnson D.A., King L.S., *AIAA J.*, **23**, No. 11, pp. 1684-1692, 1985
11. Prandtl L., *Nachr. Akad. Wiss., Göttingen, Math.-Phys.* Kl., pp. 6-19, 1945
12. Kolmogorov A.N., *Izv. Akad. Nauk. SSR, Seria fizicheska,* **VI**, No. 1-2, pp. 56-58, 1942
13. Kline S.J., Morkovin M.V., Sovran G., Cockrell D.J., *Computation of Turbulent Boundary Layers* - 1968 AFOSR - IFP-Stanford Conf., Vol. I, Stanford Univ., 1969
14. Wolfshtein M., *Int. J. Heat Mass Transfer*, **12**, pp. 301-318, 1969
15. Norris L.H., Reynolds W.C., Rept. No. FM-10, Stanford Univ., Dept. Mech. Eng., 1975
16. Rodi W., Paper AIAA - 91 - 0216, 1991
17. Bradshaw P., Ferriss D.H., Atwell N.P., *J. Fluid Mech*, **28**, pp.593-616, 1967
18. Spalart P.R., Allmaras S.R., *La Recherche Aèrospatiale*, **1**, pp 5-21, 1994
19. Rotta I.C., *Zeitschr. f. Physik*, **131**, pp. 51-57, 1951
20. Launder B.E., Spalding D.B., *Comp. Meth. in Appl. Mech. Eng.*, **3**, pp. 269-289, 1974
21. Wilcox D.C., Traci R.E., AIAA Paper 76-351, 1976
22. Wilcox D.C., *Turbulence Modeling for CFD*, DCW Industries, California, 1993
23. Kline S.J., Cantwell B.J., Lilley G.M., *1980 - 81 AFOSR-HTTM-Stanford Conference on Complex Turbulent Flows*, Stanford University, CA, 1981
24. Patel V.C., Rodi W., Scheuerer G., *AIAA J.*, **23**, pp. 1308-1319, 1985
25. Menter F.R., *AIAA J.*, **32**, pp. 1598-1601, 1994
26. Rodi W., Scheuerer G., *J. Fluids Eng.*, **108**, pp. 174-179, 1986
27. Durbin P.A., *J. Theoret. Comp. Fluid Dyn.,* **3**, pp. 1-13, 1991
28. Bardina J.E., Huang P.G., Coakley T.J., AIAA paper 97-2121, 1997

INSTABILITY AND TRANSITION IN BOUNDARY LAYERS: DIRECT NUMERICAL SIMULATIONS

Hermann F. Fasel
The University of Arizona, Department of Aerospace and Mechanical Engineering
1130 N. Mountain Ave., Tucson, AZ 85721, USA; e-mail: faselh@email.arizona.edu

Abstract: The fundamental aspects of DNS for investigating transition in boundary layers are discussed. Emphasis of this paper is on the so-called spatial simulation model. Several key examples of successful applications of DNS for advancing the understanding of transition physics are presented.

Key words: instability, transition, boundary layer, direct numerical simulation

1. INTRODUCTION

Research in laminar-turbulent transition in boundary layers is intricately connected with the name of Ludwig Prandtl and the Göttingen "school." "Entstehung der Turbulenz" ('Origin of Turbulence'), as Ludwig Prandtl called it, was one of his research interests early on. After the formulation and validation of his boundary layer theory, laminar-turbulent transition in wall-bounded flows caught his particular interest. For wall boundary layers, both flow states, laminar and turbulent, could be observed and investigated in detail in experiments. However, the transition itself remained elusive. Based on theoretical/mathematical considerations, Prandtl concluded at first [**1**] that small-amplitude waves should always be damped due to the effects of viscosity and that therefore large-amplitude (nonlinear) waves were required to cause transition.

Early water-tunnel experiments in Göttingen seemed to support this conjecture as the onset of the turbulent flow depended strongly on the disturbance level at the inflow (see the experiments on the generation of turbulent spots, "Turbulenzherd"). This may be an indication that, at first, Prandtl may have doubted the applicability of the Orr-Sommerfeld [**2**,**3**] linear stability theory approach for transition in boundary layers. However, in a newly constructed water tunnel, intermittent waves with

G.E.A. Meier and K.R. Sreenivasan (eds.), IUTAM Symposium on One Hundred Years of Boundary Layer Research, 257-267,

relatively long wave lengths would be observed that were slowly amplified in the downstream direction. Likely motivated by these observations, Prandtl and his students continued to probe the relevance of the small disturbance theory for boundary layer transition. Tollmien [**4**], building on the work of Tietjens [**5**] and Heisenberg [**6**], was finally successful in solving the Orr-Sommerfeld equation for a Blasius boundary layer and established the famous diagram defining regions where the flow is stable or unstable. He also determined a critical Reynolds number below which all small disturbances decay. Schlichting [7] later repeated and essentially verified Tollmien's calculations and computed the amplitude distributions of the disturbance waves which, today, are typically referred to as Tollmien-Schlichting waves. Finally, 14 years later in their "landmark" experiments, Schubauer and Skramstad [**8**] were able to generate and measure such instability waves and thus validated the theory by Tollmien and Schlichting. Ever since, linear stability theory (LST) has been a fundamental tool for investigating stability and transition in boundary layers.

With the rapid increase of computing power of digital computers in the 1970's, Direct Numerical Simulations (DNS) developed into an additional powerful tool for transition research. In DNS, the Navier-Stokes equations are solved directly, without restricting assumptions with respect to the base flow or the form or amplitude of the disturbance waves. In this paper, some of the fundamental issues regarding the DNS for investigating stability and transition in boundary layers are discussed. Typical results of transition simulations are presented. The examples are selected to demonstrate the remarkable progress made in this field during the last 30 years.

2. APPROACHES FOR DIRECT NUMERICAL SIMULATIONS

For DNS of transition, two fundamentally different approaches were pursued, the "Temporal Model" and the "Spatial Model."

Temporal approach: In the temporal model the boundary layer base flow is assumed to be independent of the downstream direction (thus neglecting effects of the growing boundary layer) and that the disturbances are periodic in the downstream direction. Due to these constraints, in a simulation the disturbances develop (grow or decay dependent on the Reynolds number and the downstream wavelength) in the time-direction when forced initially. The assumption of periodicity allows simulations of very small downstream domains (typically one disturbance wavelength) and allows efficient use of pseudo-spectral methods. Thus the temporal model is directly related to the formulation of the linear theory eigenvalue problem where for the disturbance ansatz the spatial wavenumber is real and the frequency complex.

To quantitatively compare results from temporal LST or temporal DNS with experiments, where amplification of the waves was not in time but in the downstream direction, typically a transformation using the phase velocity is employed. However, even for the linear development this can be very inaccurate. Especially when nonparallel effects are important and/or the nonlinear development is of interest, quantitative comparison is impossible. Despite of these shortcomings temporal DNS was employed successfully for exploring numerous transition phenomena. For detailed discussions of the temporal approach see [**9**].

Spatial approach: In the spatial model the downstream extent of the computational domain is much larger so that the spatial disturbance development can be investigated. The spatial simulation model corresponds to the traveling wave ansatz of linear stability theory when the spatial wave number is complex and the frequency is real. This is of course the "correct" model for the "convective" nature of the instability mechanism observed in the Schubauer and Skramstad experiments [**8**]. However, the physical realism of the spatial model comes with a price, as spatial DNS is much more demanding both in necessary computational resources and in the development of the computer codes. These difficulties arise from the fact that a much larger extent of the computational domain is required (e.g. many wavelengths in the downstream direction) and that at the outflow, boundary conditions are required such that the disturbances are convected out of the computational domain to prevent contamination of the solution within the computational domain upstream of the outflow boundary [**10**].

The advantages and disadvantages of spatial versus temporal simulation approaches are discussed in more detail in [**11**]. In spite of the difficulties, this author has insisted from the very beginning on the development of spatial DNS models. This decision was founded on the belief that the open questions of stability and transition could only be answered by a realistic simulation model. In the following some of the fundamental issues of spatial DNS will be discussed and a few typical results that are representative for the development of spatial DNS will be presented.

3. NUMERICAL METHODS

Direct Numerical Simulations of transition are based on the numerical solution of the complete Navier-Stokes equations within a specified computational domain, and using boundary and initial conditions that are consistent with the physical transition problem to be investigated. The governing equations together with boundary and initial conditions have to be solved using numerical methods, such as finite difference, spectral, or finite element methods. Here, for the sake of brevity, only a discussion is presented. More detailed information regarding numerical procedures are presented in a recent survey paper [**12**].

Forms of the Navier-Stokes equations: The Navier-Stokes equations can be cast into different forms, all of which have certain advantages/disadvantages for developing numerical methods for spatial DNS. The choice of a particular form of the equations depends on the flow geometry to be investigated, the numerical method used for solving the governing equations (e.g. finite difference, spectral, finite element etc.), and last but not least on the boundary conditions used. For compressible transition simulations, the Navier-Stokes equations in primitive variables and the so-called "conservative" form was successfully used by us [**13,14**] and others [**15**]. For incompressible simulations, in addition to the primitive variable formulation, the Navier-Stokes equations can be used in vorticity-transport form and in combination with Poisson-type equations (for the velocity components, for the streamfunction for 2-D or for the vector potential for 3-D). More recently, a formulation based on vertical velocity/vertical vorticity [**12**] has been used successfully for spatial transition simulations. In our research we had most success with a vorticity-velocity formulation that we pioneered starting with the original (2-D) demonstration that DNS can be employed for investigating stability and transition in boundary layers [**16**].

Boundary conditions: Specification and implementation of "proper" boundary conditions are one of the greatest difficulties when developing a Navier-Stokes code for spatial DNS. The problems arise from the fact that the outflow boundary has to be "transparent" to downstream propagating disturbances, so that possible upstream reflections that would contaminate the solution in the computational domain are minimized. The difficulties regarding the outflow boundary treatment are discussed in detail in [**10**]. In addition to the outflow boundary conditions, "numerical" or "indirect" boundary conditions are required at the wall. Using the primitive variable formulation for incompressible simulations, "numerical" conditions have to be imposed for the wall pressure so that the resulting computed velocity fields are divergence free. For the vorticity-velocity formulation, "numerical" conditions for the wall vorticity have to be specified and implemented so that the velocity field is divergence free and that vorticity is conserved.

Numerical methods for solving governing equations: For solving the Navier-Stokes equations together with the boundary conditions several numerical approaches are available. Discretizing of the time derivative determines not only the temporal accuracy but also if the method is explicit or implicit, which has implications with regard to numerical stability. For spatial discretization we have employed finite-difference approximations and we found that higher accuracy (4th order and higher, standard or compact) is a prerequisite for efficient transition simulations, in particular when the late stages of the transition are of interest. In the spanwise direction, when the base flow is 2-D, pseudo-spectral (Fourier) approximations lead to highly efficient simulation codes. For solving the Poisson equations that result from the incompressible Navier-Stokes equations, multi-grid methods have been employed with great success [**17**,**18**,**19**] while for simple rectangular domains and Cartesian grids, direct methods are more efficient [**20**,**21**]. Overall, the question which numerical method should be chosen for a given method is complicated by the fact that the architecture of the supercomputer to be employed has a major impact on the computational efficiency of a particular method.

4. RESULTS FROM DNS OF INSTABILITY AND TRANSITION

In this section some typical results from DNS of transition of boundary layers will be presented. The examples are not chosen to elaborate on physical issues of the transition process, but rather to indicate some of the major stepping stones in the development of spatial DNS for transition research. For reasons of brevity and availability of graphical results, the examples presented here were taken almost exclusively from work that this author was directly or indirectly involved in. There are numerous other examples in the literature that could have served for elucidating the considerable progress made in DNS during the last 30 years.

To appreciate the progress made in this field since the modest beginnings in the early 1970s, results from likely the first successful application of DNS for transition research [**16**] are presented in Fig. 1. These simulations, although only two-dimensional and with a grid of only 8379 points, required the then most powerful supercomputers (CDC 6600/7600).

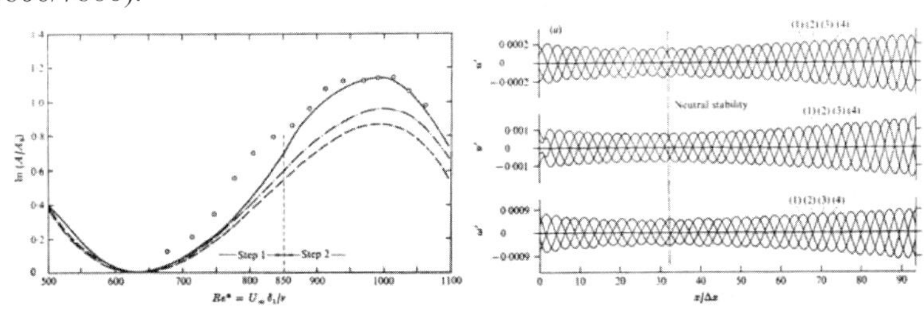

Figure 1. Development of T-S waves in a flat-plate boundary layer. DNS by Fasel [**16**]. *Left:* comparison of disturbance amplification with linear stability theory and experimental measurements. Shown are amplitude curves from ——DNS, – – –linear theory, —·—non-parallel theory, ◦ experiments. *Right:* For Step 1 in figure on the left, downstream development of perturbations (u′, v′, ω′) for four instants in time.

Careful validation of the developed computer codes by comparison with linear stability theory and experiments confirmed the validity of the Navier-Stokes approach. Although very modest by today's standards, these first attempts already clearly indicated the considerable potential of DNS for transition research, and thus laid the groundwork for the future astounding developments. Even with modest 2-D simulations, fundamental unresolved issues of instability and transition could be successfully addressed, such as for example the importance of the so-called non-parallel effects, where our Navier-Stokes calculations [**22**] confirmed the non-parallel theory by Gaster [**23**] and thus resolved the ambiguities of several competing theories at that time [**24,25**].

The next major milestone was the 3-D simulation of the experiments by Klebanoff *et al.* [**26**], with astonishing agreement between experiments and simulations (Fig. 2) [**27**]. However, this agreement was possible only after a slight adverse pressure gradient was imposed in the DNS so that the amplification curve for the small amplitude case of the experiments was matched. From this we conjectured that a mild adverse pressure gradient may have existed in the experiments. This example also points towards an additional role of DNS for interpreting and explaining laboratory transition experiments.

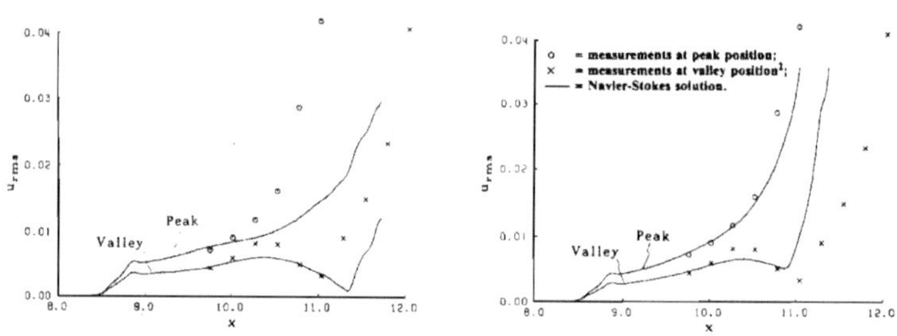

Figure 2. 3-D DNS of fundamental breakdown [**26**] matching experiments by Klebanoff et al. [**25**]. Amplification curves (rms of total streamwise disturbance) at the peak and valley positions, *left:* zero pressure gradient, *right:* small adverse pressure gradient.

An additional important step forward was the use of spatial DNS for transition in supersonic boundary layers [**13**], an area where laboratory experiments are notoriously difficult and expensive. Furthermore, real flight conditions are practically impossible to establish in laboratory experiments. In contrast, once a code is validated using laboratory data, the effect of free-flight conditions can be investigated with spatial DNS. As an example simulations for a M=1.6 boundary layer are presented here where the role of various breakdown scenarios was investigated. This led to the discovery of a new and likely very relevant breakdown scenario, the so-called "oblique" breakdown (Fig. 3) [**13**,**14**,**28**]. Another progress was the demonstration that spatial DNS can not only be used for investigating transition physics but that it can be an important tool for exploring transition control, both passive and active [**29**,**30**].

Subsequently, DNS was employed for investigating transition phenomena of ever increasing complexity. Examples are: Boundary layers subjected to adverse pressure gradients [**17**], highly nonlinear, 3-D stages of breakdown (Fig. 4) [**18**], stability and transition in 3-D boundary layers (cross-flow instability) [**31**,**32**], by-pass transition due to Klebanoff modes [**21**], and boundary layers separating due to strong adverse pressure gradients (Fig. 5) [**33**,**34**,**35**]. Finally, an example of probably one of the computationally most intensive DNS to-date with 400 million grid points for the latest stages of breakdown is presented in Figure 6 [**36**].

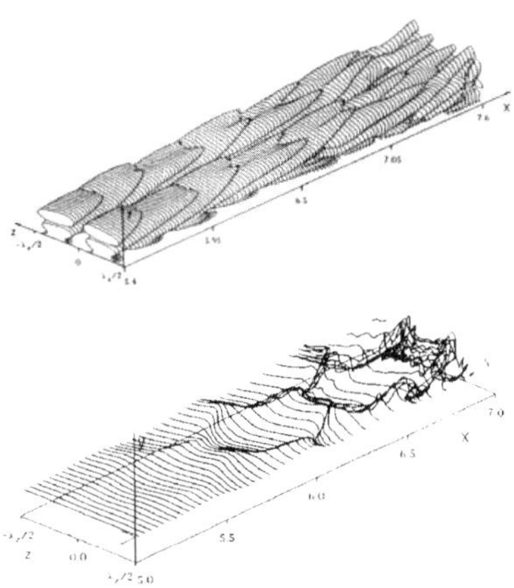

Figure 3. Typical flow structures for oblique breakdown in a supersonic flat-plate boundary layer at M=1.6 [**13**]. Structures visualized by streamwise vorticity (left) and by timelines (right).

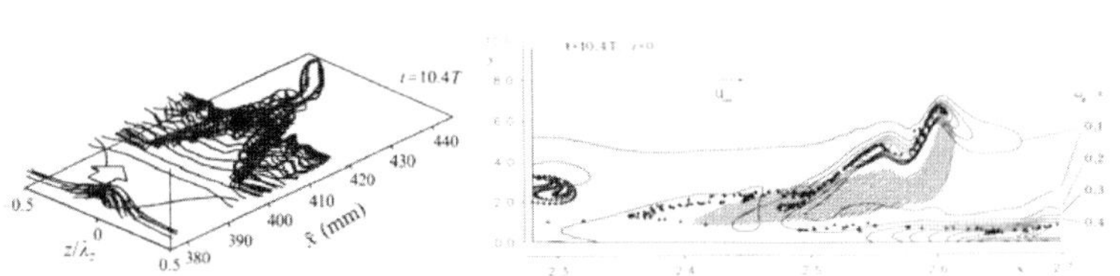

Figure 4. 3-D DNS of fundamental breakdown [**18**]. Shown at one time-instant is a lamba-vortex structure visualized by timelines (left) and by marker particles overlayed with isolines of spanwise vorticity (right).

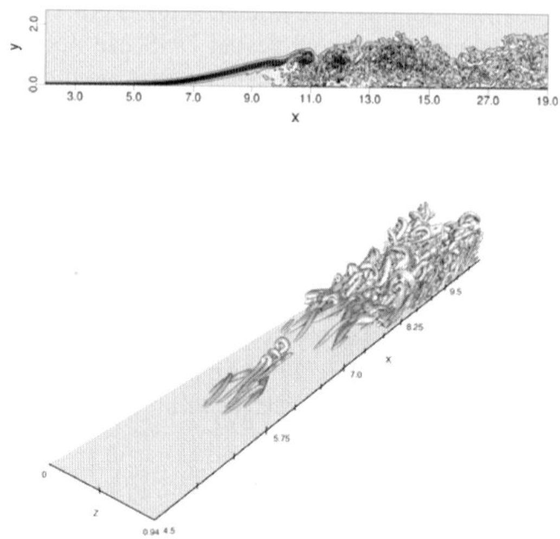

Figure 5. 3-D DNS (using 80 million grid points) of transitional boundary layer subject to strong adverse pressure gradient [**35**].Top*:* flow separation and transition of the unforced flow. Instantaneous contours of spanwise vorticity (spanwise average). *Bottom:* separation bubble is significantly reduced in size due to the application of vortex generator jets (VGJs). Vortical structures visualized over one spanwise period using λ_2-criterion.

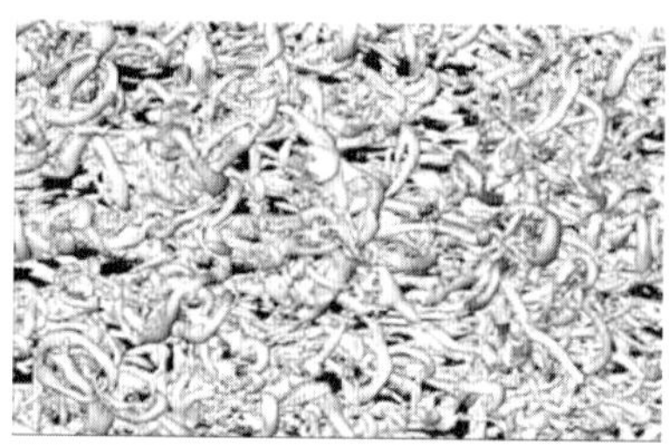

Figure 6. Instantaneous view of vortical structures (visualized using λ_2-criterion) during the very late stage of turbulent breakdown in a flat-plate boundary layer. DNS by Meyer et al. [**36**].

5. CONCLUSIONS

The examples presented here, and many others published in the literature, convincingly demonstrate that spatial DNS is a powerful tool for transition research. One of the main attractions of DNS is the fact that detailed temporal and spatial information can be extracted from the simulations that are difficult, if not impossible, to obtain from experiments, especially for supersonic/hypersonic transition. Because of the detailed data available from DNS, insight into the physics of the transition processes in boundary layer flows can be obtained that is not possible with any other means. Without doubt, DNS has contributed significantly to the high level of understanding of the transition physics that we have today. However, the computationally most intensive DNS of recent years have also exposed a possible "curse" of DNS as a tool for future transition investigations: The amount of data generated by such simulations and the information that can be extracted form these data are enormous and will increase even more rapidly in the future. Therefore, the question arises if we truly "understand" transition once we are able to simulate it in all details. In contrast to the often widespread belief that simulations will eventually replace theory in transition research, this author believes that theoretical models will be needed even more in the future than in the past in order to exploit the massive information supplied by DNS. Brilliant minds like L. Prandtl would have enthusiastically embraced tools like DNS if available then, but not instead of "theory" but rather as an additional tool for developing and testing the theoretical models that are required for a true fundamental understanding of the relevant physical processes of transition.

ACKNOWLEDGEMENTS

This paper is dedicated to my former thesis advisor, Prof. R. Eppler, on the occasion of his 80th birthday. I was motivated to work on the topic of stability and transition in boundary layers by his interest in transition prediction for the design of laminar airfoils for sailplanes. In the context of the present symposium honoring the work of L. Prandtl it is worthwhile to mention that R. Eppler made extensive use of L. Prandtl's boundary layer theory for his highly successful development of airfoils. He has also personally met L. Prandtl in 1942 during the selection process for the Lilienthal prize when Prandtl acted as chair of the selection committee. I also would like to acknowledge the contributions of the many doctoral students that I advised in DNS of stability and transition. Finally, the funding I received for transition research, first from the Deutsche Forschungsgemeinschaft (DFG) and later from ONR and AFOSR is gratefully acknowledged.

REFERENCES

1. Prandtl, L., 1921, "Bemerkungen über die Entstehung der Turbulenz," *ZAMM I*, pp. 431-436.
2. Orr, W. M. F., 1907, "The stability or instability of the steady motions of a perfect liquid and of a viscous liquid. Part I: A perfect liquid; Part II: A viscous liquid," *Proc. Roy. Irish Acad.* **27**, pp. 9-68 and pp. 69-138.
3. Sommerfeld, A., 1908, "Ein Beitrag zur hydrodynamischen Erklärung der turbulenten Flüssigkeitsbewegungen," *Atti del 4. Congr. Internat. Dei Mat. III*, Roma, pp. 116-124.
4. Tollmien, W., 1929, "Über die Entstehung der Turbulenz," *1. Mitt. Nachr. Ges. Wiss. Göttingen, Math. Phys. Klasse*, pp. 21-44.

5. Tietjens, O., 1922, "Beiträge zur Entstehung der Turbulenz," Diss., Göttingen; also *ZAMM* **5**, pp. 200-217, 1925.
6. Heisenberg, W., 1924, "Über Stabilität und Turbulenz von Flüssigkeitsströmungen," *Ann. d. Phys.* **74**, pp. 577-627.
7. Schlichting, H., 1933, "Zur Entstehung der Turbulenz bei der Plattenströmung," *Nachr. Ges. Wiss. Göttingen, Math. Phys. Klasse*, pp. 182-208.
8. Schubauer, G. B. and Skramstad, H. K., 1943, "Laminar boundary layer oscillations and stability of laminar flow," National Bureau of Standards, Research Paper 1772; also *J. Aeronaut. Sci.* **14**, p.69, 1947.
9. Kleiser, L. and Zang, T., 1991, "Numerical Simulation of Transition in Wall-Bounded Shear Flows," *Ann. Rev. Fluid Mech.* 23, pp. 495-537.
10. Kloker, M., Konzelmann, U., and Fasel, H., 1993, "Outflow Boundary Conditions for Spatial Navier-Stokes Simulations of Transition Boundary Layers," *AIAA J.* **31** (4), pp. 620-628.
11. Fasel, H., 1990, "Numerical Simulation of Instability and Transition in Boundary-Layer Flows," IUTAM-Symp., Toulouse, France, 1989, In *Laminar-Turbulent Transition,* Arnal, D., and Michel, R. (eds.), Springer, pp. 587-598.
12. Rempfer, D., 2003, "Low-Dimensional Modeling and Numerical Simulation of Transition in Simple Shear Flows," *Ann. Rev. Fluid Mech.* **35**, pp. 229-265.
13. Fasel, H., Thumm, A., and Bestek, H., 1993, "Direct Numerical Simulation of Transition in Supersonic Boundary Layers; Oblique Breakdown," In *Transitional and Turbulent Compressible Flows*, *ASME-FED* **151**, pp. 77-92.
14. Eissler, W. and Bestek, H., 1993, "Spatial Numerical Simulations of Nonlinear Transition Phenomena in Supersonic Boundary Layers," In *Transitional and Turbulent Compressible Flows*, *ASME-FED* **151**, pp. 69-76.
15. Zhong, X., 2001, "Leading-Edge Receptivity to Free Stream Disturbance Waves for Hypersonic Flow over a Parabola," *J. Fluid Mech.* **441**, pp. 315-367.
16. Fasel, H., 1976, "Investigation of the Stability of Boundary Layers by a Finite Difference Model of the Navier-Stokes Equations," *J. Fluid Mech.* **78**, pp. 355-383.
17. Kloker, M., 1993, "Direkte numerische Simulation des laminar-turbulenten Strömungsumschlages in einer stark verzögerten Grenzschicht," Dissertation, University of Stuttgart.
18. Rist, U. and Fasel, H., 1995, "Direct Numerical Simulation of Controlled Transition in a Flat-Plate Boundary Layer," *J. Fluid Mech.* **298**, pp. 211-248.
19. Postl, D., Wernz, S., and Fasel, H., 2004, "Case 3: Direct Numerical Simulation of the Cray X1," In *CFD Validation of Synthetic Jets and Turbulent Separation Control,* Proceedings of Langley Research Center Workshop, March 29-31, Williamsburg, VA.
20. Swarztrauber, P., 1977, "The Methods of Cyclic Reduction, Fourier Analysis and the FACR Algorithm for the Discrete Solution of Poisson's Equation on a Rectangle," *SIAM Rev.* **19** (3), pp. 490-501.
21. Meitz, H.L., 1996, "Numerical Investigation of Suction in a Transitional Flat-Plate Boundary Layer," Dissertation, University of Arizona.
22. Fasel, H. and Konzelmann, U., 1990, "Non-Parallel Stability of a Flat-Plate Boundary Layer Using the Complete Navier-Stokes Equations," *J. Fluid Mech.* **221**, pp. 311-347.
23. Gaster, M., 1974, "On the Effect of Boundary-Layer Growth on Flow Instability," *J. Fluid Mech.* **66**, pp. 465-480.
24. Saric, W. and Nayfeh, A., 1975, "Non-parallel Stability of Boundary Layer Flows," *Phys. Fluids* **18**, pp.945-950.
25. Bouthier, M., 1973, "Stabilité Linéaire des Écoulements Presque Paralléles. Part 1," J. Méc. **12**, pp.75-95.
26. Klebanoff, P., Tidstrom, K., and Sargent, L., 1962, "The Three-Dimensional Nature of Boundary-Layer Instability," *J. Fluid Mech.* **12**, pp. 1-34.

27. Fasel, H., Rist, U., and Konzelmann, U., 1990, "Numerical Investigation of the Three-Dimensional Development in Boundary-Layer Transition," *AIAA J.* **28**, pp. 29-37.
28. Pruett, D. and Chang, C.-L., 1995, "Direct Numerical Simulation of High-Speed Boundary-Layer Flows-Part II: Transition on a Cone in Mach 8 Flow," *Theor. Comp. Fluid Dyn.* **7** (5), pp. 397-424.
29. Kral, L. and Fasel, H., 1994, "Direct Numerical Simulation of Passive Control of Three-Dimensional Phenomena in Boundary-Layer Transition Using Wall Heating," *J. Fluid Mech.* **264**, pp. 213-254.
30. Kral, L. and Fasel, H., 1991, "Numerical Investigation of Three-Dimensional Active Control of Boundary-Layer Transition," *AIAA J.* **29** (9), pp. 1407-1417.
31. Müller, W., Bestek, H., and Fasel, H., 1996, "Nonlinear Development of Travelling Waves in a Three-Dimensional Boundary Layer," IUTAM-Symp., Manchester, UK, 1995, In *Fluid Mechanics and its Applications* **35**, Duck, P.W., and Hall, P. (eds.), Kluwer.
32. Wassermann, P. and Kloker, M., 2003, "Mechanisms and Passive Control of the Development of Secondary Perturbations on a Swept Wing," *J. Fluid Mech.* **456**, pp. 49-84.
33. Augustin, K., Rist, U. and Wagner, S., 2003,"Investigation of 2D and 3D Boundary-Layer Disturbances for Active Control of Laminar Separation Bubbles," AIAA Paper 2003-0613.
34. Marxen, O., Rist, U., and Wagner, S., 2004, "Effect of Spanwise-Modulated Disturbances on Transition in a Separated Boundary Layer," *AIAA J.* **42** (5), pp. 937-944.
35. Postl, D., Gross, A., and Fasel, H., 2004, "Numerical Investigation of Active Flow Control for Low-Pressure Turbine Blade Separation," AIAA Paper 2004-0750.
36. Meyer, D., Rist, U., and Wagner, S., 2003, "Direct Numerical Simulation of the Development of Asymmetric Perturbations at Very Late Stages of the Transition Process," In *Recent Results in Laminar-Turbulent Transition*, Wagner, S., Kloker, M., Rist, U. (eds.) , NNFM **86**, Springer, Heidelberg, pp. 63-74.

WALL MODELING FOR LARGE-EDDY SIMULATION OF TURBULENT BOUNDARY LAYERS

Parviz Moin and Meng Wang
Center for Turbulence Research, Stanford University
Building 500, Stanford, CA 94305, U.S.A.
{moin, wangm}@stanford.edu

Abstract: Large-eddy simulation (LES) of wall bounded flows is extremely expensive at high Reynolds numbers if one attempts to resolve the small but dynamically important eddies in the near-wall region. To alleviate this problem LES can be combined with a wall model. In recent years at the Center for Turbulence Research we have explored two classes of wall models: those based on turbulent boundary-layer equations and those based on control theory. An overview of these modeling approaches and their applications are provided in this article.

Key words: Wall modeling, large-eddy simulation.

1. INTRODUCTION

The rapid advance in computing power has made the large-eddy simulation (LES) methodology increasingly practical for engineering applications. Unlike direct numerical simulation in which all the flow scales are resolved, LES attempts to resolve only the energy containing scales, thus allowing much larger grid spacing in free-shear flows and in the outer layer of wall-bounded flows. Near a no-slip wall, however, the eddies scale with the distance from the wall and move increasingly closer to the wall as the Reynolds number increases. These eddies are dynamically important and their effects on the outer flow must be included. To resolve these eddies in LES, the Reynolds number scaling of

G.E.A. Meier and K.R. Sreenivasan (eds.), IUTAM Symposium on One Hundred Years of Boundary Layer Research, 269-278,

the required number of grid points is nearly the same as for direct numerical simulation [1, 2]. This has been a major roadblock in extending LES to high Reynolds-number wall-bounded flows.

The combination of LES with wall modeling is designed to alleviate the stringent near-wall resolution requirement. In this method, the dynamics of the near-wall eddies is modeled, and its effect on the outer flow is provided to the LES as a set of approximate boundary conditions. This allows the LES to be conducted on a relatively coarse grid which scales with the outer flow scales, making the computational cost only weakly dependent on the Reynolds number.

The use of a wall model in LES was pioneered by Deardorff [3] in a channel flow at infinite Reynolds number. Most wall models are analogous to the wall functions commonly used in Reynolds-averaged Navier-Stokes (RANS) approaches. They provide an algebraic relationship between the instantaneous local wall stresses and the tangential velocities at the first off-wall velocity nodes. This approach was first employed in a channel flow simulation by Schumann [4], who assumed that the streamwise and spanwise velocity fluctuations are in phase with the respective surface shear stress components. A number of modifications to Schumann's model have been made by, for example, Grötzbach [5] and Werner and Wengle [6] to eliminate the need for *a priori* prescription of the mean wall shear stress and to simplify computations, and by Piomelli et al. [7] to empirically account for the phase shift between the wall stress and near-wall tangential velocity due to the tilting of near-wall eddies.

The algebraic wall stress models mentioned above all imply the logarithmic (power) law of the wall for the mean velocity, which is not valid in complex flows with pressure gradients and separation. In recent years a more robust approach based on boundary-layer approximations [8, 9] has received much attention. In this so called TBLE model, or two-layer model, the turbulent boundary-layer equations with a RANS type eddy viscosity are solved numerically on an embedded near-wall mesh to compute the wall stress. Reasonable success has been documented by Balaras et al. [8] in a plane channel, square duct, and rotating channel, and by Cabot and Moin [9, 10] in a channel flow and backward-facing step flow. See [9] and [11] and the references therein for a comprehensive review of the various LES wall models.

Our most recent efforts have been focused on extending the wall modeling approach to more challenging complex flows and enhancing its robustness. Wang and Moin [12] successfully used the TBLE model with a dynamically adjusted RANS eddy viscosity to perform LES of the flow past an airfoil trailing-edge with incipient separation. Catalano et al. [13] applied a simplified version of the TBLE model to compute the flow over a circular cylinder at super-critical Reynolds numbers, and obtained encouraging but mixed results. To compensate for large numerical and subgrid scale (SGS) modeling errors on coarse meshes and at very high Reynolds numbers, a new wall modeling approach based on optimal control theory was proposed by Nicoud et al. [14, 15]. A brief review of these recent developments is given in the following sections.

2. TBLE MODEL FOR COMPLEX FLOWS

With the TBLE model, the equations solved in the thin wall-layer are [8, 9]

$$\frac{\partial}{\partial x_2}\left(\nu+\nu_t\right)\frac{\partial u_i}{\partial x_2}=F_i, \quad i=1,3, \tag{1}$$

for the two wall-parallel velocity components, where in general

$$F_i=\frac{1}{\rho}\frac{\partial p}{\partial x_i}+\frac{\partial u_i}{\partial t}+\frac{\partial}{\partial x_j}u_i u_j. \tag{2}$$

The eddy viscosity ν_t is of RANS type, and we have employed the simple mixing-length eddy viscosity model with near-wall damping: $\nu_t/\nu=\kappa y_w^+\left(1-e^{-y_w^+/A}\right)^2$. The pressure is assumed constant across the wall layer and is taken from the LES at the edge of the wall layer. Equations (1) and (2) can be solved numerically using an algorithm similar to that for the LES. The computational cost is small relative to that of the LES because of the simplifications in these equations and the absence of a Poisson equation for pressure. Two simpler variants of the above wall model, with $F_i=0$ and $F_i=\frac{1}{\rho}\frac{\partial p}{\partial x_i}$, have also been considered. They are particularly easy to implement because Eq. (1) reduces to an ordinary differential equation, which can be integrated to give a closed-form expression for the wall shear stress [12]. In particular, the simplest case with $F_i=0$ implies the instantaneous log law for the wall-parallel velocities when $y_w^+\gg 1$.

The TBLE model and its simplified forms have been tested extensively in channel flow and flow over a backward-facing step [9]. The results are found to be insensitive to the type of wall

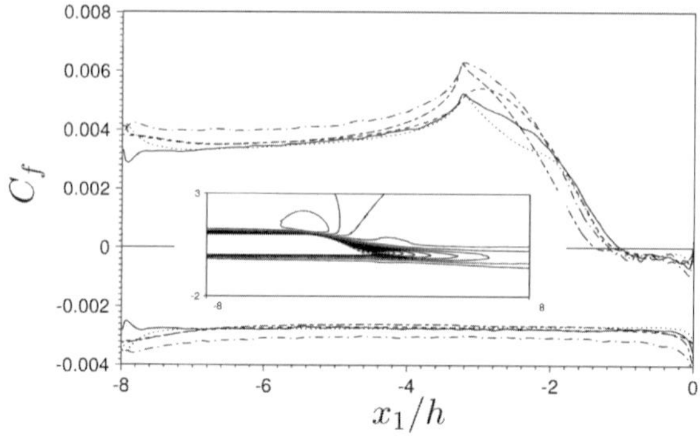

Figure 1. Distribution of the mean skin friction coefficient along an airfoil trailing edge, computed using LES with wall models given by Eq. (1). —·— TBLE model with $\kappa = 0.4$; ——— TBLE model with dynamic κ; - - - - simple model with $F_i = 0$; — - — simple model with $F_i = \frac{1}{\rho}\frac{\partial p}{\partial x_i}$; ········ full LES. Lower curves are for the flat pressure side. Insert: Trailing-edge shape and mean streamwise velocity contours.

stress model in the channel flow. In the massively separated flow behind a step, however, the full TBLE model better predicts the wall stresses. The most accurate solutions are obtained when the mixing-length eddy viscosity is reduced from the standard RANS value using a dynamic procedure which matches the total stresses between the wall layer and the outer layer.

In [12], the above wall models are analyzed and applied to the flow past a model airfoil with an asymmetric trailing-edge, shown in the insert of Fig. 1, at chord Reynolds number of 2.15×10^6. The results further demonstrate the need to reduce the value of the RANS eddy viscosity if the wall-layer equations contain non-linear convective terms and hence carry Reynolds stress. This is important for all flows, particularly attached flows. As illustrated in Fig. 1, the skin friction coefficient computed using the full TBLE model with the standard (von Kármán) constant $\kappa = 0.4$ shows overprediction in most regions relative to the full LES with resolved wall-layers [16]. By using a reduced, variable model coefficient the C_f is much improved. It gives better overall agreement with the full LES result, compared with the two simpler wall models with $F_i = 0$ and $F_i = \frac{1}{\rho}\frac{\partial p}{\partial x_i}$. The dynamically computed κ is found to be only a small fraction of the standard value of 0.4. On average, on the flat surfaces, less than 20% of the Reynolds stress is modeled by the mixing-length eddy

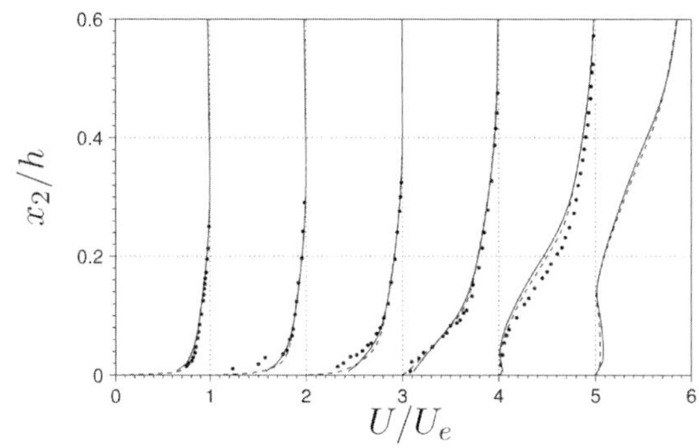

Figure 2. Profiles of the normalized mean velocity magnitude as a function of vertical distance to the upper surface, at (from left to right) $x_1/h = -3.125$, -2.125, -1.625, -1.125, -0.625, and 0. ——— LES with TBLE model; - - - - full LES; • Experiment [17]. Individual profiles are separated by a horizontal offset of 1 with the corresponding zero lines located at 0, 1, ..., 5.

viscosity. The rest is directly accounted for by the the nonlinear terms in the wall layer equations.

A comparison of the velocity predictions using LES with the TBLE wall model and those from the full LES [16] shows excellent agreement, as illustrated in Fig. 2. The agreement with experimental data [17] is also quite good. In particular, the separated region, which starts at $x_1/h \approx -1.125$, is predicted correctly. The velocity profiles computed using the two simpler wall models (not shown, see [12]) are less accurate but acceptable. In these simulations, the use of wall modeling has resulted in a 5/6 reduction in grid size and 90% reduction in CPU cost.

Another challenging test case considered is the high Reynolds number flow over a circular cylinder [13]. To take the best advantage of wall modeling, the super-critical flow regime, in which the boundary layer on the cylinder becomes turbulent prior to separation, is considered. The simpler variant of the TBLE model with $F_i = \frac{1}{\rho}\frac{\partial p}{\partial x_i}$ is used in the calculations.

Simulations at three Reynolds numbers ($Re_D = 5 \times 10^5$, 1×10^6, and 2×10^6) have been performed. LES results are shown to capture correctly the delayed boundary-layer separation and reduced drag coefficients after the drag crisis. The mean pressure distributions and overall drag coefficients are predicted reasonably well at $Re_D = 5 \times 10^5$ and 10^6, as exemplified in Fig. 3. However, the Reynolds-number dependence of the drag

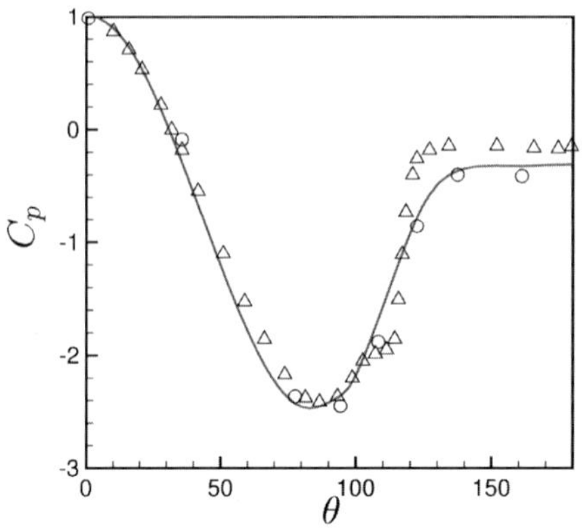

Figure 3. Mean pressure distribution on the circular cylinder. ——— LES with wall model ($F_i = \frac{1}{\rho}\frac{\partial p}{\partial x_i}$) at $Re_D = 10^6$; ○ Experiment of Warschauer & Leene at $Re_D = 1.26 \times 10^6$ (from [18], spanwise averaged); △ Experiment of Flachsbart at $Re_D = 6.7 \times 10^5$ [19].

coefficient has not been captured, and the solutions become increasingly inaccurate at higher Reynolds numbers. In a sense, even the good predictions observed are fortuitous because neither the grid resolution nor the wall model allows an adequate description of the boundary-layer transition, which is extremely complex and sensitive to external disturbances in the Reynolds number range considered. Furthermore, the grid used near the cylinder surface, particularly before separation, is quite coarse judged by the need to resolve the outer boundary-layer scales. This is in contrast to the trailing-edge flow case [12] for which the model works well. At such marginal resolution, the SGS modeling errors and numerical errors tend to dominate the LES in the near-wall region, which cannot be corrected by a wall model based purely on physics. A control-based model, to be discussed in the following section, can account for these errors and hence may provide a better alternative.

3. CONTROL-BASED WALL MODELING

Cabot [10] experimented with the use of wall stresses from a well resolved computation as the approximate boundary conditions in a coarser mesh LES. The results indicate that even the "correct" stresses cannot produce satisfactory solutions. This illustrates that a wall model cannot improve a coarse grid LES by modeling the near-wall physics alone. SGS modeling errors, which dominate the near-wall region as the turbulent integral scale shrinks relative to the filter width, and numerical errors associated with the low-order finite difference methods on the coarse near-wall grids must both be taken into account.

A pertinent question to ask is then what stresses the wall model should produce in order to obtain an accurate coarse grid LES solution. To address this issue, a sub-optimal control method has been used by Nicoud et al. [14] to determine the wall stress conditions required to force the LES solution to a target velocity profile. This strategy is implemented in a plane channel with $Re_\tau = 4000$ on a very coarse grid of 32^3. The cost function is selected to be the difference between the plane-averaged LES velocity and the logarithmic profile, integrated across the channel. In the initial formulation [14], the controls are defined as the two wall shear-stress components, and the normal velocity at the boundary is set to zero (no transpiration). Transpiration velocity was later included as an additional control parameter [15] in an attempt to better approximate the effect of near-wall turbulence structures. Note that in the latter case the net transpiration velocity is kept at zero.

As demonstrated in Fig. 4, the control-based wall models produce mean velocity profiles in very good agreement with the log-law. The addition of the transpiration control improves the mean velocity over the case when only wall stress controls are considered. Also shown in the figure is the mean velocity profile obtained using the simple wall stress model of Piomelli et al. [7], which is typical of current wall stress models. The root-mean-square (rms) velocity fluctuations obtained from the LES with optimized boundary conditions are shown in the right plot of Fig. 4. The agreement with the fully resolved LES data of Kravchenko et al. [20] is poor for the streamwise component in the near-wall region, and the inclusion of the transpiration velocity control is seen to exacerbate the overprediction. It is noted that the sub-optimal wall stresses are optimized to compensate for numerical and SGS modeling errors so as to achieve the correct mean velocity profile. It is therefore not surprising that unphysical velocity fluctuations may be present.

In an attempt to improve the turbulence intensity, the cost function is enhanced to include a component targeting the reference rms velocity fluctuations of Kravchenko et al. [20]. Modest improvement is observed in the case with transpiration velocity, but the streamwise rms velocity remains overpredicted near the wall [15]. Furthermore, the objective of matching the rms velocity fluctuations appears to compete with the objective to

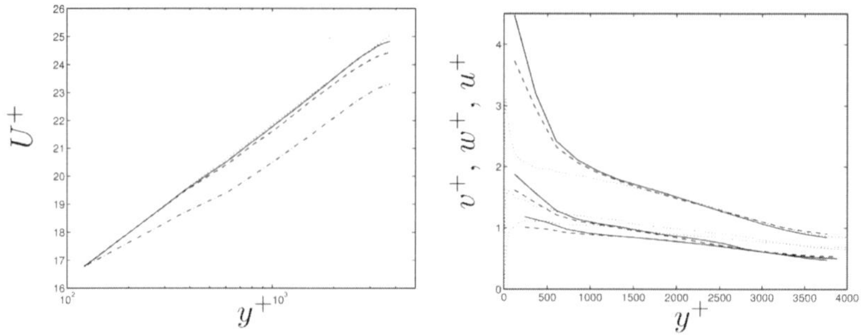

Figure 4. Channel flow statistics at $Re_\tau = 4000$ obtained on a coarse-grid using LES with control-based wall models. Left plot: mean velocity profiles (- - - - control includes wall stresses only; ——— control includes wall stresses and transpiration; —·— no control, uses wall stress model of Piomelli et al. [7]; ········ $U^+_{\text{ref}} = 2.41 \ln y^+ + 5.2$). Right plot: root-mean-square of velocity fluctuations (- - - - control includes wall stresses only; ——— control includes wall stresses and transpiration; ········ reference profiles from [20].)

match the mean velocity, resulting in a slightly less accurate mean profile.

The sub-optimally controlled simulations are computationally expensive. To develop a practical wall model, the data generated from the simulations are subjected to a linear regression scheme to derive a simple, inexpensive model which predicts the wall stresses from the near-wall velocity field. When used in actual simulations, this so called linear stochastic estimate model is shown to reproduce accurately the results of the sub-optimal control calculations. Furthermore, correct mean velocity profiles are predicted over a range of Reynolds numbers from $Re_\tau = 180$ to $Re_\tau = 20000$ using the same grid and numerical methods, even though the model is derived for one particular Reynolds number. However, the linear stochastic estimate model is found to be sensitive to the numerical methods, grid, and SGS model employed. This suggests that while the physics may be captured in a more universal manner, a wall model will always have to adapt itself to the computational environment in which it is used.

4. CONCLUSIONS AND FUTURE DIRECTIONS

Significant progress has been made in recent years in the development of wall-layer models for cost-effective LES at high

Reynolds numbers. The major achievement is the extension of wall modeling capability for flows with strong pressure gradients and separation. The TBLE model, which is a generalization of the classic algebraic models, has enjoyed much success because of its ability to incorporate more near-wall physics. Nonetheless, more challenging tests are needed to fully evaluate the robustness of the method, particularly at very high Reynolds numbers and on very coarse grids.

Wall modeling using control theory is a promising new technique. The pioneering work of Nicoud et al. [14] indicates that in order to obtain the correct mean velocity profile, the wall stresses must not only model the physics but also compensate for numerical and SGS modeling errors. Because of this, control-based models can potentially be more robust than models based purely on physics. While providing valuable insight into wall modeling, the original method is not useful as a predictive tool since a target profile (log law) must be specified *a priori*. In addition, the computational cost is very high due to the need to iterate on the state and adjoint equations.

The future prospect of control-based wall modeling depends on its predictive capability and computational efficiency. Our current efforts are directed toward these goals. To make the model predictive, one approach under consideration is to define the cost function in terms of RANS velocity predictions in the near-wall region as the target profiles for LES. Computational cost may be reduced by taking advantage of the fact that both the control and cost function are defined in a small region near the wall and are computed over short times. This can lead to simplifications in the equations, which will be cheaper to evaluate and can be more readily implemented in complex geometries.

ACKNOWLEDGMENTS

The work described in this article has been supported by the U.S. Air Force Office of Scientific Research and Office of Naval Research. We would like to thank J.S. Baggett, W. Cabot, J. Jiménez, F. Nicoud, and J. Templeton for their contributions and many useful discussions.

REFERENCES

1. Chapman DR. "Computational Aerodynamics Development and Outlook," *AIAA J.*, **17**, pp. 1293-1313, 1979.
2. Baggett JS, Jiménez J, Kravchenko AG. "Resolution Requirements in Large-Eddy Simulations of Shear Flows," *Annual Research Briefs*,

Center for Turbulence Research, NASA Ames/Stanford Univ., 1997, pp. 51-66.
3. Deardorff JW. "A Numerical Study of Three-Dimensional Turbulent Channel Flow at Large Reynolds Numbers," *J. Fluid Mech.*, **41**, pp. 453-480, 1970.
4. Schumann U. "Subgrid Scale Model for Finite Difference Simulations of Turbulent Flows in Plane Channels and Annuli," *J. Comp. Phys.*, **18**, pp. 376-404, 1975.
5. Grötzbach G. "Direct Numerical and Large Eddy Simulation of Turbulent Channel Flows," in *Encyclopedia of Fluid Mechanics*, edited by Cheremisinoff NP, Gulf Pub. Co., 1987, Chap. 34, pp. 1337-1391.
6. Werner H, Wengle H. "Large Eddy Simulation of Turbulent Flow over and around a Cube in a Plane Channel," *Proc. Eighth Symp. Turb. Shear Flows*, 1991, pp. 1941-1946.
7. Piomelli U, Ferziger J, Moin P, Kim J. "New Approximate Boundary Conditions for Large Eddy Simulations of Wall-Bounded Flows," *Phys. Fluids*, **1**, pp. 1061-1068, 1989.
8. Balaras E, Benocci C, Piomelli U. "Two-Layer Approximate Boundary Conditions for Large-Eddy Simulation," *AIAA J.*, **34**, pp. 1111-1119, 1996.
9. Cabot W, Moin P. "Approximate Wall Boundary Conditions in the Large-Eddy Simulation of High Reynolds Number Flow," *Flow Turb. Combust.*, **63**, pp. 269-291, 1999.
10. Cabot W. "Near-Wall Models in Large-Eddy Simulations of Flow Behind a Backward-Facing Step," *Annual Research Briefs*, Center for Turbulence Research, NASA Ames/Stanford Univ., 1996, pp. 199-210.
11. Piomelli U, Balaras E., "Wall-Layer Models for Large-Eddy Simulations," *Annu. Rev. Fluid Mech.*, **34**, pp. 349-374, 2002.
12. Wang M, Moin P. "Dynamic Wall Modeling for LES of Complex Turbulent Flows," *Phys. Fluids*, **14**, pp. 2043-2051, 2002.
13. Catalano P, Wang M, Iaccarino G, Moin P. "Numerical Simulation of the Flow around a Circular Cylinder at High Reynolds Numbers," *Int. J. Heat Fluid Flow*, **24**, pp. 463-469, 2003.
14. Nicoud F, Baggett JS, Moin P, Cabot W. "Large Eddy Simulation Wall Modeling Based on Suboptimal Control Theory and Linear Stochastic Estimation," *Phys. Fluids*, **13**, pp. 2968-2984, 2001.
15. Baggett JS, Nicoud F, Mohammadi B, Bewley TR, Gullbrand J, Botella O. "Sub-Optimal Control Based Wall Models for LES – Including Transpiration Velocity," *Proc. Summer Program*, Center for Turbulence Research, NASA Ames/Stanford Univ., 2000, pp. 331-342.
16. Wang M, Moin P. "Computation of Trailing-Edge Flow and Noise Using Large-Eddy Simulation," *AIAA J.*, **38**, pp. 2201-2209, 2000.
17. Blake WK. *A Statistical Description of Pressure and Velocity Fields at the Trailing Edge of a Flat Strut*, David Taylor Naval Ship R & D Center Report 4241, Bethesda, Maryland, 1975.
18. Warschauer KA, Leene JA. "Experiments on Mean and Fluctuating Pressures of Circular Cylinders at Cross Flow at Very High Reynolds Numbers," *Proc. Int. Conf. on Wind Effects on Buildings and Structures*, Saikon, Tokyo, 1971, pp. 305-315 (see also [19]).
19. Zdravkovich MM. *Flow around Circular Cylinders. Vol. 1: Fundamentals*, Oxford University Press, 1997, Chap. 6.
20. Kravchenko, AG, Moin P, Moser R. "Zonal Embedded Grids for Numerical Simulations of Wall-Bounded Turbulent Flows," *J. Comput. Phys.*, **127**, pp. 412-423, 1996.

REVISITING THE TURBULENT SCALE EQUATION

F. R. Menter, Y. Egorov

ANSYS Germany, Staudenfeldweg 12, D-83624 Otterfing, Germany, E-mail: florian.menter@ansys.com, yury.egorov@ansys.com

Abstract: The paper revisits two-equation turbulence models with an emphasis on the turbulent length scale equation. Rotta's derivation of the *kL* model is re-evaluated and some of the assumptions made during the original formulation of the *kL* model will be relaxed. As a result, the second derivative of the velocity field appears in the length scale equation instead of the third derivative as in the original model. First results, including aspects of Scale-Adaptive Simulation (SAS) capabilities of the new model formulation will be presented.

Key words: Turbulent scales, two-equation models, SAS, von Karman length scale.

1. INTRODUCTION

Two-equation turbulence models reflect the principle that the minimum information required for modelling turbulence are two independent scales obtained from two independent transport equations. They also form the basis of higher-order models like Reynolds stress or algebraic stress models. Even one-equation models using the eddy viscosity as a single variable can be derived from two-equation models using equilibrium assumptions [1].

Three main families of two-equation models have emerged over the years: The first family is based on the k-ω model as proposed by Kolmogorov [2] and later extended by Saffman [3] and Wilcox [4]. The second family is based on the k-ε model proposed by Launder and Spalding [5], and later modified by others. The third family builds on the integral length scale equation of Rotta [6], and is typically formulated as a k-kL model.

Instead of using heuristic arguments for determining the basic form of the turbulent scale equation, Rotta [6] derived an exact transport equation for the integral length scale and modelled it on a term-by-term basis. The distinguishing factor of the model proposed by Rotta is the appearance of a natural length scale in the source terms of the kL-equation, involving the third derivative of the velocity field. As will be discussed in the course of this paper, the availability of a natural length scale is an attractive feature,

G.E.A. Meier and K.R. Sreenivasan (eds.), IUTAM Symposium on One Hundred Years of Boundary Layer Research, 279-290.

as it allows a more subtle reaction of the model to resolved flow features. However, the third velocity derivative has turned out to be problematic and was never actually used in any of the *k-kL* model variants.

In the following, the scale equation proposed by Rotta [6] will be revisited. It will be argued that the assumption, which resulted in the elimination of the second derivative of the velocity field, was too restrictive and can be avoided. Without this assumption, the equations contain the von Karman length scale as the natural length scale. The resulting formulation is attractive, both as a standard RANS model, but also as a model with the ability to partially resolve the turbulent spectrum. The latter concept is termed Scale-Adaptive Simulation (SAS), and was introduced in [7]. It refers to models, which can operate in RANS and scale-resolving mode, without a need for explicit grid information. These formulations offer a great potential for numerous technical flows, as they avoid the grid-sensitivity of DES models in the RANS regime [7], while serving a similar function. It will be shown that the SAS concept arises naturally from the formulation of the turbulence length scale equation.

2. DISCUSSION OF ROTTA'S *kL*-EQUATION

The equation for the turbulence length scale derived by Rotta [6] is based on the definition of the integral length scale *L*, using two-point correlations of the velocity fluctuations u_i (summation over equal indices):

$$kL = \frac{3}{16}\int_{-\infty}^{\infty} R_{ii}(\vec{x}, r_y)\,dr_y \qquad R_{ii} = \overline{u_i(x)u_i(\vec{x}+r_y)} \tag{1}$$

The resulting exact equation for the variable $\Phi = kL$ reads for a shear layer in *y*-direction:

$$\begin{aligned}
&\frac{\partial \Phi}{\partial t} + U_j \frac{\partial \Phi}{\partial x_j} + \frac{3}{16}\int_{-\infty}^{\infty}\left[\frac{\partial U(\vec{x}+r_y)}{\partial x} - \frac{\partial U(\vec{x})}{\partial x}\right] R_{ii}\,dr_y = \\
&-\frac{3}{16}\frac{\partial U(\vec{x})}{\partial y}\int_{-\infty}^{\infty} R_{21}\,dr_y - \frac{3}{16}\int_{-\infty}^{\infty}\frac{\partial U(\vec{x}+r_y)}{\partial y} R_{12}\,dr_y \\
&+\frac{3}{16}\int_{-\infty}^{\infty}\frac{\partial}{\partial r_k}\left(R_{(ik)i} - R_{i(ik)}\right)dr_y + \nu\frac{3}{8}\int_{-\infty}^{\infty}\frac{\partial^2 R_{ii}}{\partial r_k \partial r_k}\,dr_y \\
&-\frac{\partial}{\partial y}\left\{\frac{3}{16}\int_{-\infty}^{\infty}\left[R_{(i2)i} + \frac{1}{\rho}\left(\overline{p'v'} + \overline{v'p'}\right)\right]dr_y - \nu\frac{\partial \Phi}{\partial y}\right\}
\end{aligned} \tag{2}$$

where U_i represent the mean flow velocity components, ν the molecular viscosity, r_y the distance between two probe locations, *p'* the fluctuating pressure and R_{ijk} are triple correlations. Rotta's modelled equation resulting from a term-by-term analysis of the above equation reads (using a somewhat simplified diffusion formulation):

$$\frac{\partial \Phi}{\partial t}+U_j\frac{\partial \Phi}{\partial x_j}=-\overline{u'v'}\left(\tilde{\zeta}_1 L\frac{\partial U}{\partial y}+\tilde{\zeta}_2 L^3\frac{\partial^3 U}{\partial y^3}\right)-\tilde{\zeta}_3\cdot k^{3/2}+\frac{\partial}{\partial y}\left[\frac{\nu_t}{\sigma_\Phi}\frac{\partial \Phi}{\partial y}\right] \quad (3)$$

The main difference to standard scale equations is the occurrence of the third derivative on the right hand side of Eq. (3). It is the result of a Taylor series expansion of the exact term:

$$\int_{-\infty}^{\infty}\frac{\partial U(\vec{x}+r_y)}{\partial y}R_{12}dr_y \rightarrow \frac{\partial U(\vec{x})}{\partial y}\int_{-\infty}^{\infty}R_{12}dr_y$$
$$+\frac{\partial^2 U(\vec{x})}{\partial y^2}\int_{-\infty}^{\infty}R_{12}r_y dr_y+\frac{1}{2}\frac{\partial^3 U(\vec{x})}{\partial y^3}\int_{-\infty}^{\infty}R_{12}r_y^2 dr_y+\ldots \quad (4)$$

Rotta [6] argues that the second term in this expansion can be neglected, as the integral is zero in homogenous turbulence. This leaves the third derivative as the leading term (in addition to the production term $\propto \partial U/\partial y$), resulting in the model formulation given in Eq. (3). The third derivative was however never used in any of the actual implementations of the *k-kL* model. There are several reasons for the omission of this term. The most important is that it is not intuitively clear why the third derivative should be more relevant than the second derivative in the determination of a natural length scale. In addition, a third derivative is a tedious quantity to compute in a general purpose CFD code. However, with the omission of the higher derivative term, the *k-kL* model lost its main distinction compared to the k-ε and the k-ω models. It was therefore not widely used in CFD simulations.

Rotta's argument for dropping the term involving the second velocity derivative is based on the observation that the function R_{12} is symmetric with respect to r_y in homogenous turbulence. The integral $\int_{-\infty}^{\infty}R_{12}r_y dy$ would therefore be zero as r_y is asymmetric. This argument appears overly restrictive, as the entire term involving the second derivative is non-homogenous by nature (in homogenous flow, the second derivative would be zero). It is therefore necessary to evaluate the integral for non-homogenous flows. Ideally, the evaluation would be based on detailed high-Reynolds number measurements in turbulent shear flows. As such data are not readily available, a generic logarithmic layer is used instead.

Figure 1 illustrates a virtual experimental set-up. Two probes are located inside the logarithmic region of a boundary layer. Probe 1 is fixed and probe 2 is located at a distance r_y in realization I (r_y should be of the order of the integral length scale). In realization II, both probes are moved away from the wall by the distance r_y. In realization III, the second probe is first located at x_2+r_y and then at x_2-r_y. The correlation $\tilde{R}_{12}(\vec{x}+\vec{r}_y)=\frac{1}{\overline{u'v'}}\overline{u'(\vec{x}\,)v'(\vec{x}\,+\vec{r}_y)}$ is measured (note that $\overline{u'v'}$ is constant in a logarithmic layer).

Since the integral length scale increases linearly with the distance from the surface, it follows that:

$$
\begin{gathered}
\tilde{R}^{I}_{12}(\vec{r}_y) < \tilde{R}^{II}_{12}(\vec{r}_y) \\
\tilde{R}^{I}_{12}(\vec{r}_y) \approx \tilde{R}^{III}_{12}(-\vec{r}_y) \qquad \tilde{R}^{II}_{12}(\vec{r}_y) = \tilde{R}^{III}_{12}(\vec{r}_y) \\
\tilde{R}^{III}_{12}(-\vec{r}_y) < \tilde{R}^{III}_{12}(\vec{r}_y)
\end{gathered} \tag{5}
$$

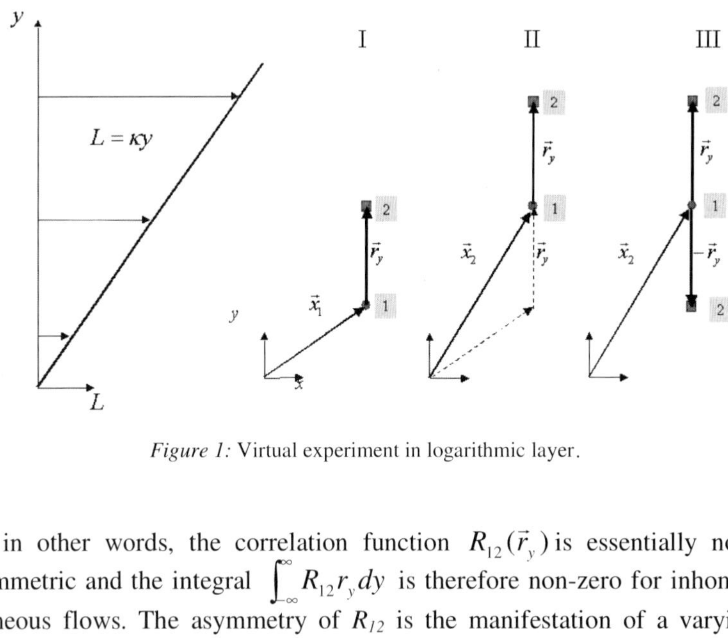

Figure 1: Virtual experiment in logarithmic layer.

or in other words, the correlation function $R_{12}(\vec{r}_y)$ is essentially non-symmetric and the integral $\int_{-\infty}^{\infty} R_{12} r_y dy$ is therefore non-zero for inhomogeneous flows. The asymmetry of R_{12} is the manifestation of a varying length scale, which is not present in homogenous flows. The exact evaluation of the integral requires experimental data at high-Reynolds numbers, which are not yet available. Simple estimates indicate however that the term is of a similar order of magnitude as the production term in a logarithmic region.

The following two model formulations are proposed for this term:

$$\frac{3}{16}\frac{\partial^2 U(\vec{x})}{\partial y^2}\int_{-\infty}^{\infty} R_{12} r_y dr_y \rightarrow \zeta_2 \overline{u'v'}\left|\frac{\partial^2 U(\vec{x})}{\partial y^2}\right| L^2 \tag{6}$$

$$\frac{3}{16}\frac{\partial^2 U(\vec{x})}{\partial y^2}\int_{-\infty}^{\infty} R_{12} r_y dr_y \rightarrow \zeta_2 \overline{u'v'}\left|\frac{\partial^2 U(\vec{x})}{\partial y^2}\right| L^2 \frac{1}{\kappa}\left|\frac{\partial L}{\partial y}\right| \tag{7}$$

where the von Karman constant $\kappa = 0.41$. The first formulation is modelled similarly to the third derivative term proposed by Rotta. The second formulation is based on the argument that the asymmetry of R_{12} is a result of a variation in the length scale, L. The integral $\int_{-\infty}^{\infty} R_{12} r_y dy$ is therefore always non-zero in regions of variations of L as shown in the above heuristic arguments concerning the integral in a logarithmic region. A blend of the two formulations is used in the new model.

3. $\sqrt{k}L$ - MODEL FORMULATION

It is not possible to present the complete model development in this paper. The goal of the present work is rather to present the basic ideas and benefits of using the second velocity derivative in the turbulent scale equation. A full engineering turbulence model will be developed at a later stage.

For reasons beyond the current discussion, the authors prefer the variable $\Phi = \sqrt{k}L$ instead of kL as used by Rotta. The model reads for incompressible flows:

$$\frac{\partial k}{\partial t} + U_j \frac{\partial k}{\partial x_j} = P_k - c_\mu^{3/4}\frac{k^2}{\Phi} + \frac{\partial}{\partial y}\left[\frac{\nu_t}{\sigma_k}\frac{\partial k}{\partial y}\right]$$

$$\frac{\partial \Phi}{\partial t} + U_j \frac{\partial \Phi}{\partial x_j} = \zeta_1 \frac{\Phi}{k} P_k - \hat{\zeta}_2 \nu_t S\left|U''\right|\frac{\Phi^2}{k^{3/2}} - \zeta_3 \cdot k + \frac{\partial}{\partial y}\left[\frac{\nu_t}{\sigma_\Phi}\frac{\partial \Phi}{\partial y}\right] \tag{8}$$

$$\nu_t = c_\mu^{1/4}\Phi \;;$$

with:

$$\hat{\zeta}_2 = \zeta_2 \max\left(c_{SAS}; \left|\frac{L'}{\kappa}\right|\right)$$

$$|L'| = \sqrt{\frac{\partial L}{\partial x_j}\frac{\partial L}{\partial x_j}}; \quad |U''| = \sqrt{\frac{\partial^2 U_i}{\partial x_j \partial x_j}\frac{\partial^2 U_i}{\partial x_k \partial x_k}}; \quad P_k = \nu_t S^2 \tag{9}$$

where S is the absolute value of the strain-rate and $c_\mu = 0.09$.

The determination of the constants follows mainly Rotta's estimates resulting in $\zeta_1 = 0.8$ (corresponding to $\tilde{\zeta}_1 = 1.3$ instead of 1.2 in [6]) and $\zeta_3 = 0.0326$ (corresponding to $\tilde{\zeta}_3 = 0.115$, or c_L=0.7 in [6]). $\zeta_2 = 3.51$ follows from the demand that the equations have to satisfy the logarithmic law of the wall. The diffusion constants are chosen as $\sigma_k = \sigma_\Phi = 2/3$ to ensure a proper behaviour of the equations at a viscous-inviscid interface. The constant c_{SAS} allows the model to adjust to a given resolved length scale of the mean flow, as will be shown below.

4. LENGTH SCALE DETERMINATION BY TWO-EQUATION MODELS

While the principle of two separate scales for the statistical description of turbulence is widely accepted, it is not always clear how two-equation models determine these two scales. This is particularly true in unsteady (URANS) simulations where some of the larger scales of the flow are resolved.

As only the strain-rate, *S*, of the mean flow is provided to the source terms of standard two-equation models, only one scale can be obtained from equilibrium of source terms. As *S* has dimension *1/T*, the source terms determine the turbulent frequency, ω. The turbulence length scale cannot be obtained from the source terms alone. It requires the inclusion of diffusion terms. Estimating the diffusion term for a generic variable as:

$$\frac{\partial}{\partial y}\left(\frac{\nu_t}{\sigma}\frac{\partial \Theta}{\partial y}\right) \propto \nu_t \frac{\Theta}{\delta^2} \tag{10}$$

where δ is the thickness of the shear layer, one obtains:

$$L \sim \delta \tag{11}$$

for standard RANS models.

In order to allow for a more subtle reaction to the underlying flow field, information on the higher derivatives of the resolved velocity field is explicitly required, as indicated by the exact equation for the length scale, Eq. (2). In Rotta's model the resulting length scale is given by:

$$L = \text{const.} \times \sqrt{\left|\frac{\partial U / \partial y}{\partial^3 U / \partial y^3}\right|} \tag{12}$$

The length scale from the current model, Eq. (8), depends on the function $\hat{\zeta}_2 = \zeta_2 \max\left(c_{SAS}; |L'|/\kappa\right)$

$$L \propto \sqrt{\delta\, L_{\nu K}} \quad \text{for} \quad c_{SAS} < |L'|/\kappa \quad \text{RANS Mode} \tag{13}$$

$$L \propto L_{\nu K} \quad \text{for} \quad c_{SAS} > |L'|/\kappa \quad \text{SAS Mode} \tag{14}$$

$$L_{\nu K} = \kappa \left|\frac{\partial U / \partial y}{\partial^2 U / \partial y^2}\right| \tag{15}$$

In order to demonstrate the effect of the von Karman length scale and the influence of the constant c_{SAS}, the following *frozen* harmonic mean flow field is provided to the turbulence model:

$$U(y) = U \cdot sin\left(y \cdot 2\pi / \lambda\right) \tag{16}$$

If periodic boundary conditions are specified for the two turbulence variables, standard two-equation models will not provide a steady-state solution to this problem, as no layer thickness δ is available. While the turbulent frequency ω will converge to a value $\omega \approx U/\lambda$, the turbulent kinetic energy, k, and the turbulent length scale, L, will grow indefinitely. The reason is that no information on the natural length scale λ is provided to the source terms of the models.

In order to demonstrate the influence of the constant c_{SAS}, two settings for c_{SAS} are tested. For $c_{SAS}=0$, the model behaves similar to standard two-equation models with k and L growing indefinitely, again because δ is not available, see Eq. (13). For $c_{SAS}=0$, the model always operates in RANS mode.

As no solution exists for $c_{SAS}=0$ and periodic boundary conditions, the difference in the model behaviour is best demonstrated for a finite layer thickness. Both variables, k and Φ, are set to zero at $y = 0$ and $y = \delta$ for this test. Figure 2 shows the distribution of L for $c_{SAS} = 0$ and $c_{SAS} = 0.54$ (a realistic value for turbulent flows, as shown later) for a variation in the shear-layer thickness from $\delta = 4\lambda$ to $\delta = 8\lambda$. While the solution for $c_{SAS}=0$ returns a length scale which scales with the square root of the layer thickness, the $c_{SAS}=0.54$ solution adjusts to the "resolved scales" and returns $L \sim \lambda$ for both simulations. The reason why the $c_{SAS}=0$ solution behaves similar to a standard RANS model is because $|L'| << 1$. The effect of the von Karman length scale is thereby reduced.

For practical flows and $c_{SAS}=0.54$, the RANS mode of Eq. (13) is typically activated in attached boundary layers, developed pipe flows, etc., whereas the SAS mode of Eq. (14) is active in regions of massive flow separation with global flow instabilities (turbulent vortex shedding, etc.). The model therefore behaves like a Detached Eddy Simulation (DES) model, but without its explicit grid dependency. It thereby avoids the main problem of DES, where grid refinement in the RANS region can activate the DES limiter and thereby compromise the RANS model performance [7].

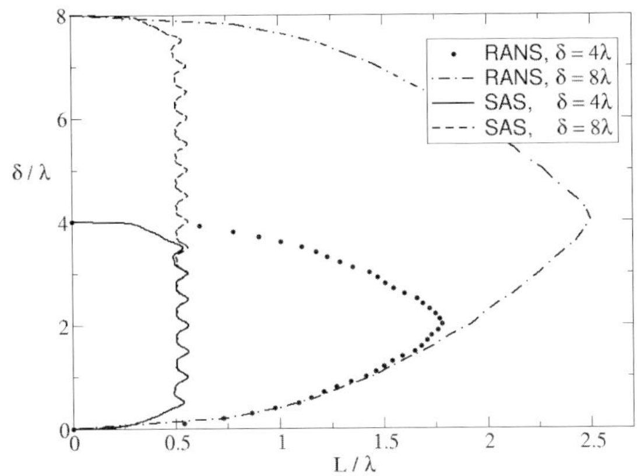

Figure 2: Length scale for sinusoidal velocity field (Eq. 16) and two different layer thicknesses.

5. TEST CASES

5.1. Self-similar flows

The current model has not yet been optimized for a wide range of flows. The purpose of the present tests is to show that the model does produce solutions even at this early stage of development, which are competitive with those of other two-equation models.

Table 1 gives the spreading rates for the three main free shear flows. The far wake has not been computed, as it is only relevant at large distances from the body. The near wake is dominated by the boundary layer computed at the trailing edge. The range of experimental data is taken from [4]. The simulations have been carried out with 1-D codes. The spreading rates are in good agreement with the experimental data, especially for $c_{sas}=0.54$. Fine-tuning of the model could also give better agreement for $c_{sas}=0$.

Table 1: Spreading rates for free shear flows

	$c_{sas}=0$	$c_{sas}=0.54$	Experiment
Mixing Layer	0.128	0.110	0.115
Plane Jet	0.117	0.094	0.100-0.110
Round Jet	0.138	0.095	0.086-0.095

5.2. Flat plate boundary layer

The purpose of the flat plate simulation is to demonstrate that the model given by Eq. (8) has a proper steady-state solution for boundary layers, both for $c_{SAS}=0$ and $c_{SAS}=0.54$. Figure 3 shows the wall shear stress coefficient and the velocity profiles for both model variants in comparison with the data of Wieghardt. The simulations have been carried out with the use of wall-function boundary conditions, as the development of a viscous sublayer model is still in progress. The agreement with experimental correlations is good for both parameter settings. Both parameter settings operate in RANS mode for this case.

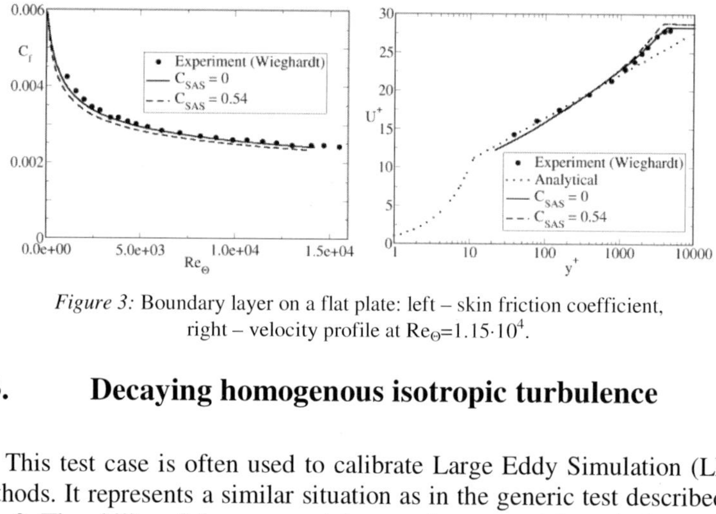

Figure 3: Boundary layer on a flat plate: left – skin friction coefficient, right – velocity profile at $Re_\Theta=1.15\cdot10^4$.

5.3. Decaying homogenous isotropic turbulence

This test case is often used to calibrate Large Eddy Simulation (LES) methods. It represents a similar situation as in the generic test described in Sec. 3. The ability of the new model to produce a partial resolution (to the grid limit) of the resolved turbulence is a result of its ability to adjust to the length scale of the resolved structures. Under those conditions, the model provides an eddy viscosity, which is of a similar order of magnitude as that produced by an LES model.

For this test case, a generic divergence-free initial turbulent flow field, based on the experimental spectrum is produced for a cubical domain [8,9] with a $32\times32\times32$ grid point resolution. The decay of these turbulent fluctuations depends on the level of eddy viscosity provided by the turbulence model, as well as on the dissipation of the numerical scheme. The present simulations are carried out using a second-order accurate central difference scheme in space and a second-order backward Euler method in time. Peri-

best example is the flow around a cylinder in cross-flow. A classical URANS model can produce three-dimensional unsteady structures [9] for this problem. However, the resolved structures are typically of the size of the cylinder diameter (or in other words of the size of the shear layer thickness) as a result of the turbulent length scale in URANS being proportional to the thickness of the shear layer and not proportional to the size of the resolved scales. The SAS model allows the original flow instability to develop into a turbulent spectrum down to the resolution limit of the grid. The behaviour of the SAS model is therefore similar to that of a Detached Eddy Simulation (DES) model [10,11]: the attached boundary layers are solved like in a RANS model and the "detached" unsteady-state flow behind the cylinder results in a LES-like solution. The advantage of the SAS model compared to DES is that the grid spacing does not explicitly appear in the equations. The interaction of the DES-limiter with the RANS model, which poses a quality-assurance problem in DES, can thereby be avoided [7].

The simulations have been carried out for a Reynolds number of Re = 3.6×10^6 and a hexahedral grid with 3.25×10^6 nodes. Figure 5 shows the variable $S^2 - \Omega^2$, where Ω is the absolute value of the vorticity for the SAS model simulation. The picture shows turbulent structures behind the cylinder, which are typically only observed for LES simulations. However, the SAS model handles the attached boundary layers like a RANS model and can therefore be applied to much higher Reynolds numbers than an LES method. The simulations are still in progress and no time-averaged quantities are available at this point.

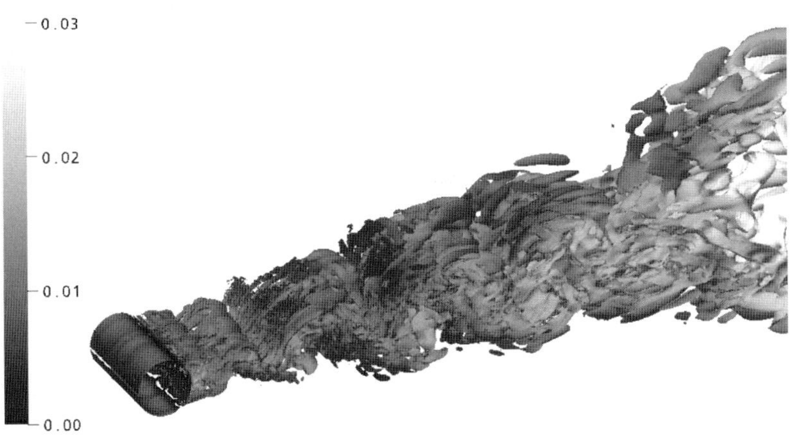

Figure 5: Circular cylinder in a cross flow at Re=$3.6 \cdot 10^6$. Isosurface of $S^2 - \Omega^2$, shadowed according to the turbulent length scale/cylinder diameter (c_{SAS} = *0.54*).

6. SUMMARY

A new turbulence model based on the theory developed by Rotta [6] for the derivation of the *k-kL* model is presented. It is argued that the expansion of one of the terms in the original *kL*-equation is based on overly restrictive arguments assuming homogenous turbulence. In inhomogeneous turbulence, the second term in the expansion can be retained in the equations, introducing the von Karman length scale as a natural scale into the equations.

The new model can be operated in classical RANS mode (enforced by $c_{SAS}=0$), or in a scale-resolving mode termed SAS. The RANS model offers attractive features compared to existing two-equation models, which will be developed further in the future. The SAS mode shows a similar behaviour as popular DES methods, but avoids the grid sensitivity of DES in the RANS regime.

ACKNOWLEDGMENT

This work was supported by the EU projects EVG1-2001-00026 (EXPRO) and AST3-CT-2003-502842 (DESIDER).

REFERENCES

1. Menter, F. R., "Eddy Viscosity Transport Equations and their Relation to the k-ε Model", *ASME Journal of Fluids Engineering,* Vol. 119, pp. 876-884, 1997
2. Kolmogorov, A. N., "Equations of Turbulent Motion of an Incompressible Fluid," *Izvestia Academy of Science,* USSR, Physics, Vol. 6, No. 1-2, 1942
3. Saffman, P. G., "A Model for Inhomogeneous Turbulent Flow", *Proc. Roy. Soc.*, London, Vol. A 317, 1970
4. Wilcox, D. C., *Turbulence Modeling for CFD*, DCW Industries, La Canada, CA, 1993.
5. Launder, B. E., Spalding, D. B., "Mathematical Models of Turbulence", Academic Press, London, 1972
6. Rotta, J. C., *Turbulente Strömungen*, Teubner Verlag, Stuttgart, 1972
7. Menter, F. R, Kuntz, M., Bender R., "A Scale-Adaptive Simulation Model for Turbulent Flow Predictions", AIAA Paper 2003-0767
8. Comte-Bellot, G., Corrsin, S., "Simple Eulerian Time Correlation of Full and Narrow-Band Signals in Grid-Generated Isotropic Turbulence", J. Fluid Mech, Vol. 48, 1971
9. Travin, A., Shur, M., Spalart, P. R., Strelets, M., "On URANS Solutions with LES-Like Behaviour", Proc. ECCOMAS 2004, Jyväskylä, 2004
10. Spalart, P. R., Jou, W-H. Strelets, M., Allmaras, S. R., 1997, "Comments on the Feasibility of LES for Wings and on a Hybrid RANS/LES Approach," *Advances in DNS/LES*, Proc. 1st AFOSR International Conference on DNS/LES, Louisiana Tech University, eds, C. Liu, Z. Liu, L. Sakell, 1997
11. Strelets, M., 2001, "Detached Eddy Simulation of Massively Separated Flows," AIAA Paper 2001-0879, 2001

INDUSTRIAL AND BIOMEDICAL APPLICATIONS

Frank Smith, Nicholas Ovenden and Richard Purvis

University College London, Mathematics Department, UCL, Gower St, London WC1E 6BT tel. (UK) 02076792839, fax (UK) 02073835519, email frank@math.ucl.ac.uk

Abstract: Theoretical models for two distinct current applications are described, one industrial on violent water-air interaction during an impact process and the other biomedical on network flow. Each involves Prandtl's boundary-layer equations, accompanied by very short-scale physical adjustments. Oblique impacts and successive bifurcations are the respective particular themes.

Key words: industrial, biomedical, impacts, networks, theory, computation, scales.

1. INTRODUCTION

It is a pleasure to pay tribute to the great leader and innovator Ludwig Prandtl. What an inspiration he and his work were and still are!

A common feature in the two areas of modelling described in the present article is the laminar boundary layer for an incompressible fluid. Among Prandtl's many other brilliant researches, his idea of the *Grenzschicht* (boundary layer) and related thin viscous layers led to the flowering of singular perturbation studies and multiple scaling in engineering and mathematics during the twentieth century and their continuation into the twenty-first. The idea has had enormous application in industry over decades, not only in aeronautics but in many other fields as well. More recently the application in biomedical studies has grown.

We focus on two distinct applications from recent work, one industrial on violent water-air interaction during an impact process and

G.E.A. Meier and K.R. Sreenivasan (eds.), IUTAM Symposium on One Hundred Years of Boundary Layer Research, 291-300.

the other biomedical on networks. Background work on the former is described fully in [**1-7**] and on the latter in [**8-14**]. Since the flow in each case is very difficult to solve in full by direct computation, the present new contribution uses the Prandtl idea of thin viscous layers, among several other features, to help improve physical understanding and to create predictions and comparisons.

Sections 2, 3 describe the industrial and the biomedical applications in turn, highlighting the short-scale physical changes which are another common feature. In each application Prandtl's boundary-layer equations,

$$u_x + v_y = 0, \qquad (1.1a)$$

$$u_t + u\,u_x + v\,u_y = -\,p_x(x, t) + Re^{-1}\,u_{yy} \qquad (1.1b)$$

play a key role in the thin layers present. The equations are written here in non-dimensional form with velocity field (u, v) in respective Cartesian coordinates (x, y), pressure p and time t. In the usual manner, the relevant dimensional scalings are a typical flow speed U*, a representative length L and the fluid density ρ. The Reynolds number Re is U*L/ν, with ν denoting kinematic viscosity, and planar motion is assumed. Section 4 provides final comments.

2. IMPACTS: OBLIQUENESS AND AIR EFFECTS

A main issue in impacts concerns when and how substantial air-water interaction first occurs near the oblique impact of a water droplet onto a flat horizontal fixed solid surface (wall) or another body of water, with air in-between, as in Fig. 1. The effects of an oblique approach can be significant in industrial terms and so these are incorporated whereas other important effects such as from gravity, surface tension and compressibility are examined in the literature cited in the previous section. The representative quantities U*, L here are taken to be the vertical component V of the droplet approach velocity and a typical droplet diameter, while ρ is ρ_l , the density of the water (or fluid 1), likewise ν is ν_l in the water, and the pressure is measured relative to the atmospheric value. The coordinates and time are centred near the area and instant of impact. The starting point is the Navier-Stokes equations, which are, with Δ denoting the Laplacian,

$$\underline{u}_t + (\underline{u}\,.\,\text{grad})\,\underline{u} = -\,(\rho_l / \rho_n)\,\text{grad}\,p + (\nu_n / \nu_l)\,Re^{-1}\,\Delta\underline{u} \qquad (2.1)$$

both in the water, with subscript n = 1, and in the air (fluid 2) with n = 2 where the air density and kinematic viscosity are ρ_2, ν_2 in turn. The continuity equation (1.1a) applies in each fluid. The impact then has rapid local interaction involving the thin air layer.

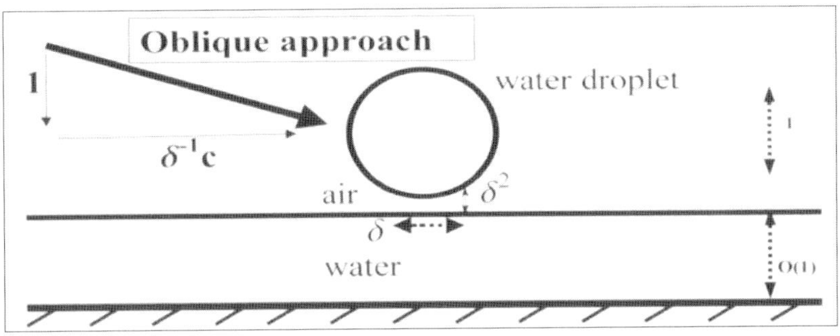

Figure 1. Diagram of oblique impact involving water droplet, air and water/solid surface.

The present theory now takes the density and viscosity ratios ρ_2 / ρ_1, μ_2 / μ_1 of the two fluids 1, 2 to be small: for dry air with pure water these two ratios are near 1/828 and 1/55 in turn, at 20 degrees C and one-atmosphere pressure. Near impact, as the aspect ratio δ of the air layer becomes small, the length scalings of that layer are $(x, y) = (\delta X, \delta^2 y_2)$ in view of the droplet's O(1) curvature, whereas the length scalings in the water are $(\delta X, \delta Y)$. An order of magnitude argument therefore suggests

$$(u, v, p) = (\delta^{-1}c + u_1, v_1, \delta^{-1}p_1) + \ldots \text{ in the water,} \tag{2.2a}$$

$$(u, v, p) = (\delta^{-1}u_2, v_2, \delta^{-1}p_2) + \ldots \text{ in the air} \tag{2.2b}$$

based on the kinematic and pressure conditions at the unknown interface and on (1.1a), (2.1), with the typical time scale $t = \delta^2 T$ being short. The size $\delta^{-1}c$ of the relative incident horizontal velocity component U/V (see also Fig. 1) of the droplet is such as to significantly affect the local interaction. The governing equations in the air are of thin-layer type. In fact the individual contributions in (2.1) for n = 2 are now of order δ^{-3} [acceleration], δ^{-3} [inertia], $\delta^{-2}(\rho_1 / \rho_2)$ [pressure gradient], $(\nu_2 / \nu_1)\, Re^{-1}\, \delta^{-5}$ [viscous] in the x direction and all are in balance if $Re \sim (\nu_2 / \nu_1)\, \delta^{-2}$ and $(\rho_2 / \rho_1) \sim \delta$, yielding

$$Re \sim (\nu_2\, \rho_1^{\,2}) / (\nu_1\, \rho_2^{\,2}) . \tag{2.3}$$

Hence the air-water interaction is controlled by the coupled system of (1.1a, b) in the air, provided (X, y_2, T, 1, u_2, v_2, P) replaces (x, y, t, Re, u, v, p), along with

$$(\partial_T + c\,\partial_x)^2 F = \pi^{-1}\,(\text{P.V.}) \int P_s(s, T)\,(X-s)^{-1}\,ds \qquad (2.4)$$

from the unsteady potential flow in the water, where the principal value integral extends from minus infinity to plus infinity. The unknown pressure P stands for p_1 evaluated at Y zero and is identical to p_2. The boundary conditions on (1.1a, b) include the kinematic condition on v_2 and a matching constraint $u_2 = c$ at the unknown scaled interface $y_2 = F$, as well as no slip at y_2 zero.

The estimate (2.3) acts as a critical Reynolds number and its value is about 10^7 for water with air. Because of the industrial setting [**1, 4-7**] where typical Re values of 10^4 to 10^5 are encountered there is much interest in the subcritical range. There (1.1a, b) and the appropriate boundary conditions reduce to the Reynolds lubrication equation

$$(F^3 P_X)_X = 12\,(F_T + c F_X/2) \qquad (2.5)$$

in normalised form. So far the working is for oblique droplet impact onto a fixed solid but essentially the same system (2.4), (2.5) applies for impact onto water [**5, 7**]. Also, equivalent equations hold in a frame moving with the horizontal velocity U of the droplet where the wall appears as an upstream-moving wall [**7**].

Computational solutions of (2.4), (2.5) were obtained by adapting a numerical method from the papers [**5, 7**], which address the case of zero c (normal impact). Grids and time steps similar to those used in these papers were applied here as well and tested satisfactorily for accuracy. The initial conditions and the far-field boundary conditions are those of an approaching parabola shape $(X-cT)^2 - T$ for F, corresponding to the lower reaches of the smooth total incoming droplet, and of negligible induced pressure (P tending to zero) which is associated with the atmospheric pressure holding outside the interaction region. The results for two different positive values of c are shown in Fig. 2 and indicate effects not dissimilar to those of inclined gravity [**7**]. A skewed touchdown (F tending to zero) is indicated generally at negative T; thus the presence of air hastens touchdown. This is in the fixed frame, note, with a relatively high incident horizontal velocity component, and the angle of approach measured from the horizontal is V/U, i.e. δ / c. In the extreme of large c the majority of the solution has $T \propto c$ and enlarged lengths $X-cT \propto c^{1/2}$, so that in a moving frame the right side of (2.5) approaches $-6c\ F_X$; (2.5) now integrates to give $F^3\ P_X$ as $-6c\ F + G(T)$. Here G must be nonzero to give zero farfield P. Substitution into (2.4) to yield an F equation then suggests that the G term drives the touchdown process, in which the minimum F tends to zero in a square-root manner in scaled time.

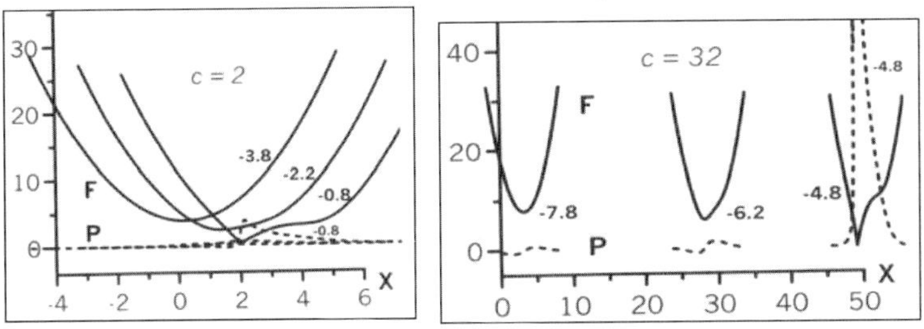

Figure 2. Droplet shapes (solid) and pressures (dashed) for two c values, at times marked.

More widely, the results also prompt thoughts on pre-existing air flow effects since we would expect such air flow also to provoke skewing, for example by means of an extra streamwise mass flux. Analysis for large negative T however suggests that, at least for zero c, skewing is present only if the incident shape is already skewed. This is because the farfield P must be zero. Solutions with such incident skewing are included in [7]. With nonzero c an extra mass flux is induced but it is a definite amount rather than arbitrary. In fact, skewing of the incident shape may well be how pre-existing air flow influences the impact process in practice, by altering the droplet shape considerably before the local interaction comes into play.

3. NETWORK FLOW

Moving on to the separate issue of networks, we describe modelling of a planar network of bifurcating tubes as in Fig. 3 starting at x = 0. The representative quantities U*, L here are, respectively, a typical axial speed induced in the incident mother flow (taken to be fully developed motion with no slip at the walls and unknown total mass flux) and a typical tube width. Steady flow is assumed over a long length scale that is nevertheless short compared with the viscous length Re, together with zero pressure upstream in the mother and prescribed pressures at the downstream ends of the network. The upstream-influence length scale axially [**11**] is O (ϵ^{-1}) where ϵ^7 is 1/Re and is small. The thin viscous layer at the lower outer wall near y = 0 then has thickness of order ϵ^2 and Prandtl's equations (1.1a, b) hold there in re-scaled terms with Re replaced by unity. The boundary conditions required are

$$u \sim \lambda_0 \,(y + A(\dot{x})) \qquad \text{as } y \to \infty, \tag{3.1a}$$

$$\text{no slip at } y = f(x), \tag{3.1b}$$

where the positive constant λ_0 stands for the scaled incident wall shear and f(x) denotes the given lower-wall shape. If we ignore upstream influence for now, the negative boundary-layer displacement A(x) can be obtained (to within a factor related to the mass flux) by the core-flow solution valid outside the wall layer; see below. In that case the viscous wall-layer problem determines the ϵ^4 scaled wall pressure p to within a constant.

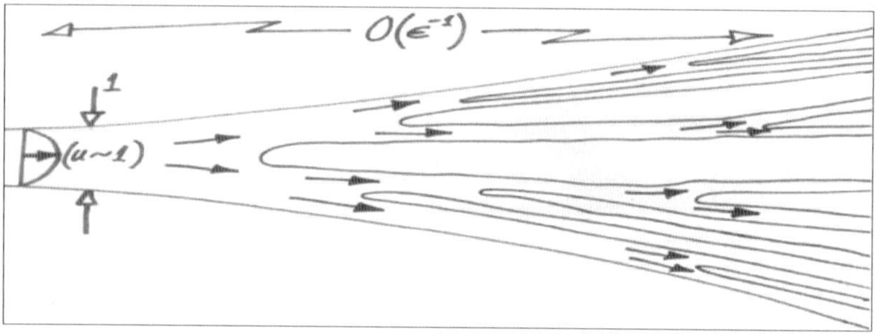

Figure 3. Schematic of a branching network with successive bifurcations

Suppose first that we have a single 1-to-2 branching. The inviscid core then within the lower daughter acts mostly as if distinct from that in the upper daughter and likewise for the viscous upper wall layer, over the present length scales. In the lower daughter core, the pressure is of order ϵ^4 and the stream-function expands as

$$\psi = \psi_0(y) + \epsilon^2 \{ A(x)\, u_0(y) + \lambda_2\, \psi_6(y) \} + \ldots \tag{3.2}$$

where ψ_0 is $\lambda_0\,(y^2/2 - y^3/3)$ and u_0 is $\lambda_0\,(y - y^2)$, corresponding to the fully developed Poiseuille flow in the absence of any bifurcation, whereas the constant λ_2 is an unknown associated with the altered mass flux. The undeveloped viscous layers on the internal dividers of the daughters have negligible impact on the flow (they are passive), implying a tangential flow condition on the given divider underside y = c_0 - ϵ^2

$T_0(x)$ say. Taking $T_0(0)$ as zero without loss of generality thus yields the classical thin-channel result

$$A(x) = T_0(x) + K_0, \quad \text{for } x > 0, \tag{3.3}$$

(since $u_0(c_0)$ is nonzero) which determines the function A(x) to within the additive constant K_0. Similarly, upstream influence present in the mother tube yields a free-interaction behaviour [**11**]

$$A(x) = K e^{\kappa x}, \quad \text{for } x < 0, \tag{3.4}$$

where κ is a known positive constant and (3.4) represents an elliptic effect. A novel feature due to the presence of the bifurcation (branching junction) however is that an axial jump in displacement can occur across the daughter entrances from 0- to 0+. The jump is admissible, and in fact necessary due to the set pressures upstream and downstream [**12, 14**]. At the outer walls in particular, where the incident velocity is close to zero, the viscous layers of (3.1a,b) allow the Bernoulli quantity $p + u^2/2$ to be conserved as required along each local inviscid streamline by means of a scaled pressure jump, in this case $\lambda_0^2 (K^2 - K_0^2)/2$. The jumps are smoothed out over a shorter axial scale by an Euler region of length O(1) in x [**12-14**], which provides some direct communication between the two daughters and the mother. The feature that K, K_0 are unequal in general allows adjustment of K_0 in order to allow the lower-daughter pressure to satisfy the downstream pressure conditions, and likewise for the upper daughter.

Second, suppose a 1-to 4 network. Then another new feature appears as follows. Again attention can be restricted to a lower part, consisting now of a daughter described essentially as in (3.2) and two granddaughters which begin at $x = x_1 > 0$. The lower of these granddaughters is also described essentially by (3.2). The upper one however must suffer higher typical pressure variations of order ϵ^2 such that

$$\psi = \psi_0(y) + \epsilon^2 \{ D(x) u_0(y) + \lambda_2 [\psi_0(y) - \psi_0(c_1)]\} + \ldots \tag{3.5}$$

where $c_1 - \epsilon^2 T_1(x)$, $c_1 + \epsilon^2 S_1(x)$ are the underside and topside respectively of the divider between these two granddaughters and

$$D(x) = - p_1(x) \int u_0^{-2}\, dy - S_1(x) + \gamma_1. \tag{3.6}$$

The integral is from c_l to y, while the ϵ^2 scaled pressure p_l and the constant γ_l are unknown. The novel feature here is that another jump must usually occur, namely in pressure across the entrance of the upper granddaughter from x_l- to x_l+. This again is admissible, as the incident velocity is nonzero at all y heights of that granddaughter, allowing the Bernoulli property to be maintained along each streamline. This active jump is also smoothed out on a shorter axial scale by an O(1) Euler region in $x - x_l$. (Overall this is another type of ellipticity.) As a result, it is found that a jump is also induced in the effective A(x) function here which although still similar to (3.3), (3.4) now has

$$A(x) = K e^{\kappa x}, \text{(discontinuity)}, T_0(x) + K_0, \text{(discontinuity)}, T_l(x) + K_l \quad (3.7)$$

The doubly discontinuous form (3.7) then drives the viscous wall-layer response by means of the constraint (3.1a). The displacement constants K_0, K_l in (3.7) are controlled not only by the outermost (lower granddaughter) imposed pressure downstream but also by the inner (upper) granddaughter pressure imposed downstream.

Third, suppose a 1-to-8 network. Again consider its lower part. Yet another new feature enters as this new generation can contain some inner bifurcations which have nonzero incident velocity throughout and so can provoke the higher $O(\epsilon^2)$ pressure (and jumps) all the way across in y as well as for long distances axially upstream and downstream, while outermost bifurcations continue the earlier established trend. One case, to focus attention, has the triply discontinuous form

$$A(x) = \text{as in (3.7), then a discontinuity, then } T_2(x) + K_2. \quad (3.8)$$

The three constants K_0, K_l, K_2 however depend on the four pressures imposed downsteam in the four great-granddaughters (of this lower part) via the higher pressure responses and pressure jump occurring in the (implied) inner bifurcation as just described. The forms alternative to (3.8) in a 1-to-8 network depend on the relative positioning of each divider, making either (3.7) or a four-times discontinuous form hold.

Larger/generalized networks produce similar effects, i.e. potentially many discontinuities in the negative displacement A(x) which, along with f(x), forces the viscous layer by means of (3.1a,b) and induces discontinuities in the wall pressure(s). The viscous layer is nonlinear in general, requiring numerical solution and admitting separation [**11-14**]. By virtue of Prandtl's transposition theorem, the solution depends only on the effective thickness (A+f) [=B say], thus giving wide application.

For small B, a linearized form applies and gives merely small discontinuities in pressure as in Fig. 4. The sample solutions in Fig. 4 show the scaled pressure induced at an outer wall: the "1 or 5 branchings" refers to the number of branching junctions as seen from that outer wall only, whereas the total system could have more branching junctions unseen from that wall (as in Fig. 3 for example). A contraction of the outermost tube width broadly leads to a favourable pressure gradient and increasing wall shear, and expansion to an adverse pressure gradient with decreasing wall shear, as expected, but the discontinuities due to the branching junctions can counteract those trends.

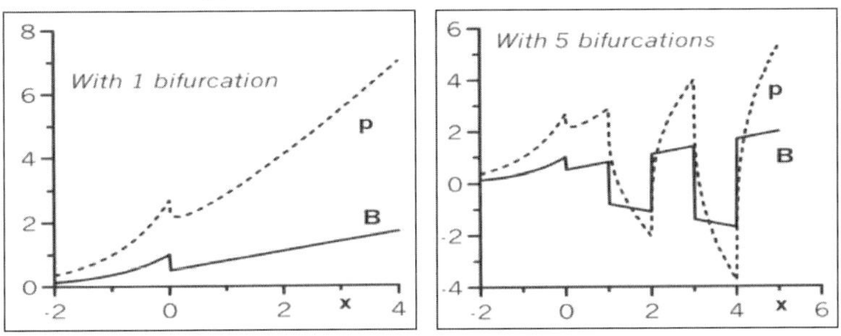

Figure 4. Outer-wall pressure in networks with 1 or 5 seen branchings, for effective thicknesses B as shown. The seen junctions are at x = 0 (left) and 0, 1, 2, 3, 4 (right).

4. FURTHER COMMENTS

Concerning the industrial application in section 2, an extension akin to the extreme for large c might admit an account of skimming (bouncing) via ground effect, an extension which is similar to using an enhanced Reynolds number based on U. Along with that, and the limit of quasi-inviscid air studied anew in [**5**], it would clearly be interesting to see what happens when the complete Prandtl system (1.1a,b) applies, even including compressibility for instance. Concerning the biomedical application in section 3, the presence of multiple jumps in the solution(s) is likewise intriguing, especially if coupled with separation in nonlinear and/or unsteady cases [**12, 14**]. Common needs for both applications are increased understanding of three-dimensional phenomena eventually, further investigation of full nonlinearity and following through on ideas from the current article, all of which is partly continuing work. We aim to report more later.

Above all, Prandtl's full boundary-layer system (1.1a,b) remains fascinating and relevant. Both extreme and linearized cases in this article exhibit the rich structure and physics arising from the system. The article also hints clearly, we hope, at the richness and diversity in terms of application areas.

ACKNOWLEDGEMENTS

Industrial: we thank EPSRC and QinetiQ for support through the Faraday Partnership for Industrial Mathematics, managed by the Smith Institute, and also David Allwright, Roger Gent, David Hammond, Richard Moser and Manolo Quero for their interest and helpful discussions. **Biomedical:** we thank EPSRC for support through the Mathematics Programme and Life Sciences interface, and also Neil Kitchen, Stefan Brew, Joan Grieve, Robert Bowles and Stephen Baigent for their interest and helpful discussions.

REFERENCES

1. Gent RW, Dart NP, Cansdale JT. "Aircraft icing", *Phil Trans Roy Soc A,* **358**, pp. 2873-2911, 2000.
2. Josserand C, Zaleski S. "Droplet splashing on a thin liquid film", *Phys Fluids*, **15**, pp. 1650-1657, 2003.
3. Wilson SK. "A mathematical model for the initial stages of fluid impact in the presence of a cushioning fluid layer", *J Eng Maths*, **25**, pp. 265-285, 1991.
4. Purvis R, Smith FT. "Large droplet impact on water layers", *Proc. 42nd Aerospace Sci Conference*, Reno, NV, USA, Jan. 5-8, 2004, paper no. 2004-0414.
5. Purvis R, Smith FT. "Air-water interactions near droplet impact", *Euro J Applied Maths*, to appear, 2004.
6. Purvis R, Smith FT. "Droplet impact on water layers: post-impact analysis and computations", *Phil Trans Roy Soc A*, in press, 2004.
7. Smith FT, Li L, Wu G-X. "Air cushioning with a lubrication / inviscid balance", *J Fluid Mech*, **482**, pp. 291-318, 2003.
8. Brada M, Kitchen ND. "How effective is radiosurgery for arteriovenous malformations?", *J Neurol Neurosurg Psychiatry*, **68**, pp. 548-549, 2000.
9. Hademenos GJ, Massoud TF and Vinuela F. "A biomathematical model of intercranial arteriovenous malformations based on electric network analysis: theory and haemodynamics", *Neurosurgery*, **38**(5), pp. 1005-1015.
10. Zhao Y, Brunskill CT, Lieber BB. "Inspiratory and expiratory steady flow analysis in a model symmetrically bifurcating airway", *J Biomech Eng*, **119**, pp. 52-65, 1997.
11. Smith FT. "Upstream interactions in channel flows", *J Fluid Mech*, **79**, pp. 631-655, 1997.
12. Smith FT, Ovenden NC, Franke P, Doorly DJ. "What happens to pressure when a flow enters a side branch?", *J Fluid Mech*, **479**, pp. 231-258, 2003.
13. Smith FT, Jones MA. "AVM modelling by multi-branching tube flow: large flow rates and dual solutions", *IMA J Maths Medicine Biol*, **20**, pp. 183-204, 2003.
14. Smith FT, Dennis SCR, Jones MA, Ovenden NC, Purvis R, Tadjfar M. "Fluid flows through various branching tubes", *J Eng Maths*, **47**, pp. 277-298, 2003.

ANALYSIS AND CONTROL OF BOUNDARY LAYERS: A LINEAR SYSTEM PERSPECTIVE[1]

John Kim
Department of Mechanical and Aerospace Engineering
University of California, Los Angeles, CA 90095-1597
jkim@seas.ucla.edu

Junwoo Lim
Pittsburgh Supercomputing Center
4400 Fifth Avenue, Pittsburgh, PA 15213
jlim@psc.edu

Abstract:
Motivated by the recent successful applications of linear controllers to nonlinear flows, several approaches of boundary-layer control are analyzed from a linear system point of view. The singular value decomposition (SVD) is applied to the linearized Navier-Stokes system in the presence of control. The performance of control is examined in terms of the largest singular values, which represent the maximum growth of disturbance energy attainable in the linear system under control. It is shown that there exists a strong similarity between the trend observed in the SVD analysis (linear) and that observed in direct numerical simulations (nonlinear), thus reaffirming the importance of linear mechanisms in the near-wall dynamics of turbulent boundary layers. The present study illustrates that a proper linear analysis, the SVD analysis in particular, can be used as a guideline for designing controllers for drag reduction in turbulent boundary layers.

1. INTRODUCTION

It has been generally accepted that nonlinearity is an essential characteristic of turbulent flows. Consequently, except for special situations in which a linear mechanism is expected to play a dominant role (e.g., rapidly straining turbulent flows to which the rapid distortion theory can be applied), the role of linear mechanisms in turbulent flows has not received much attention. Even for transitional flows, a common notion is that the most a linear theory can provide is insight into the early stages of transition to turbulence.

[1]Portions of this paper have been published in Lim and Kim [1].

G.E.A. Meier and K.R. Sreenivasan (eds.), IUTAM Symposium on One Hundred Years of Boundary Layer Research, 301-312,

But several investigators have recently shown that linear mechanisms play an important role even in fully turbulent, and hence fully nonlinear, flows. Examples of such studies include: 'optimal' disturbance[2] in turbulent boundary layers [2, 3]; transient growth due to non-normality of the Navier-Stokes system [4]; energy amplification in the linearized Navier-Stokes system [5]; essentially linear feedback controllers for drag reduction in turbulent channel flows [6, 7]; applications of a linear control theory to transitional and turbulent channel flows [8–12]; and a numerical experiment [13] demonstrating that near-wall turbulence could not be maintained in turbulent channel flow when a linear mechanism was artificially suppressed, thus further illustrating the essential role of a linear process in the nonlinear flow.

In this paper, some of the above-mentioned work on the role of linear mechanisms in wall-bounded shear flows are reviewed from a perspective that controlling linear processes in turbulent boundary layers is a viable route for control of turbulent boundary layers, especially for the purpose of viscous drag reduction. We then analyze turbulent channel flows from a linear system point of view. After recasting the linearized Navier-Stokes equations into a state-space representation, we apply the singular value decomposition (SVD) analysis to the linear system, and show that the SVD analysis presents the same trend observed in nonlinear simulations.

2. LINEAR MECHANISM AND LINEAR CONTROL

The linear mechanism by which disturbances can grow in a shear flow has been pointed out by several investigators ([14–17], to name a few). Landahl [15] referred to it as a lift-up process, in which streamwise-independent streamwise disturbances can grow linearly in time. From consideration of the vorticity equation, it is a vortex tilting term (from spanwise vorticity associated with mean shear to wall-normal vorticity); sometimes it is also referred to as the linear coupling term [13], since it couples the Orr-Sommerfeld equation for wall-normal velocity with the Squire equation for the wall-normal vorticity in the linearized Navier-Stokes equations (see Section 3 below). More importantly, it enhances non-normality of the linearized Navier-Stokes operator, although the Orr-Sommerfeld operator itself is slightly non-normal (*i.e.*, not self-adjoint). The transient growth due to this non-normality in transitional flows has received much attention lately, since it is now rec-

[2]This is actually the worst disturbance from the point of controlling disturbances, and we use this commonly used term in quotes to avoid confusion with the optimal control in linear control theory.

ognized that some subcritical transition to turbulence can be related to this mechanism (see, for example, [4, 18, 19]).

Butler and Farrell [2, 3] were the first to recognize the relevance of this linear mechanism in fully developed turbulent, and hence nonlinear, flows. Their so-called 'optimal' disturbances, which have the largest transient growth due to non-normality of the linearized Navier-Stokes operator, have strong resemblance to the organized turbulence structures observed in the near-wall region of turbulent boundary layers. Although they had to impose the eddy turnover time, which represents a nonlinear process in turbulent flows, as an additional constraint in their linear analysis (otherwise, the 'optimal' disturbance turned out to be much larger than the commonly observed near-wall structures; see Section 3 for further discussion), it was the first demonstration – to the best of our knowledge – that the same linear mechanism plays an important role in a fully developed turbulent flow. Kim and Lim [13] further illustrated the importance of this linear mechanism through a numerical experiment, in which modified nonlinear Navier-Stokes equations without the linear coupling term were used to show that a fully developed turbulent channel flow could not be maintained without the linear coupling term. This demonstrated that, although a nonlinear process is a critical element in the overall self-sustaining mechanism of near-wall turbulence in wall-bounded turbulent shear flows as suggested in [20, 21], the aforementioned linear mechanism plays an important role, without which near-wall turbulence could not be maintained. A controller that can suppress or mitigate this linear mechanism would be very effective in controlling near-wall turbulence (and hence controlling the skin-friction drag and surface heat transfer) in turbulent boundary layers. In fact, Kim and Lim's [13] numerical experiment, referred to as a virtual flow in [13], can be viewed as a turbulent flow with a perfect controller with which the linear coupling term is completely eliminated.

Lee *et al.* [6] developed a nonlinear adaptive controller, using a neural network representing an adaptive inverse model of the Navier-Stokes equations. Once properly trained, this inverse model neural network was able to predict an optimal control input for a desired output. Applying this controller to a turbulent channel flow for skin-friction drag reduction resulted in about 20% drag reduction. A subsequent examination of the neural network suggested, however, the possibility of using a simpler linear network. This linear network resulted in almost identical drag reduction, suggesting that the flow dynamics of interest, *i.e.*, those relevant to high skin-friction drag, can be approximated by a linear model. This result further substantiates the notion that a linear mechanism plays an important role in the near-wall flow dynamics.

Bewley *et al.* [22] applied an adjoint-based optimal control, in which a control objective was minimized over a finite time by iteratively solving the

Navier-Stokes equations together with their adjoint equations. This approach led to flow laminarization when applied to a low Reynolds-number turbulent channel flow. However, this approach requires solving the Navier-Stokes equations forward in time and the adjoint equations backward in time, which makes it too difficult, if not impossible, to implement in practice for real-time control. Lee *et al.* [7] approximated this adjoint-based approach by taking an adjoint of only the linear part of the discretized Navier-Stokes equations in their search for a suboptimal state. This approach led to a feedback control law very similar to that derived from the linear network mentioned above, resulting in a similar drag reduction. It is worth noting that the two control schemes derived from two different approaches – one as a linear approximation of a nonlinear neural network and the other as a linear approximation of the nonlinear adjoint equations – yielded very similar feedback control laws. It appears that whatever physics relevant to skin-friction drag reduction in turbulent boundary layers could be adequately approximated by a linear model.

Much advance has been made recently in applying linear optimal control theories to flow control [8–12]. In this approach, an optimal controller, which is designed to minimize a control objective for a linear system, is developed for a linearized Navier-Stokes system and then applied to transitional and turbulent flows, which are governed by the nonlinear Navier-Stokes equations. Standard linear control theoretic approaches, such as linear quadratic regulator (LQR) and linear quadratic Gaussian (LQG) synthesis, have been applied. It has been observed that these linear controllers work surprisingly well in achieving the control objective – minimizing disturbances – in spite of the fact that the flow under consideration is strictly nonlinear. These results suggest, once again, that the essential flow dynamics, especially those concerned with near-wall turbulence, are well represented in the linear system. This led Lim and Kim [1] to analyze turbulent channel flow from a linear system of view, which will be discussed in the next section.

3. SVD ANALYSIS OF LINEAR SYSTEM

We analyze a turbulent channel flow, as an example of wall-bounded turbulent shear flows, from a linear system perspective. The flow is statistically homogeneous in the streamwise (x) and spanwise (z) directions, which allows us to represent the governing equations in the Fourier space and the linearized Navier-Stokes equations decouples for each wave number, thus making the analysis simple. By representing the wall-normal velocity, v, and the wall-normal vorticity, ω, in terms of Fourier modes in the streamwise and the spanwise directions, the linearized incompressible Navier-Stokes equations can be written in the following operator form for each wave number pair,

$$\frac{\partial}{\partial t}\begin{bmatrix} v \\ \omega \end{bmatrix} = [\mathcal{A}]\begin{bmatrix} v \\ \omega \end{bmatrix}, \tag{1}$$

where the operator $\mathcal{A}$ represents the linearized Navier-Stokes system defined as

$$[\mathcal{A}] = \begin{bmatrix} L_{os} & 0 \\ L_c & L_{sq} \end{bmatrix}.$$

Here L_{os}, L_{sq} and L_c represent the Orr-Sommerfeld, Squire, and linear coupling operators, respectively, and are defined as

$$\begin{aligned} L_{os} &= \Delta^{-1}\left(-\mathrm{i}k_x U\Delta + \mathrm{i}k_x\frac{d^2U}{dy^2} + \frac{1}{Re}\Delta^2\right), \\ L_{sq} &= -\mathrm{i}k_x U + \frac{1}{Re}\Delta, \\ L_c &= -\mathrm{i}k_z\frac{dU}{dy}. \end{aligned} \tag{2}$$

Here, k_x and k_z are the streamwise and spanwise wavenumbers, respectively, $\Delta = \partial^2/\partial y^2 - k_x^2 - k_z^2$, and U is the time-averaged mean velocity on which the linearized form is based. Reynolds number, Re, is based on the wall-shear velocity, $u_\tau = \sqrt{\tau_w/\rho}$, and the channel half-width, h, where $\tau_w = \nu dU/dy|_w$ is the mean shear stress at the wall, and ν and ρ denote the kinematic viscosity and the density, respectively. All flow variables are non-dimensionalized by u_τ and h unless stated otherwise.

Equation(1) with control input can be written in the following state-space representation:

$$\frac{d\mathbf{x}}{dt} = \mathbf{A}\mathbf{x} + \mathbf{B}\mathbf{u}, \tag{3}$$

$$\mathbf{u} = -\mathbf{K}\mathbf{x}, \tag{4}$$

where $\mathbf{x}$ and $\mathbf{u}$ are defined as

$$\begin{aligned} \mathbf{x} &\equiv \begin{bmatrix} v_1 & \ldots & v_{N-1} & \omega_1 & \ldots & \omega_{N-1} \end{bmatrix}^T, \\ \mathbf{u} &\equiv \begin{bmatrix} v_0 & v_N \end{bmatrix}^T. \end{aligned}$$

The vector $\mathbf{x}$ represents a 'state' of the system, and it consists of the wall-normal velocity and wall-normal vorticity at each collocation point.[3] The subscripts denote collocation points in the wall-normal direction (0 and N correspond to the upper and lower wall, respectively). The other vector $\mathbf{u}$ represents 'control,' which is blowing and suction at the wall in the present study. Equation (3) represents a state equation inside the flow domain, which is being forced by the control input, $\mathbf{u}$, at the boundary of the domain. In linear optimal control theory, $\mathbf{K}$ is obtained such that the controlled system is linearly stable and a certain cost function is minimized (see, for example, [10] for further details).

By combining Eqs. (3) and (4), the system equation for controlled cases is given as $d\mathbf{x}/dt = (\mathbf{A} - \mathbf{BK})\mathbf{x}$. For uncontrolled cases, $\mathbf{K}$ is zero and the system equation simply becomes $d\mathbf{x}/dt = \mathbf{Ax}$. The traditional eigenvalue analysis, which predicts whether a linear system is stable or unstable based on the eigenvalues of the system, is inadequate in explaining the transient growth of the kinetic energy of certain disturbances in an otherwise stable system. Instead, the transient growth can be analyzed by applying the SVD analysis to the system operator, by which the amplification factor of the 'optimal' disturbance can be determined. It is assumed that the SVD analysis is also applicable for examining the performance of controllers for turbulent boundary layers. Effective controllers must reduce the singular values, which represent the transient energy growth, of the controlled system. Note that the reduction of singular value is related to the reduction of non-normality of the flow system, which is partially responsible for sustaining near-wall turbulence structures (which are in turn responsible for high skin-friction drag in turbulent boundary layers).

To analyze the transient energy growth through the SVD analysis, we consider the ratio of the kinetic energy of a disturbance at a given time (τ) to the disturbance energy at $t = 0$,

$$G(\tau) = \sup_{\mathbf{x}(\cdot,0)\neq 0} \frac{||\mathbf{x}(\tau)||^2}{||\mathbf{x}(0)||^2}, \tag{5}$$

where

$$||\mathbf{x}||^2 \equiv \int_{-1}^{1} \left[v^* v + \frac{1}{k_x^2 + k_z^2} \left(\frac{\partial v^*}{\partial y} \frac{\partial v}{\partial y} + \omega^* \omega \right) \right] dy,$$

and τ is the given time mentioned above. The quantity $||\mathbf{x}||^2$ represents the kinetic energy of $\mathbf{x}$ and can be expressed as $||\mathbf{x}(t)||^2 = \mathbf{x}^*(t)\mathbf{Q}\mathbf{x}(t)$, where the Hermitian matrix $\mathbf{Q}$ is defined in terms of an inner product in discrete space. The matrix $\mathbf{Q}$ can be further decomposed in the form $\mathbf{Q} = \mathbf{F}^*\mathbf{F}$, where $\mathbf{F}^*$ is

[3] We use the collocation points to represent a state vector, but other state vectors, such as a Gallerkin projection, can be used as well.

the Hermitian conjugate of $\mathbf{F}$. The solution of the system equation is simply given by $\mathbf{x}(t) = \exp[(\mathbf{A} - \mathbf{BK})t\,]\,\mathbf{x}(0)$. It follows that

$$\begin{aligned} ||\mathbf{x}(t)||^2 &= \mathbf{x}^*(t)\mathbf{F}^*\mathbf{F}\mathbf{x}(t) \\ &= ||\mathbf{F}\mathbf{x}(t)||_2^2 \\ &= ||\mathbf{F}\exp[(\mathbf{A} - \mathbf{BK})t\,]\mathbf{x}(0)||_2^2\,, \end{aligned} \tag{6}$$

where $||\cdot||_2$ represents the 2-norm (Euclidian Norm). Combining Eqs. (5) and (6), we obtain the growth ratio at $t = \tau$ as

$$\begin{aligned} G(\tau) &= \sup_{\mathbf{X}(\cdot,0)\neq 0} \frac{||\mathbf{F}\exp[(\mathbf{A} - \mathbf{BK})\tau\,]\mathbf{x}(0)||_2^2}{||\mathbf{F}\mathbf{x}(0)||_2^2} \\ &= \sup_{\mathbf{X}(\cdot,0)\neq 0} ||\mathbf{F}\exp[(\mathbf{A} - \mathbf{BK})\tau\,]\mathbf{F}^{-1}||_2^2\,. \end{aligned} \tag{7}$$

The 2-norm of a matrix can be easily computed from the SVD of the matrix. Typical SVD subroutines provide a diagonal matrix Σ and two orthogonal matrices $\mathbf{U}$ and $\mathbf{V}$, with which the original matrix can be expressed in the following form:

$$\mathbf{U}^*\mathbf{A}\mathbf{V} = \Sigma. \tag{8}$$

The column vectors of $\mathbf{V}$ and $\mathbf{U}$ are referred to as right and left singular vectors, respectively. The diagonal elements of $\boldsymbol{\Sigma}$ are the singular values, each of which represents the 2-norm ratio of corresponding column vectors of $\mathbf{V}$ and $\mathbf{U}$.

The relevant singular values (σ's) for the present analysis are the singular values of $\mathbf{F}\exp[(\mathbf{A} - \mathbf{BK})\tau]\mathbf{F}^{-1}$, representing the amplification of initial kinetic energy over time τ. In naturally evolving wall-bounded shear flows, only a few singular values are larger than one, as shown in Fig. 1, implying that only a few particular disturbances can have the transient growth. The largest σ represents the maximum energy growth ratio at τ, and the corresponding column vectors of $\mathbf{U}$ and $\mathbf{V}$ are the flow field at τ and the initial flow field, respectively. In other words, the initial flow field V_1 evolves in time τ to become U_1 with the growth ratio $G(\tau) = \sigma_1$, where σ_1 is the largest singular value. Note that the singular vectors are orthogonal to each other (both $\mathbf{U}$ and $\mathbf{V}$ are orthogonal matrices), and each singular vector can be expressed in terms of a combination of the eigenvectors of the system.

The singular vector V_1 corresponding to the largest $G(\tau)$ for all wavenumber pairs is the 'optimal' disturbance. The term 'optimal' was originally chosen in the sense that this disturbance would have the largest (optimal) transient growth [2]. In a turbulent (hence nonlinear) flow environment, the evolution of this disturbance represents the most probable – at least linearly – scenario to grow and survive in the disruptive environment. In a turbulent flow environment, the given time scale, τ, plays an important role in deter-

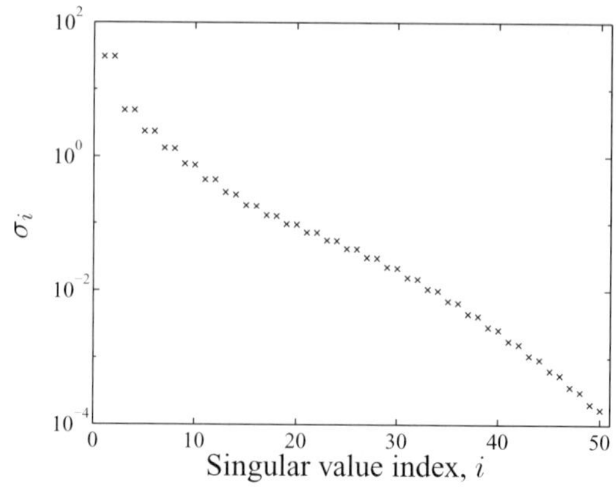

Figure 1: The first 50 singular values of a turbulent channel (Re_τ=180) for τ^+=80, k_x=0 and k_z=10.5.

mining the 'optimal' disturbance. The time scale that was 'globally' optimal for the maximum energy growth was found to be relatively large ($\sim \mathcal{O}(Re)$) and it was argued that such an 'optimal' disturbance could not attain its potentially maximum state, as nonlinear activities constantly disrupt the linear process [3]. Butler and Farrell [3] used the eddy turn-over time in the near-wall region, approximately $t^+ = 80$ (the superscript $^+$ denotes a variable non-dimensionalized by u_τ and ν/u_τ), for their τ, which resulted in the 'optimal' disturbance similar to those observed in turbulent boundary layers.

Singular values corresponding to a turbulent channel flow with various different control schemes have been examined, and the results are shown in Fig. 2. In addition to the channel flow with a linear optimal controller (*i.e.*, the control gain matrix **K** determined by an LQR synthesis), results from the opposition control of Choi *et al.* [23] (see Lim and Kim [1] for the structure of **K** corresponding to opposition control) and those from Kim and Lim's [13] virtual flow (*i.e.*, the operator **A** with $L_c = 0$) are also shown in Fig. 2. From the distribution of large singular values corresponding to different control approaches, the efficiency of each controller can be predicted, assuming that the present SVD analysis is still valid for the actual nonlinear system. Note that no singular values corresponding to the virtual flow is larger than one, indicating that there would be no transient growth in this case.

In Fig. 3, mean skin-friction drag history from numerical simulations of a turbulent channel flow (*i.e.*, nonlinear system) with various controllers are shown. Note that the case without the linear coupling term (virtual flow) resulted in complete laminarization, consistent with the SVD analysis. Other cases are also consistent with the SVD analysis, demonstrating that the SVD analysis is indeed a viable tool in predicting the performance of a controller in the nonlinear turbulent channel flow. It should be noted that these simulations were performed at an extremely low Reynolds number, and extending these results to much higher Reynolds number flows must be done with some care.

Lim and Kim [1] also applied the SVD analysis to opposition control with different detection planes in order to explain the observed behavior, and the results were consistent with the reported results: that is, the largest singular value is minimum with the detection plane located at $y^+ \approx$ 10-15, and larger than that for the uncontrolled case with the detection plane located beyond $y^+ \approx 20$. They also showed that the efficiency of opposition control would decrease as the Reynolds number increases, consistent with the LES results by Chang *et al.* [24], thus indirectly validating the application of the present SVD analysis, at least for opposition control, to higher Reynolds numbers.

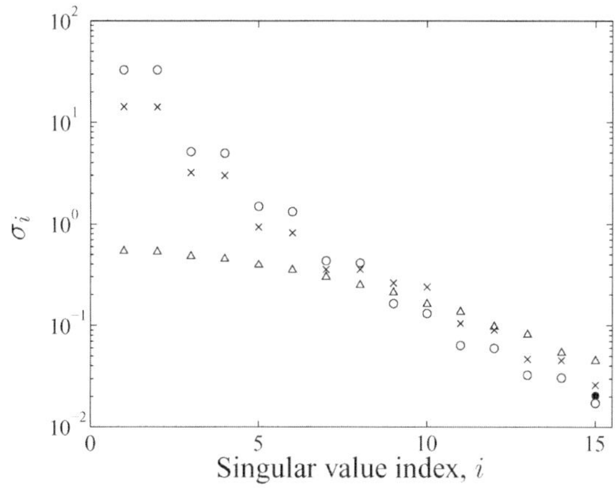

Figure 2: Singular values in a turbulent channel with different controllers: ∘, no control; •, opposition control; ×, LQR control; △, virtual flow. This is for the case of ($k_x = 0$, $k_z = 6.0$), corresponding to $\lambda_z^+ \approx 100$, and Re_τ=100. A similar trend can also be observed in other wavenumbers.

4. CONCLUDING REMARKS

We have shown that boundary layers can be analyzed from a linear system perspective. The SVD analysis can provide useful information regarding the controller's capability of attenuating the transient growth of disturbances in turbulent boundary layers. The trends observed from the SVD analysis were similar to those observed in DNS or LES of drag-reduced turbulent flows, illustrating that the linear system model can describe an important part of the near-wall dynamics and that it can be used as a guideline for various control designs for drag reduction. It could be used, for example, in optimizing control parameters without actually performing expensive nonlinear computations. Other issues, such as the effects of using the evolving mean flow as control applied to a nonlinear flow system (also known as gain scheduling), and high Reynolds number limitations, can also be addressed through the SVD analysis.

It is worth mentioning, however, that the linearized Navier-Stokes equations are not sufficient in general to describe many features of turbulent boundary layers, including the self-sustaining mechanism of near-wall turbulence, in which a nonlinear mechanism plays an essential role [20, 21]. This limitation notwithstanding, it is shown here that much can be learned from a proper linear analysis of nonlinear flows, especially for the wall-bounded turbulent shear flows in which a linear mechanism plays an important, if not dominant, role. In this regard, and to some extent paradoxically, wall-bounded turbulent shear flows are more amenable to a theoretical analysis than, say, homogeneous isotropic turbulent flows, where no dominant linear mechanism is present.

In 1904, Prandtl resolved the d'Alembert's Paradox by concluding that viscous stress cannot be ignored in the wall region due to the presence of the strong velocity gradient near the wall. At this celebration of the 100th anniversary of his seminal lecture, we conclude that the linear mechanism in turbulent boundary layers cannot be ignored due to the presence of the strong mean velocity gradient near the wall.

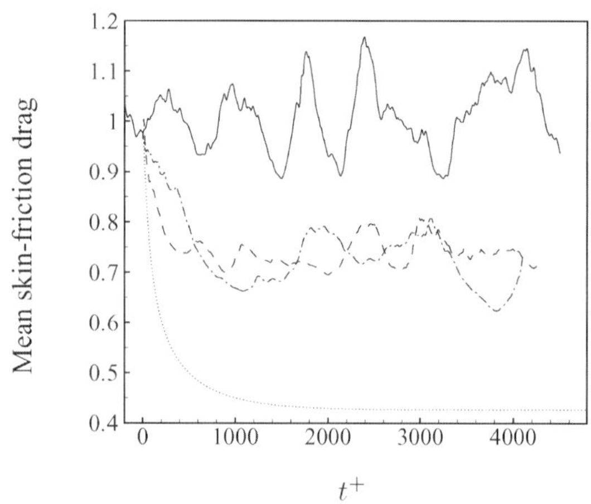

Figure 3: Mean skin-friction drag history with various control methods (Re_τ = 100). ——— , no control; - - - - , opposition control ($y_d^+ \approx 15$), – · – · – , LQR control (minimizing $\mathbf{x}^*\mathbf{x}$); ········ , virtual flow. Note that the same control parameters used in Fig. 2 are used here.

ACKNOWLEDGMENTS

This work has been supported in part by the Air Force Office of Scientific Research (Program Managers: Dr. Marc Jacobs, Dr. Berlinda King, Lt. Col. Sharon Heise, and Dr. Thomas Beutner). Computer time provided by NSF NPACI Centers (Pittsburgh Supercomputing Center in particular for JL) is also gratefully acknowledged.

REFERENCES

[1] J. Lim and J. Kim. A singular analysis of boundary layer control. *Phys. Fluids*, **16**(6):1980–1988, 2004.

[2] K. M. Butler and B. F. Farrell. Three-dimensional optimal perturbations in viscous shear flow. *Phys. Fluids A*, **4**(8):1637–1650, 1992.

[3] K. M. Butler and B. F. Farrell. Optimal perturbations and streak spacing in wall-bounded turbulent shear flow. *Phys. Fluids A*, **5**(3):774–777, 1993.

[4] S. C. Reddy and D. S. Henningson. Energy growth in viscous channel flows. *J. Fluid Mech.*, **252**:209–238, 1993.

[5] B. Bamieh and M. Dahleh. Energy amplification in channel flows with stochastic excitation. *Phys. Fluids*, **13**(11):3258–3269, 2001.

[6] C. Lee, J. Kim, D. Babcock, and R. Goodman. Application of neural networks to turbulence control for drag reduction. *Phys. Fluids*, **9**(6):1740–1747, 1997.

[7] C. Lee, J. Kim, and H. Choi. Suboptimal control of turbulent channel flow for drag reduction. *J. Fluid Mech.*, **358**:245–258, 1998.

[8] S. S. Joshi, J. L. Speyer, and J. Kim. A systems theory approach to the feedback stabilization of infinitesimal and finite-amplitude disturbances in plane Poiseuille flow. *J. Fluid Mech.*, **332**:157–184, 1997.

[9] K. Lee, L. Cortelezzi, J. Kim, and J. L. Speyer. Application of reduced-order controller to turbulent flows for drag reduction. *Phys. Fluids*, **13**(5):1321–1330, 2001.

[10] J. Lim. *Control of wall-bounded turbulent shear flows using modern control theory*. Ph.D. dissertation, University of California, Los Angeles, 2003.

[11] H. Högberg, T. R. Bewley, and D. S. Henningson. Linear feedback control and estimation of transition in plane channel flow. *J. Fluid Mech.*, **481**:149–175, 2003.

[12] H. Högberg, T. R. Bewley, and D. S. Henningson. Relaminarization of Re_τ=100 turbulence using gain scheduling and linear state-feedback control. *Phys. Fluids*, **15**(11):3572–3575, 2003.

[13] J. Kim and J. Lim. A linear process in wall-bounded turbulent shear flows. *Phys. Fluids*, **12**(8):1885–1888, 2000.

[14] T. Ellingsen and E. Palm. Stability of linear flow. *Phys. Fluids*, **18**(4):487–488, 1975.

[15] M. T. Landahl. Wave breakdown and turbulence. *SIAM J. App. Math.*, **28**:735–756, 1975.

[16] L.S. Hultgren and L.H. Gustavsson. Algebraic growth of disturbances in a laminar boundary layer. *Phys. Fluids*, **24**(6):1000–1004, 1981.

[17] L.H. Gustavsson. Energy growth of three-dimensional disturbances in plane poiseuille flow. *J. Fluid Mech.*, **224**:241–260, 1991.

[18] L. N. Trefethen, A. E. Trefethen, S. C. Reddy, and T. A. Driscoll. Hydrodynamic stability without eigenvalues. *Science*, **261**:578–584, 1993.

[19] E. Reshotko. Transient growth: A factor in bypass transition. *Phys. Fluids*, **13**(5):1067–1075, 2001.

[20] J. M. Hamilton, J. Kim, and F. Waleffe. Regeneration mechanisms of near-wall turbulence structures. *J. Fluid Mech.*, **287**:317–348, 1995.

[21] F. Waleffe. Transition in shear flows. Nonlinear normality versus non-normal linearity. *Phys. Fluids*, **7**(12):3060–3066, 1995.

[22] T. R. Bewley, P. Moin, and R. Temam. Dns-based predictive control of turbulence: an optimal benchmark for feedback algorithms. *J. Fluid Mech.*, **447**:179–225, 2001.

[23] H. Choi, P. Moin, and J. Kim. Active turbulence control for drag reduction in wall-bounded flows. *J. Fluid Mech.*, **262**:75–110, 1994.

[24] Y. Chang, S. S. Collis, and S. Ramakrishnan. Viscous effects in control of near-wall turbulence. *Phys. Fluids*, **14**(11):4069–4080, 2002.

THE DEVELOPMENT (AND SUPPRESSION) OF VERY SHORT-SCALE INSTABILITIES IN MIXED FORCED-FREE CONVECTION BOUNDARY LAYERS

P.W. Duck[1], J.P. Denier[2] and J. Li[2]
[1]*Department of Mathematics, The University of Manchester*
[2]*School of Mathematical Sciences, The University of Adelaide*

Abstract: The two-dimensional boundary-layer flow over a cooled/heated flat plate is investigated. A cooled plate (with a free-stream flow and wall temperature distribution which admit similarity solutions) is shown to support non-modal disturbances, which grow algebraically with distance downstream from the leading edge of the plate. In a number of flow regimes, these modes have diminishingly small wavelength, which may be studied in detail using asymptotic analysis. Corresponding non-self-similar solutions are also investigated. It is found that there are important regimes in which if the temperature of the plate varies (in such a way to break self-similarity), then standard numerical schemes exhibit a breakdown at a finite distance downstream. This breakdown is shown to be related to very short-scale disturbance modes, which manifest themselves by means of the spontaneous formation of an essential singularity at a finite downstream location. We show how these difficulties can be overcome by treating the problem in a quasi-elliptic manner, in particular by prescribing suitable downstream (in addition to upstream) boundary conditions.

1. INTRODUCTION AND FORMULATION

With the exception of [10], all results to date concerning the stability of mixed forced-free convection boundary layers have focused upon wave-like (normal mode) disturbances. Recently, however, there has been renewed interest in the problem of non-modal disturbances (both two and three-dimensional) in boundary-layer flows, a problem first studied in [7]. Ref. [8] considered the instability of a flat plate boundary layer to three-dimensional non-modal (in the streamwise direction) disturbances. Such disturbances are ascribed to the interplay between inviscid algebraic growth and viscous dissipation and it is demonstrated that, although viscous dissipation provides for an algebraic decay, it is unable to overcome the inviscid algebraic growth. This is in stark contrast to the equivalent problem in a parallel flow where algebraic decay resulting from viscous dissipation does serve to damp out the inviscid algebraic growth, thus resulting in a flow which is stable

G.E.A. Meier and K.R. Sreenivasan (eds.), IUTAM Symposium on One Hundred Years of Boundary Layer Research, 313-323,

to such disturbances. The existence of algebraically-growing disturbances (within the context of the boundary-layer flow) provides a driving mechanism for the linear amplification stage observed in experiments on by pass transition. Ref. [9] considered the question of optimal (three-dimensional) disturbances within the boundary-layer flow, thus strengthening the link between algebraic disturbances and by-pass transition for further discussion on algebraically growing disturbances see [1] for spatial growth and [11] for temporal growth).

We should emphasise here that Refs. [8], [9] and [1] have demonstrated algebraically growing three-dimensional disturbances; the base flows they consider are stable to two-dimensional disturbances. Here we are concerned with the Reynolds number independent (in)stability of a class of mixed forced-free convection boundary-layer flows, by investigating two-dimensional perturbations of a class analogous to those of Ref. [7], but unlike Ref. [7] it is found that these can grow algebraically downstream.

Consider then the two-dimensional motion of a steady, incompressible Boussinesq fluid flowing over a heated/cooled flat plate. Fluid velocities are scaled on a typical free-stream speed U_∞ and the distances along (x) and normal to (y) the plate expressed relative to a characteristic length L. The pressure $\tilde{p}$ is non-dimensionalized using $\rho_0 U_\infty^2$ where ρ_0 is the density of the fluid at the temperature of the plate (T_0) and the fluid temperature $\tilde{T}$ is written relative to the difference $T_0 - T_\infty$, where T_∞ is the temperature of the free-stream fluid. The Prandtl number is denoted by σ (taken to be equal to 0.72 in what follows) and the Reynolds number Re takes its usual form equal to $U_\infty L/\nu$ where ν represents the kinematic viscosity of the fluid. Lastly we define $G = Gr\, Re^{-2}$ where $Gr = g\beta L^3(T_0 - T_\infty)/\nu^2$ is the Grashof number; g denotes the acceleration due to gravity, and β is the coefficient of volume expansion.

When $Re \gg 1$ the velocity, pressure and temperature fields within the boundary layer can be written as

$$(\tilde{u}, \tilde{v}, \tilde{T}, \tilde{p}) = (u, \mathrm{Re}^{-1/2} v, T, -\frac{1}{2}u_e^2(x) + \mathrm{Re}^{-1/2} G_{0p}) + ...,$$

where the quantities on the right-hand-side are functions of x and the boundary-layer variable $Y = Re^{-1/2} y$. The governing (boundary-layer-type) equations then become (to leading order in powers of Re)

$$\frac{\partial u}{\partial x} + \frac{\partial v}{\partial Y} = 0, \quad u\frac{\partial u}{\partial x} + v\frac{\partial v}{\partial Y} = u_e\frac{\mathrm{d}u_e}{\mathrm{d}x} - G_0\frac{\partial p}{\partial x} + \frac{\partial^2 u}{\partial Y^2},$$

$$\frac{\partial p}{\partial Y} = T, \quad u\frac{\partial T}{\partial x} + v\frac{\partial T}{\partial Y} = \frac{1}{\sigma}\frac{\partial^2 T}{\partial Y^2}, \tag{1}$$

subject to the boundary conditions $u = v = 0$, $T = T_w(x)$ on the horizontal flat plate situated at $Y = 0$ and that $u \to u_e(x)$, $T \to 0$ as $Y \to \infty$. In the above, the apparently small term $G_0 = Re^{-1/2}G$ has been retained. In this formulation G_0 is defined with respect to a typical wall temperature and so is a constant for a given flow. The extreme cases in which $G_0 \ll 1$ or $G_0 \gg 1$ are respectively referred to as forced and free convection boundary layers, whereas our interest here is with the intermediate regime for which $G_0 = O(1)$, the mixed forced-free convection problem.

We consider a general similarity-like solution to the boundary-layer equations (1) by setting

$$u = x^m f'(\eta, x), T = x^{(5m-1)/2} g(\eta, x), p = x^{2m} q(\eta, x), \tag{2}$$

$$v = -\frac{1}{\sqrt{2}}\left((m+1)x^{\frac{m-1}{2}} f + (1-m)\eta x^{\frac{m-1}{2}} f' - 2x^{\frac{m-1}{2}} f x\right),$$

where we have set $u_e = x^m$, $T_w = x^{(5m-1)/2} g_w(x)$. The "similarity" variable is given by $\eta = Y/(\sqrt{2}x^{(1-m)/2})$. Under this transformation the boundary-layer equations become

$$f''' + 2m(1-(f')^2) + (m+1) f f'' = 2x\left(f'\frac{\partial f'}{\partial x} - f''\frac{\partial f}{\partial x}\right)$$

$$+ G_0\left(4mq + (m-1)\eta q' + 2x\frac{\partial q}{\partial x}\right), \tag{3a}$$

$$\frac{1}{\sigma} g'' - (5m-1) g f' + (m+1) f g' = 2x\left(f'\frac{\partial g}{\partial x} - g'\frac{\partial f}{\partial x}\right) \tag{3b}$$

$$q' = \sqrt{2} g, \tag{3c}$$

(where a prime denotes differentiation with respect to η). This system must be solved subject to the boundary conditions

$$f = 0, f' = 0, g = g_w(x) \text{ on } \eta = 0, \tag{4a}$$

$$f' \to 1, g \to 0, q \to 0 \text{ as } \eta \to \infty. \tag{4b}$$

Note that although we have introduced similarity-like variables, this certainly does not preclude us from studying non-similarity-type flows. Indeed, the latter are an important component of this work. However, since the spatial development of the solution will, close to the leading edge, $x = 0$, take on a locally similar form, the variables introduced above enable us to progress the solution downstream in an entirely natural manner.

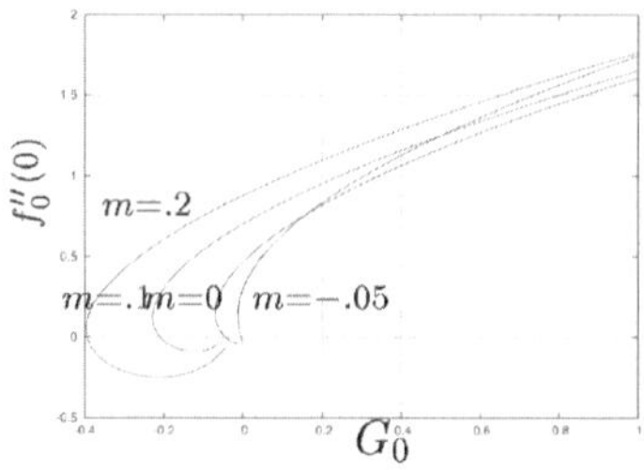

Figure 1. Variation of $f_0''(0)$ with G_0

2. SIMILARITY FLOWS

In this section we study first the form of similarity solutions of (3), and then a particular class of disturbance to these base flows. The similarity forms (denoted here by $f_0(\eta)$, $g_0(\eta)$, $q_0(\eta)$) may be obtained from (3) by merely setting $x = 0$ and solved subject to (4). Results for $f''(0)$ for $m = -.05$, 0, .1 and .2, over a range of G_0 are shown in figure 1. These base-flow (similarity) results are reminiscent of the well-known [6] distributions encountered in classical Falkner-Skan distributions of wall shear versus Hartree parameter (effectively our parameter 'm'). In particular the non-uniqueness in the solution for negative values of G_0 mimics the non-uniqueness for the classical Falkner-Skan equations found for negative values of the Hartree parameter. The question as to which solution is physically relevant has been considered in Ref. [10].

We now consider stability issues related to these similarity solutions by considering perturbations to the basic boundary-layer flow of the form (analogous to those of Ref. [7])

$$(f, g, q) = (f_0(\eta), g_0(\eta), q_0(\eta)) + x^{\lambda}(f_1(\eta), g_1(\eta), q_1(\eta)) + O(x^{2\lambda}), \quad (5)$$

where it is implicitly assumed that $x \ll 1$. Note that the leading-order terms in this expansion are functions of η alone; this is only appropriate if T_w is constant and so we will, without loss of generality, set $T_w = 1$ for the remainder of this section.

Taking the $O(x^{\lambda})$ terms when (5) is substituted into (3) we obtain

$$f_1''' + (m+1) f_0 f_1'' - 2(2m+\lambda) f_0' f_1' + (m+1+2\lambda) f_1 f_0''$$
$$= G_0\left(2(2m+\lambda)q_1 + (m-1)\eta q_1'\right), \quad (6a)$$

$$\frac{1}{\sigma} g_1'' - (5m - 1 + 2\lambda) g_1 f_0' + (m+1) f_0 g_1' = -(m + 1 + 2\lambda) f_1 g_0' + (5m-1) g_0 f_1', \tag{6b}$$

$$q_1' = \sqrt{2} g_1, \tag{6c}$$

which are subject to the homogeneous boundary conditions

$$f_1 = 0, f_1' = 0, g_1 = 0 \quad \text{on} \quad \eta = 0, \tag{7a}$$

and

$$f_1' \to 0, g_1 \to 0, q_1 \to 0 \text{ as } \eta \to \infty. \tag{7b}$$

This system represents an eigenvalue problem for λ as a function of G_0 (and the Prandtl number σ); details of the numerical scheme used to solve this system can be found in Ref. [4].

In all the cases (i.e. for all values of the pressure gradient parameter m) studied, for heated boundary layers, namely $G_0 > 0$, only negative values for λ were found, which, in the context of the present paper are of limited interest, although these are of relevance in the far-downstream behaviour of flow disturbances, in the sense of [7].

We focus on regimes which admit positive values of the eigenvalue λ, and therefore flows which can support downstream algebraically-growing disturbances. Figure 2a presents results for the case $m = 0$, in particular distributions of (the logarithm of) λ are shown as a function of the wall shear (for later cases, this turns out to be advantageous, rather than as a function of G_0, given the aforementioned non-uniqueness). In this case just one (real)

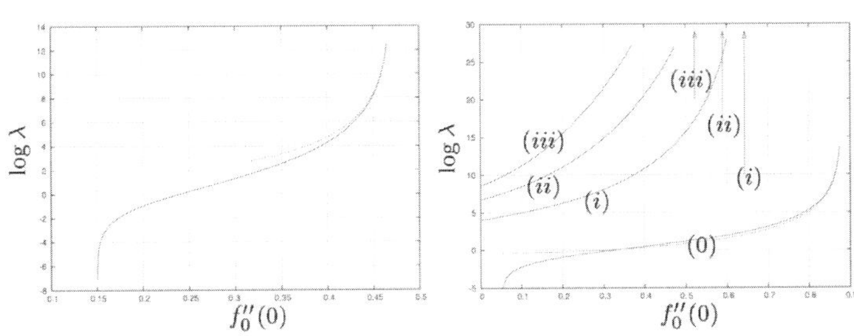

Figure 2. Plot of (a) the positive eigenvalue for $m = 0$ (asymptotic values shown as broken line) and (b) the positive eigenvalues for $m = 0.2$ (asymptotic values/locations shown as broken line/arrows)

positive eigen-value was found over the range shown, and the following are the key observations: (i) as $G_0 \to 0-$, $\lambda \to \infty$; (ii) as the 'nose' of the $f_0''(0)$ versus G_0 curve is approached, i.e. as $G_0 \to -.0699 \ldots, f''(0) \to 0.149..$, then so $\lambda \to 0$; (iii) for the lower branch solution, i.e. for $0.149\ldots > f_0''(0) > 0$ only negative values of λ were encountered; (iv) for $f_0''(0) < 0$ (that is, for reversed flow solutions) a large (probably infinite number) of positive values of λ were encountered (these are *not* shown) and (v) the vast majority of eigenvalues were real. Note that (i) will be considered in detail below, (ii) is an inevitable consequence of the non-uniqueness, whilst (iv) is a reflection of the 'ellipticity' of the flow in the case of flow reversal.

A second set of results is presented (figure 2b), which is for $m = 0.2$. Although in this case the vast majority of eigenvalues appeared to be real (again), the results are intriguingly qualitatively different from the results for the previous choice of m. In this case, although the unboundedness in λ as $G_0 \to 0-$ is clearly present again (this mode terminating with $\lambda = 0$ at the nose of the distribution curve, figure 1, as G_0 is reduced, additional ($\lambda > 0$) modes form, with infinite magnitude arise at other (negative) values of G_0. Figure 2b shows the first four modes found; many others appeared to arise at progressively more negative values of G_0.

From these results, it is quite clear that there are two (distinct) limits leading to $\lambda \to \infty$, the first as $G_0 \to 0-$, the second at discrete, non-zero critical values of G_0 (which seem to occur only for $m > 0$); these two limits are both important, and have been analysed in detail by [4]. In particular, in the latter case, the critical values of G_0 (or alternatively $f_0''(0)$) at which modes appear can be predicted, and these are indicated by arrows in figure 2b.

3. THE NON-PARALLEL EVOLUTION OF THE BOUNDARY LAYER

We now turn our attention to the question of the non-parallel development of the buoyant boundary layer, in particular to the question of how the boundary layer responds when the wall temperature is variable (or more precisely, decreases with distance downstream from the leading edge of the plate). Our initial interest in this problem arose because of a desire to understand the phenomena of velocity overshoot in buoyant mixed forced-free convection boundary layers and whether this overshoot could be controlled through a judicious choice of wall temperature. The development of overshoot is intimately linked with the generation of short wave instabilities in this class of boundary-layer flow. These short wave instabilities, which should not to be confused with those described in this paper, manifest themselves in the form of finite amplitude waves travelling at the maximum speed of the basic flow; full details can be found in [3] and [2].

To tackle the question of spatially developing (i.e. non-self-similar) flows, we revert to a consideration of system (3) - (4b). The self-similarity is broken by the choice of wall temperature which varies downstream as

$$g_w(x) = e^{-x} + \gamma(1 - e^{-x}), \tag{8}$$

where here γ is treated as a parameter, controlling the downstream evolution of the flow. Consistent with the form (8), the initial profiles at $x = 0$ may be taken as the similarity solutions obtained from the previous section. A second-order finite-difference/Crank-Nicolson scheme (coupled with Newton iteration) was employed to march the solution downstream.

Figure 3 shows the downstream development of the wall-shear stress ($f''(\eta = 0)$), for the case $m = 0$, $\gamma = -0.1$, $G_0 = 0.5$. These results were obtained using four streamwise grid sizes, namely $\Delta x = 10^{-2}$, 10^{-3}, 10^{-4} and 10^{-5} (for a fixed transverse gridsize, $\Delta\eta = 5 \times 10^{-3}$). It is immediately apparent that these results suffer a spontaneous breakdown, characterised by sudden oscillations (which on close inspection are of a streamwise point-to-point nature). Other flow quantities (e.g. the wall temperature gradient) exhibit the same type of behaviour. The genesis of this is highly grid dependent. There is clearly no sign of flow reversal occurring, nor of any other 'suspicious' base-flow behaviour immediately prior to this event. Computations with γ=0.25 but with all other parameters unchanged from the previous example led to solutions that proceeded downstream, unabated, with the far-downstream form being (asymptotically) approached.

In order to explore the reason for the "breakdown" (or lack of it) a procedure based upon (5) developed for the similarity states was adopted. We seek local solutions of the form

$$\begin{aligned}(f, g, q) = &(f_0(\eta; x), g_0(\eta; x), q_0(\eta; x)) \\ &+ \varepsilon(f_1(\eta; x), g_1(\eta; x), q_1(\eta; x)) \exp\Theta(x) + O(\varepsilon^2),\end{aligned} \tag{9}$$

where the amplitude ε is assumed small.

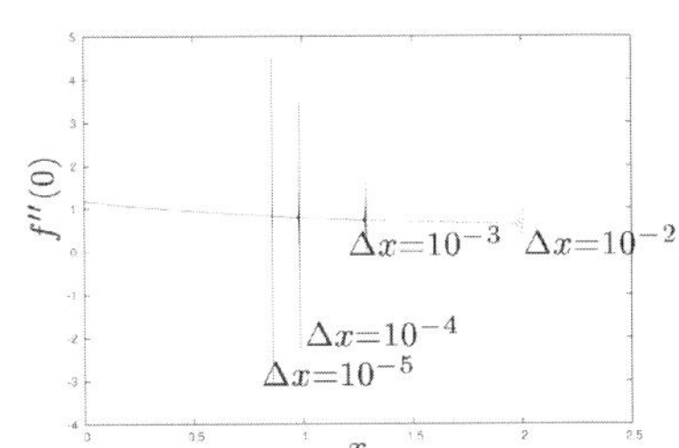

Figure 3. Spatial development of wall-shear stress, $m = 0$, $G_0 = 0.5$, $\gamma = -.1$.

From the onset it must be stressed that the approach to be adopted here may be regarded as somewhat heuristic, but nonetheless extremely useful in understanding the difficulties experienced in the numerical marching computations, detailed above, and does become increasingly valid in the short-wavelength limit, $|x\Theta_x| \to \infty$. Taking the $O(\varepsilon)$ terms when (9) is substituted into (3) yields

$$f_1'''+(m+1)f_0f_1''-2(2m+x\Theta_x)f_0'f_1'+(m+1+2x\Theta_x)f_1f_0''$$
$$-G_0\left(2(2m+x\Theta_x)q_1+(m-1)\eta q_1'\right)= \qquad (10a)$$
$$2x\left(f_0'\frac{\partial f_1'}{\partial x}+f_1'\frac{\partial f_0'}{\partial x}-f_0''\frac{\partial f_1}{\partial x}-f_1''\frac{\partial f_0}{\partial x}\right),$$

$$\frac{1}{\sigma}g_1''-(5m-1+2x\Theta_x)g_1f_0'+(m+1)f_0g_1'+(m+1+2x\Theta_x)f_1g_0'$$
$$+(5m-1)g_0f_1'=2x\left(f_0'\frac{\partial g_1}{\partial x}+f_1'\frac{\partial g_0}{\partial x}-g_0'\frac{\partial f_1}{\partial x}-g_1'\frac{\partial f_0}{\partial x}\right), \qquad (10b)$$

$$q_1'-\sqrt{2}g_1=0, \qquad (10c)$$

subject to (6); as before, primes denote differentiation with respect to η. We now make the assertion that both the base flow f_0, g_0, q_0 and perturbation quantities f_1, g_1, q_1 are slowly varying in the streamwise direction, thereby permitting the neglect of the right-hand-side terms in (10). If we then write $\lambda = x\Theta_x$, we recover (6). This system was solved in precisely the same manner as that employed previously, except the analysis was performed at each streamwise location (i.e. on the corresponding local base flow profile). Results for $\gamma = -0.1$ are shown in figure 4. Here we have only shown the 'unstable' eigenvalue; it is immediately apparent that a large (infinite) eigenvalue forms at a finite downstream location, which therefore suggests that infinitely short wavelength disturbances are responsible for the numerical marching difficulties experienced with $\gamma = -0.1$ and $\gamma = 0.1$. Similar eigenvalue searching procedures were adopted for the case $\gamma = 0.25$, but these failed to detect any positive values of λ, an observation entirely consistent with the lack of difficulties encountered with the marching scheme in this case.

We can make further analytical progress on understanding this phenomenon by supposing that $\lambda \to \infty$ as $x \to x_0$. We then seek a disturbance whose wavelength (which must be determined by an asymptotic balance) is $O(|x-x_0|^3)$. So in (9) $\Theta(x) = \Theta(x) = \dfrac{\lambda_0}{(x-x_0)^2}$, which leads to

$$f'_{00} f_1'' + \left[\frac{\sqrt{2} g_0^{crit} g'_{00}}{f'_{00}} - f'''_{00} \right] f_1 = 0. \tag{11}$$

(subject to $f_1(0) = 0, f_1' (\eta \to \infty) \to 0$) where f_{00} and g_{00} correspond to the *local* base-flow solution (evaluated at $x = x_0$). The system was solved at each streamwise location, and when the eigenvalue G_0^{crit} corresponded to the actual value of G_0, this location was then deemed to be the point where infinite spatial (λ) eigenvalues appeared, i.e. $x = x_0$. This location is clearly marked by a vertical arrow on figure 4. Inspection of this figure indicates consistency between the numerical results obtained from (10) and the predicted origin of these modes (x_0) obtained from (11).

4. THE SUPPRESSION OF THE INSTABILITY

The difficulties associated with the failure of marching schemes are a serious restriction on the usefulness of the procedure. In order to overcome these difficulties, system (3) was treated quasi-elliptically. As before, boundary conditions at the leading edge were imposed (using the appropriate similarity solution), whilst Neumann boundary conditions were imposed (at a finite x location) *downstream* (this treatment worked well in the study of [5] and proved very effective in the present study); this condition is completely consistent with the imposed conditions, notably (8), which is expected to lead to a similarity form far downstream. Second-order central differencing was used in both the η and x directions. In order to solve the resulting nonlinear set of algebraic equations, Newton iteration was employed, i.e. the entire flowfield was calculated simultaneously.

Results using this quasi-elliptic procedure are shown in figure 5; this corresponds to the case computed earlier with the marching routine, as illustrated in figure 3, namely $\gamma = -0.1$, $G_0 = 0.5$ and $m = 0$. The elliptic-type procedure has no difficulty in computing solutions. It is clear that the imposition of (reasonable) downstream conditions leads to a complete

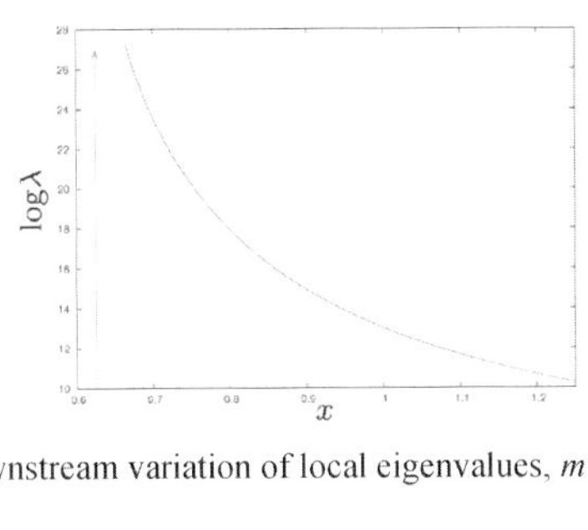

Figure 4. Downstream variation of local eigenvalues, $m = 0$, $G_0 = 0.5$, $\gamma = -.1$

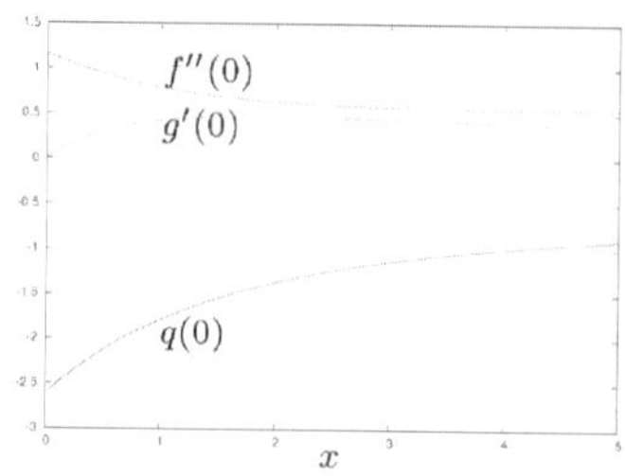

Figure 5. Spatial development of wall-quantities, $m = 0$, $G_0 = 0.5$, $\gamma = -.1$

suppression of the small-scale instabilities, over which marching schemes have no control. This same procedure also yielded completely regular results for other cases which encountered difficulties with the numerical marching scheme of the previous section.

5. CONCLUSIONS

Flows of the type considered in this paper exhibit a number of interesting phenomena, most of which are linked to the occurrence of algebraically-growing (in the downstream direction) eigensolutions. Instabilities of this type have received a great deal of attention recently, in particular with regard to transient growth in boundary layers ([8], [1], [9]). However there is one important distinction between the current work, and these previous studies, insofar as the present problem exhibits algebraic instabilities in a two-dimensional context, whilst in the aforementioned boundary-layer studies, three-dimensionality was an inherent necessity for the occurrence of such modes. In the present study, one requirement is that the similarity forms are algebraically unstable and for this to occur it is necessary that the wall temperature be lower than that of the freestream, i.e, the requirement is that $G_0 < 0$.

Non-similar flows are especially intriguing. The results of section clearly reveal the further subtleties associated with flows of this type, especially the occurrence of eigensolutions which form spontaneously through the formation of essential singularities. These in turn lead to exceedingly challenging numerical tasks. The present work also reveals that heated plates can also be susceptible to these numerical difficulties. However, the quasi-elliptic treatment of the non-self-similar flows bypasses the difficulties associated with the triggering of (very) short wavelength disturbances; the imposition of downstream boundary conditions appears to render the problem well posed, and this leads to sensible (credible) solutions extending from the leading edge of the plate, to far downstream.

ACKNOWLEDGEMENTS

JPD and JL gratefully acknowledge the financial support of the Australian Research Council through grant DP0210877.

REFERENCES

[1] Andersson, P., Berggren, M. & Henningson, D.S. 1999 Optimal disturbances and bypass transition in boundary layers. *Phys. Fluids* **11**, 134.

[2] Denier, J.P. & Bassom, A.P. 2003 The non-parallel evolution of nonlinear short waves in buoyant boundary layers. *Stud. Appl. Math.*, **110**, 139–156.

[3] Denier, J.D. & Mureithi, E.W. 1996. Weakly nonlinear wave motions in a thermally stratified boundary layer, *J. Fluid Mech.* **315**, 293–316.

[4] Denier, J.D., Duck, P.W. & Li, J. 2005 On the growth (and suppression) of very short-scale disturbances in mixed forced-free convection boundary layers. *J. Fluid Mech.* **526**, 147–170.

[5] Duck, P.W., Stow, S. & Dhanak, M.R. 1999 Non-similarity solutions to the corner boundary-layer equations (and the effects of wall transpiration). *J. Fluid Mech.* **400**, 125.

[6] Hartree, D.R. 1937 On an equation occurring in Falkner and Skan's approximate treatment of the equations of the boundary layer. *Proc. Camb. Phil. Soc.* **33**, 223.

[7] Libby, P.A. & Fox, H. 1964 Some perturbation solutions in laminar-boundary layer theory. *J. Fluid Mech.* **17**, 433.

[8] Luchini, P. 1996 Reynolds-number-independent instability of the boundary layer over a flat surface. *J. Fluid Mech.* **327**, 101.

[9] Luchini, P. 2000 Reynolds-number-independent instability of the boundary layer over a flat surface: optimal perturbations. *J. Fluid Mech.* **404**, 289.

[10] Steinrück, H. 1994 Mixed convection over a cooled horizontal plate: non-uniqueness and numerical instabilities of the boundary-layer equations. *J. Fluid Mech.* **278**, 251.

[11] Trefethen, L.N., Trefethen, A.E., Reddy, S.C. & Driscoll, T.A. 1993 Hydrodynamic stability without eigenvalues. *Science* **261**, 578.

COMPUTATIONAL STUDIES OF BOUNDARY-LAYER DISTURBANCE DEVELOPMENT

Christopher Davies

Cardiff School of Mathematics, Cardiff University, Senghennydd Road, Cardiff, CF24 4AG, UK. E-mail: DaviesC9@cardiff.ac.uk

Abstract: Numerical simulations can play a complementary role to physical experiments in the development of theoretical explanations of boundary-layer disturbance behaviour. We briefly review a computationally advantageous vorticity-based formulation of the Navier-Stokes equations that can be used to conduct such simulations. The formulation has the merit that it facilitates the imposition of no-slip conditions by means of fully-equivalent integral constraints on the vorticity. Various features of an associated numerical discretization scheme are discussed.

Keywords: Boundary layer, disturbances, no-slip conditions, wall compliance, velocity-vorticity formulation, numerical methods

1. INTRODUCTION

One hundred years after Prandtl first introduced the concept of the boundary layer into fluid dynamics, the behaviour of disturbances in laminar boundary layers still remains an active area of research. Even for the simplest incompressible boundary layers, such as Blasius flow past an aligned flat plate, there are open questions about the generation, spatio-temporal evolution and possibilities for control of the disturbances that trigger the transition between laminar and turbulent flow. Amongst various reasons why the study of such boundary-layer disturbances continues to be interesting and demanding, the following are particularly important: (i) it is often a significant mathematical challenge to determine even approximate analytical solutions of an appropriately chosen set of governing equations; (ii) if such solutions can in fact be obtained, there may be no straightforward and unambiguous way of matching them to behaviour that is readily observed in physical experiments at Reynolds numbers for which laminar flow can be maintained. Comparison between mathematical theory and experiment can still be problematic when computer-based experiments are

G.E.A. Meier and K.R. Sreenivasan (eds.), IUTAM Symposium on One Hundred Years of Boundary Layer Research, 325-334,

employed in order to complement the data available from physical experiments. Resolution requirements, limited computational resources and the numerical stability properties of feasible algorithms all conspire to limit the range of Reynolds numbers that can be effectively explored by means of computer simulation.

To have any hope of observing the operation of a given theoretically predicted mechanism in the results obtained from a computer experiment, it is clearly necessary to take the theory into account when setting up the finest scale numerical resolution and the overall size of the computational domain. For example, if it is anticipated that there will be disturbances that are governed by triple-deck theory [1], then the spatial grid used in the simulations will need to be fine enough to resolve a thin viscous sublayer that lies immediately adjacent to the wall. If δ is the boundary-layer thickness, then this viscous sublayer will only be of vertical extent $\delta\ R_\delta^{-1/4}$, where R_δ is the local Reynolds number based on δ. But the computational domain will also need to be large enough to include structures that have lengths that are of order $\delta\ R_\delta^{1/4}$, in both the streamwise and the vertical direction. This is to capture any effects attributable to an upper deck, where disturbances take on an inviscid and irrotational character. For large values of the Reynolds number, numerical simulations that meet such requirements can become very computationally demanding. Thus it is important to develop methods for conducting simulations that are as efficient as possible.

2. DISTURBANCE EQUATIONS

Within the confines of such a short paper it is not possible to give a fair and systematic review of the merits of the various formulations of the Navier-Stokes equations that might be used to determine the development of disturbances in an incompressible boundary layer. Instead we will just provide a summary statement of one particular vorticity-based formulation [2] and highlight its advantages for computational purposes.

Velocity-vorticity formulation

We will consider the governing equations for the flow perturbation variables only, by first decomposing the total velocity and vorticity fields into the form $\mathbf{U} = \mathbf{U}^B + \mathbf{u}$, $\boldsymbol{\Omega} = \boldsymbol{\Omega}^B + \boldsymbol{\omega}$, where the superscript B is used to distinguish the boundary-layer meanflow, which is assumed to be already known. The boundary layer is taken to be located entirely above a surface that is positioned at $z = \eta(x, y, t)$.

The components of the perturbation flow variables $\mathbf{u} = (u, v, w)$, $\boldsymbol{\omega} = (\omega_x, \omega_y, \omega_z)$ are divided into two sets. Namely, the primary variables $\{\omega_x, \omega_y, w\}$ and the secondary variables $\{\omega_z, u, v\}$. This terminology is adopted because the secondary variables can be explicitly defined in terms of the primary variables by performing an integration along the wall-normal z-direction, across the boundary layer, in the following manner:

$$u = -\int_z^\infty \left(\omega_y + \frac{\partial w}{\partial x}\right) dz \tag{1}$$

$$v = \int_z^\infty \left(\omega_x - \frac{\partial w}{\partial y}\right) dz \tag{2}$$

$$\omega_z = -\int_z^\infty \left(\frac{\partial \omega_x}{\partial x} + \frac{\partial \omega_y}{\partial y}\right) dz. \tag{3}$$

The governing equations for the disturbances can then be written in the form:

$$\frac{\partial \omega_x}{\partial t} + \frac{\partial N_z}{\partial y} - \frac{\partial N_y}{\partial z} = \frac{1}{R}\nabla^2 \omega_x \tag{4}$$

$$\frac{\partial \omega_y}{\partial t} + \frac{\partial N_x}{\partial z} - \frac{\partial N_z}{\partial x} = \frac{1}{R}\nabla^2 \omega_y \tag{5}$$

$$\nabla^2 w = \frac{\partial \omega_x}{\partial y} - \frac{\partial \omega_y}{\partial x}, \tag{6}$$

where R is the Reynolds number and the convective quantity $\mathbf{N} = (N_x, N_y, N_z)$ is given by $\mathbf{N} = \boldsymbol{\Omega}^B \times \mathbf{u} + \boldsymbol{\omega} \times \mathbf{U}^B + \boldsymbol{\omega} \times \mathbf{u}$. The two vorticity transport equations and the single Poisson equation stated immediately above specify a velocity-vorticity system of three governing equations for the three unknown primary variables $\{\omega_x, \omega_y, w\}$. The secondary variables, which are only required in order to evaluate the components of the quantity $\mathbf{N}$, can all be eliminated because each of them can be expressed solely in terms of the primary variables. It may be shown [2] that the stated velocity-vorticity system remains fully equivalent to the primitive variables formulation of the Navier-Stokes equations, provided that fairly weak conditions can be imposed on the primary variables for $z \to \infty$.

No-slip conditions

A very obvious advantage of the velocity-vorticity system (4)-(6) is that there is a reduction in the number of variables and governing equations, compared with other formulations of the Navier-Stokes equations for three-dimensional incompressible flow. But another, arguably more significant, advantage is that the no-slip conditions can be imposed in an elegant and mathematically self-consistent fashion. If the perturbation fluid velocity is specified as $\bar{\mathbf{u}} = (\bar{u}, \bar{v}, \bar{w})$ on the bounding surface located at $z = \eta(x, y, t)$, then the no-slip conditions can be written, using the definitions of the secondary variables, in the form:

$$\int_\eta^\infty \omega_y \, dz = -\bar{u} - \int_\eta^\infty \frac{\partial w}{\partial x} \, dz \tag{7}$$

$$\int_\eta^\infty \omega_x \, dx = \bar{v} + \int_\eta^\infty \frac{\partial w}{\partial y} \, dz. \tag{8}$$

Each of these two integral conditions can be associated with the vorticity transport equation for the corresponding component of the perturbation vorticity. They thus provide a very natural means of imposing constraints on the vorticity evolution that arise directly from the no-slip conditions. Moreover, the no-slip conditions are employed once, and once only, to provide such constraints. The remaining condition to be applied at the surface $z = \eta(x, y, t)$, namely the normal velocity matching condition

$$w(x, y, \eta) = \bar{w}, \tag{9}$$

may be imposed on the solution of the Poisson equation (6) in a relatively straightforward manner.

The coupling between the no-slip conditions and the vorticity evolution that is facilitated by the integral conditions (7), (8) remains in force even when a linearized version of the velocity-vorticity system is employed. This is not the case for some other possible velocity-vorticity formulations, where the introduction of artificial vorticity boundary conditions does not always guarantee a well-posed connection between the vorticity and the no-slip conditions.

Pressure calculation

Within the context of the velocity-vorticity formulation, the pressure usually plays a purely passive role and so it does not have to be computed. However, if there is pressure driven wall motion, such as is the case when the bounding surface at $z = \eta(x, y, t)$ is compliant rather than rigid, then the pressure will need to be calculated. The perturbation pressure can be found from the relation

$$p = \int_z^\infty \left(\frac{\partial w}{\partial t} + N_z + \frac{1}{R} \left[\frac{\partial \omega_y}{\partial x} - \frac{\partial \omega_x}{\partial y} \right] \right) dz - \left(\frac{1}{2} |\mathbf{u}|^2 + \mathbf{u}.\mathbf{U}^B \right) \tag{10}$$

This expression for the pressure is obtained, in effect, by integrating the wall-normal momentum equation. It may be observed that, once the primary variables are known, the pressure can be independently evaluated for each distinct x-y position by simply performing an integration along the wall-normal direction. Computing the pressure in such a way has the advantage that there is a component of the surface pressure that can be readily identified as the rate of change of the normal-momentum due to all of the fluid that lies immediately above a given point on the surface, namely the part of the pressure that is contributed by the first term that appears in the integral in equation (10). When the surface consists of a compliant wall this pressure term can be carefully combined with a corresponding inertial term from the governing equation for the compliant wall motion so as to yield a numerically stable scheme for simulating interactively coupled wall and fluid motions [2-6].

Approximations and simplifications

The full velocity-vorticity system (4)-(6) may be readily modified in order to artificially eliminate various effects due to three-dimensionality, nonlinearity and meanflow inhomogeniety. Such deliberate idealization can often be very helpful in identifying important physical mechanisms and facilitating comparisons with theoretically predicted behaviour. To study purely two-dimensional disturbances, the spanwise perturbation velocity v and spatial variations along the spanwise direction must both be neglected. This gives a system comprised of a single vorticity transport equation and a Poisson equation that, taken together, determine the evolution of the primary variables $\{\omega_y, w\}$. The streamwise velocity component u, which is the only non-vanishing secondary variable, remains explicitly defined by the integral relation (1), while the streamwise no-slip condition is ensured by the corresponding integral constraint (7). Some further details for the two-dimensional case can be found in reference [7]. Studies that neglect nonlinear perturbation effects may be performed by simply dropping the product $\boldsymbol{\omega} \times \mathbf{u}$ in the calculation of the convective quantity $\mathbf{N}$. The effects of spatial inhomogeniety in the meanflow can also be quite simply removed.

3. NUMERICAL METHODS

A full description of a numerical scheme for discretizing the velocity-vorticity system of governing equations is included in [2], for the particular case of disturbances developing in the three-dimensional von

Kármán boundary layer over a rotating disc. Thus we will not attempt to give here a very extensive account of our numerical methods. However, we will discuss a few aspects of the time discretization in a little more detail. This is mainly in order to draw attention to an efficient procedure that can be used to impose the integral constraints on the vorticity. These constraints, it should be recalled, are fully equivalent to the no-slip conditions. Their satisfactory imposition is thus an important requirement for the success of any numerical scheme.

Summary of the discretization scheme

The main features of the discretization are that: (i) The streamwise variation is discretized using finite-differences which are typically of at least fourth-order accuracy. (ii) Fourier expansions are deployed for the variation in the spanwise direction. (iii) Chebyshev expansions are used for the discretization in the wall-normal direction across the boundary layer. An algebraic co-ordinate transformation is deployed to map the semi-infinite physical domain onto a finite computational domain. (iv) To avoid an overly restrictive numerical stability limit on the size of the timestep, the temporal discretization of the vorticity transport equations is taken to be implicit for viscous terms that involve second derivatives in the wall-normal direction, but all other terms may be treated explicitly. (v) The integral constraints on the vorticity can also be treated in an explicit fashion. This allows the solution of the two discretized vorticity transport equations and the discretized Poisson equation to be fully decoupled within the time-stepping procedure. (Some additional comments about how this can be achieved are given later.) (vi) The wall-normal discretization is formulated so as to involve only pentadiagonal matrix operations. This facilitates the direct solution of the discretized transport equations using a Thomas algorithm. The Poisson equation can be solved by combining the Thomas algorithm with either an iterative streamwise line marching procedure or a direct method that involves a fast sine-transform along the streamwise direction. (vii) A pseudo-spectral transform technique is used to compute the nonlinear and other product terms in the transport equations that appear via the convective quantity **N**.

Decoupling and the imposition of integral constraints

We will now descibe how, at each timestep, it is possible to effectively decouple the solution of the discretized vorticity transport equations from the solution of the discretized Poisson equation for the normal velocity component. This decoupling contributes to the efficiency of the numerical scheme and helps to simplify its implementation. By examining a specific time-stepping scheme, we can also illustrate the role of the vorticity integral constraints in determining the evolution of the vorticity. For the sake of simplicity we will consider a relatively straightforward second-order Adams-Bashforth/Crank-Nicolson time-stepping scheme that involves a predictor and a corrector stage. More sophisticated and/or higher-order time-stepping procedures can be readily implemented in a similar fashion. We will only describe the two-dimensional case, where there is just a single vorticity transport equation to consider for the spanwise perturbation vorticity ω_y. The three-dimensional case can be tackled by treating the second vorticity equation for the streamwise vorticity ω_x, and its associated integral constraint, in essentially the same manner. In fact the discretized equation for the streamwise vorticity can be cast in exactly the same general form as for the spanwise vorticity. The only real difference is that it needs to be solved subject to a slightly altered form of vorticity integral constraint. To help make this more apparent, and to simplify the notation in what follows, we will use ω rather than ω_y to denote the spanwise perturbation vorticity.

If all terms in the spanwise vorticity transport equation except the viscous term that involves a second-order wall-normal derivative are dealt with in an explicit fashion then we can obtain a discretized vorticity equation that may be formulated as

$$\left(\frac{1}{\Delta t} - \frac{1}{2R}\frac{\partial^2}{\partial z^2}\right)\hat{\omega}^l = \left(\frac{1}{\Delta t} + \frac{1}{2R}\frac{\partial^2}{\partial z^2}\right)\omega^{l-1} + \frac{3}{2}M^{l-1} - \frac{1}{2}M^{l-2}\,, \tag{11}$$

where l is the timestep label and $\hat{\omega}^l$ is the predictor stage value of the spanwise perturbation vorticity. The quantities M^{l-1}, M^{l-2} both combine, for each of the previous two timesteps, the convective terms and those of the viscous terms that are treated explicitly. Equation (11) can thus be viewed as a second-order differential equation along the wall-normal z-direction, with $\hat{\omega}^l$ as the unknown quantity that must be determined. It can be solved, uniquely, by imposing two conditions on $\hat{\omega}^l$. The first condition is that $\hat{\omega}^l \to 0$ for $z \to \infty$. For the second condition we can use a discretized version of the integral constraint (7) that may be written as

$$\int_0^\infty \hat{\omega}^l\,dz = -\int_0^\infty \left\{\omega^{l-1} + 3\frac{\partial w}{\partial x}^{l-1} - \frac{\partial w}{\partial x}^{l-2}\right\} dz\,. \tag{12}$$

Here, again for the purposes of simplicity, we have chosen to assume that the bounding surface is flat and stationary, so that $\bar{u} = \eta = 0$ in equation (7). Equation (12) is obtained by imposing the integral constraint at the mid-point of the time-step, but with the deployment of two different second-order accurate time-averaging schemes. A centred average is used for the perturbation vorticity, but a backwards average is used for the normal velocity. This means that, as for the discretized transport equation (11), the right-side of (12) only involves quantities that have already been determined from previous time-steps.

The condition that $\hat{\omega}^l \to 0$ for $z \to \infty$ may be readily imposed, in a conventional fashion, on the numerical solution of the discretized vorticity equation (11). (For example, we chose to use a wall-normal mapping together with a form of Chebyshev expansion for which the condition was automatically satisfied.) The integral constraint would, at first sight, appear to be more troublesome. However, if the variation of the dependent variables along the wall-normal z-direction is discretized using spectral series expansions, then it need not present any particular difficulties. The non-local nature of the integral constraint, in that it necessarily couples together the values of $\hat{\omega}^l$ at positions across the whole of the boundary layer, is not very radically different from the distinctively global character of the derivative operators that arises for any spectral discretization of a differential equation.

Once the predictor stage value of the vorticity $\hat{\omega}^l$ has been determined, the Poisson equation (6) can be solved to determine the normal-velocity w^l. In two dimensions this takes the form

$$\nabla^2 w^l = -\frac{\partial \hat{\omega}^l}{\partial x} \, . \tag{13}$$

Since the right-hand side is already known, the numerical procedure that is required to obtain the solution of the discretized version of the Poisson equation is decoupled from the procedure that is used to find the solution of the discretized vorticity transport equation. For the three-dimensional case, it becomes necessary to simply add another predetermined term, involving the predictor stage value of the streamwise perturbation vorticity, to the right-hand side of the Poisson equation.

After the normal-velocity w^l has been computed, the perturbation vorticity may be recalculated to obtain the corrected value ω^l. It is not strictly necessary to implement a corrector stage, but it can help to improve the numerical stability of the time-stepping scheme. For instance, the following discretized version of the vorticity transport equation may be employed for the corrector

$$\left(\frac{1}{\Delta t} - \frac{1}{2R}\frac{\partial^2}{\partial z^2}\right)\omega^l = \left(\frac{1}{\Delta t} + \frac{1}{2R}\frac{\partial^2}{\partial z^2}\right)\omega^{l-1} + \frac{1}{2}\hat{M}^l + \frac{1}{2}M^{l-1}. \quad (14)$$

A numerical solution to this differential equation needs to determined subject to the condition that $\omega^l \to 0$ for $z \to \infty$ and the integral constraint

$$\int_0^\infty \omega^l \, dz = -\int_0^\infty \frac{\partial w^l}{\partial x} \, dz \, . \quad (15)$$

The quantity $\hat{M}^l$ that appears in the discretized vorticity transport equation can be evaluated using the known values of the vorticity from the predictor stage and the wall-normal velocity w^l. As before, it may be seen that all the quantities on the right-hand sides of the discretized vorticity transport equation and the integral constraint have predetermined values. So once more, the numerical solution procedure that is needed to update the vorticity may be decoupled from the procedure that is used to solve the Poisson equation for the wall-normal velocity.

4. CONCLUDING REMARKS

Because fluid boundary layers can be viewed as being, essentially, concentrations of vorticity that are formed adjacent to solid walls, it seems fairly natural to try to use a vorticity-based simulation method to study the development of boundary-layer disturbances. But the implementation of such an approach is not without difficulty. Although the no-slip boundary condition at a solid wall surface can be interpreted as providing a localized source for the diffusion of vorticity, there is no natural boundary condition that can be applied to the vorticity itself. To cope with this lack of a vorticity boundary condition, it has been deemed to be necessary, within many practical vorticity-based simulation schemes, to adopt rather *ad hoc* methods in order to specify the vorticity at solid surfaces. We have shown that, contrary to such procedures, there is no need to utilize any artificial boundary conditions that determine the wall vorticity. It is in fact possible to formulate completely rigorous and elegant constraints on the vorticity that are fully equivalent to no-slip conditions. Crucially, these constraints are not entirely local in nature. Though they can be applied independently for each distinct position along a solid wall surface, they involve integrals of the vorticity across the whole of the boundary layer, rather than just the value of the vorticity at the wall.

Unfortunately, due to space limitations in the present paper, it is not possible to present any of the numerical simulation results that have been obtained using our vorticity-based scheme that incorporates the no-slip conditions by means of vorticity integral constraints. Instead, we will have to be content with just listing various different forms of boundary-layer disturbance that have been successfully studied to date. These have included: (i) disturbances generated by highly localized suction slots; (ii) disturbances generated and controlled by interactive MEMS devices; (iii) various types of disturbances evolving over compliant surfaces; (iv) disturbances associated with the absolute instability of the von Kármán boundary layer over a rotating disk; (v) weakly nonlinear and modulated two-dimensional Tollmien-Schlichting waves, as well as their strongly nonlinear development leading to spike formation. The interested reader should refer to references [2-10] for further information about these studies. A general review is presented in reference [11].

REFERENCES

1. Rothmeyer, A. P., Smith F.T. "Incompressible Triple-Deck Theory". In *The Handbook of Fluid Dynamics* (ed. R.W. Johnson), Chap. 23, CRC Press, 1998.
2. Davies, C., Carpenter P.W. "A novel velocity-vorticity formulation of the Navier-Stokes equations with applications to boundary layer disturbance evolution", *J. Comp. Phys.*, **172**, pp. 119-165, 2001.
3. Davies, C., Carpenter P.W. "Numerical simulation of the evolution of Tollmien-Schlichting waves over finite compliant panels", *J. Fluid Mech.*, **335**, pp. 361-392, 1997.
4. Davies, C., Carpenter, P.W. "Global behaviour corresponding to the absolute instability of the rotating-disc boundary layer", *J. Fluid Mech.*, **486** pp. 287-329, 2003.
5. Ali, R. "Receptivity and transition in boundary layers over rigid and compliant surfaces". PhD thesis, University of Warwick, 2003.
6. Davies, C. "Convective and absolute instabilities of flow over compliant walls". In *Flow past highly compliant boundaries and in collapsible tubes* (eds. Carpenter, P.W. & Pedley, T.J.), Chap. 4, pp. 69-93, Kluwer, 2003.
7. Bowles, R.I., Davies, C. & Smith, F.T. "On the spiking stages in deep transition and unsteady separation", *J. Eng. Math.*, **45** pp. 227-245, 2003.
8. Carpenter, P.W., Lockerby, D.A., Davies, C. "Numerical simulation of the interaction of microactuators and boundary layers", *AIAA J.*, **40** pp. 67-73, 2002.
9. Houten, S. "Finite amplitude disturbances in a boundary layer". PhD thesis, Keele University, 2004.
10. Houten, S., Healey, J.J., Davies, C. "Nonlinear evolution of Tollmien-Schlichting waves at finite Reynolds numbers". In *Laminar-Turbulent Transition* (eds. Fasel, H.F. & Saric, W.S.), pp. 181-186. Springer, 2000.
11. Davies, C. "Numerical simulation of boundary-layer disturbance evolution", *Phil. Trans. R. Soc. London A*, (in press), 2004.

HYPERSONIC REAL-GAS EFFECTS ON TRANSITION

Hans G. Hornung
Graduate Aeronautical Laboratories, California Institute of Technology, Pasadena, CA 91125, USA. hans@galcit.caltech.edu

Abstract: Some of the results of an extensive research program into the effects on transition of the vibrational and chemical relaxation processes that occur in high-enthalpy flows are presented. Relaxation effects are found to influence transition significantly, with increases of the transition Reynolds number by up to a factor of five. The mechanism responsible for this transition delay is shown to be the damping of the acoustic second mode instability by relaxation processes. Transition is also found to be further delayed by up to a factor of two by suitable wall porosity.

Key words: Hypervelocity flow, relaxation effects, transition delay, transition control.

1. INTRODUCTION

In incompressible boundary layer flows the viscous instability, which was first discovered (with some surprise) by Ludwig Prandtl [1] and his co-workers in 1921, is usually responsible for the path to transition. One of the important differences between low-speed and hypersonic flows is that the dominant instability mode in the latter is the second or Mack [2] mode, in which the boundary layer acts as a wave guide for acoustic noise, where selected frequencies are trapped and amplified, eventually leading to transition.

The second important difference is that in high-enthalpy hypersonic (hypervelocity) flows the relaxation processes associated with vibrational excitation and dissociation provide mechanisms for damping acoustic waves and may therefore be expected to affect the second mode. These relaxation

G.E.A. Meier and K.R. Sreenivasan (eds.), IUTAM Symposium on One Hundred Years of Boundary Layer Research, 335-344,

effects can exist quite subtle, however, since they can also affect the mean structure and therefore the stability properties of the boundary layer.

A large part of the experimental work on the problem of stability and transition at high Mach number has been done in cold hypersonic facilities. In such facilities, the test gas is expanded from a reservoir at relatively low temperature (of order 1000 K), so that the high Mach number is produced, not so much by raising the speed, but mainly by lowering the speed of sound. Important experimenters in this regime are Demetriades, Stetson and Kendall. (see the review by Reshotko **[3]**). Together with the linear stability analysis by Mack **[2]**, they provide a substantial basis for understanding the path to transition in cold hypersonic flow. They are, however, not able to capture the phenomena that occur in hypervelocity flows because of the vibrational excitation and dissociation.

Some of the specific problems of hypervelocity boundary layer stability have been addressed computationally by a number of authors. They include Malik and Anderson **[4]**, who considered equilibrium vibration and dissociation, and Stuckert and Reed **[5]** who assumed vibrational equilibrium but finite-rate chemistry. Both found that the new effects caused the boundary layer to be destabilized. However, more recent work by Johnson *et al.* **[6]** found that non-equilibrium chemistry had a strong damping effect, in agreement with recent experimental evidence. The apparent contradiction between the results of these investigations is not too surprising in view of the complicated manner in which the rate processes can influence the stability problem, and the large number of parameters involved in it.

In this paper we present the results of an extensive experimental program of research conducted over the last decade, in which the focus is specifically on the regime where relaxation processes associated with vibrational excitation and dissociation are important. In the laboratory, such flows can only be maintained for very short times, since they require the gas to be expanded from a reservoir at very high temperature and very high pressure, conditions at which it can be contained only for a period of typically 2 ms. This, and the aggressive environment of the high temperatures and pressures make it impossible to use many techniques that are available to experiments in longer-duration, cold facilities. It is therefore necessary to approach the problem with more indirect methods that use such simple evidence as the location of transition in a careful exploration of the parameter space. Fortunately, the frequencies of the most strongly amplified modes are typically 1-3 MHz, so that the short test time is not a serious limitation.

2. TRANSITION ON A SLENDER CONE

Much of the work on transition in hypersonic flow has been performed on the simplest possible shape, namely the slender cone. The flow over a slender cone has the advantages that the pressure gradient is zero, and that it is free of side effects. The first experiments to be performed in the newly completed T5 hypervelocity free-piston shock tunnel in 1991-1993 were designed for a 5 deg half-angle cone also, in order to be able to compare the new high enthalpy results with those from the cold hypersonic wind tunnels.

The first series of experiments (for details see Germain and Hornung [7]) explored the behavior of the transition location on the cone as a function of the total enthalpy of the flow in air and nitrogen. The transition location was determined from the distinct rise in heat flux. An example of how this is done is shown in Figure 1.

In hypervelocity flow simulation it is important to reproduce the actual speed of the flow, so that the vibrational excitation and dissociation are reproduced correctly. This is often done at the expense of reproducing the Mach number. This is the case in the T5 experiments also, where the free-stream Mach number is typically 5.5, but the speed ranges up to 6 km/s. Thus, the boundary layer edge temperature in a free-flight situation is very different than in the T5 experiments, but the temperature profile in the *inner part* of the boundary layer is almost the same in both cases. It is therefore more meaningful to compare any two flows in terms of the Reynolds number evaluated at the reference condition rather than that based on the edge conditions. The reference temperature is given by

$$\frac{T^*}{T_e} = 0.5 + \frac{\gamma - 1}{2}\frac{\sqrt{Pr}}{6}M_e^2 + 0.5\frac{T_w}{T_e},$$

where the wall and boundary layer edge conditions are identified by the subscripts w and e respectively.

The results of experiments in nitrogen and air flows are plotted in Figure 2 in the form of the Reynolds number at transition, evaluated at the reference temperature and based on the distance from the cone tip to the transition location, versus the total enthalpy of the flow. Two new features are brought out by this plot. First, a significant increase in transition Reynolds number (evaluated at reference conditions) with total enthalpy increase is observed, and second, this increase is slightly larger in air than in nitrogen. This led us to suspect that transition is significantly influenced by high-enthalpy real-gas effects, and that it might be interesting to explore what happens in other gases, such as helium, which behaves like a perfect gas in our total enthalpy range, and carbon dioxide, which exhibits strong vibrational and dissociational effects in this range. The first experiment, with helium, showed that, even at 15 MJ/kg, the transition Reynolds number was the same as the low enthalpy value. A dramatically larger transition Reynolds number was observed in carbon dioxide flows, also shown in Figure 2, (for details see Adam and Hornung **[8]**).

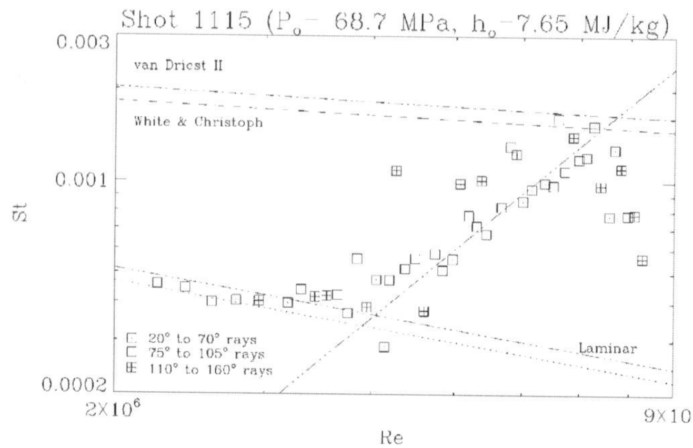

Figure 1. Plot of Stanton number against boundary-layer edge Reynolds number for one experiment on the cone. As may be seen, the Stanton number follows the theoretical laminar flow line (dotted line) at low Reynolds numbers, and rises up toward the turbulent level (as given by two turbulence models) at high Reynolds number. The transition Reynolds number is determined by the intersection of a straight line fit of the transitional data with the laminar line.

It is clear from these results that a dramatic transition delay, which is completely absent at low speeds, is evident at high enthalpy, and that the magnitude of the phenomenon and the enthalpy at which it sets in are different for different gases.

It was at this point that Graham Candler and his group became interested in testing our results by making linear stability computations at the conditions of our experiments. Their results agreed with the trends observed in the nitrogen and air flows, and illustrated dramatically how strongly thermochemical non-equilibrium effects can influence the growth rate of disturbances. An example of their results is shown in Figure 3. These results also establish the acoustic Mack mode as being responsible for the path to transition in the T5 experiments.

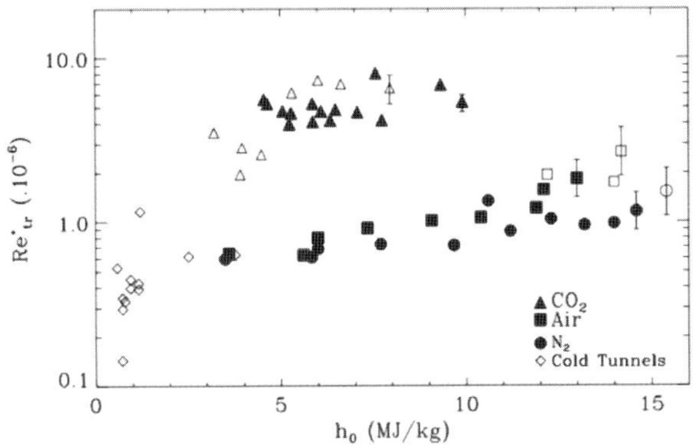

Figure 2. Transition Reynolds number evaluated at reference conditions as function of total enthalpy. Open symbols correspond to cases where the flow was laminar all the way to the end of the cone. The cold tunnel data are from papers by Demetriades **[10]** and DiCristina **[11]**. The carbon dioxide results are superimposed on this plot as triangular symbols. Note the large transition delay relative to the nitrogen results.

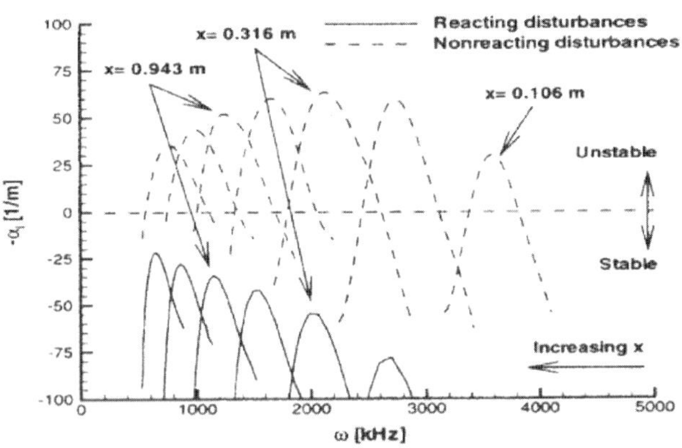

Figure 3. Johnson *et al.*'s results of linear stability calculations with thermochemical nonequilibrium at the conditions of T5 shot 1150 at 4.0 MJ/kg in carbon dioxide, showing growth rate of disturbances as functions of disturbance frequency at several distances along the cone. To examine the damping effect of finite rate processes, the dashed curves show that the disturbances are amplified when relaxation is turned off.

Just as these experiments had been completed, Norman Malmuth of Rockwell Science Center and Sasha Fedorov of Moscow Institute of Physics and Technology started to discuss with us the possibility of controlling transition in hypersonic flow. They had shown theoretically that the acoustic mode could also be damped by wall porosity, see Fedorov, *et al.* **[11]**. This led to the research project described in the next section.

3. PASSIVE CONTROL OF TRANSITION

Simply stated, the acoustic disturbances are trapped and amplified in the boundary layer, which acts like a wave guide for them. It has been known (early work on the subject included that of Kirchhoff and Rayleigh) that acoustic disturbances are absorbed in wall porosity by viscous action and heat conduction. Fedorov and Malmuth quantified the damping rate in a hypersonic boundary layer and suggested types of porosity for optimum results. At the conditions of the cone experiments in T5, small-diameter, deep, blind holes that are closely spaced were predicted to produce suitable damping. The proportions of the configurations chosen are related to the boundary layer thickness in the schematic sketch of Figure 4.

With such a fine distribution of blind holes, the sheer number of holes required (some 15 million) appears to be prohibitive. After finding a company (Actionlaser, in Sydney, Australia) that was able to make holes at the required spacing and diameter in a stainless steel sheet of 0.5 mm thickness, we decided to wrap such a sheet around the aluminum cone and remake the intermediate part of the tip to provide a flush transition from the (non-porous) tip to the porous surface. In order to provide a control experiment in every shot by making half of the cone surface porous and half non-porous, the porous sheet was formed into a half cone and welded to a similar half cone sheet without holes. The resulting hollow cone was then slipped over the aluminum cone at a low temperature (190 K) to take advantage of the difference in the thermal expansion coefficients of stainless steel and aluminum, which thus provides an interference fit of approximately 0.1 mm. At the same time, disassembly is still possible by cooling to liquid nitrogen temperature.

As may be imagined, the process of getting this model manufactured and assembled required a considerable effort in development work. To do this, several attempts had to be made in the rolling of the sheet into an accurate conical shape, and in the extremely fine and accurate welding of the sheets. An impression of part of this task is given by the micrographs of Figure 5.

The cone was then instrumented with thermocouple heat flux gauges as in the previous experiments. The same procedure for determining the transition location was applied, this time separately on the smooth and on the porous side.

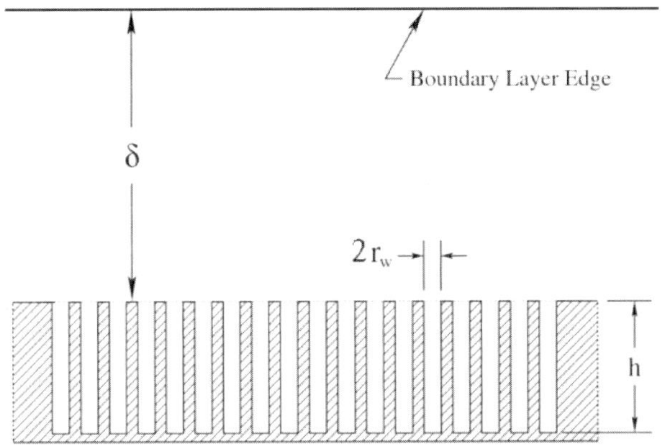

Figure 4. Showing the approximate proportions of the hole diameter, spacing and depth in relation to the laminar boundary layer thickness. With a typical boundary layer thickness of one mm, This makes the desirable hole depth 0.5 mm and the hole diameter and spacing 0.05 mm and 0.1 mm respectively.

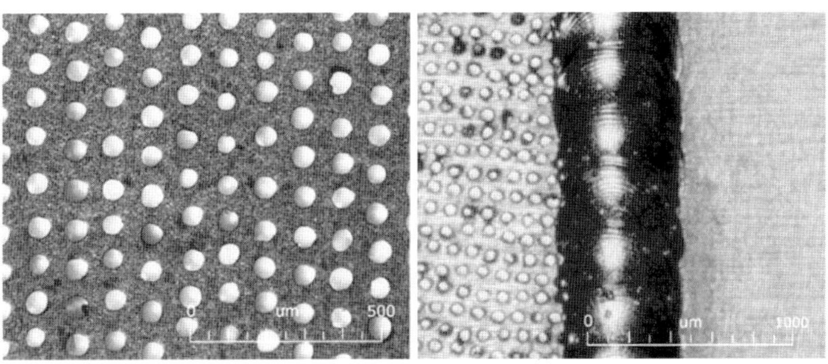

Figure 5. Left: Magnified image of the stainless steel Actionlaser perforated sheet. At this scale the grain boundaries of the metal can be resolved. Note that the length of the half-millimeter scale bar is equal to the depth of the holes. Right: Micrograph of the weld joining the porous and solid sides along a generator of the cone. The weld is 0.5 mm wide.

The results obtained in nitrogen flow are shown in Figure 6. They confirm approximately the results of the previous experiments and exhibit a dramatic transition delay on the porous side of the cone. The increase of the

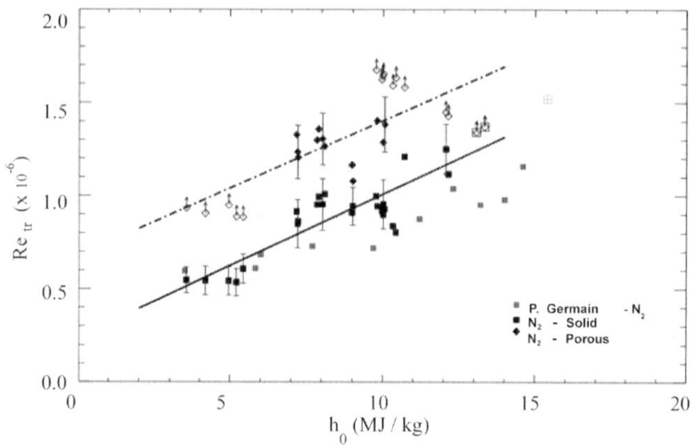

Figure 6. Plot of transition Reynolds number vs total enthalpy for the N2 data. Dark squares show the results from the non-porous side of the cone. Gray squares show the nitrogen data from Figure 4 for comparison. The filled diamonds show the values from the porous side of the cone. As may be seen, transition is very significantly delayed on the porous side. The open diamonds symbolize situations in which the boundary layer was laminar on the porous side all the way to the end of the cone. In these cases, the Reynolds number plotted is that based on the length of the cone. The lines are linear fits to the points to guide the eye. (For detail, see Rasheed *et al.* **[12]**).

transition Reynolds number is typically 400,000 which is as much as 80% at the low-enthalpy end of the range. Both at the low and at the high end of the range, transition could not even be achieved on the porous side, the boundary layer remaining laminar all the way to the end of the cone. The effect is shown dramatically in Figure 7, which shows a shadowgraph that includes the boundary layers on both sides of the cone.

In carbon dioxide flows we observed a very different effect of the porosity. At low enthalpy, where the transition Reynolds number on the solid surface is comparable with that in nitrogen, transition is delayed by the porosity, but at approximately 3 MJ/kg, a crossover occurs, and the porosity causes transition to be advanced at higher enthalpy by as much as 50%. It turns out that the high Reynolds number that is reached with the flow still laminar on the solid surface in carbon dioxide makes the holes act like roughness elements that cause the boundary layer to be tripped. The Reynolds number based on the hole diameter at the crossover point is approximately 200, which is in agreement with the critical value for tripping according to Reda **[13]**.

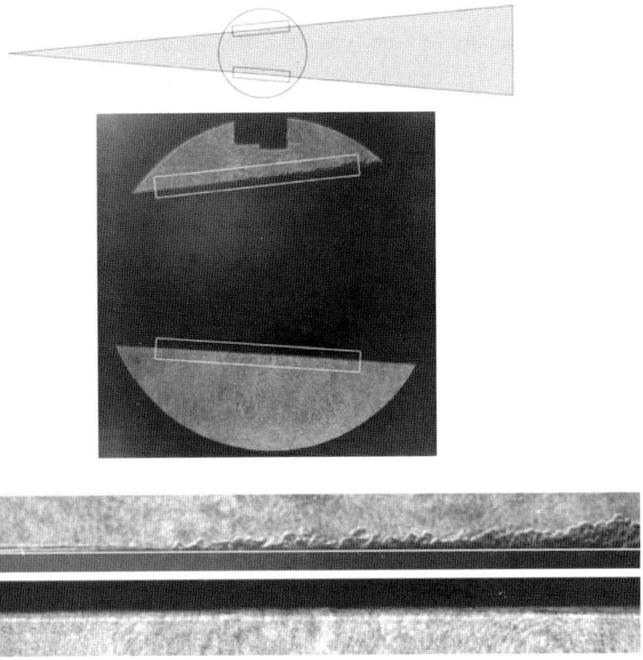

Figure 7. The schematic at the top shows the location of the viewing window relative to the cone. The next frame down shows a shadowgraph taken through this window of nitrogen flow at 9.8 MJ/kg and at a reservoir pressure of 48.2 Mpa. At the top surface, which is the smooth side of the cone, the boundary layer changes from laminar at the left to turbulent at the right, while, at the bottom (porous side), it is laminar all the way to the end of the picture. The white rectangular boxes in the main image are shown enlarged at the bottom for a more detailed view

4. CONCLUSIONS

An extensive series of experiments performed in the high-enthalpy shock tunnel T5 demonstrate conclusively that the relaxation processes of vibrational excitation and dissociation can have very dramatic stabilizing effects on transition in flows over slender cones. The mechanism by which this damping occurs is through the influence of relaxation on acoustic waves. In addition it was demonstrated that transition could be delayed very significantly by suitable blind porosity of the surface. Both results establish the acoustic instability mode as the dominant path to transition in this regime.

ACKNOWLEDGEMENTS

The work described in this paper was supported by AFOSR Grants F49610-92-J-0110, F49620-93-1-0338, and F49620-98-1-0353.

REFERENCES

1. Prandtl L. "Bemerkungen ueber die Entstehung der Turbulenz", *ZAMM*, **1**, pp. 431-436 (1921).
2. Mack LM. "Boundary-Layer Stability Theory, Special Course on Stability and Transition of Laminar Flow", *AGARD* Report Number 709 (1984).
3. Reshotko E. "Boundary-Layer Stability and Transition", *Ann. Rev. Fluid Mech.*, **8**, pp. 311-349 (1976).
4. Malik MR, Anderson EC. "Real Gas Effects on Hypersonic Boundary-Layer Stability", *Physics of Fluids A*, **3**, pp. 803-821 (1991).
5. Stuckert G, Reed H. "Linear Disturbances in Hypersonic, Chemically Reacting Shock Layers", *AIAA Journal*, **32**, pp. 1384-1393 (1994).
6. Johnson HB, Seipp T, Candler GV. "Numerical Study of Hypersonic Reacting Boundary Layer Transition on Cones", *Physics of Fluids*, **10**, pp. 2676-2685 (1998).
7. Germain P, Hornung HG. "Transition on a Slender Cone in Hypervelocity Flow", *Experiments in Fluids*, **22**, pp. 183-190 (1997).
8. Demetriades A. "Hypersonic Viscous Flow over a Slender Cone. Part III: Laminar Instability and Transition", *AIAA Paper* 74-535, 1974 (7th Fluid and Plasma Dynamics Conference, June 17-19, Palo Alto, CA, USA).
9. DiCristina V. "Three Dimensional Laminar Boundary Transition on a Sharp 8 deg Cone at Mach 10", *AIAA Journal*, **8**, pp. 852-856 (1970).
10. Adam P, Hornung HG. "Enthalpy Effects on Hypervelocity Boundary Layer Transition: Ground Test and Flight Data", *J. Spacecraft and Rockets*, **34**, pp. 614-619 (1997).
11. Fedorov AV, Malmuth ND, Rasheed A, Hornung HG. "Stabilization of hypersonic boundary layers by porous coatings", *AIAA Journal* **34**, pp. 605-610 (2001).
12. Rasheed A, Hornung HG, Fedorov AV, Malmuth ND. "Experiments on passive hypervelocity boundary-layer control using an ultrasonically absorptive surface", *AIAA Journal,* **40**, pp. 481-489, (2002).
13. Reda DC. "Roughness-Dominated Transition on Nosetips, Attachment Lines and Lifting-Entry Vehicles", *AIAA Paper* 2001-0205, (39th AIAA Aerospace Sciences Meeting and Exhibit, January 8-11, 2001, Reno, NV, USA).

STABILIZATION OF HYPERSONIC BOUNDARY LAYER BY MICROSTRUCTURAL POROUS COATING

Anatoly A. Maslov
Institute of Theoretical and Applied Mechanics SB RAS, Institutskaya 4/1, 630090 Novosibirsk, Russia; phone: +7 (3832) 30 38 80, fax: +7 (3832) 34 22 68, e-mail: maslov@itam.nsc.ru

Abstract: The present paper overviews the experimental studies of hypersonic laminar boundary layer stability and transition. The progress in development of new methods of measurements and data processing allow measuring high frequency disturbances and investigating nonlinear processes in hypersonic boundary layers. Experiments confirmed the theoretical prediction of Fedorov and Malmuth, that the porous coating strongly stabilizes the second mode of disturbances and marginally destabilizes the first mode. The bispectral analysis of the hot-wire data showed that the harmonic resonance, which is known to be the primary nonlinear mechanism for the breakdown to turbulence in the hypersonic boundary layer on solid surfaces, is completely absent on the porous surface. Experiments revealed that the porous coating significantly delays the transition.

Key words: Boundary layer, new measurement technique, linear and nonlinear stability, laminar-turbulent transition, control, porosity effect.

1. INTRODUCTION

For small freestream disturbances and negligible surface roughness, the laminar-turbulent transition is due to amplification of unstable modes in the boundary layer [**1**]. For essentially two-dimensional supersonic and hypersonic flows, the initial phase of transition is associated with excitation and amplification of the first and/or second modes.

The first mode is an extension to high speeds of the Tollmien-Schlichting (TS) waves, which represent viscous instability at low Mach numbers. The inviscid nature of the first mode begins to dominate when the Mach number increases, since compressible boundary layer profiles

G.E.A. Meier and K.R. Sreenivasan (eds.), IUTAM Symposium on One Hundred Years of Boundary Layer Research, 345-354,

contain a generalized inflection point [**2**]. This mode can be stabilized by wall cooling, suction, and favorable pressure gradient.

The second mode results from an inviscid instability driven by a region of supersonic mean flow relative to the disturbance phase velocity. This instability belongs to the family of trapped acoustic modes propagating in a wave-guide between the wall and the sonic line [**2-3**]. Once the second mode arrives, it becomes the dominant instability since its growth rate tends to exceed that of the first mode. For insulated surfaces, this occurs for Mach number higher than 4. In contrast to the TS instability, wall cooling destabilizes the second mode. This effect can be important in the transition of hypersonic flows. Since the relative temperature of the hypersonic vehicle surface is relatively small, natural cooling eliminates the TS instability, while the second mode remains unstable and may provoke an early transition. Increasing the laminar run requires diminishing the second-mode instability.

In high-speed flows, the second mode is associated with disturbances of relatively high frequency corresponding to the ultrasonic band. Fedorov and Malmuth [**4**] assumed that a passive, ultrasonically absorptive coating (UAC) of fine porosity may suppress these fluctuations and, at the same time, may not trip the boundary layer due to roughness effects; *i.e.*, the passive UAC may stabilize the second and higher modes by a disturbance-energy-extraction mechanism.

This hypothesis was confirmed by the inviscid [**4**] and viscous [**5**] linear stability analyses. The present paper overviews the experimental studies that validate the original concept of Fedorov and Malmuth.

2. EXPERIMENTAL APPARATUS

2.1 Porous UAC characteristics

Characteristics of a porous coating must meet certain requirements to effectively suppress the boundary-layer instability. Based on Fedorov's preliminary estimation, two types of porous UACs were chosen: a regular porous UAC and a random porous UAC.

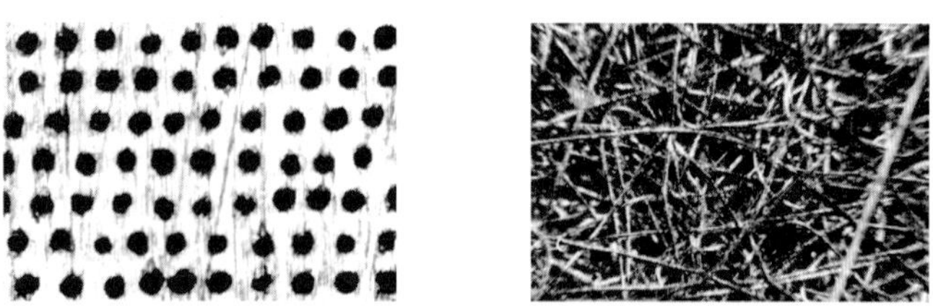

Figure 1. Regular (left) and random (right) prototypes of porous UACs.

The regular porous UAC is a stainless steel perforated sheet, which has equidistant cylindrical holes of depth 450 μm with the average spacing of 100 μm. The average hole diameter is 50 μm on the face side and 64 μm on the back side (see Fig. 1, left); i.e., holes are slightly conical with a taper angle of 0.9°.

The random porous UAC is a felt-metal coating. It is composed of stainless steel fibers 30 μm in diameter. To provide coating integrity, the fibers are hard sintered randomly on a solid stainless steel sheet 0.245 mm thick. After that, they are rolled to a porosity of 75%. The porous layer has a thickness of 0.75 mm, so that the total thickness of the felt metal sheet is 1 mm. Magnified images of the porous surface are shown in Fig. 1, right.

The perforated sheets are flash mounted on the cone surface.

2.2 Models and equipment

The experiments were conducted in the T-326 hypersonic blow-down wind tunnel with an open-jet test section and the AT-303 adiabatic wind tunnel at the Institute of Theoretical and Applied Mechanics of the Siberian Branch of the Russian Academy of Sciences (ITAM). The measurements were performed at Mach number of 6-12. The T-326 run time can be as long as 30 min, and the AT-303 run time is about 0.1 sec.

The models were the 7° half-angle sharp cones 500 mm long. Half of the base part was covered by the UAC. The model was equipped with a 3-D generator providing a high-frequency glow discharge in a small chamber. Artificial disturbances generated by the glow discharge were introduced into the boundary layer through an orifice 0.4 mm in diameter located at the distance of 69 mm from the model nose. The generator construction is similar to that used for excitation of artificial wave packets in supersonic boundary layers [**6**].

The model was installed in the test section of the wind tunnels at zero angle of attack.

2.3 Measuring system

A constant-current hot-wire anemometer (CCA) and atomic layer thermo pile (ALTP) probes were used for pulsation measurements.

A CCA custom built at ITAM was used to measure mass flow fluctuations. The hot-wire probes were made of tungsten wire 5 μm in diameter and 1 mm long, which was welded to pointed stings. The overheat ratio was 0.5; the frequency response of hot-wire anemometer was 600 kHz. This technique allows investigating the behavior of linear disturbances but cannot be applied to nonlinear researches.

For investigation of high frequency disturbances, a new technique and gages have been developed in recent years. A revolutionary step in disturbance measurements was made in the group of Knauss (IAG, Stuttgart University), where a new type of sensors, ALTP, was developed **[7]**. One of the advantages of ALTP sensors is their wide frequency bandwidth. Due to the atomic-layered structure, their pass band can reach several megahertz. In addition to the above-mentioned wider frequency range in comparison to hot-wire probes, the ALTP sensors have another advantage. Having high rigidity, they can be applied in high-enthalpy and high-turbulent compressible flows, where wire breakage is the reason for hot-wire inapplicability.

An example of ALTP application to hypersonic boundary layer stability experiments is illustrated in Fig. 2. The oscillation spectra were obtained in the boundary layer of the cone at M=12 on solid surface. The measurements were conducted in the AT-303 short-duration wind tunnel. The maximum at a frequency of 200-300 kHz corresponds to the second mode. The second and third harmonics are visible in spectra too.

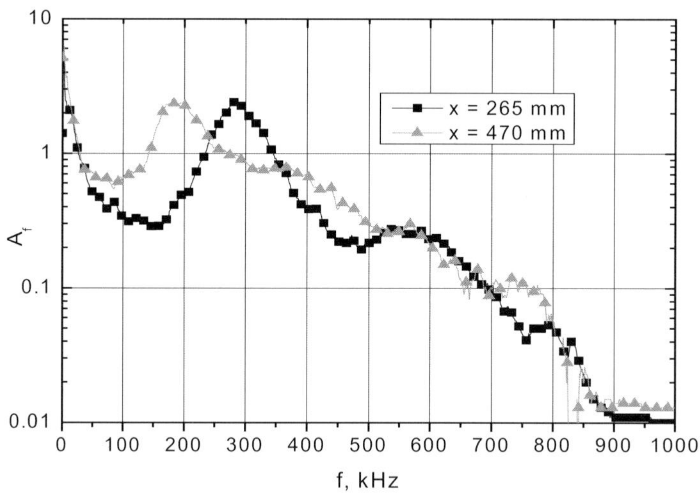

Figure 2. Oscillation spectra in the boundary layer of the cone at M=12.

In experiments with artificial disturbances, a special generator is used. The high-frequency glow discharge system consists of clock and high-voltage generators. The clock generator signal is used to trigger the high-voltage generator and synchronize hot-wire measurements with the high-voltage generator initiation. To obtain the amplitude and phase of artificially excited disturbances, the discrete Fourier transform is used **[6]**.

3. LINEAR DISTURBANCE DEVELOPMENT

Hot-wire measurements [**8-9**] showed that boundary layers on solid and porous surfaces (both regular and random) are laminar in the range of hot-wire measurements. The mean flow profiles are similar in both cases and agree with the self-similar solution of the boundary-layer equations.

The hot-wire measurements of "natural" disturbances showed that the disturbance spectra on the solid surface are typical for hypersonic boundary layers, with the second mode being dominant [**10-11**]. Examples of the mass flow pulsation spectra are sown in Fig. 3. The second mode dominates at frequency of 280 kHz.

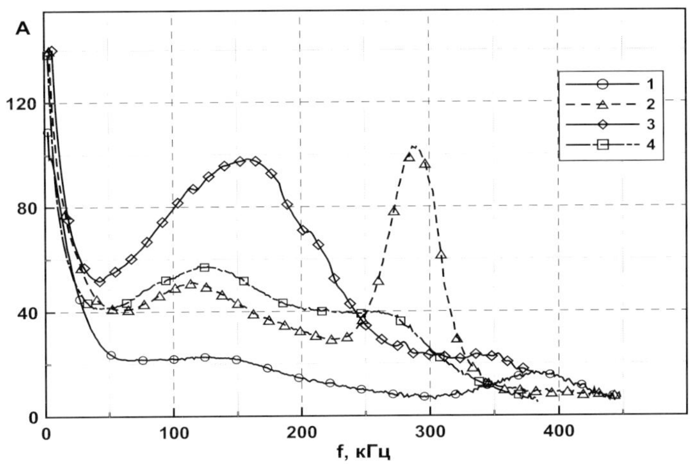

Figure 3. Mass flow pulsation spectra. 1 – initial cross section at $Re_{eX}=2.84\cdot 10_6$, solid surface; 2 – solid surface; 3 - felt metal UAC; 4- regular structure UAC ($Re_{eX} = 4.5\cdot 10^6$).

On the porous surface, the second mode is so strongly suppressed that it is not observable in the measurement region, while the first mode becomes unstable. The roughness of the fibers and increasing in receptivity to acoustic waves may explain the dramatic increasing of the first mode disturbances (at a frequency of approximately 100 kHz) on the random porous surface. Disturbances of the first mode on the regular UAC surface are slightly larger than on the solid surface. The surface on the regular porous UAG sheet is smooth enough and destabilization occurs owing to porosity.

To investigate the second mode stabilization effect, artificial wave packets were generated in the boundary layer at a frequency corresponding to the second mode instability.

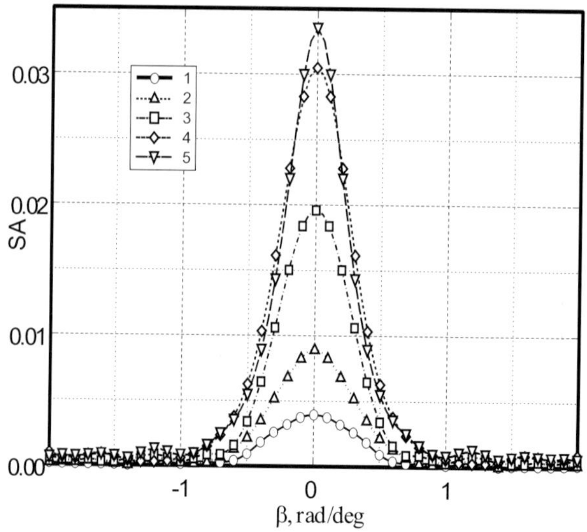

Figure 4. Transversal wave spectra (M=6, T-326 wind tunnel). 1-5 – different Reynolds numbers.

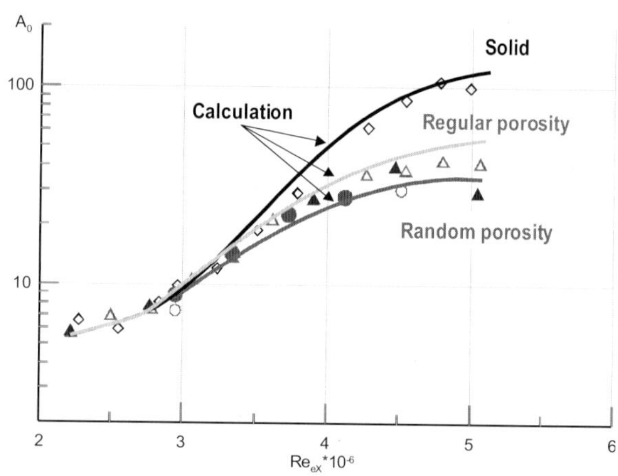

Figure 5. Comparison of the theoretical amplification curves (solid lines) with experimental data (symbols) for the two-dimensional component of an artificially excited wave packet (f= 280 kHz, M=6, T-326 wind tunnel).

The transversal wave spectra (the β-spectra resulting from the Fourier transform of the transverse distributions) are plotted in Fig. 4. Two-dimensional waves of the wave packets are dominant and unstable on the solid surface. The porous coating essentially reduced the second-mode growth rate. Measurements of the UAC effect on two- and three-dimensional disturbances showed that the porous coating strongly stabilizes the second mode and marginally destabilizes the first mode.

The theoretical amplification curves (solid lines) are compared with experimental data (symbols) for the two-dimensional component of an artificially excited wave packet at a frequency of 280 kHz in Fig. 5. The calculations were by Fedorov for test conditions on the basis of the nonparallel linear stability theory.

A comparison of the theoretical amplification curves with experimental data showed that the theoretical growth rates are remarkably close to experimental data, which confirms the theoretical model.

4. NONLINEAR DISTURBANCE DEVELOPMENT

The nonlinear aspects of stabilization of the second mode disturbances by means of passive, ultrasonically absorptive coatings with a regular microstructure are studied using the bispectral analysis [**12**]. The experimental data are hot-wire measurements made in artificially excited wave packets introduced into a hypersonic boundary layer on both solid and porous surfaces. The bispectral measurements show that the subharmonic and harmonic resonances of the second mode are significantly modified [**13**]. The harmonic resonance, which is quite pronounced in the late stages of the hypersonic boundary layer on solid surfaces, is completely absent on the porous surface. The degree of nonlinear phase locking associated with the subharmonic resonance and identified on the solid surface is substantially weakened on the porous surface. This nonlinear interaction persists further downstream on porous surface than on the solid surface; however in contrast to case of the solid surface, there are no strongly preferred interaction modes. The spectral measurements show that the first mode is moderately destabilized on the porous surface. The bispectral measurement presented here identify a nonlinear interaction, with is associated with the destabilized first mode; however this is observed to be a very weak nonlinear interaction that has no adverse effect on UAC performance.

Let us consider an example of bispectral analysis application to investigation of stability of the boundary layer on the porous surface [**13**]. At the downstream station close to transition, we have confirmation of second mode stabilization with the use of UAC (Fig. 6). The range and intervals of the contour levels in both parts of the figure are the same. It is, thus, quite clear that degree of nonlinear phase locking is substantially reduced on the porous surface compared to the solid surface. Generation of the harmonic associated with the second mode, f=290 kHz, is completely suppressed on the porous surface. On the solid surface (Fig. 6, left) the nonlinear phase locking is also quite large between the second mode and its subharmonic, f=(140kHz, 140kHz), and between the first and second modes, f=(210kHz, 90kHz). There are also increased levels of nonlinear phase locking between triads in the complete range of frequencies. This situation may be contrasted with the state of the boundary layer on the porous surface, which is observed to remain quite laminar.

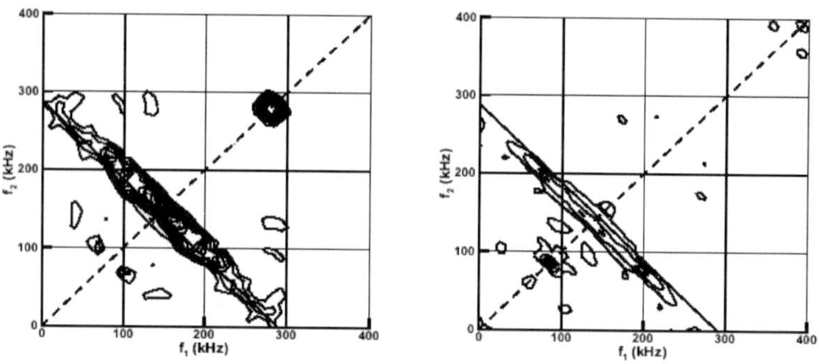

Figure 6. Bicoherence spectra at x=286 *mm*. Solid (left) and porous (right) surfaces.

The bispectral measurements on the porous surface (Fig. 6, right) show that the nonlinear phase locking involves a limited range of frequencies, 50-230kHz. The energy transfer between the wave triads in this frequency range is, therefore, enhanced. However, on the porous surface, in contrast to the solid surface, there are no preferred mode interactions; it is, therefore, obvious that UAC is effective in substantially weakening the nonlinear interactions involving the second mode, its subharmonic and the first mode.

5. TRANSITION MEASUREMENTS

The knowledge on influence of porous coatings on the transition of a laminar boundary layer to a turbulent state is extremely important for practical application. The only experiment performed in this aspect was made at the California Institute of Technology [**14**]. The concept was verified in the GALCIT T-5 shock tunnel by testing a 5° half-angle sharp one cone 1 m long. Half of the cone surface in these tests was solid, and the other half had a porous sheet regular structure. The model was instrumented by thermocouples, and the transition onset point was determined from the Stanton number distributions measured simultaneously on both sides of the model for each run. The experiments were performed for the freestream Mach number M=(4.59–6.40). This study revealed that the porous coating significantly delays the transition. For most runs, the boundary layer on the porous surface was laminar up to the model base, while the transition on the untreated solid surface was observed halfway along the cone. These experiments can be considered as another qualitative confirmation of the theoretical prediction of Fedorov and Malmuth [**5**].

6. CONCLUSIONS

The progress in hypersonic boundary layer measurements is associated with development of new methods of measurements and data processing. Application of artificial disturbances, new generation of hot-wire anemometers and ALTP sensors allows one to measure high frequency disturbances and investigate nonlinear processes in hypersonic boundary layers. For identification of these nonlinear aspects, application of the bi-spectral analysis is very effective.

Measurements of the UAC effect on 2-D and 3-D disturbances showed that the porous coating strongly stabilizes the second mode and marginally destabilizes the first mode. A comparison of the theoretical amplification curves with experimental data showed that the theoretical growth rates are remarkably close to experimental results, which confirms the theoretical model of Fedorov and Mulmuth.

A nonlinear analyses shows that the harmonic resonance, which is quite pronounced in the latter stages of the hypersonic boundary layer transition on solid surfaces, is completely absent on the porous surface. The leading role belongs to the subharmonic resonance. The porous coating significantly delays the transition.

Stability calculations for the cooled wall case performed by Fedorov indicate that a combination of cooling and UAC leads to strong stabilization of the hypersonic boundary layer. For actual hypersonic vehicles, the wall temperature ratio is small, which eliminates the first-mode instability. By diminishing 3-D effects (which helps to avoid cross-flow vortices), reducing the TPS roughness (that helps to avoid bypass mechanism) and stabilizing the second mode with the help of a thin porous coating, it is feasible to achieve a long laminar run on hypersonic vehicle surfaces.

ACKNOWLEDGMENTS

The author acknowledges all the co-authors of the overviewed works. He is especially grateful to Dr. A. Fedorov and Dr. N. Malmuth for initiation the experimental work, theoretical support, explanations and consultations. The author also thanks Dr. A. Shiplyuk, who carried out all the experiments and linear data processing. He thanks Prof. N. Chokani for initiation of nonlinear approach, presentation of the codes, processing and interpretation of the results of bispectra analysis. He appreciates Dr. Knauss suggestion of ALTP application to stability measurements, development of measurement techniques and methods, and cooperation in experiments.

Portions of this work were sponsored by the Boeing, AFOSR and Russian Foundation of Basic Research (RFBR) under the grant 02-01-00141.

REFERENCES

1. Reshotko E. "Boundary layer instability, transition and control", *AIAA Paper 94-0001*, 1994.
2. Mack LM. "Boundary-layer stability theory", Special Course on Stability and Transition of Laminar Flow, AGARD Rep No. 709, pp. 1-81,1984.
3. Gushchin VR, Fedorov AV. "Asymptotic analysis of inviscid perturbations in a supersonic boundary layer", *Zhurnal Prikl. Mekh. i Tekh. Fiz.*, **no. 1,** pp. 69-75 1989 (in Russian).
4. Malmuth ND, Fedorov AV, Shalaev V, Cole J, Khokhlov A. "Problems in high speed flow prediction relevant to control" *AIAA Paper 98-2695*, 1998.
5. Fedorov AV, Malmuth ND, Rasheed A, Hornung HG. "Stabilization of hypersonic boundary layers by porous coatings", *AIAA Journal*, **vol. 39, no. 4,** pp. 605-610, 2001.
6. Kosinov AD, Maslov AA, Shevelkov SG. "Experiments on the stability of supersonic laminar boundary layers", *J. Fluid Mech.*, **vol. 219,** pp. 621-633, 1990.
7. Maslov AA, Bountin DA, Shiplyuk AN, Smorodsky B, Knauss H, Gaisbauer U, Wagner S, Betz J, "ALTP sensor application for boundary layer measurements", *Proc. of the ICMAR Conference, Part II,* Novosibirsk, Russia, June 28 – July 3, 2004, pp. 137-146.
8. Fedorov A, Shiplyuk A, Maslov A, Burov E, Malmuth N. "Stabilization of a hypersonic boundary layer using an ultrasonically absorptive coating", *J. Fluid Mech.*, **vol. 479,** pp. 99-124, 2003.
9. Fedorov A, Kozlov V, Shiplyuk A, Maslov A, Sidorenko A, Burov E, Malmuth ND. "Stability of hypersonic boundary layer on porous wall with regular microstructure", *AIAA Paper 2003-4147*, 2003.
10. Stetson KF, Thompson ER, Donaldson JC, Siler LG. "Laminar boundary layer stability experiments on a cone at Mach 8. Part 1: Sharp cone". *AIAA Paper 83-1761*, 1983.
11. Stetson KF, Kimmel RG. "On hypersonic boundary-layer stability", *AIAA Paper 92-0737*, 1992.
12. Chokani N. "Nonlinear spectral dynamics of hypersonic laminar boundary layer flow", *Physics of Fluids*, **vol. 12,** pp. 3846-3851,1999.
13. Shiplyuk A, Buntin D, Maslov A, Chokani N. "Nonlinear aspects of hypersonic boundary layer stability on a porous surface", *AIAA Paper 2004-0255*, 2004.
14. Rasheed A, Hornung, HG, Fedorov AV, Malmuth ND. "Experiments on passive hypervelocity boundary layer control using an ultrasonically absorptive surface", *AIAA Journal,* **vol. 40, no. 3,** pp. 481-489, 2002.

THE ASYMPTOTIC STRUCTURE OF HIGH-REYNOLDS NUMBER BOUNDARY LAYERS

Peter A. Monkewitz [1] and Hassan M. Nagib [2]

[1] *Laboratory of Fluid Mechanics (LMF), Swiss Federal Institute of Technology Lausanne (EPFL), CH-1015 Lausanne, Switzerland ; e-mail: peter.monkewitz@epfl.ch*

[2] *Dept. of Mechanical and Aerospace Engineering, Illinois Institute of Technology (I I T), Chicago, IL 60616-3793, USA ; e-mail: nagib@iit.edu*

Abstract: The methodology of matched asymptotic expansions or "generalized boundary layer theory" for large Reynolds number Re_τ (based on friction velocity and outer length scale), pioneered by Prandtl [**1**], is used to extract from measured or computed mean velocity profiles in turbulent channel and pipe flows their limiting behavior at infinite Re_τ. After fitting an "outer expansion" in terms of suitable functions of the outer wall-normal coordinate η to the data, the construction of composite expansions is used "in reverse" to extract the "inner expansion". Its leading term of order $O(Re_\tau^0)$ represents the near-wall solution for infinite Re_τ which is found to be identical for channels and pipes. For large values of the inner wall-normal coordinate y_+, this limiting inner expansion for the streamwise velocity is furthermore shown to be well described by Prandtl's famous "log-law".

Key words: Mean velocity profiles in turbulent channel and pipe flows, Matched asymptotics, Infinite Re limit

1. INTRODUCTION

Ever since Prandtl "invented" boundary layers [**1**] one hundred years ago – thereby starting the branch of applied mathematics now called singular perturbation which has been enormously fruitful far beyond fluid boundary layers – and formulated the "law of the wall" for the near-wall mean velocity profile in wall-bounded turbulent flows (later complemented by different "laws of the wake"), the debate on the appropriate form of these laws has been going on (see e.g. [**2-4**]). Particular issues are the values of the "fitting parameters" in the different laws, the most notorious parameter being the Kármán constant κ in the "log-law" $U_+ = \kappa^{-1}\ln(y_+) + B$, where " $_+$ " denotes "inner" or "wall" variables, non-dimensional with the friction velocity $u_\tau = (\tau_w/\rho)^{1/2}$ and kinematic viscosity ν, where τ_w is the wall shear stress and ρ the density. The debate has been singularly complicated by the limited accuracy and reliability of high-Re experimental data and by the limitation of DNS to relatively low Reynolds numbers. As a consequence, the literature supporting one or the other law and/or specific values of fitting parameters is vast and replete with controversies. The aim of the present

G.E.A. Meier and K.R. Sreenivasan (eds.), IUTAM Symposium on One Hundred Years of Boundary Layer Research, 355-362,

study is to use the methodology of matched asymptotic expansions (MAE; cf. for instance [**5**]) to try to structure this debate and to contribute to the clarification of some of the issues, in particular the question of the existence and possible universality of the infinite-Re limit of the inner part of these wall-bounded flows.

In the following, we concentrate first on fully developed turbulent channel and pipe flow, in order to avoid the additional difficulties associated with non-parallel effects, i.e., the broken translation invariance in the mean flow direction found in the boundary layer, mixing layer and wake, for instance.

2. MAE METHODOLOGY FOR DATA ANALYSIS

The quantity chosen for the following analysis is the viscous shear stress (dU_+/dy_+) normalized by the total stress $(1 - \eta)$, where $\eta = y_+/Re_\tau$ is the outer variable and $Re_\tau = u_\tau L/\nu$ the Reynolds number based on friction velocity and L, the channel half-width or the pipe radius. With this, the time-averaged momentum equation reduces for both the channel and the pipe to

$$S_M + S_R = 1, \text{ with} \tag{1}$$

$S_M \equiv (dU_+/dy_+)(1 - \eta)^{-1}$ and $S_R \equiv (\overline{-u_+v_+})(1 - \eta)^{-1}$ the normalized molecular and Reynolds shear stresses, respectively. The normalization with the total stress is chosen to eliminate the Reynolds number contained in the coordinate $\eta = y_+ / Re_\tau$ from equ. (1). This opens the possibility of a near-wall behaviour $S_M \sim S_M(y_+)$ and $S_R \sim S_R(y_+)$ which is Reynolds number independent for all but the lowest Reynolds numbers. Note that at $\eta = 1$ S_M = "0/0" which makes the evaluation of S_M from real data near the centerline an experimental implementation of l'Hopital's rule.

The relative novelty of the present approach is the identification of the measured or computed mean velocity profiles with composite expansions which allows to extract the "inner" expansion from the original data after an approximate "outer" expansion is determined by fitting the data far from the wall, where the contribution from the inner expansion is negligible. The composite expansion of the multiplicative type (see e.g. [**5**]) of order (m,n) is given by

$$f^{(m,n)comp} = f^{(m)inner} \cdot f^{(n)outer} / [f^{(n)outer}]^{(m)inner}, \tag{2}$$

where $[f^{(n)outer}]^{(m)inner}$ is the m-term inner expansion of the n-term outer expansion. For S_M, for instance, it is taken to be of the form

$$S_M^{outer} = \Sigma\ C_i(Re_\tau)\, f_i(\eta)\,; \quad f_i(\eta) = [1-(1-\eta)^2]^i\,, \tag{3}$$

where the functions f_i have the appropriate symmetry about the channel center $\eta = 1$. Furthermore, the f_i are chosen such that $f_i \propto y^i$ near the wall $\eta \to 0$. Therefore, the coefficient C_{-1} is directly related to the Kármán constant κ by $C_{-1} = 2(\kappa\, Re_\tau)^{-1}$, as the term i = -1 yields, upon integration, the logarithm of the "log-law" $U_+ = \kappa^{-1} \ln(y_+) + B$.

After fitting (3) to the data near the channel or pipe centerline, one can extract the infinite Reynolds number limit (the leading term) of the inner expansion simply by choosing m=1 in equ. (2). To fully exploit the result, the infinite Reynolds number limit of the inner expansion, i.e. the 1-term inner expansion, is then fitted by a 4/5 order Padé approximant $P_{4/5}$ which asymptotically matches the leading order inner expansion of (3). This implies that $P_{4/5} \sim (\kappa\, y_+)^{-1} + O(y_+^{-2})$ for $y_+ \to \infty$. Once a analytical fit for both the inner and outer expansions of S_M is at hand, one can for instance integrate both fits, match them and use the construction (2) to obtain an explicit composite expansion for the mean velocity profile.

3. RESULTS AND SOME CONCLUSIONS

The normalized molecular stress S_M for several different channel data sets obtained with hotwire, LDV, PIV & CFD [**6-9**] is shown in Fig. 1, together with the fitted "outer expansion" of the form (3). Note that the derivatives dU_+/dy_+ of the data have been calculated with the simplest finite difference scheme without additional filtering and that the same expansion is used for all the data. In other words, the Reynolds number dependence is built into the coefficients $C_i(Re_\tau)$, such that for instance $S_M^{outer}(\eta=1) \propto Re_\tau^{-1}$.

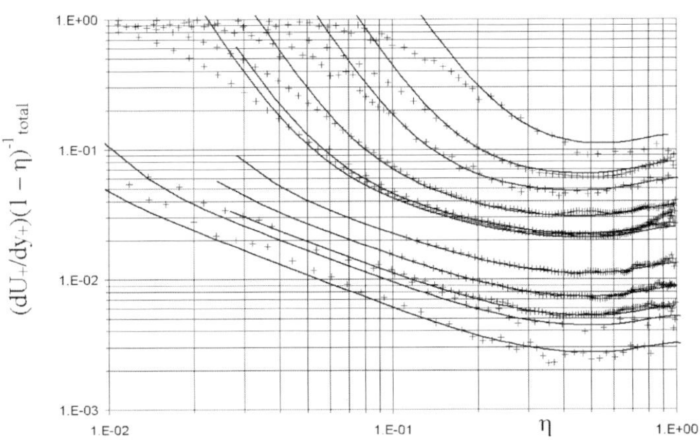

Figure 1. Viscous shear stress data (+) in 2D channels, normalized by the total stress, versus wall-normal coordinate η (nondim. with channel half-width) for Re_τ ranging from 118 to 4780 (from [**6-9**]). ——: Fitted "outer solution" with $\kappa = 0.38$.

One observes on Fig. 1 that the outer expansion is only a good fit to the data near the centerline $\eta = 1$; It actually diverges for $\eta \to 0$ which is common in singular perturbation problems. For $\eta > 0.25$ all the data are within 10% of the fit for all $Re_\tau \geq 250$. At this point, a comment regarding C_{-1} , i.e. the Kármán constant κ, is necessary. One may expect that optimizing the fit (3) would automatically yield C_{-1}, but unfortunately the quality of the fit is remarkably insensitive to the value of κ. Therefore, we have elected to make an a priori choice of the value of κ and to optimize the fit using the C_i with $i \neq -1$. For the 2 values of κ used in this paper (0.38 and 0.43), the corresponding outer fits are within 1% of each other for $\eta > 0.25$. Because of the asymptotic matching with the inner solution, the behaviour of the latter for $y_+ \to \infty$ is of course $S_M^{inner} \sim (\kappa\, y_+)^{-1}$ with the **same** chosen κ. This insensitivity of the fits is thought to be the main reason for the long lasting controversies over the value of κ. Here, one may also ask whether the term i=-1 in (3), which leads to the "log-law", is required at all. As a matter of fact, an excellent outer fit (3) can be constructed with $C_{-1} = 0$, without a "log-law". However, with such a choice, the "1-term inner expansions" for different data sets (for different Re_τ) do no longer collapse. This shows that in the framework of MAE the "log-law" is a necessary ingredient, albeit it appears in its pure form only in the limit of infinite Reynolds number. We feel that this is a stronger, but more subtle argument in favor of the log-law than the usual "best fit arguments".

For the channel data of Fig. 1, the obtained "1-term inner expansion", i.e., the infinite Re limit of the inner solution, is shown in Fig. 2a for $\kappa = 0.38$ and in Fig. 2b for $\kappa = 0.43$ (Note that on the outer scale η, the

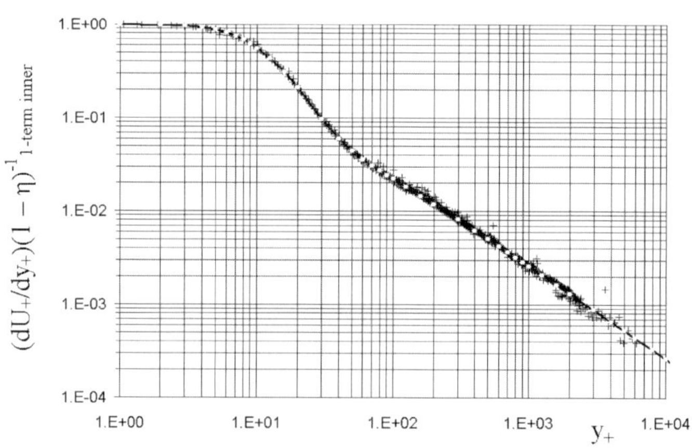

Figure 2a. 1 – term "inner expansion" with $\kappa = 0.38$ of the viscous shear stress (+), normalized by the total stress, versus inner wall-normal coordinate y_+ for the channel data of Fig. 1. White dots : 4/5-order Padé fit of limiting "inner solution".

entire layer shown in this figure has zero thickness!). On both graphs, the Padé approximants $P_{4/5}$ (different for the two κ!) are indicated by white dots. The collapse of the data is such that all the data are within 10% of the Padé fit.

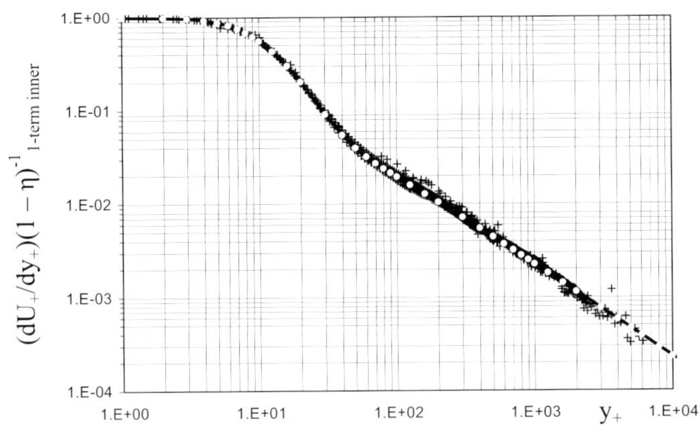

Figure 2b. Same as Fig. 2a, but with κ = 0.43. White dots: Padé fit **different** from that in Fig. 2a.

The analogous exercise has been carried out for various pipe data [**9-11**] for low to moderate Reynolds numbers – For the higher Reynolds numbers [e.g. **10**] the data do not extend close enough to the wall to allow a valid comparison with the full limiting inner expansion. The result for the "1-term inner expansion" with κ = 0.38 is shown in Fig. 3.

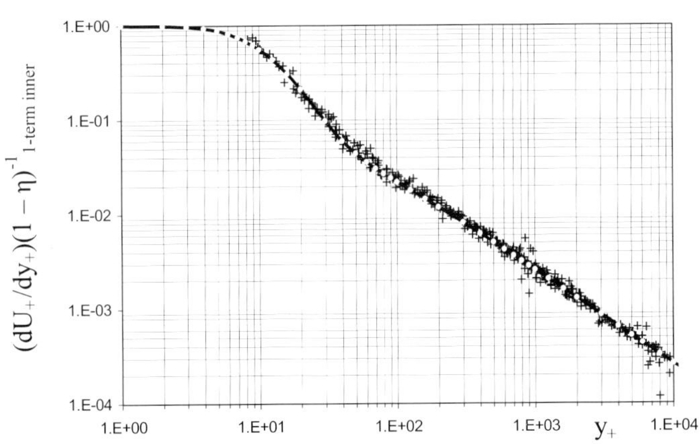

Figure 3. 1 – term "inner expansion" with κ = 0.38 of the viscous shear stress (+), normalized by the total stress, versus inner wall-normal coordinate y_+ for pipe data with Re_τ ranging from 851 to 8509 (from [**9-11**]). White dots : **same** Padé fit as in Fig. 2a.

Comparing Figs. 2 and 3, it is seen that, within experimental error, the resulting limiting inner expansion for the channel and the pipe are **identical** (for the same κ !). As briefly described in section 2, the inner and outer analytical fits of the normalized stress for the above channel data with both κ = 0.38 and κ = 0.43 have been integrated and combined **without further adjustments** into a composite expansion for U_+ which is shown in Figs. 4a and 4b, respectively.

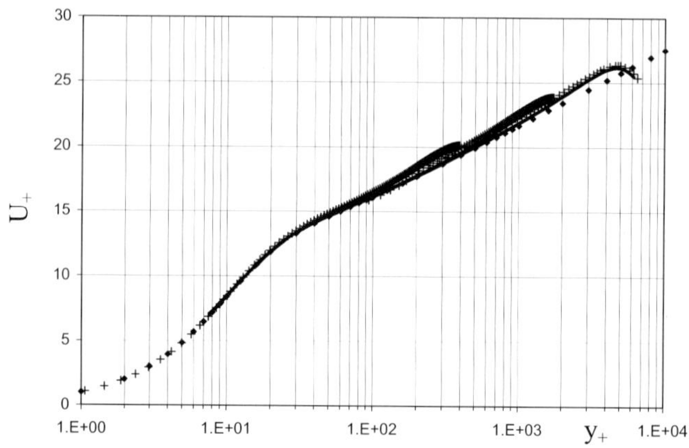

Figure 4a. Mean velocity profiles (+) in 2D channels versus y_+ for $Re_\tau = 390$, 1747 and 4783 (from [**7-9**]). ——: "Composite expansion" with κ = 0.38 . ♦: Integral of the Padé approximant of the 1-term "inner solution" of Fig. 2a with the pure "log-law".

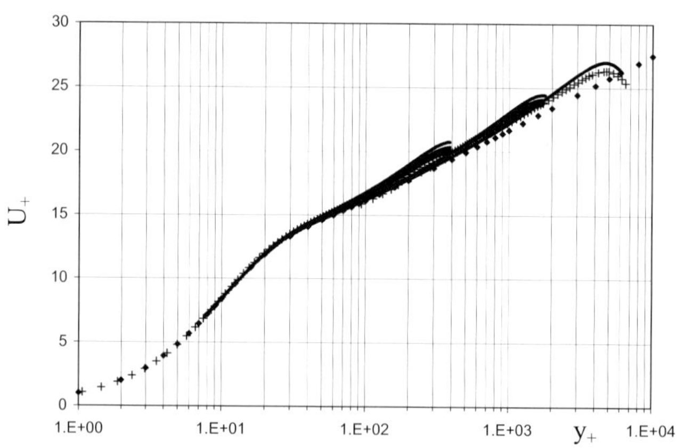

Figure 4b. Same as Fig. 4a, but with κ = 0.43.

It is worth pointing out that this determination of U_+ amounts to a consistency check between the fits for S_M and S_R and the pressure drop known from experiment or computation, since the friction factor based on the centerline velocity is, by definition, equal to $2/U_+(\eta=1)$.

One notes on the above figures of $U_+(y_+)$ that near the centerline the correspondence between the composite expansion (solid line) and the data (+) is better for $\kappa = 0.38$ (Fig. 4a) than for $\kappa = 0.43$ (Fig. 4b), i.e. that the friction factor is more accurately reproduced with $\kappa = 0.38$. This suggests, that $\kappa = 0.38$ is the "better" value for the Kármán constant, which is also the value suggested by Österlund et al. [**12**] for the boundary layer with zero pressure gradient. However, the reader needs to be warned that this conclusion concerning κ may be biased for the following reason: None of the available channel and pipe data are at high enough Reynolds number to safely neglect the $O(Re_\tau^{-1})$ correction to $U_+^{(1)inner}$. While the correction associated with the factor $(1-\eta) \equiv (1-y_+ Re_\tau^{-1})$ in the definition of S_M poses no problem, the "noise" in the data has made it impossible to determine $S_M^{(2)inner}$, i.e. the $O(Re_\tau^{-1})$ correction to $P_{4/5}$, and hence the **complete** $U_+^{(2)inner}$. Therefore, the possibility cannot be excluded that a proper accounting of the $O(Re_\tau^{-1})$ corrections discussed above would lead to somewhat different "best" values of the Kármán constant. To improve the situation, high quality data down to $y_+ = O(100)$ are badly needed for $Re_\tau > 10^4$ or preferably $Re_\tau >> 10^4$!

Nevertheless, our procedure of decomposing the normalized mean velocity derivative S_M into outer and inner solutions, then reconstructing the complete velocity profile U_+ from those two solutions and comparing the reconstructed profile to the original mean velocity data appears to be a more discriminating test for the value of κ than curve fitting the raw data.

In conclusion, we have used the methodology of MAE, which has grown out of Prandtl's original boundary layer ideas, to show that his celebrated "log-law" in its original "pure" form is indeed the proper law of the (near-)wall, **if it is interpreted as a limiting law for infinite Reynolds number**. Furthermore we have shown that, based on the available data, this limiting law is the same in channels and pipes. To obtain better fits at finite Reynolds numbers, this limiting law can in principle be augmented in a systematic manner by the next terms in the large-Re_τ expansion, but the available data have proven too noisy to do this in practice. Hence, there is still plenty of room left for the miscellaneous modified "log-laws" and power laws which have been developed to fit particular data sets at finite Reynolds numbers.

In a next step, finally, we plan to adapt the present methodology to (streamwise evolving) flat-plate boundary layer data described in Österlund et al. [**12**] and Nagib et al. [**13**].

ACKNOWLEDGEMENTS

The financial support of the US AFOSR and the Swiss ERCOFTAC Leonhard Euler Center is gratefully acknowledged. We also thank all those who have generously shared their data with us, in particular Ron Adrian, Ken Christensen, Bob Moser, Lex Smits and El-Sayed Zanoun.

REFERENCES

1. Prandtl L, "Über die Flüssigkeitsbewegung bei sehr kleiner Reibung", *Verhandlungen des III. Internationalen Mathematiker Kongresses*, Heidelberg, p. 484, 1904.
2. Barenblatt G I, "Scaling laws for fully developed turbulent shear flows. Part 1. Basic hypotheses and analysis", *J. Fluid Mech.*, **248**, p. 513, 1993.
3. Panton R L, "Evaluation of the Barenblatt-Chorin-Prostokishin power law for a boundary layer", *Phys. Fluids*, **14**, p. 1806, 2002.
4. Buschmann M, Gad-el-Hak M, "Generalized logarithmic law and its consequences", *AIAA J.*, **41**, p. 40, 2003. "Debate concerning the mean-velocity profile of a turbulent boundary layer", *AIAA J.*, **41**, p. 565, 2003.
5. van Dyke M, *Perturbation Methods in Fluid Mechanics*, Parabolic Press, 1964.
6. Fischer M, "Turbulente wandgebundene Strömungen bei kleinen Reynoldszahlen", *Ph.D. thesis*, Universität Erlangen-Nürnberg, 1999.
7. Christensen K T, "Experimental investigation of acceleration and velocity fields in turbulent channel flow", *Ph.D. thesis*, University of Illinois, Urbana-Champaign, 2000.
8. Moser R D, Kim J, Mansour N N, "Direct numerical simulation of turbulent channel flow up to Re_τ= 590", *Phys. Fluids*, **11**, p. 943, 1999.
9. Zanoun M, "Answers to some open questions in wall-bounded laminar and turbulkent shear flows", *Ph.D. thesis*, Universität Erlangen-Nürnberg, 2003.
10. Zagarola M V, Smits A J, "Mean-flow scaling of turbulent pipe flow", *J. Fluid Mech.*, **373**, p. 33, 1998.
11. Perry A E, *Data base for turbulent flow in a smooth pipe*, AGARD working group, October 1997.
12. Österlund J M, Johansson A V, Nagib H M, Hites M H, "A note on the overlap region in turbulent boundary layers", *Phys. Fluids*, 12, p. 1, 2000.
13. Nagib H, Christophorou C, Monkewitz P A, "High Reynolds number turbulent boundary layers subjected to various pressure-gradient conditions", IUTAM 2004, Göttingen, Germany, 2004.

INSTABILITIES NEAR THE ATTACHMENT-LINE OF A SWEPT WING IN COMPRESSIBLE FLOW

Jörn Sesterhenn and Rainer Friedrich
FG Strömungsmechanik, TU München
Boltzmannstr. 15, D-85748 Garching, Germany
joern.sesterhenn@lrz.tum.de, r.friedrich@lrz.tum.de

Abstract: We report on the numerical investigation of the swept leading edge flow for a compressible fluid for several Reynolds numbers and nose radii under supersonic conditions. The classical Görtler–Hämmerlin attachment line instability was recovered for this flow, but for the case of a finite nose radius, it was not found to be the dominant instability.

Key words: compressible flow, attachment line –, centrifugal –, cross-flow instability, direct numerical simulation

1. INTRODUCTION

The flow conditions along the leading edge of a swept wing are of crucial importance for the transition process to set in. Delay of transition is desirable for several reasons. For the operation of an aircraft the costs are dominated by energy consumption. Keeping the flow laminar greatly reduces these costs. For space vehicles, the heat load due to turbulent flow is also substantially higher. This increases the weight and cost of thermal protection systems.

The swept leading edge flow is inherently three–dimensional. Consider a plane spanned by the normal to the wing surface and the direction of the free stream. When the distance from the wall is increased the velocity vector is not confined within this plane. This is evident from figure 1 where the three velocity components at a location close to, but not on the attachment–line, are depicted in streamline coordinates over the wall normal direction.

This three–dimensional nature of the boundary layer profile is capable of hosting a wealth of possible instability mechanisms. The best known instabilities in this situation are the attachment-line instability and the cross-flow instability. Less known, but closely related to the latter, is a centrifugal instability. These instabilities are now briefly revisited [1, 2].

G.E.A. Meier and K.R. Sreenivasan (eds.), IUTAM Symposium on One Hundred Years of Boundary Layer Research, 363-372.

1.1 Attachment–line instability

The attachment–line instability is a viscous, linear mechanism. Its base flow is a stagnation-point flow with a sweep component, and thus termed stagnation-line flow. The non–swept base flow was discovered by Hiemenz, a student of Prandtl [3], and the swept version is commonly termed swept-Hiemenz flow. This flow is linearly unstable below a critical Reynolds number. The first investigation of this flow is due to Görtler and Hämmerlin [4, 5] and reported in the predecessor of this conference. The instability mode shows a linear dependency of the spanwise velocity component with the distance from the stagnation line and is named Görtler–Hämmerlin–mode. Interest in this mechanism has not declined since then [6, 7, 8, 9, 10, 11]. Compressibility is found to stabilise the flow according to M^2 [12]. Leading edge curvature was found to stabilise the flow as well [13].

1.2 Crossflow instability

The crossflow instability is of non-viscous nature. Its mechanism is based on an inflection point in the velocity profile. This can again be seen in figure 1. The crossflow component v vanishes, both at the wall as well as in the free stream. In between it acquires a maximum. Such profiles are known to be dynamically unstable. The mechanism is named after the typical crossflow velocity. This instability is under vivid investigation as well [14, 15, 16].

1.3 Centrifugal instability

Itoh [17] reported a further instability in the vicinity of the leading edge. The mechanism is akin to the one of the crossflow–instability but is locally confined to a region of high streamline curvature.

The appearance of this instability resembles the crossflow–instability as well, but exhibits a different phase relationship. Whereas the latter has a significant phase shift with increasing distance from the wall, the former has almost constant phase.

In this paper we describe numerical investigations of compressible leading edge flow with different Reynolds numbers and nose radii. Other parameters of the flow, as Mach number, wall temperature or surface roughness are not varied.

2. BASE FLOW

2.1 Parameters

The base flow on a curved leading edge in compressible flow is described by several parameters.

The proper length scale is the viscous length, which is about a third of the boundary layer thickness. It is formed by the velocity gradient and the viscosity and reads $\delta = \sqrt{\nu / \frac{\partial u_\infty}{\partial x}}$. x denotes the chord direction, z the spanwise direction and y is perpendicular to both. Below, we will also refer to n as the normal direction to the body surface and s as the coordinate along the body surface starting at the leading edge. In the flat plate case the velocity gradient is an arbitrary parameter and in the curved case approximated by the velocity gradient of a potential flow at the wall of a circular cylinder $\frac{\partial u_\infty}{\partial x} = 2u_\infty / R$. The nose radius may be expressed in terms of this length scale, being a first dimensionless parameter of the problem. If we omit blowing and suction or the use of trip-wires, we are left with three more dimensionless parameters. The Reynolds number is $\frac{w_\infty \delta}{\nu}$. It is constructed with the sweep velocity w_∞ parallel to the leading edge, the viscous length scale and a reference viscosity. The reference viscosity is taken to be the viscosity on the stagnation line for an adiabatic wall. The adiabatic wall temperature T_r is available from [18]. In our investigation, the wall temperature T_w was always adiabatic thus, the dimensionless parameter $\tau = \frac{T_w - T_r}{T_o - T_\infty}$ was kept zero. Additionally a sweep Mach number $M = w_\infty / c_\infty$ can be defined.

Please note that the sweepback angle is hidden in the Reynolds number by the ratio of the two velocity components. It is thus no longer a formal parameter of the problem.

It is known from the literature [6, 11] that below a critical Reynolds number of $Re = 583$ the flow is linearly stable. This value is raised by a finite nose radius [13] and compressibility [12].

Practical situations, e.g. for an airplane or fighter, typically involve Reynolds numbers of $Re = 400 - 1000$, nose radii of $R/\delta = 300 - 1000$ and sweep Mach numbers of $M = 0.3 - 1.5$.

2.2 Model assumptions and numerical scheme

We have introduced the assumption of periodicity in spanwise direction for our computations. Thus we neglect spanwise boundary layer growth. This approximation corresponds to a temporal DNS of a boundary layer transition and the parallel flow assumption in classical linear stability analysis. Therefore all results have to be interpreted with the same precaution as the classical results. By this assumption we avoid the trouble of specifying the boundary conditions in that direction and are able to compute the base flow in two dimensions only. This makes the investigation of the relevant parameter space more affordable. In future, we want to perform three—dimensional computations in representative regions of the parameter space and quantify the validity of the periodicity assumption.

At the outflow, non-reflecting boundary conditions are employed. At the inlet we consider in this paper supersonic conditions with a moving bow shock which is prescribed by use of the Rankine–Hugoniot conditions. The body surface is taken to be a parabolic leading edge.

We use a characteristic-type numerical scheme which was developed for DNS of compressible transitional and turbulent flow by Sesterhenn [19]. Its features are low numerical dissipation and dispersion and 5th, respectively 4th order accuracy in space and time. Shocks are treated explicitly with a shock-fitting procedure as demonstrated in Fabre et al. [20]. Since the shock location is initially unknown and may change in time due to incoming flow perturbations and wave reflections in the shock layer, the grid has to be time-dependent. In the present simulations we typically use grids with $350 \times 128 \times 8$ points in wall-normal, tangential and spanwise direction. Eight points in spanwise direction are able to resolve the principal mode. Up to 64 points in that direction were employed for comparison and showed no change in the behaviour of the solution with respect to the results presented here.

2.3 Description of the base flow

In the sequel, we consider a flow of $M_\infty = 8$ over a swept wing of $\Lambda = 30°$, leading to a sweep Mach number $M = 1.25$. The wing surface was treated as adiabatic. The spanwise velocity component at the attachment–line has the typical Blasius–like profile, whereas the wall normal velocity increases almost linearly with wall distance.

The velocity away from the attachment–line is depicted in fig.1.

We show the velocity components in a coordinate system which uses the wall normal n/δ, the potential streamline and the cross–product of these two

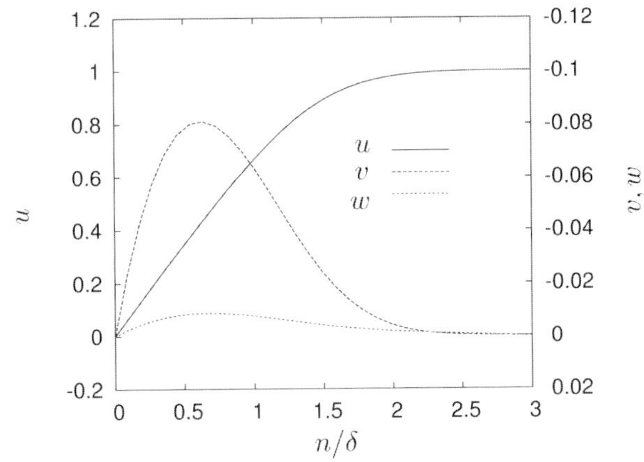

Figure 1. Velocity profile in streamline coordinates. u is the velocity in direction of the streamline, v the crossflow velocity and w the wall–normal one.

directions for its base. The u profile looks like a Blasius boundary layer but there is a crossflow component which leads to an inflection point of the overall profile. The local streamline curvature is depicted in figure 2. In that diagram a cut normal to the stagnation line is presented. We show contour lines of the local streamline curvature. The flow is from top to bottom. The upper border is the location of a detached bow shock and the lower border is the body surface.

The local curvature is maximal away from the leading edge and the locus of maximal streamline curvature is at $(x/\delta, y/\delta) \approx (100, \pm 200)$. Additionally

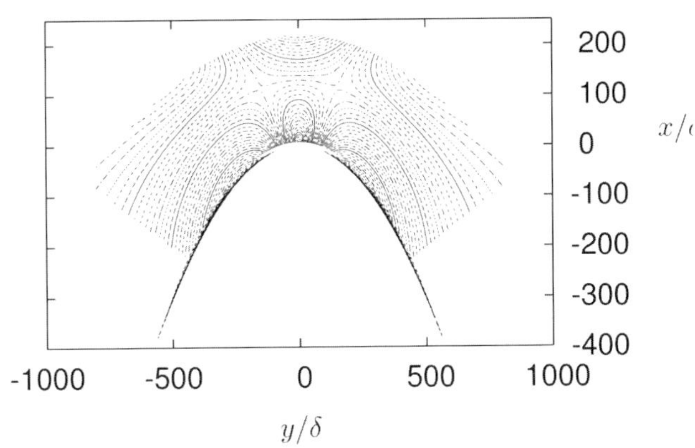

Figure 2. Local streamline curvature depicted in a plane perpendicular to the attachment line.

the primary instability vortices are shown. They are visible as the two thin stripes along the body surface. They will be discussed later. The streamlines are concave and exhibit no inflection point.

3. PERTURBED FLOW

3.1 Form of the Perturbations

The steady base flow was perturbed in two ways: randomly in the vicinity of the leading edge and coherently by an entropy perturbation upstream of the shock. Both perturbations were introduced with an amplitude such that the response was linear. This was checked afterwards.

The first method instantaneously triggered the crossflow or centrifugal instability. Very close to the attachment line the growth of an instability was observed, but it could not easily be identified as the attachment-line instability due to the fact that it was rapidly superseded by the other growth mechanism. For identification, tailored perturbations were introduced which generate a vortex pair travelling along the boundary layer edge. Thus it was hoped to favour the attachment line instability.

3.2 Results

Computations were performed for the parameters indicated in the following table 1. The extension of the computational domain in z-direction fixes

Re	**600**	**642**	**700**	**750**	**800**
R/δ	**377**	**409**	**446**	**472**	**504**

Table 1: Overview of Reynolds numbers and nose radii

the wavelength of the possible growing modes. Therefore different depths in this direction had to be investigated. For time being the computational domain is only large enough to host one principal wavelength and its higher harmonics. The most unstable mode for the incompressible case has a wavelength of $\lambda \approx 23\delta$. Thus we have chosen to vary the dimensionless wavenumber $\alpha = 2\pi\delta/\lambda = 2\pi/30 - 2\pi/20$.

Random perturbations lead to an unconditional growth of kinetic energy of the principal mode when measured globally in the full computational domain or in the full boundary layer. The growth rate was measured between 2 and 6, based on the dimensionless time $\left(\frac{\partial u_\infty}{\partial_x}\right)^{-1}$, given above. Flow visualisations revealed that this was due to vortices roughly inclined as the potential streamlines. They did not extend along the full body surface but were locally confined to the region of the strongest streamline curvature of the flow. A cut in a plane parallel to the leading edge and normal to the body surface is presented in figure 3. The plane is located near the locus of maximal streamline curvature at a distance of $s = 189\delta$ from the stagnation line. Contour lines of the v velocity component are shown. They resemble very closely the known pattern of the crossflow instability. Figure 4 shows the phase of the Fourier component of this mode. It shows a strong phase shift which is typical for crossflow modes. Atypical, and rather indicating a centrifugal instability is the local confinement of the the vortices.

In order to identify the attachment line instability, separate measurements of the kinetic energy of the perturbation were performed at the locus of maximal streamline curvature and in the vicinity of the leading edge. They revealed that the main kinetic energy is found in the strong vortices described above. At the leading edge a conditionally unstable mode was found. Its growth rates were measured to be a factor of ten less. They are reported in table 2. The critical Reynolds number is shifted towards $Re = 635$ and the wave number is lowered as compared to the flat incompressible case.

In a second test series, the flow was perturbed by an entropy spot ahead of the detached bow shock. Upon interaction with the shock, entropy, vorticity and acoustic disturbances are generated behind the shock. The vorticity perturbations have the form of two conterrotating vortices [20], and it was hoped that they strongly favour the weaker attachment–line instability.

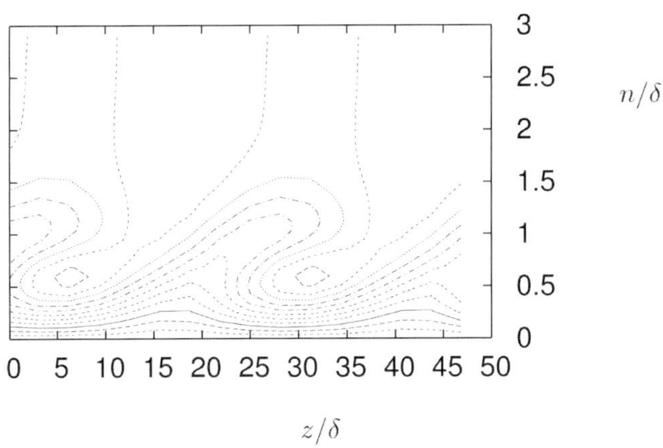

Figure 3. Contour lines of the v–velocity at $s = 189\delta$. The distance of the contourlines is ten percent of the reference velocity behind the shock.

α/Re	600	642	700	750	800
$2\pi/20$	−1.0672	−1.201	−1.3842	−1.694	−1.4605
$2\pi/22$	−0.6381	−0.580	−0.702	−0.656	−0.7515
$2\pi/24$	−0.1085	−0.2501	−0.880 (?)	**0.0804**	**0.3103**
$2\pi/26$	−1.3554	−0.1333	**0.6291**	**0.5018**	**0.7563**
$2\pi/28$	−1.3447	**0.2719**	**0.562**	**0.7268**	**0.9164**
$2\pi/30$	−1.4227	−0.357 2	**0.4062**	**0.5409**	**0.8382**

Table 2: Dependence of growth rate on Reynolds– and wave number

These attempts proved unsuccessful and it was not possible to excite the attachment–line instability strong enough to temporarily exceed the other instabilities.

4. CONCLUSIONS

For the compressible swept leading edge flow at a sweep Mach number of $M = 1.25$ with adiabatic wall conditions at a parabolic leading edge with a nose radius of $300 - 500$, the attachment–line instability was observed. It is substantially weaker than a cross-flow or centrifugal instability which is locally confined to the locus of maximal streamline curvature. The critical Reynolds number is increased whereas the corresponding wavenumber decreases. The dominating instability is unconditionally unstable in the investigated parameter range. The exact nature of this instability is uncertain. The phase relationship indicates a crossflow instability whereas the local

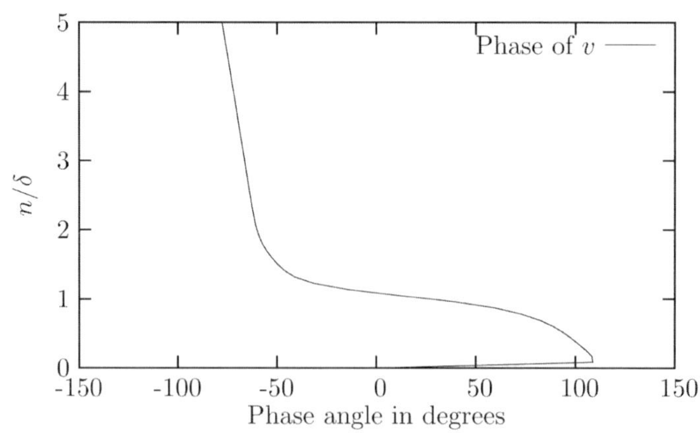

Figure 4. Phase relation of the observed instabilities

confinement of the instability to the place of maximal streamline curvature indicates a centrifugal instability.

For further investigations a global stability solver based on the current DNS-code is being developed since the DNS data is difficult to analyse and expensive to obtain.

References

[1] H. Reed and W. Saric. Stability of three-dimensional boundary layers. *Annual Review of Fluid Mech.*, 21:235–284, 1989.

[2] W. S. Saric, H. L. Reed, and E. B. White. Stability and transition of three-dimensional boundary layers. *Annual Review of Fluid Mech.*, 35:413–440, 2003.

[3] K. Hiemenz. *Die Grenzschicht an einem in den gleichförmigen Flüssigkeitsstrom eingetauchten, geraden Kreiszylinder*. PhD thesis, Göttingen, 1911. *Dingl. Polytechn. J* **326**,321(1911).

[4] H. Görtler. Dreidimensionale Instabilität der ebenen Staupunktströmung gegenüber wirbelartigen Störungen, in *50 Jahre Grenzschichtforschung*, pages 304–314. Vieweg, Braunschweig, 1955.

[5] G. Hämmerlin. Zur Instabilitätstheorie der ebenen Staupunktströmung, in *50 Jahre Grenzschichtforschung*, pages 315–327. Vieweg, Braunschweig, 1955.

[6] P. Hall, M. Malik, and D.I. Poll. On the stability of an infinite swept attachment–line boundary layer. *Proc. R. Soc. Lond.*, A(395):229–245, 1984.

[7] P.R. Spalart. Direct numerical study of leading edge contamination. Technical Report CP-438, Fluid Dyn. of 3D Turb. Shear Flows and Transition, AGARD, 1988.

[8] Ronald D. Joslin. Simulation of nonlinear instabilities in an attachment–line boundary layer. *Fluid Dynamics Research*, 18:81–97, 1996.

[9] Fabio P. Bertolotti. On the connection between cross-flow vortices and attachment–line instabilities. In *IUTAM Symposium on Laminar–Turbulent Transition*, pages 625–630, Sedona, USA, September 1999.

[10] V. Theofilis, A. Fedorov, D. Obrist, and U. Dallmann. The extended Görtler-Hämmerlin model for linear instability of three-dimensional incompressible swept attachment-line boundary layer flow. *JFM*, 2002. Under consideration for publication.

[11] Dominik Obrist and Peter Schmid. On the linear stability of swept attachment-line boundary layer flow. Part 1. Spectrum and asymptotic behaviour. *Journal of Fluid Mechanics*, 493:1–29, 2003.

[12] Anne Le Duc, Jörn Sesterhenn, and Rainer Friedrich. On instabilities in compressible attachment–line boundary layers. *Physics of Fluids*, 2003. submitted.

[13] Ray-Sing Lin and Mujeeb R. Malik. On the stability of attachment–line boundary layers. part 2. the effect of leading edge curvature. *J. Fluid Mech.*, 333:125–137, 1997.

[14] Christian Mielke. *Numerische Untersuchungen zur Turbulenzentstehung in dreidimensionalen kompressiblen Grenzschichtströmungen*. PhD thesis, ETH Zürich, 1999.

[15] M.R. Malik, Fei Li, M.M. Choudhari, and C.-L. Chang. Secondary instability of crossflow vortices and swept wing boundary layer–transition. *J. Fluid Mech.*, 399: 85–115, 1999.

[16] W.S. Saric and H.L. Reed. Crossflow instabilities – theory & technology. In *AIAA Paper*, number 2003–0771, 2003.

[17] N. Itoh. Instability of three-dimensional boundary layers due to stream-line curvature. *Fluid Dyn. Res.*, 14:353–66, 1994.

[18] Eli Reshotko and Ivan E. Beckwith. Compressible laminar boundary layer over a yawed infinite cylinder with heat transfer and arbitrary Prandtl number. Technical Report 1379, National Advisory Committee for Aeronautics, 1958.

[19] Jörn Sesterhenn. A characteristic–type formulation of the Navier–Stokes equations for high order upwind schemes. *CAF*, 30(1):37–67, 2001.

[20] David Fabre, Laurent Jacquin, and Jörn Sesterhenn. Linear interaction of a cylindrical entropy spot with a shock. *Physics of Fluids*, 13(8):2403–2422, August 2001.

STRUCTURE FORMATION IN MARGINALLY SEPARATED AERODYNAMIC AND RELATED BOUNDARY LAYER FLOWS

Alfred Kluwick and Stefan Braun

Institute of Fluid Mechanics and Heat Transfer, Vienna University of Technology, Resselgasse 3/E322, A-1040 Vienna, Austria; email: alfred.kluwick@tuwien.ac.at

Abstract: The present study deals with near critical marginally separated and triple deck flows. Asymptotic analysis then shows that unsteady three-dimensional disturbances of a steady two-dimensional critical state are governed by a nonlinear diffusion equation of Fisher's type in both cases. Solutions may exhibit finite-time blow-up but it is found that they can be extended to larger times, nevertheless. This in turn leads to the formation of characteristic flow structures localized in space and time which are universal in the sense that they are essentially independent of the initial form of the disturbances.

Key words: separation bubble, laminar-turbulent transition, finite-time blow-up

1. INTRODUCTION

Without doubt, boundary layer theory represents one of the corner stones of modern fluid mechanics. According to the original concept going back to the seminal paper [10] by Prandtl (1904) "Über die Flüssigkeitsbewegung bei sehr kleiner Reibung" the calculations of viscous wall bounded flows in the limit of large Reynolds numbers can be carried out in successive steps dealing with essentially inviscid (external) and viscous dominated (boundary layer) flow regions. This hierarchical structure in general leads to difficulties if boundary layer separation occurs. However as shown by a number of authors starting in the late 1960ies, e.g. [7], these do not signal a breakdown of the boundary layer equations and can be overcome if inviscid and viscous regions are allowed to interact already in leading rather than higher order. It is then found that there exist two different routes leading to separation of a laminar boundary layer under steady, two-dimensional flow conditions.

Firstly, a firmly attached laminar boundary layer may be forced to separate due to the presence of a large adverse pressure gradient acting over a short distance. The interaction region exhibits a triple deck structure and viscous effects are of importance inside a thin layer adjacent to the wall (lower deck) only. Here the flow is governed by the (nonlinear) boundary layer equations of an incompressible fluid.

Secondly, the formation of a short separation bubble may be caused by the presence of an adverse pressure gradient acting over a distance of order one on the typical boundary layer length scale. Also in such cases of

G.E.A. Meier and K.R. Sreenivasan (eds.), IUTAM Symposium on One Hundred Years of Boundary Layer Research, 373-382,

so-called marginal separation the interaction region splits into three layers with different physical properties. Viscous effects on the interaction mechanism are again confined to a thin wall layer where the flow is governed by the boundary layer equations of an incompressible fluid which, however, are linearized with respect to the separation profile. Solutions exist if the wall shear (or equivalently the negative perturbation displacement thickness) satisfies nonlinear solvability conditions of integro-differential form. These typically contain a single controlling parameter Γ characterizing for example the angle of attack of a slender airfoil, [11], [12], or the turning angle of a wall jet, [15] and have the remarkable property that (real) solutions exist up to a critical value Γ_c of Γ only and form two branches for a range of $\Gamma < \Gamma_c$.

Multiplicity of solutions and critical values of the controlling parameter beyond which two-dimensional steady state solutions do not exist are not a characteristic feature of marginally separated flows only. Similar phenomena are known to occur also in situations where triple deck theory applies, i.e. in situations described before as route one towards separation. Examples displaying such a branching behaviour include flows past flared cylinders, [6], subsonic flows past expansion ramps and subsonic trailing edge flows, [8].

The investigation of separated flows which are characterized by the occurrence of a critical state which terminates the regime of possible steady two-dimensional flows is of significant importance both from a practical as well as a theoretical point of view, for example in connection with leading and trailing edge stall. It is, however, severely hampered by the fact that the solution of the interaction equations – necessarily generalized to include unsteady and three-dimensional effects – represents a formidable numerical task. As noted by Braun and Kluwick [2], [3] further analytical progress is possible in the case of near critical marginally separated subsonic boundary layer flows. The asymptotic analysis capturing the flow behaviour in the limit $|\Gamma - \Gamma_c| \to 0$ is outlined in section 2 with emphasis on the finding that the resulting evolution equation is generic in the sense that it covers all known forms of marginally separated flows. In section 4 the analysis then is extended to near critical triple deck flows and on the basis of the simplified analytical results section 4 addresses the question if and how near critical both marginally separated and triple deck flows develop structures which are intrinsingly connected with the passage through Γ_c.

2. NEAR CRITICAL MARGINALLY SEPARATED FLOWS

In this section we investigate unsteady, three-dimensional disturbances of a steady two-dimensional marginally separated flow. Incompressible boundary layers have been treated first in [3] where it is shown that the wall shear A satisfies the solvability condition

$$A^2 - x^2 + \Gamma = -\lambda \int_{-\infty}^{x} \frac{1}{\sqrt{x-\xi}} \left[\frac{\partial P(\xi, z, t)}{\partial \xi} + \int_{-\infty}^{\xi} \frac{\partial^2 P(\zeta, z, t)}{\partial z^2} \mathrm{d}\zeta \right] \mathrm{d}\xi$$

$$-\gamma \int_{-\infty}^{x} \frac{1}{(x-\xi)^{1/4}} \frac{\partial (A-h)}{\partial t} \mathrm{d}\xi - \kappa \int_{-\infty}^{x} \frac{v_w}{(x-\xi)^{1/4}} \mathrm{d}\xi \,. \tag{1}$$

In addition the pressure disturbance $P(x, z, t)$ and $A(x, z, t)$ are related through the interaction law

$$P = -\frac{1}{2\pi} \int_{-\infty}^{\infty} \int_{-\infty}^{\infty} \frac{1}{\sqrt{(x-\xi)^2 + (z-\eta)^2}} \frac{\partial^2 (A-h)}{\partial \xi^2} \mathrm{d}\xi \mathrm{d}\eta \,. \tag{2}$$

Herein x, z and t denote Cartesian coordinates in the streamwise and spanwise directions and the time. All quantities are suitably nondimensionalized and scaled. Furthermore, $h(x, z, t)$ and $v_w(x, z, t)$ account for the effects of controlling devices such as wall bounded obstacles and suction strips.

Using the arguments of [12] the relationships (1), (2) can readily be shown to describe also subsonic flows of perfect gases past adiabatic walls if the coordinates and the time are appropriately redefined while in the case of external supersonic flow the interaction law (2) has to be replaced by

$$P = -\frac{1}{\pi} \int_{-\infty}^{x} \int_{z-(x-\xi)}^{z+(x-\xi)} \frac{1}{\sqrt{(x-\xi)^2 - (z-\eta)^2}} \frac{\partial^2 (A-h)}{\partial \xi^2} \mathrm{d}\xi \mathrm{d}\eta \,. \tag{3}$$

Finally, we note that the condition (1) also holds for marginally separated wall jets where the interaction law is of local form

$$P = -\frac{\partial^2 (A-h)}{\partial x^2} \,. \tag{4}$$

As pointed out before, the steady, two-dimensional versions of the problems resulting from equations (1)-(4) for $\partial / \partial t = \partial / \partial t \equiv 0$ have the remarkable property that solutions exist up to a critical value of only. Furthermore, two branches of solutions which differ, among others, by the length of the separated flow region can be determined in a range $\Gamma < \Gamma_c$. A

PSfrag replacements

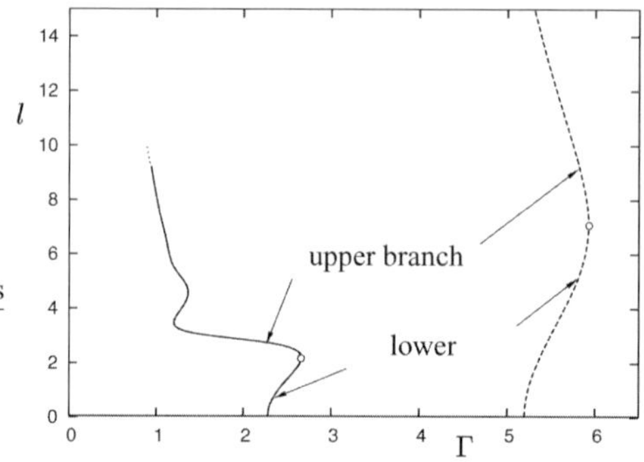

Figure 1. Nonuniqueness of the separation bubble-length l for marginally separated subsonic/supersonic boundary layer flows (solid line) and triple deck ramp flow (dashed line).

representative example is considered in Fig. 1 which displays the length l of the separated flow region inside a boundary layer with external subsonic or supersonic flow as a function of Γ.

To obtain insight into the changes of the flow behaviour associated with the passage of Γ through Γ_c we concentrate on almost critical basic states of steady, two-dimensional marginally separated flows, i.e. $|\Gamma_c - \Gamma| \ll$. Asymptotic analysis extending earlier work dealing with steady three-dimensional flows by Braun & Kluwick [2] then shows that the wall shear A can then be expanded as

$$A(x, z, t) \sim A_c(x) + \varepsilon^2 a_1(x, Z, T) + \cdots , \quad \varepsilon = |\Gamma_c - \Gamma|^{1/4} \qquad (5)$$

where $Z = \varepsilon z$ and $T = \varepsilon^2 t$. Furthermore, it is found that the first order correction $\varepsilon^2 a_1$ of the critical wall shear stress distribution $A_c(x)$ can be written in the form

$$a_1(x, Z, T) = c(Z, T)\, r(x) \qquad (6)$$

where $r(x)$ denotes the right eigenfunction of a singular linear integral operator which depends on the unperturbed flow properties only. Finally, Fourier transformation of the shape function $c(Z, T)$ with respect to the lateral coordinate

$$c(Z, T) = \int_0^\infty [a(k, T)\cos(kZ) + b(k, T)\sin(kZ)]\, \mathrm{d}k \qquad (7)$$

shows that the scaling $Z = \varepsilon z$ implies the nonlinear partial differential equation

$$\frac{\partial c}{\partial T} - \nu \frac{\partial^2 c}{\partial Z^2} + \mu c^2 - \operatorname{sgn}(\Gamma_c - \Gamma)\,\delta = g(Z,T)\ , \tag{8}$$

which can be transformed to a parameter free form known as (forced) Fisher's equation for all three versions of the interaction law (2), (3) and (4). It contains positive constants ν, μ, δ and the forcing term $g(Z, T)$ which accounts for the effects of controlling devices. In the case of below critical flows $\Gamma < \Gamma_c$ the stationary points $c = \pm c_s$, $c_s = \sqrt{\delta/\mu}$ correspond to steady two-dimensional upper and lower branch solutions.

3. NEAR CRITICAL TRIPLE DECK FLOWS

The observation that triple deck solutions may be nonunique has been made first in a study of supersonic flow past flared cylinders [6]. More recently, the loss of uniqueness has been demonstrated also for subsonic trailing edge flows, [8], and subsonic flows past expansion ramps, [9]. In a number of other problems, e.g. [5], [14] and [13] the numerical results obtained so far appear to support the possible occurrence of multiple solutions which, however, has not been confirmed conclusively yet.

To be specific let us consider the expansion ramp problem in more detail. The fundamental lower deck problem can then be expressed in the form

$$\begin{aligned}
\frac{\partial u}{\partial t} + u\frac{\partial u}{\partial x} + v\frac{\partial u}{\partial y} + w\frac{\partial u}{\partial z} &= -\frac{\partial p}{\partial x} + \frac{\partial^2 u}{\partial y^2}\ ,\\
\frac{\partial w}{\partial t} + u\frac{\partial w}{\partial x} + v\frac{\partial w}{\partial y} + w\frac{\partial w}{\partial z} &= -\frac{\partial p}{\partial z} + \frac{\partial^2 w}{\partial y^2}\ ,\\
\frac{\partial u}{\partial x} + \frac{\partial v}{\partial y} + \frac{\partial w}{\partial z} &= 0
\end{aligned} \tag{9}$$

with boundary conditions

$$\begin{aligned}
y = 0:&\quad u = w = 0\ ,\quad v = v_w(x,z,t)\ ,\\
y \to \infty:&\quad u \sim y + A(x,z,t) + \cdots\ ,\\
&\quad w \sim \frac{D(x,z,t)}{y} + \cdots\ ,\quad D = -\int_{-\infty}^{x} \frac{\partial p}{\partial z}\, d\xi\ ,\\
x \to -\infty:&\quad u \sim y + \cdots
\end{aligned} \tag{10}$$

and the interaction law given by (2). Here x, y and z denote Cartesian coordinates in the streamwise direction, normal to the wall and in spanwise direction, u, v, w the corresponding velocity components, t the time, $p = P$ the pressure, A the displacement function, h the shape of the wall contour and v_w the distribution of suction velocities subject to the usual triple deck scalings and transformations.

Solutions for steady two-dimensional flow with

$$h = h_\infty(x) = \frac{\Gamma}{2}\left(x + \sqrt{x^2 + r^2}\right) , \quad r = 0.1 \tag{11}$$

and $v_w = v_{w\infty}(x) = 0$ have recently been obtained by Zametaev (private communication). According to the numerical calculations the flow remains attached if the scaled ramp angle $\Gamma = -Re^{1/4}\,\lambda^{-1/2}$ is less than 5.2. Here Re denotes the Reynolds number formed with the distance of the corner from the leading edge of the flat plate, the free stream velocity and the free stream value of the kinematic viscosity and $\lambda \approx 0.332$ is the Blasius constant. With increasing values of the ramp angle the length l of the separation zone increases monotonically up to the critical value $\Gamma_c \approx 5.93$ beyond which no steady two-dimensional solution can be found, Fig. 1. In addition – and similar to marginally separated flows – there exist a second branch of solutions where l increases further as Γ is continuously reduced below Γ_c. This suggest that an analysis similar to the one carried out for marginally separated boundary layers should be possible also in the context of classical triple deck theory. To this end the field quantities are expanded in the form

$$\begin{aligned}
u &= u_c(x,y) + \varepsilon^2 u_1(x,y,Z,T) + \cdots ,\\
v &= v_c(x,y) + \varepsilon^2 v_1(x,y,Z,T) + \cdots ,\\
w &= \varepsilon^3 w_2(x,y,Z,T) + \cdots ,\\
p &= p_c(x) + \varepsilon^2 p_1(x,Z,T) + \cdots ,\\
A &= A_c(x) + \varepsilon^2 A_1(x,Z,T) + \cdots ,
\end{aligned} \tag{12}$$

where $\varepsilon = |\Gamma - \Gamma_c|^{1/4}$, $Z = \varepsilon z$ and $T = \varepsilon^2 t$. Controlling devices are taken into account by allowing for small deformations of the critical ramp geometry $h = h_\infty(x) + \varepsilon^4 h_2(x, Z, T)$ and suction velocities $v_w = v_{w\infty}(x) + \varepsilon^4 v_{w2}(x, Z, T)$. Substitution of (12) into the triple deck equations (9), (10) and (2) including unsteady three-dimensional effects then show that the perturbations $r_1^T = (u_1, v_1, A_1)$ of the steady two-dimensional triple deck solution for the critical ramp angle can be decomposed as

$$\boldsymbol{r}_1(x,y,Z,T) = c(Z,T)\boldsymbol{r}(x,y) \tag{13}$$

in generalization of the result (6) holding for marginally separated flows. Herein $r(x, y)$ represents the right eigenvector of a singular operator matrix which depends on the unperturbed flow quantities only and the shape function $c(Z, T)$ satisfies equation (8). It thus appears that the results derived for marginally separating boundary layers have a much larger range of applicability as originally thought.

4. STRUCTURE FORMATION

The reduction of the full interaction equations for unsteady three-dimensional near critical marginally separated and triple deck flows to equation (8) leads to a significant simplification of the problem under investigation. It thus allows, among others, a systematic study of the application of controlling devices and, even more important, the conditions causing the formation of finite-time singularities. A thorough discussion of equation (8) with special emphasis on aspects of transition in marginally separated boundary layers has been given in [3], [4]. It thus suffices to summarize the main results derived therein noting that these cover the behaviour of near critical triple deck flows as well. In this connection the observation that equation (8) can be solved in closed form in the case of two-dimensional uncontrolled, i.e. unforced flow where the full interaction equations have still to be solved numerically even in the more simple case of marginally separated flows is important. In agreement with earlier numerical computations of such flows based on the full interaction equations these solutions show the possibility of finite-time blow-up. Remarkably, however, they also show that solutions do not terminate at the blow-up time, $T = T_s$ say, but can be continued to larger times where the flow recovers if $\Gamma < \Gamma_c$ Furthermore, it is found that the flow properties for $T \rightarrow T_s$ are universal in the sense that they do not depend on initial conditions. This suggests that finite-time singularities represent processes leading to the formation of intrinsic flow structures which are localized in time and space and, from a physical point of view, describe isolated bubble bursts rather than an abrupt change of the global flow. It can be shown that the phenomenon of

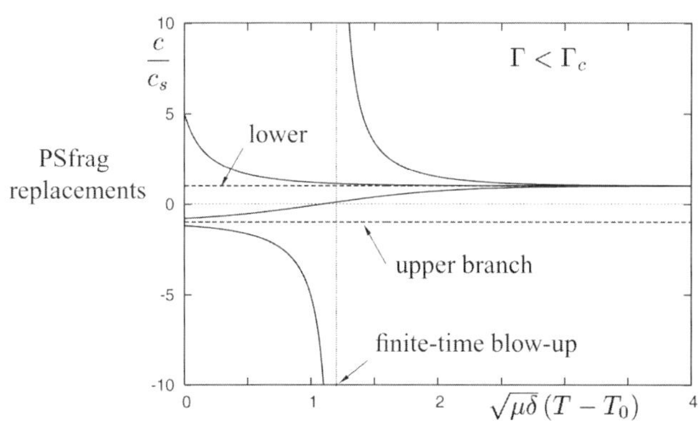

Figure 2. Perturbations of *stable* lower and *unstable* upper branch solutions, [3].

bubble bursting requires a certain finite perturbation level in below-critical situations $\Gamma < \Gamma_c$, Fig. 2, but is triggered by even infinitesimally small disturbances if $\Gamma > \Gamma_c$ where also self sustained oscillations with periodically repeated bubble bursts are possible, Fig. 3.

No closed form solutions of equation (8) are available for unsteady three-dimensional flows where the diffusion term $\partial^2_c/\partial_z^2$ comes into operation. However, using asymptotic methods it is possible to derive analytical expressions which describe the flow behaviour associated with finite-time blow-up in the limit $|T - T_s| \to 0$. The resulting flow structure is again universal and localized in time and space, [3]. It describes the focusing of disturbances for $T < T_s$ and the emergence of a pair of singularities moving in the spanwise direction for $T > T_s$, Fig. 4. They can be interpreted as vortical structures qualitatively similar to those observed in direct numerical simulations and experimental studies of transitional laminar separation bubbles.

PSfrag replacements

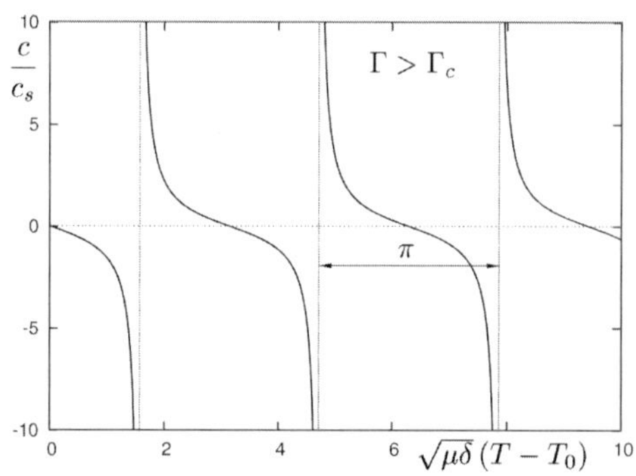

Figure 3. Self-sustained bubble bursting, [3].

PSfrag replacements

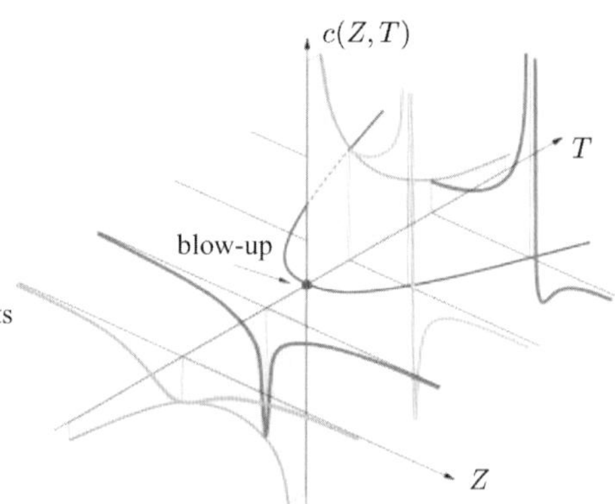

Figure 4. Blow-up structure for near-critical unsteady three-dimensional flow, [4].

5. CONCLUSIONS

A phenomenon known from interactive boundary layer theories is the existence of socalled critical states characterized by a value Γ_c of the relevant controlling parameter Γ of the problem under consideration such that no solutions exist for $\Gamma > \Gamma_c$ under steady two-dimensional flow conditions which are associated with the formation of two solution branches in a range $\Gamma \leqslant \Gamma_c$. In all known cases the critical state contains a separated flow region but the field quantities vary smoothly in the whole computational domain so that there is no sign of any irregularity similar for example the occurrence of Goldstein's separation singularity in the classical boundary layer approach which heralds the breakdown of the theory. It has, therefore, been speculated by a number of authors that the passage of Γ through be associated with a substantial (global) change of the flow behaviour.

In the present study it is argued that there exists an alternative possibility leading to a more gradual modification of the flow properties as Γ passes through Γ_c. Indeed, by performing an asymptotic analysis for near critical flow, i.e. $|\Gamma - \Gamma_c| \rightarrow 0$ it is possible to derive an evolution equation which governs unsteady and three-dimensional disturbances of the critical state. Its solutions may exhibit finite-time blow-up but it is found that they can be extended to larger times leading in turn to the formation of characteristic flow structures. These are localized in space and time and can be interpreted as (repeated) bubble bursting and the emergence of pairs of vortices propagating in lateral direction, both phenomena which are commonly observed in transitional separation bubbles. Probably the most important result in this connection is the observation that the evolution equation of near critical states is of exactly the same form for marginally separated flows and triple deck flows. This ties in nicely with a recent publication by Borodulin et al [1] in which it was argued that bursting processes in transitional laminar boundary layers share common universal properties that do not depend on the specific problem under consideration.

REFERENCES

1. Borodulin V.I., Gaponenko V.R., Kachanov Y.S., Meyer D.G.W., Rist U., Lian Q.X., Lee C.B. Late-stage transitional boundary-layer structures. Direct numerical simulation and experiment. *Theoret. Comput. Fluid Dynamics* **15**, 317–337, 2002.

2. Braun S., Kluwick A. The effect of three-dimensional obstacles on marginally separated laminar boundary layer flows. *J. Fluid Mech.* **460**, 57–82, 2002.
3. Braun S., Kluwick A. Unsteady three-dimensional marginal separation caused by surface mounted obstacles and/or local suction. To appear in *J. Fluid Mech.*
4. Braun S., Kluwick A. Blow-up and control of marginally separated boundary layer flows. To appear in *New developments and applications in rapid fluid flows* (ed. J.S.B. Gajjar & F.T. Smith), *Phil. Trans. R. Soc. Lond.* A **358**, 3113–3128, 2000.
5. Gajjar J.S.B., Türkyilmazoglu M. On the absolute instability of the triple-deck flow over humps and near wedged trailing edges. *Phil. Trans. R. Soc. Lond.* A
6. Gittler Ph., Kluwick A. Triple-deck solutions for supersonic flows past flared cylinders. *J. Fluid Mech.* **179**, 469–487, 1987.
7. Kluwick A. *Recent advances in boundary layer theory*, Wien New York, Springer, 1998.
8. Korolev G.L. Contribution to the theory of thin-profile trailing edge separation. *Izv. Akad. Nauk SSSR: Mekh. Zhidk. Gaza* **4**, 55–59 (Engl. transl. *Fluid Dyn.* **24**, 534–537), 1989.
9. Korolev G.L. Nonuniqueness of separated flow past nearly flat corners. *Izv. Akad. Nauk SSSR: Mekh. Zhidk. Gaza* **3**, 178–180 (Engl. transl. *Fluid Dyn.* **27**, 442–444), 1992.
10. Prandtl L. Über Flüssigkeitsbewegung bei sehr kleiner Reibung. *Verh. III. Intern. Math. Kongr. Heidelberg, 1904*, 484–491, Leipzig, Teubner, 1905.
11. Ruban A.I. Asymptotic theory of short separation regions on the leading edge of a slender airfoil. *Izv. Akad. Nauk SSSR, Mekh. Zhidk. Gaza* **1** (Engl. transl. *Fluid Dyn.* **17**, 33–41), 1981.
12. Stewartson K., Smith F.T., Kaups K. Marginal separation. *Stud. in Appl. Math.* **67**, 45–61, 1982.
13. Türkyilmaz I. An investigation of separation near corner points in transonic flow. *J. Fluid Mech.* **508**, 45–70, 2004.
14. Turkyilmazoglu M. Flow in the vicinity of the trailing edge of Joukowski-type profiles. *Proc. R. Soc. Lond.* A **458**, 1653–1672, 2002.
15. Zametaev V.B. Existence and nonuniqueness of local separation zones in viscous jets. *Izv. Akad. Nauk. SSSR, Mekh. Zhidk. Gaza* **1** (Engl. transl. *Fluid Dyn.* **21**, 31–38), 1986.

HIGH REYNOLDS NUMBER TURBULENT BOUNDARY LAYERS SUBJECTED TO VARIOUS PRESSURE-GRADIENT CONDITION

Hassan M. Nagib [1], Chris Christophorou [1] and Peter A. Monkewitz [2]

[1] *Dept. of Mechanical, Materials and Aerospace Engineering, Illinois Institute of Technology (I I T), Chicago, IL 60616-3793, USA ; e-mail: nagib@iit.edu*

[2] *Laboratory of Fluid Mechanics (LMF), Swiss Federal Institute of Technology Lausanne (EPFL), CH-1015 Lausanne, Switzerland ; e-mail: peter.monkewitz@epfl.ch*

Abstract: Mean velocity distributions in the overlap region, over the range of Reynolds numbers $10{,}000 < Re_\theta < 70{,}000$, under five different pressure-gradient conditions are accurately described by a log law. The pressure-gradient conditions include adverse, zero, favorable, strongly favorable, and a complex gradient. The wall-shear stress was measured using oil-film interferometry, and hot-wire sensors were used to measure velocity profiles. Parameters of the logarithmic overlap region developed from these higher Reynolds number boundary layers continue to be consistent with our recent findings and to remain independent of Reynolds number. The best estimate of the log-law parameters from the zero-pressure gradient boundary layers is $\kappa = 0.384,\ B = 4.127$. However, the Kármán "coefficient" (κ) is found to vary considerably for the non-equilibrium boundary layers under the various pressure gradients. The results highlight the variation with pressure gradient not only in the outer region of the boundary layer but also within the inner region. A slightly modified version of the almost century old Prandtl-Kármán skin friction relation provides an exceptional agreement with all three sets of data (Hites, Österlund, and Christophorou) for zero pressure gradient conditions.

Key words: Turbulent boundary layers, high Reynolds numbers, pressure gradient, wall-shear stress, Kármán constant, logarithmic law.

1. INTRODUCTION

Two independent experimental investigations of the behavior of turbulent boundary layers with increasing Reynolds number (Re_θ) were recently completed [1]. The experiments were performed in two

G.E.A. Meier and K.R. Sreenivasan (eds.), IUTAM Symposium on One Hundred Years of Boundary Layer Research, 383-394,

facilities, the MTL wind tunnel at KTH and the NDF wind tunnel at IIT. While the KTH experiments were carried out on a flat plate, the model used in the NDF was a long cylinder with its axis aligned in the flow direction. Both experiments were conducted in a zero-pressure gradient, covered the range of Reynolds numbers based on the momentum thickness from *2,500* to *27,000*, and utilized oil-film interferometry to obtain an independent measure of the wall-shear stress. Contrary to the conclusions of some earlier publications, careful analysis of the data revealed no significant Reynolds number dependence for the parameters describing the overlap region using the classical logarithmic relation. The parameters of the logarithmic overlap region were found to be constant and were estimated to be: $\kappa = 0.38$, $B = 4.1$. These two experiments have been recently extended to Reynolds numbers based on momentum thickness exceeding *70,000* on a flat plate in the NDF.

2. EXPERIMENTAL ARRANGEMENT

The current experiments were also carried out in the National Diagnostic Facility (NDF) at IIT on a 10 m long and 1.5 m wide flat plate using free-stream velocities ranging from 30 to 85 m/s. Again, hot-wire anemometry and oil-film interferometry were used to measure the velocity profiles and the wall-shear stress, respectively; the hot-wire data were collected on an equally spaced dense grid. The design of the experiments, and in particular the location and spacing of the velocity profiles in the downstream direction and the hot-wire sensor in the wall-normal direction, were carefully selected to facilitate the evaluation of wall-normal and streamwise derivatives. The arrangement of the NDF was designed to allow the adjustment of the test-section ceiling to impart various pressure gradients. Several conditions were investigated so far, including: Adverse Pressure Gradient (APG), Zero Pressure Gradient (ZPG), Favorable Pressure Gradient (FPG), Strongly Favorable Pressure Gradient (SFPG), and a Complex Pressure Gradient (CPG) as displayed in Figure 1.

3. RESULTS

One of the cornerstones of our approach to measurements of turbulent wall-bounded flows is the independent and accurate measurement of the wall shear stress with oil-film interferometry. We believe the only wall-bounded flow that may not require such measurements is the fully developed pipe flow, where the careful measurement of pressure gradient

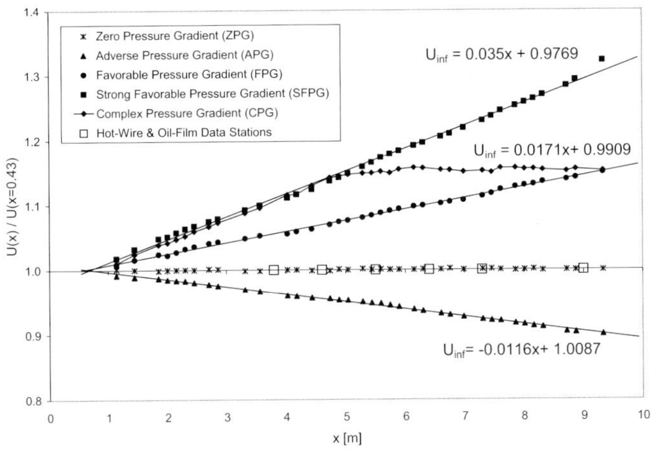

Figure 1. Variation of free-stream velocity ratio along *10-m* plate for various pressure-gradient conditions.

can lead to an accurate determination of the friction velocity. For the non-equilibrium boundary layers under various pressure-gradient conditions, one can be dramatically misled by other indirect techniques for the determination of wall-shear stress. Oil film interferometry is the most reliable method for accurate and direct measurement of mean skin friction (~1.5%) and it can also measure its direction [2]. The keys to achieving such accuracy are the independent determination of the oil viscosity as a function of temperature, the steady and monitored temperature of the oil and the surface during the experiment, and the processing of the images using advanced digital acquisition and processing. Until another even more reliable technique is developed it should be accepted as the standard for such measurements. Figure 2 compares our ZPG measurements with the recent two sets of measurements of Hites and Österlund [1]. The value of the Kárman constant extracted using the correlation used by Professor Don Coles in the 1968 Stanford Olympics on boundary layers in his outstanding contribution [3] and again proposed by Fernholz [4] for all three sets of data is approximately 0.384. We will refer to this relation as the Coles-Fernholz relation, and as readily observed from its form in Equation 1 it is based on the logarithmic velocity profile.

$$C_f = 2*[\frac{1}{\kappa}\ln(\mathrm{Re}_\theta) + C]^{-2} \tag{1}$$

Many relations are found in the literature describing the dependence of the skin friction for zero pressure gradient boundary layers on various Reynolds numbers. All of them are based on some integral form of the boundary layer equations of Prandtl complemented with experimental correlations. They include the Prandtl-Kármán relation which is used extensively in the chemical engineering literature. It is usually given as:

$$(C_f)^{-0.5} = 4\log((C_f)^{-0.5}\,\mathrm{Re}_x) - 0.4 \tag{2}$$

It is quite remarkable that Equation 2 is used for pipe flows as well as quite often for zero-pressure gradient boundary layers, although the experimental correlation utilized in deriving it is Nikuradse's pipe flow data! In the aerospace and mechanical engineering literature (e.g., Schlichting [5] and White [6]) a number of relations are used and they include the so called Prandtl approximate relation and the $1/7^{\mathrm{th}}$ power law:

$$C_f = 0.058/\mathrm{Re}_x{}^{0.2} \tag{3}$$

$$C_f = 0.027/\mathrm{Re}_x{}^{(1/7)} \tag{4}$$

Another relation based on an analytical fit to the Prandtl-Kármán relation is found in White's book on viscous flows [6]:

$$C_f = 0.455/[\ln(0.06\,\mathrm{Re}_x)]^2 \tag{5}$$

A most interesting relation based on a correlation of experimental data is due to Schultz-Grunow [7]:

$$C_f = 0.37(\log(\mathrm{Re}_x))^{-2.584} \tag{6}$$

It is very curious that the exponent of Equation 6 is nearly equal to the reciprocal of *0.38*. The most recent significant contribution to such a relation is found in the work of Professor William George and his associates [8]. Using a different asymptotic approach, which utilized the external velocity in the outer scaling, they conclude that the velocity profile is power-law like and develop a set of equations including the following for the correlation of the skin friction:

$$\frac{u_*}{U_\infty} = \frac{C_{o\infty}}{C_{i\infty}}(\delta^+)^{-\gamma_\infty}\exp[A/\ln(\delta^+)^\alpha] \tag{7}$$

The Prandtl-Kármán relation is based on integral boundary layer theory and an empirical coefficient. It is quite ironic that the version most commonly used (Equation 2), even for boundary layers, contains the constant 0.4 which is based on Nikuradse's pipe data. The only modification we make to this relation to fully agree with our data is changing this constant to a value of 2.12. This modification also renders the relation more compatible with Nikuradse's less known *flat plate* boundary layer data [9]. One of the difficulties in utilizing this relation is its implicit form. A much simpler explicit relation has been in the literature for nearly four and a half decades and is based on the logarithmic velocity profile; Equation 1. As demonstrated by Figures 2 and 3, the data of Österlund, Hites and Christophorou lead to the values $\kappa = 0.384$ and $C = 4.127$. In contrast, since the late 1960's, the most commonly used version of this relation utilized $\kappa = 0.41$ and $C = 5$. For convenience in using this relation we also found based on the results of Österlund, and Christophorou the correlation:

$$\theta = 0.022\, x^{0.835}. \qquad (8)$$

Therefore, we find that, a slightly modified version of the nearly a century old Prandtl-Kármán skin friction relation provides an exceptional agreement with zero pressure-gradient boundary (ZPG) layer data. In fact many of the commonly used skin friction relations for ZPG layers provide correct predictions for all practically encountered Reynolds numbers if they are underpinned by the same accurate measurements; some of them were even used by Coles [3]. As we concluded recently [10], the differences between the various commonly used relations, including those based on the log law or a power law velocity profile can only be resolved by theoretical arguments or the detailed measurements of velocity profiles over a wide range of Reynolds numbers as shown later here. We should also point out that in the equilibrium ZPG layers, the resulting skin friction is independent of whether the momentum thickness Reynolds number is varied by changing the free-stream velocity or the downstream distance; i.e., no effects of "initial conditions." Had the data of Schultz-Grunow [7] been used some fifty years earlier in the Prandtl-Kármán relation, we would have been in a much better shape in our ability to predict resistance on various surfaces. We also find that the seldom referenced data of Rolf Karlsson [11] deserves far more attention and are in excellent agreement with our data.

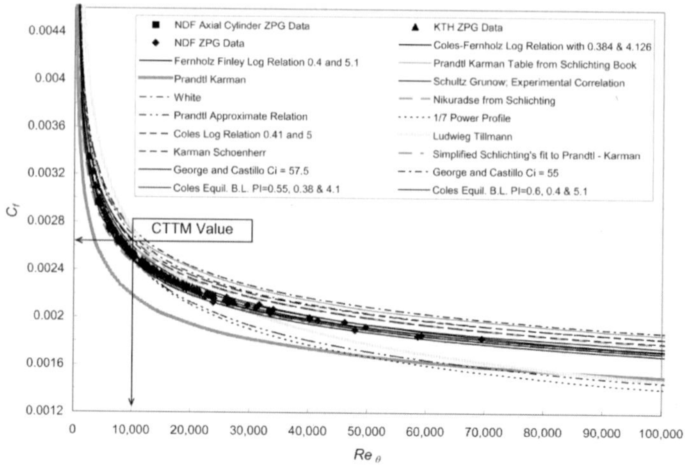

Figure 2. Variation of skin friction coefficient with momentum-thickness Reynolds number in zero-pressure gradient boundary layers compared to commonly used correlations.

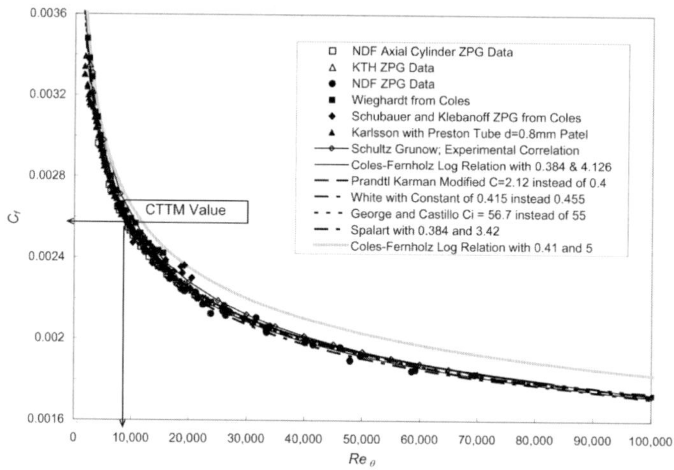

Figure 3. Variation of skin friction coefficient with momentum-thickness Reynolds number in zero-pressure gradient boundary layers compared to other selected experiments, modified version of commonly used correlations, and the relation using the most popular Coles values from 1968 [3].

Oil-film data for boundary layers under various pressure-gradient conditions are included in Figure 4. The behaviour of non-equilibrium layers like SFPG (note amount of change indicated by vertical arrows) is most revealing when contrasted to equilibrium ZPG data. In absence of direct skin friction measurements, the Clauser plot approch is commonly used based on the assumption of a constant value of the Kármán constant (κ). Such an approach leads to drastically different and obviously incorrect values for skin friction as displayed in Figure 5 with closed symbols; i.e., from this alone we can conclude that κ is not a constant.

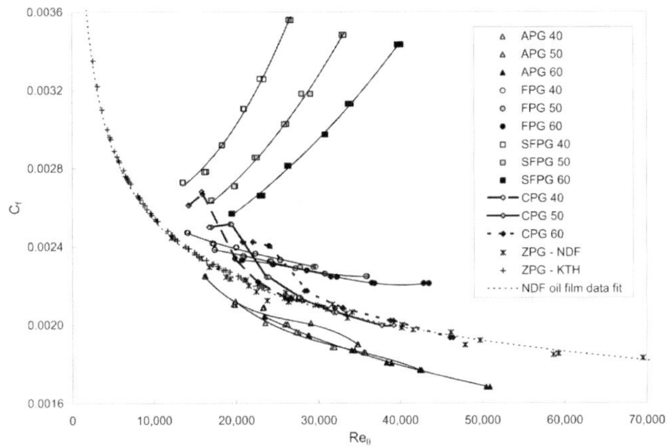

Figure 4. Variation of skin friction coefficient with momentum-thickness Reynolds number for constant upstream reference free-stream velocities of *40, 50* and *60* m/sunder the various pressure gradients of Figures 1 and 8.

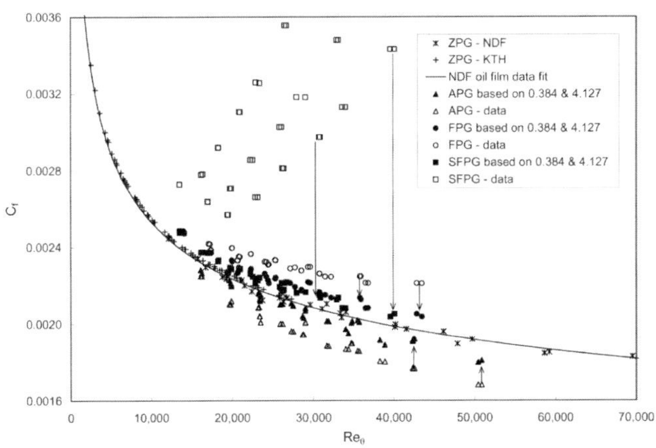

Figure 5. Variation of skin friction coefficient with momentum-thickness Reynolds number developed without use of oil-film data for pressure-gradient cases and assuming a constant value of Kármán constant κ of *0.384*; use of other popular values of κ leads to similar results.

As in our earlier work [1], profiles of the mean and rms streamwise component of the velocity and their spatial derivatives are used to examine the effects of the pressure gradient on the inner and outer layers as well as their overlap region. Figures 6 and 7 display most of the ZPG data and confirm the value of κ extracted from the oil-film data for the same conditions.

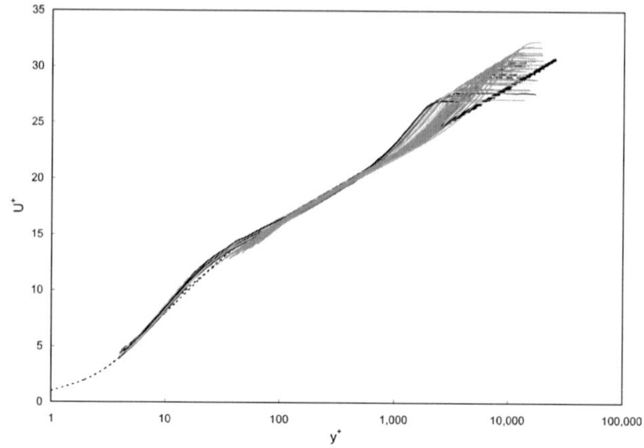

Figure 6. Inner-scaled velocity profiles measured with hot wire probe at $x = 3.8, 4.6, 5.5, 6.4, 7.3$ and 9 m for free-streams of $30, 40, 50$ and 60 m/s; figure also displays KTH data with $Re_\theta > 8{,}000$, two profiles for each case of NDF data, excluding runs with 50 and 60 m/s at $x = 7.3$ and 9 m, and a logarithmic law with $\kappa = 0.384$ and $B = 4.127$.

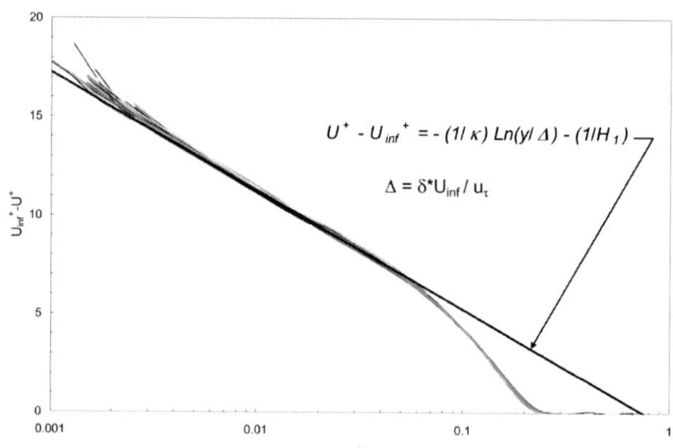

Figure 7. Outer scaled velocity profiles measured with hot wire probe at $x = 3.8, 4.6, 5.5, 6.4, 7.3$ and 9 m for free-streams of $30, 40, 50$ and 60 m/s; figure displays two profiles for each case and excludes runs with 50 and 60 m/s at $x = 7.3$ and 9 m, and a logarithmic law with $\kappa = 0.384$ and $H_1 = 1.31$.

The pressure-gradient parameter for the five different conditions is given in Figure 8, and sample velocity profiles are displayed in Figure 9. The results demonstrate that the pressure gradient causes significant changes not only in the outer region of the boundary layer but also within the inner region; i.e., the buffer layer. The effect of these changes on Coles' outer layer parameter and the behavior of the maximum turbulence stress have also been documented. In particular, the velocity profiles reveal the dependence of the log-layer parameter κ and B on the pressure gradients in these non-equilibrium boundary layers.

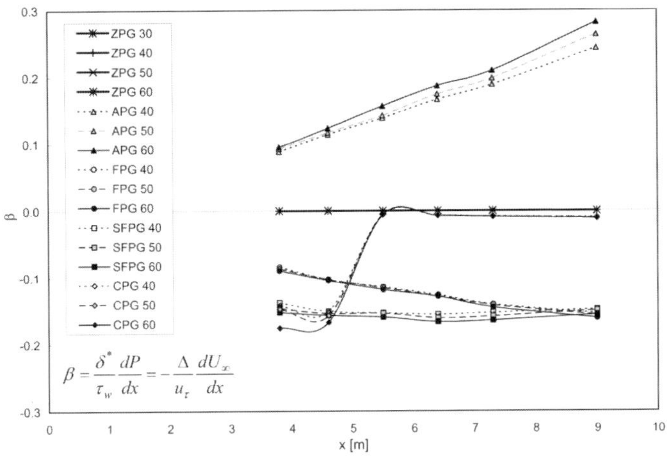

Figure 8. Variation of pressure gradient parameter β with downstream distance for conditions of Figure 1.

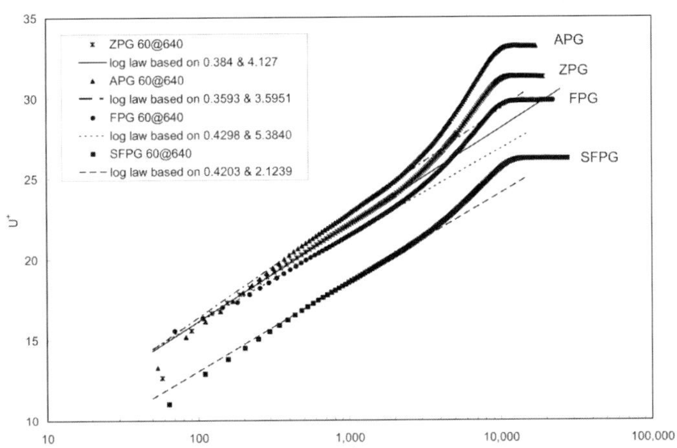

Figure 9. Velocity profiles measured with hot wire probe at $x = 6.4$ m and free-stream velocity of *60* m/s for several pressure gradients based on u_τ from oil-film measurements; figure displays actual data points throughout profile.

In order to examine the validity of the power law, the diagnostic functions for it and the log law are calculated by differentiating the hot-wire data, and sample results are shown in Figures 10 and 11 for *15,800* $< Re_\theta <$ *34,000*. It is quite clear at these higher Reynolds numbers that the overlap region is very accurately represented by a log law and not by the power law; see comments that followed our earlier work [1] in the same journal. The collapse of the velocity profiles of Figure 7 using u_τ to non-dimensionalize the velocity defect, and the far inferior collapse when the free-stream velocity is used in its place (not shown here) are additional strong evidence in support of this conclusion.

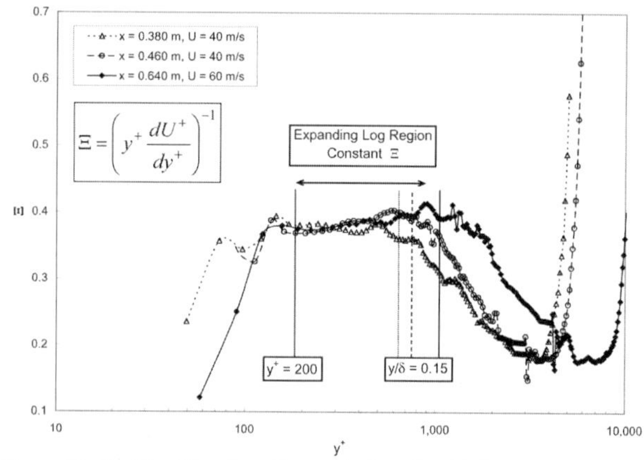

Figure 10. Profile of log-law diagnostic function Ξ for zero-pressure gradient boundary layers, 15,800 $< Re_\theta <$ *34,000*, demonstrating excellent log-like region above $y^+ \sim$ *200* and up to $y/\delta \sim 0.15$.

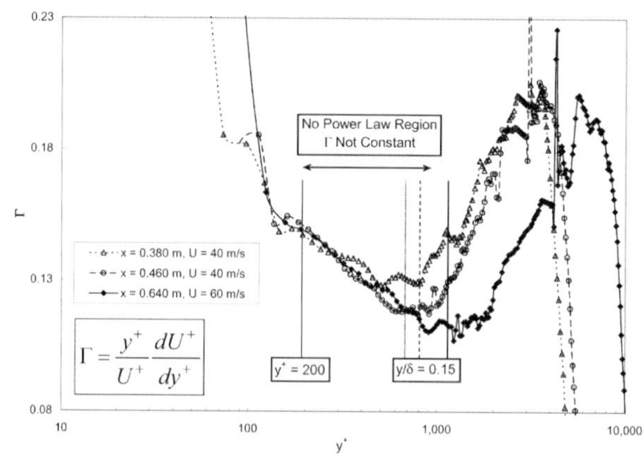

Figure 11. Profile of power-law diagnostic function Γ for zero-pressure gradient boundary layer, 15,800 $< Re_\theta <$ *34,000*, demonstrating lack of power-law Region up to $y/\delta \sim 0.15$.

Using data analysis similar to the profiles of Figure 10 for all pressure gradients from present experiment, we find that, for these non-equilibrium boundary layers, values of the Kármán "coefficient" κ vary between *0.34* & *0.38* for APG, are near *0.41* for FPG, vary between *0.36* & *0.49* for SFPG, and change between *0.38* & *0.42* for CPG.

4. CONCLUSIONS

One of the cornerstones of our approach to measurements of turbulent wall-bounded flows is the independent and accurate measurement of the wall shear stress with oil-film interferometry. We believe the only wall-bounded flow that may not require such measurements is the fully developed pipe flow, where the careful measurement of pressure gradient can lead to an accurate determination of the friction velocity.

The zero-pressure gradient boundary layer data clearly and irrefutably demonstrate that, for high Reynolds numbers, the log law is the correct representation of the overlap region and the power law is not a valid representation of the same region. In contrast to the constant value of κ for ZPG of *0.384*, the variation of the Kármán "coefficient" from generally accepted values is revealed by pressure gradient measurements in these non-equilibrium wall-bounded shear flows. In general, the value of κ tends toward the ZPG value as the boundary layers develop to an equilibrium state.

ACKNOWLEDGEMENTS

The results presented here are based on work primarily funded by the Air Force Office of Scientific Research, USAF, under grant number F49620–01–1-0445, monitored by Dr. Tom Beutner. We wish to especially thank Prof. Don Coles, of Caltech, Dr. Jens Österlund, of FOI in Sweden, Prof. K. R. Sreenivasan of the Abdus Salam ICTP in Trieste, Italy, Prof. Arne Johansson of KTH in Stockholm, Prof. Bill George of Chalmers University in Sweden, and Dr. Philippe Spalart of Boeing in Seattle, for the many stimulating discussions during the course of this research. The contributions of Kapil Chauhan to the processing and interpretation of the data have been extensive and are greatly appreciated.

REFERENCES

1. J. M. Österlund, A.V. Johansson, H. M. Nagib, and M. H. Hites, A note on the overlap region in turbulent boundary layers, Phys. Fluids, 12:1–4, 2000.
2. J-D. Ruedi, H. Nagib, J Österlund, P. Monkewitz. 2003. Evaluation of three techniques for wall-shear measurements in threedimensional flows. Exp Fluids 35:389–396.
3. D. Coles, The Young Person's Guide to the Data, Proceedings of Turbulent Boundary Layers – 1968, AFOSR-IFP-Stanford Conference, Vol. II, 1968.
4. H. H. Fernholz, Ein halbempirisches Gezetz fuer die Wandreibung in kompressiblen turbulenten Grenzschichten bei isothermer und adiabater Wand. ZAMM, 51: 148–149, 1971.
5. H. Schlichting, Boundary Layer Theory, McGraw-Hill Book Company, Seventh Edition, 1979.
6. F. M. White, Viscous Fluid Flow, McGraw-Hill Book Company, Third Edition, 2005.
7. F. Schultz-Grunow, Neues Widerstandsgesetz für glatte Platten, Luftfahrtforschung 17, 239, 1940; also NACA TM 986, 1941.
8. W. K. George, L. Castilio, and P. Knecht. The zero pressure gradient turbulent boundary layer. Technical Report TRL-153, Turbulence Research Laboratory, SUNY at Buffalo, 1996.
9. J. Nikuradse, Turbulente Reibungsschichten an der Platte, Published by ZWB, R. Oldenbourg, Munich and Berlin, 1942.
10. H. Nagib, C. Christophorou, J-D Reudi, P. Monkewitz, J. Österlund and S. Gravante, Can We Ever Rely on Results from Wall-Bounded Turbulent Flows without Direct Measurements of Wall Shear Stress?, AIAA -2004-2392, 24th AIAA Aerodynamic Measurement Technology and Ground Testing Conference Portland, OR 28 June–1 July, 2004.
11. R. I. Karlsson, Studies of Skin Friction in Turbulent Boundary Layers on Smooth and Rough Walls, Doctoral Thesis, Chalmers University of Technology, Göteborg, Sweden, 1980.

ANALYSIS OF ADVERSE PRESSURE GRADIENT THERMAL TURBULENT BOUNDARY LAYERS AND CONSEQUENCE ON TURBULENCE MODELING

T. Daris, H. Bézard
SNECMA Motors, Rd Pt R. Ravaud, 77550 Moissy, France. Tel: +33 (0) 1 60 59 82 63
thomas.daris@snecma.fr
ONERA, 2. av. E. Belin, 31055 Toulouse, France. Tel: +33 (0) 5 62 25 28 28
herve.bezard@onera.fr

Abstract: An analysis of the fully turbulent region of slightly heated boundary layers submitted to adverse pressure gradients is presented. Both moderate and strong pressure gradients are considered, leading to two different regions: the logarithmic region and the half-power region. New forms of the temperature profile for both regions are proposed and assessed through experimental data. This analysis is used to evaluate the performance of different existing four-equation turbulence models in APG boundary layers.

Key words: Turbulence, heat flux, pressure gradient, logarithmic law, half-power law, turbulence modeling, boundary layer.

1. INTRODUCTION

The frame of this work is the slightly heated incompressible turbulent boundary layer at zero pressure gradient (ZPG) or submitted to an adverse pressure gradient (APG). It follows the analysis of Huang and Bradshaw [**6**] but gives a deeper view in the effect of pressure gradient, with emphasis on the thermal behavior. First, the governing equations of the flow are recalled. Then the different regions of the boundary layer are

G.E.A. Meier and K.R. Sreenivasan (eds.), IUTAM Symposium on One Hundred Years of Boundary Layer Research, 395-404,

analyzed. Finally the consequence on the turbulence modeling is presented.

2. FLOW EQUATIONS

If temperature differences are small between the wall and the outer flow, temperature can be considered as a passive scalar and the temperature field does not influence the velocity field. The flow is described by the momentum equation for the mean longitudinal velocity u and by the energy equation written with the mean temperature T:

$$\begin{cases} \rho \dfrac{Du}{Dt} = -\dfrac{dp}{dx} + \dfrac{\partial}{\partial y}\left(\mu \dfrac{\partial u}{\partial y} - \rho \overline{u'v'} \right) \\ \rho C_p \dfrac{DT}{Dt} = \dfrac{\partial}{\partial y}\left(\lambda \dfrac{\partial T}{\partial y} - \rho C_p \overline{v'T'} \right) \end{cases} \quad (1)$$

where $-\overline{u'v'}$ and $-\overline{v'T'}$ are the shear stress and the turbulent heat flux, μ and λ are the dynamic viscosity and the thermal conductivity of the flow and C_p is the specific heat coefficient. μ and λ are connected through the Prandtl number $Pr = \mu C_p / \lambda$ (≈ 0.7 for air). For convenience, the cinematic viscosity $\nu = \mu / \rho$ and the thermal diffusivity $\alpha = \lambda / (\rho C)_p$ will be used, so that $Pr = \nu / \alpha$.

By analogy the turbulent Prandtl number is defined as $Pr_t = \nu_t / \alpha_t$, where ν_t and α_t are the eddy viscosity and diffusivity, first introduced by Boussinesq and defined as:

$$-\overline{u'v'} = \nu_t \frac{\partial u}{\partial y} \quad \text{and} \quad -\overline{v'T'} = \alpha_t \frac{\partial T}{\partial y} \quad (2)$$

The present work will be restricted to the fully turbulent inner part of the boundary layer where convection is negligible. This excludes the outer part of the boundary layer. Hence, Eq. (1) becomes after integration:

$$\begin{cases} -\overline{u'v'} + \nu \dfrac{\partial u}{\partial y} - \dfrac{1}{\rho}\dfrac{dp}{dx} = cst = \dfrac{\tau_w}{\rho} \\ -\overline{v'T'} + \alpha \dfrac{\partial T}{\partial y} = cst = \dfrac{q_w}{\rho C_p} \end{cases} \quad (3)$$

where τ_w is the wall shear stress and q_w is the wall heat flux. Eq. (3) shows that the total shear stress for ZPG and the total heat flux (viscous + turbulent) are constant in the inner region, which is confirmed by experiments.

Dimensional analysis yields to the definition of the friction velocity u and of the friction temperature T_τ, first introduced by Squire [**14**], and to the dimensionless variables y^+, u^+, T^+, usually called "wall-variables":

$$u_\tau = \sqrt{\frac{\tau_w}{\rho}};\ T_\tau = \frac{q_w}{\rho C_p u_\tau};\ y^+ = \frac{y u_\tau}{\nu};\ u^+ = \frac{u}{u_\tau};\ T^+ = \frac{T_w - T}{T_\tau} \qquad (4)$$

The other quantities can be expressed likewise:

$$p^+ = \frac{\nu}{\rho u_\tau^3}\frac{dp}{dx}\ ;\ \overline{u'v'}^+ = \frac{\overline{u'v'}}{u_\tau^2}\ ;\ \overline{v'T'}^+ = \frac{\overline{v'T'}}{u_\tau T_\tau} \qquad (5)$$

Using the wall-variables and considering only the fully turbulent part of the inner region where viscous diffusion can be neglected, Eq. (3) becomes:

$$\overline{u'v'}^+ = 1 + p^+ y^+ \quad \text{and} \quad \overline{v'T'}^+ = 1 \qquad (6)$$

This expression shows that the turbulent shear stress depends on the pressure gradient, on the contrary to the turbulent heat flux.

3. LOGARITHMIC REGION

In this section the case of moderate APG for which $p^+y^+ \ll 1$ will be considered. As a consequence, if the pressure gradient value increases, the extension in terms of y^+ of the concerned region will decrease.

A particular case is the ZPG boundary layer ($p^+=0$) for which $-\overline{u'v'}^+ = 1$. Dimensional analysis yields to the expression of the velocity and temperature gradients of the form:

$$\begin{cases} \dfrac{\partial u^+}{\partial y^+} = \dfrac{1}{\kappa y^+} & \Rightarrow u^+ = \dfrac{1}{\kappa} ln\left(y^+\right) + C \\ \dfrac{\partial T^+}{\partial y^+} = \dfrac{1}{\kappa_T y^+} & \Rightarrow T^+ = \dfrac{1}{\kappa_T} ln\left(y^+\right) + C_T \end{cases} \qquad (7)$$

which leads by integration to the well-known "logarithmic law" for the velocity and temperature profiles, with κ and κ_T known as the von Kármán constants. Log-laws are valid in experiments for $y^+>30$ approximately and the constants are: $\kappa \approx 0.41$, $C \approx 5.2$, $\kappa_T \approx 0.48$, $C_T \approx 3.8$ for air. It is straightforward to show from Eq. (2) that $\nu_t^+ = \kappa y^+$ and $\alpha_t^+ = \kappa_T y^+$ so that $Pr_t = \kappa/\kappa_T$ (0.85) in the logarithmic region.

In case of APG, the log-law remains for the velocity profile, as shown in Fig. 1 for different experiments (ZPG [**11**], APG [**7-12**]). However this is not the case for the temperature profile as seen in Fig. 2. The ZPG experimental temperature profile [**4**] exhibits the log-law with $\kappa_T = 0.48$, but the APG profiles [**1-10-9**] deviates from the law and exhibit higher values of κ_T depending on the pressure gradient. This behavior suggests that κ_T depends on the pressure gradient. In case of a moderate APG where $p^+y^+ \ll 1$, an idea is to make a Taylor expansion of the different quantities in terms of p^+y^+. For κ and κ_T at first order, it reads:

$$\kappa = \kappa_0 + \kappa_1 p^+ y^+ \quad \text{and} \quad \kappa_T = \kappa_{T_0} + \kappa_{T_1} p^+ y^+ \tag{8}$$

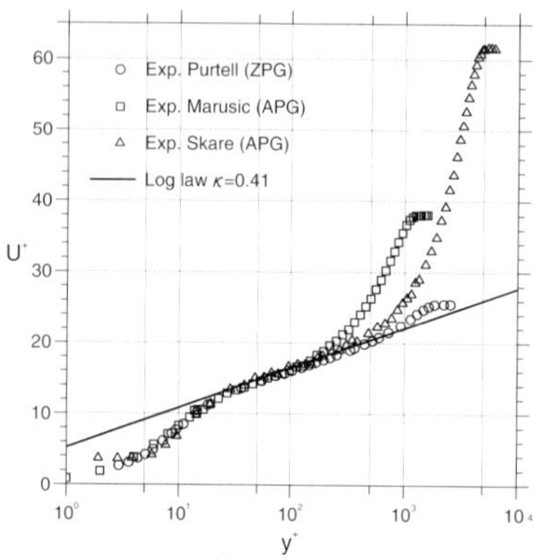

Figure 1. Logarithmic law for the velocity profile. ZPG and APG boundary layer experiments.

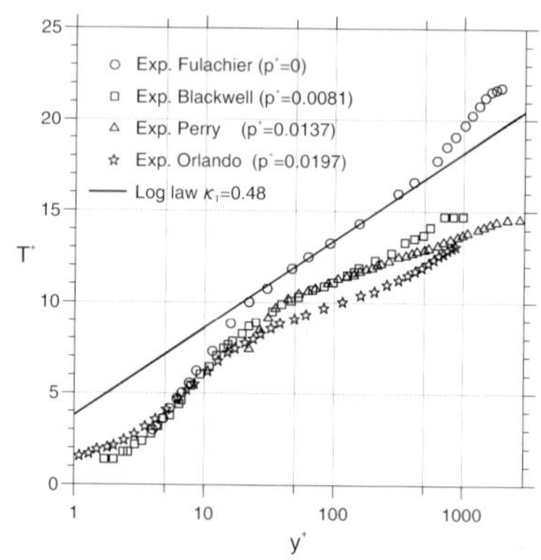

Figure 2. Logarithmic law for the temperature profile. ZPG and APG boundary layer experiments.

Based on experiments, for the velocity profile, the constants are $\kappa_0 = 0.41$, and $\kappa_1 = 0$, because κ does not change with the pressure gradient. For the temperature profile, $\kappa_{T0} = 0.48$ but κ_{T1} is more difficult to get from the experiments. However, as already pointed out by Huang and Bradshaw [**6**], experiments show that the turbulent Prandtl number is constant in the logarithmic region and independent of the pressure gradient.

Some statements have to be made here: there are no strong physical reasons for the turbulent Prandtl number to be constant in a large part of the boundary layer and independent of the pressure gradient. Indeed the Reynolds stresses are related through the continuity equation while the turbulent heat transport is not, so that there is no reason for $\overline{u'v'}$ to be proportional to $\overline{v'T'}$ in a large region of the boundary layer with the same coefficient of proportionality. In fact the turbulent Prandtl number varies when approaching to the wall or on the contrary when going to the edge and varies when considering other flows such as jet or mixing layers. However the flow addressed here is a slightly heated incompressible boundary layer where temperature can be considered as a passive scalar and depends strongly on the fluid motion. Moreover the thermal and dynamical boundary layers are assumed to develop together. Things may be different if high heat transfer flows were considered with strong differences between the developments of the respective boundary layers and this represents a limitation of the present approach.

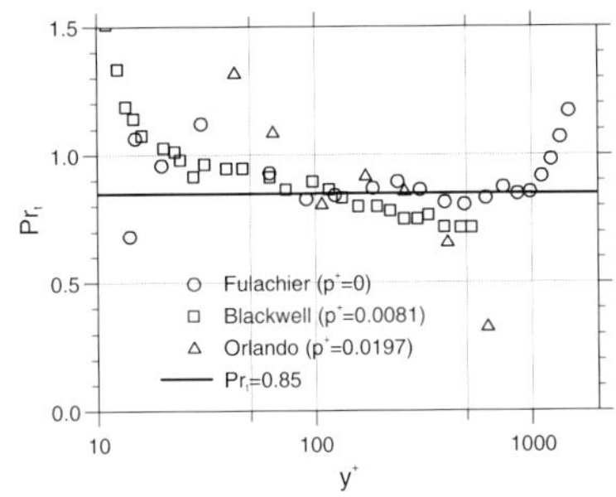

Figure 3. Evolution of turbulent Prandtl number in the boundary layer. ZPG and APG experiments.

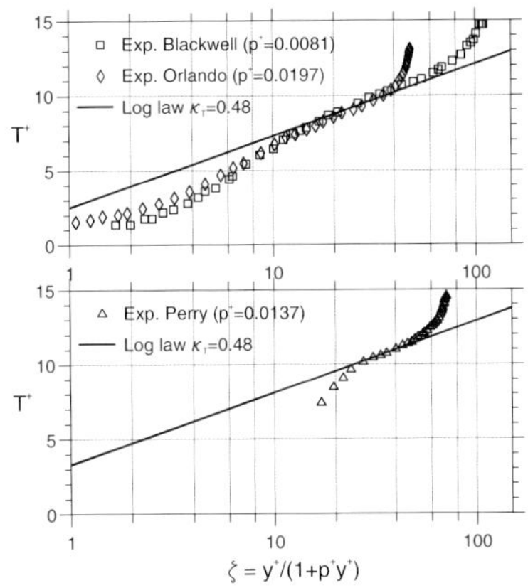

Figure 4. Experimental validation of the log-law for the temperature profile in APG boundary layers.

The evolution of the turbulent Prandtl number in ZPG and APG boundary layers is presented in Fig. 3. It can be reasonably accepted that Pr_t is constant for $30 < y^+ < 500$, *i.e.* in the logarithmic region and even above. The turbulent Prandtl number can be developed at first order in p^+y^+ and from the definition it reads:

$$Pr_t = \frac{\nu_t^+}{\alpha_t^+} = \frac{-\overline{u'v'}^+ \dfrac{\partial T^+}{\partial y^+}}{\overline{v'T'}^+ \dfrac{\partial u^+}{\partial y^+}} = \frac{1+p^+y^+}{1}\,\frac{\kappa_0+\kappa_1 p^+y^+}{\kappa_{T_0}+\kappa_{T_1}p^+y^+} \tag{9}$$
$$= \frac{\kappa_0}{\kappa_{T_0}}\left[1+\left(1+\frac{\kappa_1}{\kappa_0}-\frac{\kappa_{T_1}}{\kappa_{T_0}}\right)p^+y^+\right]$$

From Eq. (9) it can be deduced that the turbulent Prandtl number is independent of the pressure gradient if $1+\kappa_1/\kappa_0-\kappa_{T1}/\kappa_{T0}=0$, which implies, as $\kappa_1=0$:

$$\kappa_{T_1} = \kappa_{T_0} \quad \Rightarrow \quad \kappa_T = \kappa_{T_0}\left(1+p^+y^+\right) \tag{10}$$

The temperature gradient expression (7) thus reads:

$$\frac{\partial T^+}{\partial y^+} = \frac{1}{\kappa_T y^+} = \frac{1}{\kappa_{T_0}\left(1+p^+y^+\right)y^+} \Rightarrow T^+ = \frac{1}{\kappa_{T_0}} ln\underbrace{\left(\frac{y^+}{1+p^+y^+}\right)}_{\zeta} + C'_T \tag{11}$$

Hence, at given pressure gradient p^+, the temperature profile in APG boundary layer flows exhibits the same logarithmic profile as in ZPG, provided that the new variable $\zeta = y^+/(1+p^+y^+)$ is used instead of y^+. This is a new result in relation to Huang and Bradshaw analysis, which can be validated by the mean of experiments. Fig. 4 presents the same APG experimental velocity profiles as in Fig. 2 but plotted as a function of the ζ variable. It can be seen that all velocity profiles exhibit the classical log-law with slope $1/\kappa_T$ and $\kappa_T = 0.48$. The constant C'_T lies between 2.5 and 3.4, depending on experiments, instead of 3.8, which could be due to an influence of the pressure gradient. There also could be a problem of measurement and a preliminary analysis of recent experiments of Houra and Nagano [**5**] tends to show that the constant is not changed compared to the ZPG case ($C'_T \approx 3.8$).

4. HALF-POWER REGION

In this section the strong APG case, first addressed by Townsend [**15**], will be considered. p^+y^+ is assumed to be much greater than 1. The region of the boundary layer concerned is located above the logarithmic region, but far enough from the outer part to still neglect convection. Eq. (6) then becomes: $-\overline{u'v'} = p^+y^+$ and $-\overline{v'T'} = 1$. It is useful to introduce the length scale or mixing length ℓ defined as: $-\overline{u'v'} = \ell^2(\partial u/\partial y)^2$. In a ZPG boundary layer the expression of ℓ is straightforward and reads: $\ell = \kappa y$. In APG boundary layers, it is found experimentally [**12**] that the length scale still remains linear with the wall distance in the inner region even above the logarithmic region. Thus it can be written that:

$$\begin{aligned} & -\overline{u'v'}^+ = p^+y^+ = \left(\kappa y^+\right)^2\left(\frac{\partial u^+}{\partial y^+}\right)^2 \\ \Rightarrow\ & \frac{\partial u^+}{\partial y^+} = \frac{\sqrt{p^+y^+}}{\kappa y^+} \quad \Rightarrow\ u^+ = \frac{2}{\kappa}\sqrt{p^+y^+} + C_r \end{aligned} \tag{12}$$

The velocity profile thus exhibits a "half-power" form function of p^+y^+ with slope $2/\kappa \approx 4.9$. This is illustrated in Fig. 5 for the Skåre *et al.* experiment [**12**]. The "half-power" region lies approximately at $20 < \sqrt{y^+} < 40$, *i.e.* at $400 < y^+ < 1600$, which is as expected above the logarithmic region. The constant C_r depends on the reduced pressure gradient p^+.

From the velocity gradient and shear stress expression, the eddy viscosity expression is straightforward and reads: $\nu_t^+ = \kappa y^+ \sqrt{p^+ y^+} = \kappa y^{+3/2} \sqrt{p^+}$. To obtain the temperature profile, the assumption of a constant turbulent Prandtl number will be made again. As already noticed in Fig. 3, this assumption seems valid at least for $30 < y^+ < 500$, which means also above the logarithmic region that ends approximately at $y^+ = 200$ for the experiments considered here. However there are needs of accurate APG experimental data to really assess this assumption, even if, as discussed before, a constant turbulent Prandtl number assumption should be regarded with caution. If Pr_t is constant, the turbulent heat flux Eq. (2) becomes:

$$\begin{aligned} \overline{v'T'}^+ &= 1 = \alpha_t^+ \frac{\partial T^+}{\partial y^+} = \frac{\nu_t^+}{Pr_t} \frac{\partial T^+}{\partial y^+} = \frac{\kappa y^{+3/2} \sqrt{p^+}}{Pr_t} \frac{\partial T^+}{\partial y^+} \\ \Rightarrow \quad T^+ &= -\frac{2\,Pr_t}{\kappa} \frac{1}{\sqrt{p^+ y^+}} + C_{T_r} \end{aligned} \qquad (13)$$

The temperature profile thus exhibits an inverse "half-power" form function of p^+y^+ with slope $-2Pr_t/\kappa \approx -4.1$. The half-power form of the temperature profile has already been exhibited in previous studies, however an analytical expression of the slope is proposed here. This is illustrated in Fig. 6 for the different experiments shown previously. It can be seen that the constant C_{Tr} decreases when the reduced pressure gradient p^+ increases, so that all the curves are shifted. To really give evidence of this region is quite hard as there is usually a lack of data points. However the experiment of Perry *et al.* [**10**] is well detailed and exhibits clearly the inverse half-power region. It lies approximately at $-0.7 < -1/\sqrt{p^+y^+} < -0.3$, *i.e.* $150 < y^+ < 800$, which is as expected located above the logarithmic region. It should be pointed out that Perry *et al.* [**10**] presented a half-power region with slope -2.8 instead of -4.1, but they located this region at $0.5 < p^+y^+ < 3$ which corresponds to $30 < y^+ < 200$, that is exactly in the logarithmic region! This explains the wrong slope value they estimated.

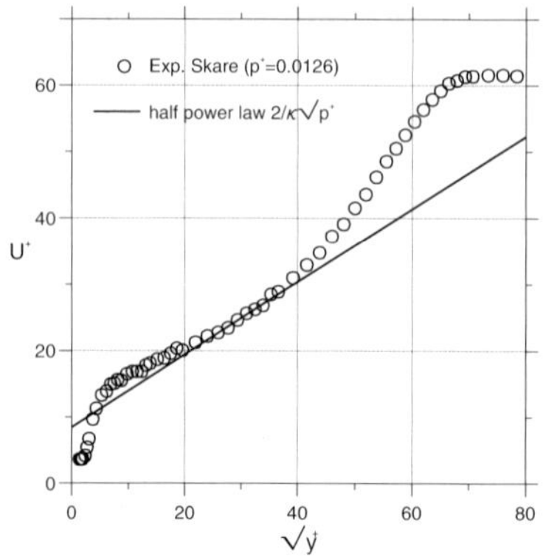

Figure 5. Half-power law for the velocity profile. APG boundary layer experiment.

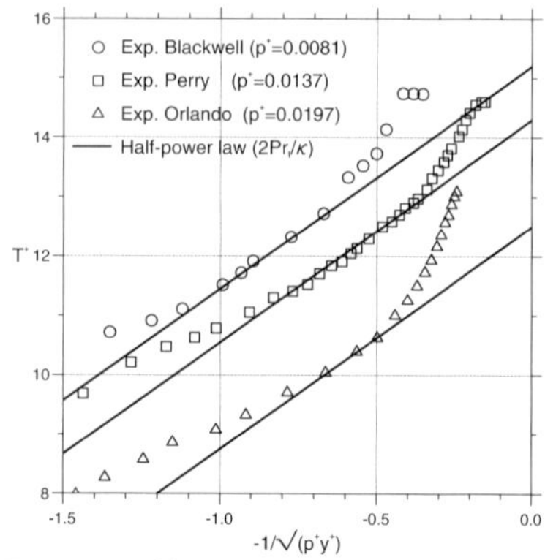

Figure 6. Half-power law for the temperature profile. APG boundary layer experiments.

5. CONSEQUENCE ON TURBULENCE MODELING

The log-law and the half-power law for both velocity and temperature profiles are physical behaviors that a turbulence model has to reproduce to improve the predictions for strong APG flows, which concern many engineering applications. It is possible to evaluate the performance of any turbulence models in both cases following the approach of Catris *et al.* [**2**], extended to the thermal part by Daris *et al.* [**3**]:

- in the logarithmic region, the turbulence transport equations are developed as Taylor expansions in terms of p^+y^+. By identification of terms of same order, it is possible to calculate the predicted values for κ_0 and κ_{T0} (terms of 0th order for ZPG) and for κ_1 and κ_{T1} (terms of 1st order for APG). A model should give the physical values, *e.g.* $\kappa_0 \approx 0.41$, $\kappa_1 = 0$, $\kappa_{T0} = \kappa_{T1} \approx 0.48$.
- in the half-power region, solutions for the mean and turbulence quantities are sought as powers of p^+y^+ such as $u^+{=}A_u(p^+y^+)^{eu}$ and $T^+{=}A_T(p^+y^+)^{eT}$. It can be shown that all models are able to reproduce the correct powers, *e.g.* e_u=1/2 and e_T=-1/2, which only means the models are dimensionally consistent. However the goal is to predict the good values for A_u and A_T, *e.g.* $A_u = 2/\kappa \approx 4.9$ and $A_T = -2Pr_t/\kappa \approx -4.1$.

Table 1 gives the predicted values for κ_0, κ_1, κ_{T0}, κ_{T1}, A_u and A_T for a variety of four-equations turbulence models: k-ε-k_T-ε_T models of Nagano-Kim [**8**] (NK) and Sommer-So-Zhang [**13**] (SSZ); k-ω-k_T-ω_T models of Huang-Bradshaw [**6**] (HB); k-$k\ell$-k_T-$k_T\ell_T$ model of Daris-Bézard [**3**] (DB),

ℓ_T being the thermal turbulent length scale. This last model has been particularly designed to reproduce APG boundary layer flows.

Table 1. Behavior of four-equations turbulence models in the log and half-power regions

Models	κ_0	κ_{T0}	κ_1	κ_{T1}	A_u	A_T
Exp./Theory	**0.41**	**0.48**	**0**	**0.48**	**4.9**	**-4.1**
NK	0.42	0.44	1	-0.58	No sol.	No sol.
SSZ	0.38	0.48	1.28	5.5	No sol.	No sol.
HB	0.40	0.47	0.19	0.56	No sol.	No sol.
DB	0.41	0.48	0.	0.48	4.9	-2.6

All models give almost the correct values for the ZPG log-law. For the APG behavior in the logarithmic region, k-ε models give poor results, which is a known result. However for the thermal point of view, it is surprising to see even opposite behaviors (κ_{T1}<0 for NK model). The HB model is very close to the theoretical values, proving superiority of k-ω models on k-ε models. However none of them, except the new DB model, is able to predict the correct behavior in strong APG for the half-power region (when no values are written means the analytical resolution of the problem is impossible and gives imaginary solutions for A_u and A_T). At the moment the new DB model is the only one which predicts the correct log-law and half-power law profiles for both velocity and temperature.

6. CONCLUSION

It has been shown that for moderate APG boundary layers, the temperature profile exhibits the classical logarithmic profile with von Kármán constant κ_T equal to 0.48 if the variable $\zeta = y^+/(1+p^+y^+)$ is used instead of y^+. The expression of the inverse half-power law for the temperature profile at strong APG has been established and the slope found to be $-2Pr_t/\kappa$. Both expressions have been validated on existing APG experiments which have shown that really accurate data is necessary to assess the presented results. Analysis of previous existing four-equation models has shown that none of them is able to reproduce correctly both log and half-power regions for APG boundary layer flows. The present analysis can also be performed on more complex models, like second order models.

REFERENCES

1. Blackwell BF, Kays WM, Moffat RJ. "The turbulent boundary layer on a porous plate: an experimental study of the heat transfer behavior with adverse pressure gradients", *Technical Report HMT-16*, Dept. Mech. Engng., Stanford University, 1972.
2. Catris S, Aupoix B. "Towards a calibration of the length-scale equation", *Int. J. Heat. Fluid Flows*, **21**(5), pp. 606-613, 2000.
3. Daris T, Bézard H. "Four-equations models for Reynolds stress and turbulent heat flux predictions", *SFT - 12th International Heat Transfer Conference*, Grenoble, France, August 18-23, 2002.
4. Fulachier L, Verollet E, Dekeyser I. "Résultats expérimentaux concernant une couche limite turbulente avec aspiration et chauffage à la paroi", *Int. J. Heat Mass Transfer*, **20**, pp. 731-739, 1977.
5. Houra T, Nagano Y. "Effect of pressure gradient on heat transfer in turbulent boundary layer", *Heat Transfer 2002 - 12th International Heat Transfer Conference*, pp. 597-602, 2002.
6. Huang PG, Bradshaw P. "The law of the wall for turbulent flows in pressure gradients", *AIAA Journal*, **33**(4), pp. 624-632, 1995.
7. Marušić I, Perry AE. "A wall-wake model for the turbulence structure of boundary layers. Part 2. Further experimental support", *J. Fluid Mech.*, **298**, pp. 389-407, 1995.
8. Nagano Y, Kim C. "A two-equation model for heat transport in wall turbulent shear flows", *J. Heat Transfer*, **110**, pp. 583-589, 1988.
9. Orlando AF, Moffat RJ, Kays WM. "Turbulent transport of heat and momentum in a boundary layer subject to deceleration, suction, and variable wall temperature", *Technical Report HMT-17*, Dept. Mech. Engng., Stanford University, 1974.
10. Perry AE, Bell JP, Joubert PN. "Velocity and temperature profiles in adverse pressure gradient turbulent boundary layer", *J. Fluid Mech.*, **173**, pp. 299-320, 1966.
11. Purtell LP, Klebanoff PS, Buckley FT. "Turbulent boundary layer at low Reynolds number", *Phys. Fluid*, **24**(5), pp. 802-811, 1981.
12. Skåre PE, Krogstad PÅ. "A turbulent equilibrium boundary layer near separation", *J. Fluid Mech.*, **272**, pp. 319-348, 1994.
13. Sommer TP, So RMC, Zhang HS. "Near-Wall Variable Prandtl-Number Turbulence Model for Compressible Flows", *AIAA Journal*, **31**(1), pp. 27-35, 1993.
14. Squire HB. "The friction temperature: a useful parameter in heat-transfer analysis", *Proc. General Discus. Heat Transfer*, Inst. Mech. Engng. and ASME, London, pp. 185-186, 1951.
15. Townsend AA. "Equilibrium layers and wall turbulence", *J. Fluid Mech.*, **11**, pp. 97-120, 1961.

THE SIGNIFICANCE OF TURBULENT EDDIES FOR THE MIXING IN BOUNDARY LAYERS

Christian J. Kähler[†]
Institut für Strömungsmechanik, TU Braunschweig, Bienroder Weg 3, 38106 Braunschweig, Germany, email: c.kaehler@tu-braunschweig.de

Abstract: In this paper high resolution turbulent boundary layer measurements will be discussed which were performed 18 m behind the leading edge of a flat plate at $Re_\Theta = 7800$. The investigation was performed in the temperature stabilized closed circuit wind tunnel at the Laboratoire de Mécanique de Lille (LML) with a stereoscopic PIV based measurement system that allows to determine at any flow velocity all three velocity components in spatially-separated planes simultaneously or separated in time [5, 6]. The aim of the investigation is the analysis of the geometrical and kinematical properties of various coherent structures and their significance for the turbulent mixing in wall bounded flows. The complete investigation can be found at: http://webdoc.sub.gwdg.de/diss/2004/kaehler/kaehler.pdf.

Keywords: coherent structure, multi-point correlation, turbulent mixing, streak, ejection.

1. INTRODUCTION

Beginning with the early channel and pipe flow measurements published by Laufer [13, 14], and the boundary layer investigations along a flat plate by Klebanoff [9], turbulent boundary layers have been examined extensively because of their technological importance, their significance for the development of fundamental turbulence models and for the validation of numerical flow simulations [17, 18, 20]. The bulk of the *quantitative* investigations has been performed with *intrusive single-point measurement techniques* [2], but also *non-intrusive flow visualization techniques* have been frequently applied, since the pioneering work performed by Kline and Runstadler [11]. Although the conclusiveness of these visualizations is often questionable [4], these investigations have improved the understanding of turbulence to a large extent, because it was possible to detect *coherent flow structures*, such as *low-speed streaks, shear-layers, stream-wise vortices* and *loop-shaped structures*, which are of fundamental importance for the turbulent mixing in wall-normal direction. Today, there is no doubt about the

---

[†] The experimental investigations have been performed at Laboratoire de Mécanique de Lille (LML), while the author was affiliated with DLR Göttingen. The main results have been published in the PhD thesis http://webdoc.sub.gwdg.de/diss/2004/kaehler/kaehler.pdf. The author would like to thank Prof. Stanislas for the experimental support and the discussions.

G.E.A. Meier and K.R. Sreenivasan (eds.), IUTAM Symposium on One Hundred Years of Boundary Layer Research, 405-414,

existence of organized flow structures, but their geometrical and kinematical properties are still under investigation, especially at large Reynolds numbers, because there is no general agreement about which structures are *fundamental* and which ones are only *secondary*, which ones are *dominant* and which ones are *irrelevant* [17, 19].

2. MULTIPOINT STATISTICS

To determine the geometrical properties of the coherent structures present in the near wall region of the turbulent boundary layer flow, the primary spatial correlations R_{uu}, R_{vv} and R_{ww} of the velocity fluctuations will be considered here. These quantities have been extensively investigated by Grant along the principal axis by using a pair of hot-wire probes [3]. Although he could confirm the high degree of order present in the flow field, he was forced to conclude that the various correlations are incompatible with the idea of energetic eddy structures because of the different size of the primary correlations in stream-wise direction and the small extend of R_{vv} along the span-wise one. However, beside experimental problems associated with the short developing zone of the boundary layer and technical problems associated with intrusive probes [3, 21] these investigations were performed at relatively large y^+. Figure 1 reveal the primary correlation of the stream-wise velocity fluctuations for three wall distances of the fixed point ($y^+ = 10$, 20, 30) measured in the *xy*-plane (stream-wise wall-normal) at $Re_\Theta = 7800$. The contours of the plots are spaced in intervals of 0.05 excluding 0.05 and 1. Clearly visible is the elliptical shape, with the principal axis slightly inclined with respect to the *x*-axis, and the variation of the correlation size from the exact location of the fixed point. The elliptical shape of the structures and their inclination can be explained by the strong gradients in wall-normal direction, which tend to elongate the flow pattern. The right column of figure 1 indicates the functional dependence of the R_{ww} correlation. Especially the different dependence of the correlation in up-stream and down-stream direction on the fixed-point should be noted as well as the inclination of the structure around the maximum of correlation and further away from this point. As this correlation appears crestfallen, two different angles of attack can be estimated when moving on the correlation ridge in up-stream and down-stream direction, and it can be seen that the latter one is larger with respect to the other one for all locations of the fixed point. The folded shape implies that the coherent structures, present in the near-wall region below $y^+ \approx 30$, differ strongly from the structures located in the logarithmic part of the boundary layer flow. The similarity between the size and of R_{uu} and R_{ww} in stream-wise direction and the elliptical shape around the maximum implies the physical relation between both fluctuations

[7]. Figure 2 shows the primary auto-correlations of the stream-wise (left column) and wall-normal (right column) velocity fluctuations measured in planes parallel to the wall (xz-plane) at $y^+ \approx 30$, 20 and 10 (top to bottom). Continuous lines indicate positive correlation whereas dotted lines assign regions with negative correlation values. Clearly visible is the characteristic elliptical shape of both correlations, with the principal axis in stream-wise direction, as well as the strong dependence of the correlation width from the wall distances. However, the most striking feature is the variation of the size and span-wise location of the weak negative correlation regions of R_{vv} because they indicate the existence of stream-wise vortices. However, it should be noted that their average length is not as long as observed by other authors, as can be seen from the stream-wise extension of the negative correlation region. For phenomena associated with the production of turbulence the cross correlation R_{vu} is more important than the primary components of the correlation tensor, as R_{vu} reflects the size and shape of the

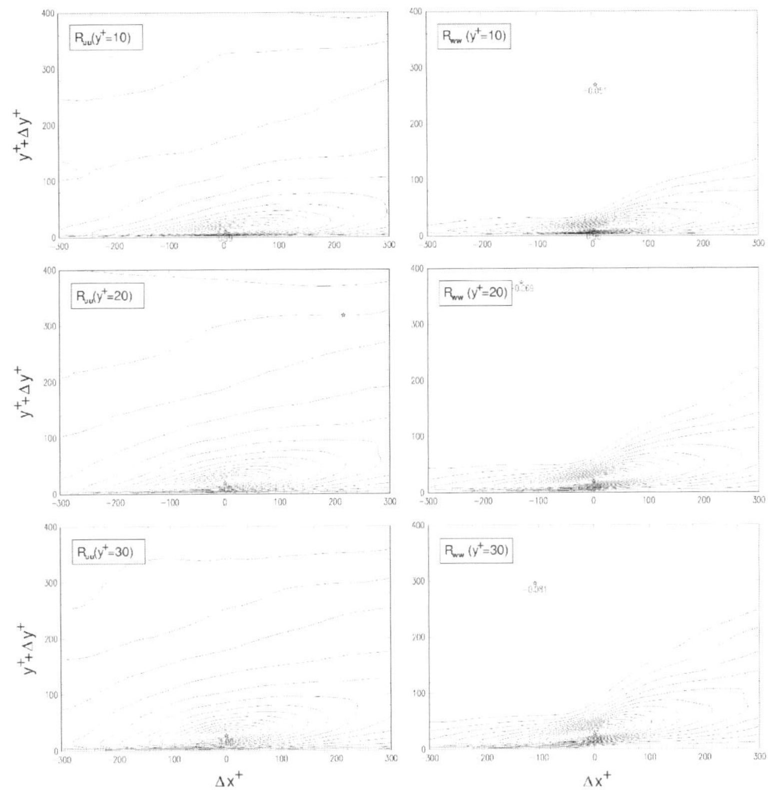

Figure 1. $R_{uu}(\Delta x^+, y^+, y^+ + \Delta y^+)$ and $R_{ww}(\Delta x^+, y^+, y^+ + \Delta y^+)$ for various y^+

structures being responsible for the transport of relatively low-momentum fluid outwards into higher speed regions and for the movement of high-momentum fluid toward the wall and into lower speed regions. This correlation was first studied by Tritton [21] along the three principal axes and at larger wall-locations. The left column in figure 3 shows the R_{vu} cross-correlation function with the v component fixed and u shifted in the two homogeneous directions. The negative sign of R_{vu} indicates that the transport of relatively low-momentum fluid outward into higher speed regions ($u < 0$ and $v > 0$) and the movement of high-momentum fluid toward the wall into lower speed regions ($u > 0$ and $v < 0$) are the predominant processes in the near-wall region, as assumed by Prandtl [15] for the derivation of the mixing length theory. In addition, the strong elliptical shape implies that the turbulent mixing in the wall-normal direction is related to the well known low-speed structures, whereby only a small part of the low-momentum structures shows a correlated motion in both stream-wise and wall-normal direction. To differentiate between high-speed structures moving towards the wall and low-speed structures moving away from the wall, the right column of the same figure reveals the conditional cross-correlation of negative wall-

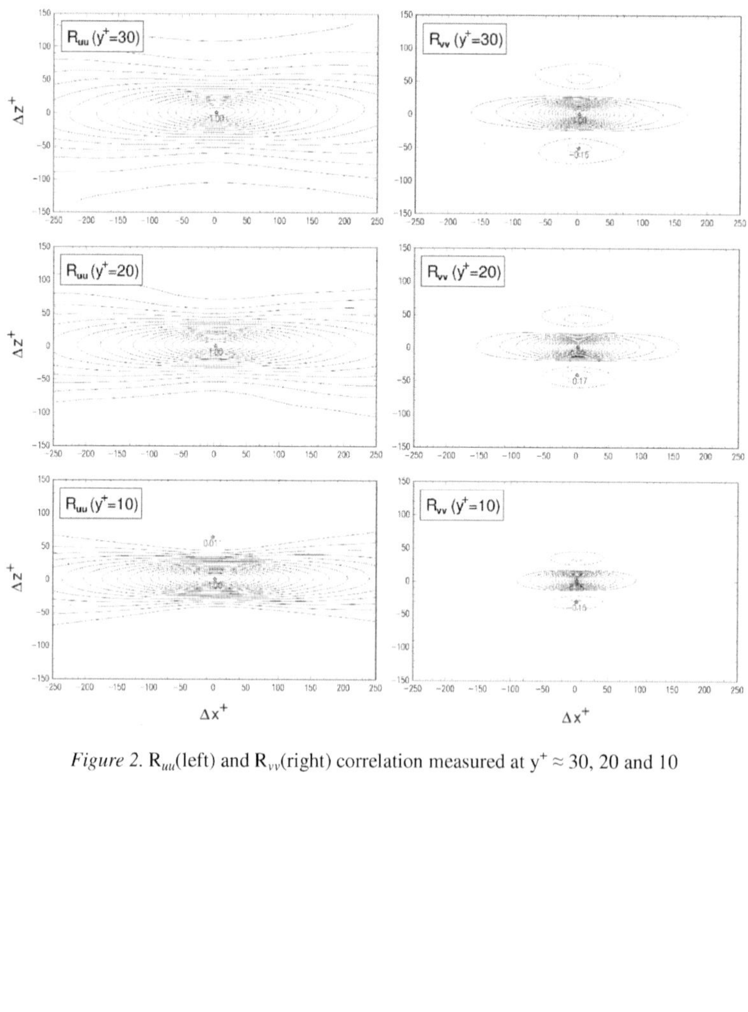

Figure 2. R_{uu}(left) and R_{vv}(right) correlation measured at $y^+ \approx 30$, 20 and 10

normal (motion towards the wall) with positive stream-wise velocity fluctuations $R_{(-v)(+u)}$ measured at $y^+ \approx 30$, 20 and 10 (top to bottom). This correlation displays the average size and shape of the coherent structures called sweeps or Q4 events in the literature.

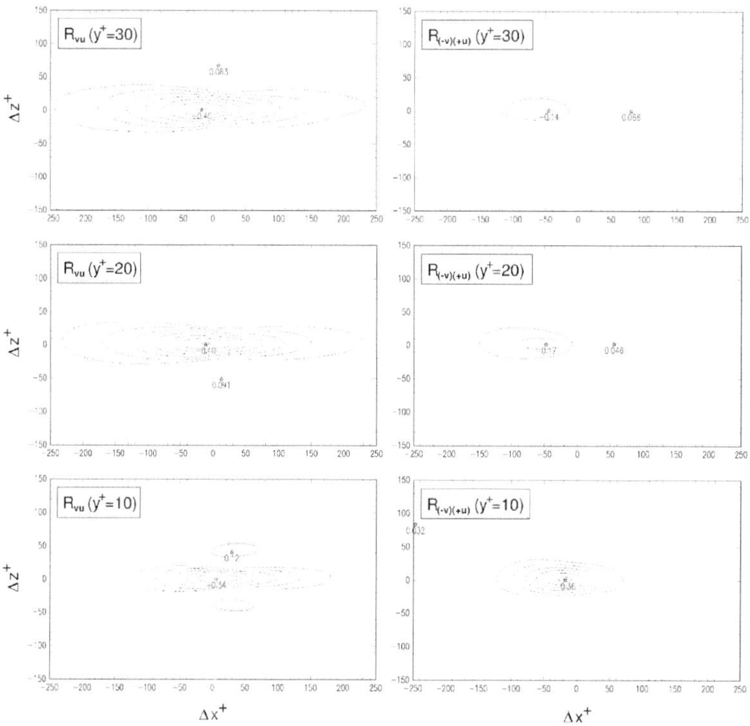

Figure 3. R_{vu} (*left*) and conditional cross-correlation of negative wall-normal with positive stream-wise velocity $R_{(-v)(+u)}$ (*right*) measured at $y^+ \approx 30$, 20 and 10

3. PROPERTIES OF COHERENT VELOCITY STRUCTURES

As the coherent structures which contribute to the correlations displayed in the previous section have a history of development which is smeared out by the statistical approach, the instantaneous velocity fields will be examined here to extract the instantaneous properties of the *dominant* coherent flow structures being present in the near-wall region. Here, only flow structures will be considered which can be labelled as *dominant* due to their occurrence according to figure 4 where the probability density function of the Reynolds stress component $-\rho uv$ measured at $y^+ = 20$ is displayed (left). However, the uvPDF(uv) distribution in the right figure indicates that their contribution to the total production of turbulence is large relative to the structures that can be characterized as *dominant* due to their large amplitude in the Reynolds stress component $-\rho uv$ see [7] for details.

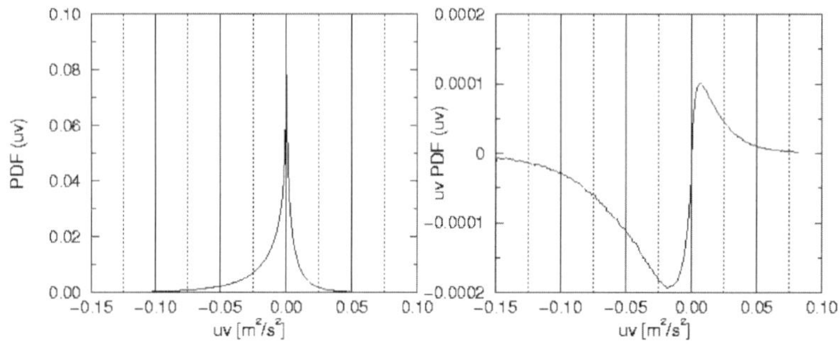

Figure 4. Distribution of uv measured at $y^+ = 20$ (*left*) and uvPDF(uv) (*right*)

3.1. Low-speed streaks

Low-speed streaks are the most striking structures in the near wall region between $y^+ \approx 5$ and $y^+ \approx 40$ - 50 which can be clearly observed in xz-planes close to the wall [10]. They appear as elongated and twisted low-speed regions, sometimes 1000 wall-units in length and on average 30 wall-units in width, with a span-wise periodicity of about 100 wall-units. It has been assumed that hairpin-vortices induce these low-speed region between the inclined legs while they are travelling downstream but a convincing experimental proof is still missing. Blackwelder and Eckelmann [1] and Kim *et al* [8] proposed that the low-speed streaks are generated between pairs of relatively weak, but highly elongated stream-wise vortices, but the existence of these vortices at high Reynolds number is still an open question [19]. Landhal and Mollo-Christensen [12] assume that the streaks might originate from a weak vertical oscillation of the fluid layers which produces strong oscillations in stream-wise direction. However, a general agreement could not be achieved. Regarding the kinematical properties, it is evident that the low-speed streaks play a dominant role in a sequence of events referred to as *bursting* phenomena. Kline *et al* [10] observed in the near-wall region of a turbulent boundary layer that extended low-speed flow structures, which move away from the wall, start to oscillate and burst finally after a certain life time into small scale turbulence. Other investigations indicate that the ejection of low-speed fluid from the wall is associated with flow structures which transfer momentum towards the wall (*sweeps* or *inrush bursts*) [16]. The connection between the bursting phenomenon near the wall and the large scale motion in the outer part is one of the key questions. Figure 5 shows a characteristic velocity field measured in the *xz*-plane at $y^+ = 10$. The flow direction is from left to right and the local mean velocity $\overline{U}$ is subtracted from the instantaneous velocity field U to display the turbulent velocity fluctuations. Predominant structures are the well known elongated flow regions that convect downstream with approximately half the local mean velocity, indicated by the vectors going from right to left. The shape, extent and span-wise separation of these slightly tilted flow regions is in quantitative agreement with the literature summarized in [17], but it should be noted that the instantaneous values of the geometrical properties can deviate strongly from the averaged ones, presented in figure 2. Another important property of the streaks is their extent in wall-normal direction. It could be shown by using the multiplane PIV technique that the streaks, which appear separated at $y^+ = 20$, belong to the same streak visible at $y^+ = 10$ [7]. Thus the decreasing length of these structures with increasing wall distance, as well as their increasing span-wise distance, is an artefact related to the statistical variation of their height. When the dynamic of the streaks is investigated with the multiplane PIV technique, it turns out that these structures tend to move away from the wall, but the spatial extent of this vertical motion is usually significantly shorter than the total length of the low-speed streaks at the same *y*-value, in agreement with the R_{uv} correlation in figure 3. This implies that no pairs of stream-wise counter-rotating vortices flank the low-speed regions over their total length, as observed in qualitative flow visualizations [1, 8].

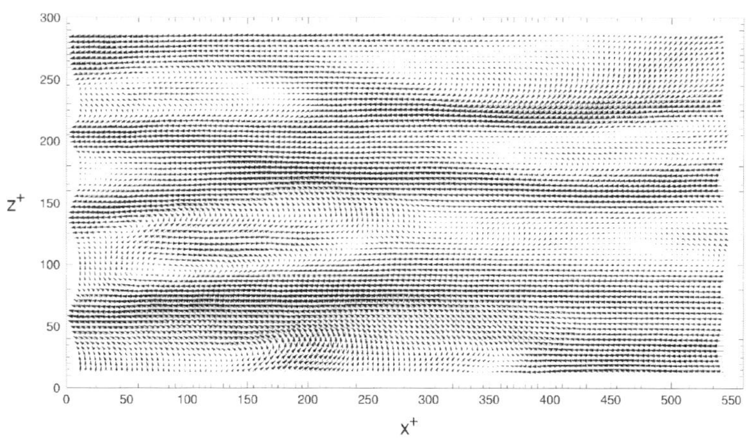

Figure 5. Velocity fluctuations measured independently at $y^+ = 10$

3.2. Sweeps

The production of turbulence caused by the lifting of the low-speed streaks is one of the basic processes identified in near-wall turbulence as already mentioned. However, due to the complexity of the turbulent motion the cause and dynamic is still the subject of controversial discussions. To investigate this exchange processes in detail, the interaction of the streaks with the surrounding fluid was investigated, whereby only the interaction regarding the high momentum flow structures which move towards the wall will be considered here. It is evident that the high-momentum flow structures must be visible in nearly each velocity field in order to compensate the low-momentum movement of the streaks, but in contrast to the low-speed streaks the variety of these structures is much larger. Usually they appear less elongated and broader than the low-speed regions because these flow structures originate statistically from flow regions which are further away from the wall. At $y^+ = 10$ these high momentum structures look sometimes similar to the streaks but more frequently they resemble small elliptically shaped islands, approximately 200 wall units in length and roughly 50 to 100 wall-units in width according to figure 5. These structures are statistically represented by the correlations shown in figure 3 (right column). If such a structure moves towards a low speed streak an interaction takes place and parts of the streaks, which are directly affected by the sweeps, are forced to move away from the wall due to continuity and this is associated with the production of Reynolds shear-stress [7]. Based on this experimental result it can be concluded that the lifting of low-speed streaks can be considered mainly as a secondary motion, induced by an interaction between a low-speed streak with a sweep, and the size and strength of the region, which moves away from the wall, is related to the local size and momentum of the sweep. In addition it can be stated that the lifting of low-speed fluid into higher momentum flow regions is accompanied by two weak stream-wise vortices because any local motion away from the wall is associated with a stream-wise vortex pair, but these vortices are produced locally and can not be considered as primary vortex structures. The similar stream-wise extent of sweeps and regions of significant outward motion supports this conjecture.

3.3. Ejection

In the past, hot wire investigations were performed to obtain quantitative information about the bursting phenomena. By using different pattern recognition techniques, a quite regular normalized velocity signal could be extracted and it was assumed that this pattern is the signature of a coherent structure which is associated with the bursting phenomena [22]. In the near-wall region below $y^+ = 30$ the pattern could be observed in nearly 65% of the total samples, and by analyzing the individual velocity signal it was found that the length of the pattern varied over quite a wide range (1:25). It could be shown by analysing the PIV results that this pattern is caused by lifting streaks [7]. As streaks appear frequently in the near-wall region, the high detection rate in [22] is not surprising and also the strong variation of the pattern length can be explained. The maximum length is given by the extension of the streaks, which can be longer than 1000 wall-units, and the lower limit of the length appears when only the cross-section of a narrow streak convects along the probe. Since the streaks are only slightly twisted, the projection of the width of the cross-section in stream-wise direction is the relevant parameter for the lower limit of the pattern length. As the minimum width is approximately 30 wall units and the maximum angle 45°, the projection of the cross-section in stream-wise direction is roughly 42 wall-units. Thus, the variation of the length can be estimated to $1000/42 \approx 24$ in agreement with [22].

4. CONCLUSION

The investigation implies that the stream-wise vortices which flank the low speed streaks are no primary vortices. They are produced locally when the streaks move away from the wall. The lift-up of the low-speed streaks on the other hand is frequently forced when a sweep-streak interaction takes place. In effect, the length of the stream-wise vortices and the region of the low-speed streak which moves away from the wall is similar to the length of the sweeps close to the wall. Finally, it could be shown that the characteristic velocity pattern observed in hot-wire investigations, is caused by low-speed streaks.

References

1. Blackwelder R.F., Eckelmann H. *Streamwise vortices associated with the bursting phenomenon*, J. Fluid Mech. **132**, 1979.
2. Fernholz H.H., Finley P.J. *The incompressible zero-pressure-gradient turbulent boundary-layer: An assessment of the data*, Prog. Aerospace Sci. **32**, 1996.
3. Grant H.L. *The large eddies of turbulent motion*, J. Fluid Mech. **4**, 1971.
4. Hama F.R. *Streaklines in a perturbed shear flow*, Phys. Fluids **5**, 1962.
5. Kähler C.J., Kompenhans J. *Fundamentals of Multiple Plane Stereo PIV*, Exp. Fluids [Suppl.], 2000.
6. Kähler C.J. *Investigation of the spatio-temporal flow structure in the buffer region of a turbulent boundary layer by means of multiplane stereo PIV*, Exp. Fluids **36**, 2004.
7. Kähler C.J. *The significance of coherent flow structures for the turbulent mixing in wall-bounded flows*, PhD thesis, University of Göttingen, Germany, 2004. See: **http://webdoc.sub.gwdg.de/diss/2004/kaehler/kaehler.pdf**
8. Kim H.T., Kline S.J., Reynolds W.C. *The production of turbulence near a smooth wall in a turbulent boundary layer*, J. Fluid Mech. **50**, 1971.

9. Klebanoff P.S. *Characteristics of turbulence in a boundary layer with zero pressure gradient*, NACA TR-1247, 1955.
10. Kline S.J., Reynolds W.C., Schraub F.A., Runstadler P.W. *The structure of turbulent boundary layers*, J. Fluid Mech. **30**, 1967.
11. Kline S.J., Runstadler P.W. *Some preliminary results of visual studies of the flow model of the wall layers of the turbulent boundary layer*, Trans. ASME. Ser. E. Vol. 2, 1959.
12. Landhal M.T., Mollo-Christensen E. *Turbulence and random processes in fluid mechanics*, Cambridge University Press, 1986.
13. Laufer J. *Investigation of turbulent flow in a 2D channel*, NACA TR-1053, 1953.
14. Laufer J. *The structure of turb. in fully developed pipe flow*, NACA TR-1174, 1954.
15. Prandtl L. *Über die ausgebildete Turbulenz*, ZAMM, **5**, 136, 1925.
16. Praturi A.K., Brodkey R.S. *A stereoscopic visual study of coherent structures in turbulent shear flow*, J. Fluid Mech. **89**, 1978.
17. Robinson S.K. *A review of vortex structures and associated coherent motions in turbulent boundary layers*, 2. IUTAM Symposium on Structures of Turbulence an Drag Reduction. Federal Institute of Technology. Switzerland Juli 25–28, 1989.
18. Rotta J.C. *Turbulent Boundary Layers in Incompressible Flow*, Reprint from "Progress in Aeronautical Science, Volume 2". Pergamon Press Oxford, 1962.
19. Smith C.R., Walker D.A. *Turbulent wall-layer vortices*, in Fluid Vortices, Green SI (ed.), Kluver, 1999.
20. Spalart P.R. *Direct simulation of a turbulent boundary layer up to* Re_Θ *=1410}*, J. Fluid Mech. **187**, 1988.
21. Tritton D.J. *Some new correlation measurements in a turbulent boundary layer*, J. Fluid Mech. **28**, 1967.
22. Wallace J.M., Brodkey R.S., Eckelmann H. *Pattern-recognized structures in bounded turbulent shear flow*, J. Fluid Mech. **83**, 1977.

UNSTABLE PERIODIC MOTION IN PLANE COUETTE SYSTEM: THE SKELETON OF TURBULENCE

Genta Kawahara [1,3], Shigeo Kida [2,3] and Masato Nagata [1,3]
[1] *Department of Aeronautics and Astronautics, Kyoto University, Kyoto 606-8501, Japan*
[2] *Department of Mechanical Engineering, Kyoto University, Kyoto 606-8501, Japan*
[3] *Advanced Research Institute of Fluid Science and Engineering, Kyoto University, Kyoto 606-8501, Japan*

Abstract: A recently discovered unstable time-periodic solution (Kawahara & Kida, *J. Fluid Mech.* **449**, 291–300, 2001) to the incompressible Navier–Stokes equation for plane Couette flow is reviewed and recomputed with higher accuracy by use of the Newton–Raphson method to track the solution by changing the Reynolds number Re and the wall-parallel size of a computational periodic box. It is found that the periodic motion, which exhibits a full regeneration cycle of near-wall coherent structures, appears through a saddle-node bifurcation and is traceable down to $Re \approx 290$. The appearance of this periodic motion would be relevant to transition to turbulence in plane Couette flow.

Key words: Plane Couette Turbulence, Coherent Structure, Periodic Motion.

1. INTRODUCTION

Since Prandtl's [1] revolutionary discovery of the boundary layer, the understanding of the effects of viscosity on wall flows has been deepened remarkably. The viscosity is known to be significant in the near-wall region of turbulent flows, where the mean flow scales with the kinematic viscosity ν and the wall friction velocity $u_\tau = \sqrt{\tau_w/\rho}$ [2], τ_w and ρ being the mean wall shear stress and mass density, respectively. The upper part of the viscous wall layer, i.e. the buffer layer ($5 \lesssim yu_\tau/\nu \lesssim 40$), is a region in which the net production of turbulence energy is positive, and turbulence energy produced therein is transferred away from the wall [3]. Moreover, turbulence activity in the buffer layer could be sustained without any interaction with outer turbulent flow [4]. In this context the buffer-layer turbulence may be considered to be a self-sustaining 'engine' of the whole wall-bounded turbulent flow.

A lot of researches on the structures of wall turbulence performed over the last four decades show that the near-wall region is dominated by coher-

G.E.A. Meier and K.R. Sreenivasan (eds.), IUTAM Symposium on One Hundred Years of Boundary Layer Research, 415-424,

ent structures which are known, at least qualitatively, to play key roles in turbulence production, energy dissipation, transport phenomena, and so on. Especially, in the buffer layer, there exist prominently the coherent structures, i.e. streamwise-velocity streaks and longitudinal vortices. The elongated low- and high-velocity streaks are arranged alternately in the spanwise direction, their spacing being around $100\,\nu/u_\tau$, while shorter quasi-streamwise vortices of both signs are typically staggered alongside them. The streaks have long been recognized to be generated by cross-flow advection of the streamwise velocity by the streamwise vortices [5], and recent studies [6–8] suggest that the sinuous instability of the streaks leads to regeneration of the vortices. It is generally believed that the streaks and the vortices are crucial ingredients of a nonlinear regeneration cycle that maintains near-wall turbulence [6,9]. Since, however, 'coherence' observed in a turbulent state is always incomplete, it is difficult to give us a strict definition to the coherent structures appearing in turbulent flows. This difficulty has prevented us from characterizing quantitatively the spatiotemporal properties of the coherent structures and from fully elucidating their dynamical or statistical roles in turbulent flows.

Recently, motivated by the discovery of the Nagata solution [10] for plane Couette system, nonlinear three-dimensional equilibrium solutions of the incompressible Navier–Stokes equation have been investigated for wall-bounded shear flows, such as plane Couette flow [10,11], plane Poiseuille flow [11–13] and autonomous wall flow [14]. The equilibrium solutions of these systems exhibit a similar structure in physical space which takes the form of wavy low-velocity streaks flanked by staggered streamwise vortices of alternating signs, and their structure closely resembles the spatial coherence observed in the buffer layer of turbulent flows. Three-dimensional time-periodic solutions to the incompressible Navier–Stokes equation have also been extracted numerically from low-Reynolds-number plane Couette turbulence [15]. One of these periodic solutions characterizes not only spatial but also temporal coherence of near-wall turbulence, i.e. the full regeneration cycle of the streaks and streamwise vortices. Although these equilibrium and periodic solutions are considered to be unstable at the Reynolds numbers where turbulence is observed, they represent saddles in phase space, in the neighbourhood of which a turbulent state would spend a substantial fraction of time. The mean and fluctuation intensity profiles in some of these equilibrium and periodic solutions are reminiscent of those in turbulent states [11,14–16]. The regeneration cycle of the near-wall coherent structures may be thought of as a chaotic oscillation about the unstable solutions. These equilibrium or periodic saddles could be a good candidate for the exact spatial or spatiotemporal description of the incomplete realization of the coherence in the buffer layer of near-wall turbulence.

In this paper we review Kawahara & Kida's [15] unstable periodic solution to the incompressible Navier–Stokes equation for plane Couette system, and recompute their solution with higher accuracy by use of the Newton–Raphson method to track it for different Reynolds numbers and different wall-parallel sizes of the computational periodic box. The original periodic solution, which is obtained by a direction-set method, is described and related with plane Couette turbulence in § 2. The periodic solution is recomputed to discuss the appearance of the periodic motion and its Reynolds-number dependence in § 3. Finally our concluding remarks are given in § 4.

2. PERIODIC SOLUTION

We first consider minimal wall turbulence [17], in which the wall-parallel size of a computational periodic box is minimized while sustaining turbulence activity. We perform direct numerical simulations of the incompressible Navier–Stokes equation, by using a spectral method, for the minimal plane Couette turbulence investigated by Hamilton, Kim & Waleffe [9] who observed qualitatively a recurrent dynamical process corresponding to the near-wall regeneration cycle. The simulation code used in this work was developed by Toh (see [12]), and its scheme is essentially the same as that in Kim, Moin & Moser [18]. The dealiased Fourier expansions are employed in the streamwise (x) and spanwise (z) directions, and the Chebyshev-polynomial expansion in the wall-normal (y) direction. The volume flux in the x-direction is set to be zero. Numerical computations are carried out on 8,448 ($= 16 \times 33 \times 16$ in x, y, and z) grid points at $Re \equiv Uh/\nu = 400$, where U is half the difference of the two wall velocities and h is half the wall separation. The numerical accuracy of this grid resolution has been checked by doubling the resolution. Hereafter we non-dimensionalize flow variables by using h as the length scale and h/U as the time scale. The Reynolds number based on u_τ and h is $Re_\tau = 34.1$. When the flow variables are normalized by ν and u_τ, we shall attach the superscript $^+$ to them. The streamwise and spanwise computational periods are $(L_x, L_z) = (1.755\pi, 1.2\pi)$ $[(L_x^+, L_z^+) = (188, 128)]$ [9].

The energy is injected through the frictional force on the moving walls and consumed at small scales over the whole flow field by viscous dissipation. The energy input rate $I = \int_0^{L_x}\int_0^{L_z}\left(\partial u/\partial y|_{y=-1} + \partial u/\partial y|_{y=+1}\right) \mathrm{d}x\mathrm{d}z/(2L_xL_z)$ and dissipation rate $D = \int_0^{L_x}\int_{-1}^{+1}\int_0^{L_z} |\boldsymbol{\omega}|^2 \mathrm{d}x\mathrm{d}y\mathrm{d}z/(2L_xL_z)$ normalized by those for a laminar state are calculated, where u is the streamwise velocity and $\boldsymbol{\omega}$ is the vorticity vector. They vary in a chaotic way in time, and their temporal averages, which are substantially larger than the corresponding ones in a laminar state (see the thin grey line with the black dots in Fig. 1), are the same because the turbulence is statistically stationary.

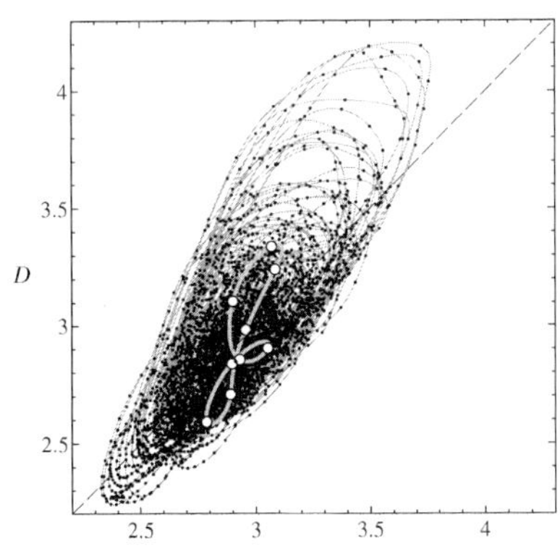

Figure 1. Projections of a turbulent and a periodic orbit on the (I, D)-plane.

In the present numerical scheme the independent variables are 31 Chebyshev coefficients for the mean streamwise and spanwise components of velocity, 7,424 (=16 × 29 × 16) Fourier–Chebyshev–Fourier coefficients for the wall-normal velocity, and 7,936 (=16 × 31 × 16) Fourier–Chebyshev–Fourier coefficients for the wall-normal vorticity. The number N of degrees of freedom of the present dynamical system is therefore 15,422. An instantaneous state of the flow field and its temporal evolution should be represented respectively as a point and its trajectory in the N-dimensional phase space spanned by all the independent variables. In Fig. 1, we plot, with a thin grey line, a projection of the turbulence trajectory over a period of 10^4 on the two-dimensional subspace spanned by I and D. Black dots are attached at intervals of $\Delta t = 2$. The orbit generally tends to turn clockwise. The energy input and dissipation rates are in balance on the dashed diagonal. The variation of the trajectory, which is confined in a finite domain, is far from periodic. On the contrary, the frequency spectrum of the total kinetic energy is continuous (figure is omitted), which suggests that the orbit may be in a chaotic state.

In a turbulent state, the spatial symmetries in the Nagata steady solution [10]: (i) the reflection with respect to the plane of $z = 0$ and a streamwise shift by a half period $L_x/2$; (ii) the 180° rotation around the line $x = y = 0$ and a spanwise shift by a half period $L_z/2$, are observed to appear approximately without being imposed on the flow. Therefore, we impose them on a time-periodic solution to be searched for, which allows us to reduce the degrees of freedom roughly to $N/4$. Note that a solution with these symmetries cannot be a travelling wave in either the x- or the z-direction.

We here employ a standard direction-set method for iteration to minimize the Euclidean distance between successive cross-points of the orbit on a Poincaré section $\mathrm{Im}(\widetilde{\omega}_{y\,0,0,1}) = -0.1875$ in phase space, where $\mathrm{Im}(\widetilde{\omega}_{y\,0,0,1})$ is the imaginary part of the Fourier–Chebyshev–Fourier coefficient of the wall-normal vorticity for the zero streamwise wavenumber, the zeroth-order Chebyshev polynomial, and the $2\pi/L_z$ spanwise wavenumber. A flow state at an arbitrary time when the turbulence trajectory is travelling more or less periodically in the phase space is chosen as the first guess for the iteration.

The iteration is continued until $\epsilon < 10^{-2}$, where ϵ is the ratio of the distance between successive cross-points to that from the origin of the earlier one (see § 3 for higher-accuracy computations). A periodic orbit thus obtained is drawn in Fig. 1 with a closed thick grey line, the period of which is $T = 65$ ($T^+ = 190$). Black dots on the turbulence trajectory are much denser near the periodic orbit, implying that the turbulent state approaches it very frequently. The approaches to the periodic orbit have been confirmed not only in this (I, D)-plane but also in other subspaces, during which the spatiotemporal structures of the turbulent flow resemble remarkably those for the periodic flow (see Fig. 2). This extraction of a periodic orbit may offer the first (to our knowledge) direct demonstration of the existence of a periodic motion embedded in a turbulent flow, at least for this highly constrained case.

A full cycle of the temporal evolution of spatial structure of the periodic solution is depicted in Fig. 2 (a–i) at nine sequential phases indicated by white dots on the periodic orbit in Fig. 1. The phase of Fig. 2 (a) corresponds to the white dot at the time of the least input and dissipation rates. Typical near-wall coherent structures are clockwise (or counter-clockwise) streamwise (x) vortices visualized by the white (or black) iso-surfaces of the Laplacian of

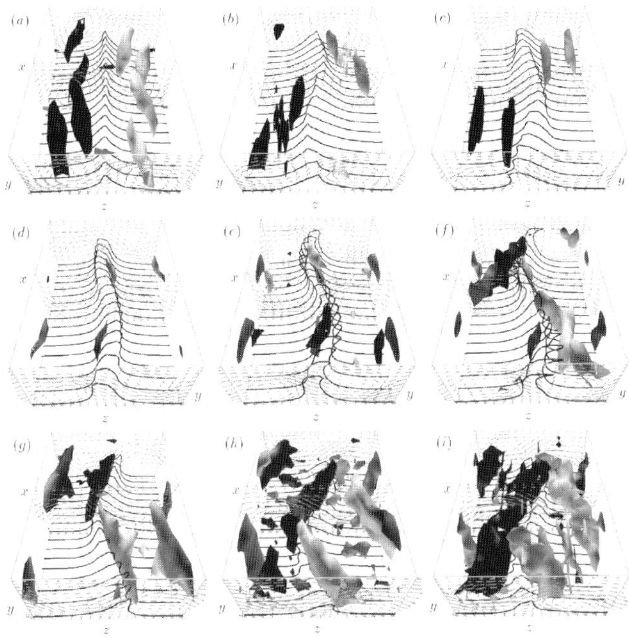

Figure 2. A full cycle of time-periodic flow. Streamwise vortices and streaks are visualized, respectively, by the iso-surfaces and iso-contours in the whole spatially periodic box ($L_x \times 2 \times L_z$) over one full cycle at nine times.

pressure, $\nabla^2 p = 0.15$, (see also the cross-flow velocity vectors) and are streamwise streaks of relatively low streamwise velocity represented by the lifted iso-contours, $u = -0.3$, in the (y, z)-planes. The dynamics of the periodic flow is described by a cyclic sequence of events which consists of (i) the formation and development of low-velocity streaks through the advection of streamwise velocity in the cross-flow induced by decaying streamwise vortices [Fig. 2 (*a*–*d*)], (ii) the bending along the streamwise direction and tilting in the spanwise (z) direction of the streaks followed by the regeneration of streamwise vortices [Fig. 2 (*e*–*g*)], and (iii) the breakdown of streaks and the violent development of streamwise vortices [Fig. 2 (*h*, *i*)]. This cyclic sequence is completely consistent with a previously reported regeneration cycle [6,9].

Figures 3 (*a*) and (*b*) compare the mean and RMS (root-mean-square) velocities for the time-periodic flow (symbols) with those for the turbulent flow (lines), where circles and solid lines indicate the streamwise component, triangles and a dotted line the wall-normal component, and squares and a dashed line the spanwise component. The mean streamwise velocity for the time-periodic flow is in very good agreement with that for the turbulent flow. It can be seen that even the RMS velocities for the time-periodic flow coincide with those for the turbulent flow. Excellent agreement in all the RMS vorticities and the Reynolds shear stress, has also been confirmed. This is expected because the turbulent state actually spends most of the time in the neighbourhood of the periodic orbit.

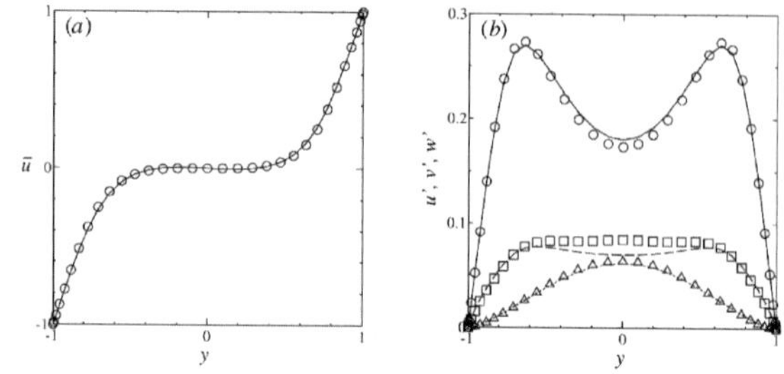

Figure 3. Comparison of (a) the mean velocity $\bar{u}$ and (b) the RMS velocities u', v', w' between a time-periodic (symbols) and a turbulent (lines) flow for $Re = 400$ and $(L_x, L_z) = (1.755\pi, 1.2\pi)$.

3. TRACKING OF SOLUTION

We next introduce the Newton–Raphson method to obtain time-periodic solutions with higher accuracy. Periodic solutions are computed as a fixed point of a Poincaré map defined on the Poincaré section $\mathrm{Im}(\widetilde{\omega}_{y\,0,0,1}) = -0.1875$. The Poincaré map, i.e. the one-period time integration of the Navier–Stokes equation, is computed by the direct numerical simulation described in § 2, and its Jacobian matrix is evaluated by a finite-difference approximation. We use the periodic solution ($\epsilon < 10^{-2}$) at the Reynolds number $Re = 400$ obtained in the preceding section as the initial guess for the Newton–Raphson iteration with accuracy $\epsilon < 10^{-7}$. The number of degrees of freedom is further reduced nearly to $N/8$ by halving the truncation order of the Chebyshev polynomials (then $16 \times 17 \times 16$ grid points in physical space) to make it easier to trace the periodic solution by changing the Reynolds number for a given box size of $(L_x, L_z) = (1.755\pi, 1.2\pi)$. The periodic solution thus obtained is slightly different from the original one for higher spatial resolution ($16 \times 33 \times 16$ grid points). For example, there is about 5% difference in the time period at $Re = 400$.

Figure 4 shows the time period of the periodic solution against Re. The periodic solution appears through a saddle-node bifurcation at the onset Reynolds number $Re = Re_{\mathrm{SN}}$ ($\simeq 321$) so that there are upper and lower solution branches at $Re > Re_{\mathrm{SN}}$. These solution branches exhibit a little complicated behaviour. If we follow the solution curve from the bifurcation point, the Reynolds number does not increases monotonically, but turns back once on both the upper and lower branches. The original solution described in § 2 lies on the lower branch. The period T normalized by h/U is around 70, while T^+ normalized by ν/u_τ^2 is around 200. The order of T^+ is comparable with the dominant time scale (≈ 400) observed in the frequency spectrum of the plane-averaged wall shear rate for minimal plane Couette turbulence at high Reynolds numbers [16]. Any bifurcation of the present periodic solution from equilibrium ones, such as Hopf or homoclinic bifurcations, has not been found.

Figures 5 (a) and (b) show the Re-dependence of the mean wall shear rate u_τ^2/ν, and half the wall separation h^+ and the spanwise period L_z^+. Note that L_z^+ (or L_x^+) can vary because of the change of u_τ even if L_z (or L_x) is fixed. In the figures there is the end point of the upper branch, at which the tracking of the solution has been terminated. Open and closed circles represent turbulent

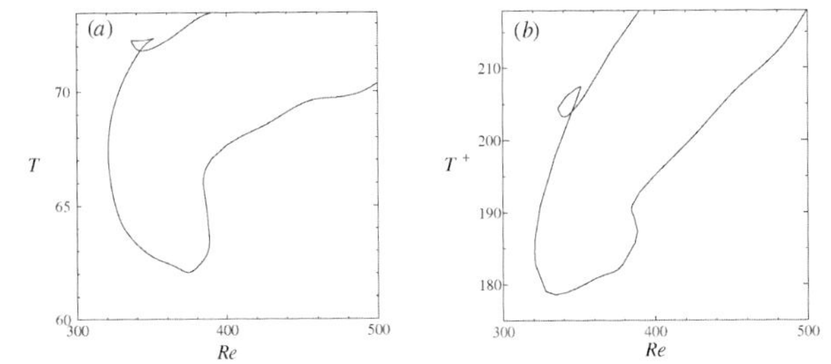

Figure 4. The time period, (*a*) T and (*b*) T^+, of the periodic solution as a function of Re for $(L_x, L_z) = (1.755\pi, 1.2\pi)$.

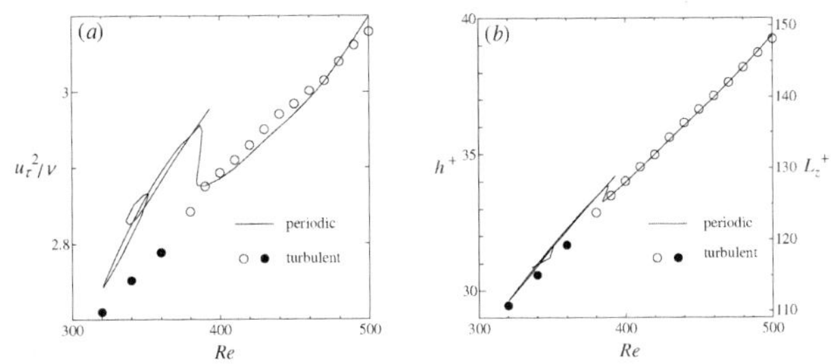

Figure 5. The Re-dependence of (*a*) the mean wall shear rate u_τ^2/ν, and (*b*) half the wall separation h^+ and the spanwise period L_z^+ for $(L_x, L_z) = (1.755\pi, 1.2\pi)$.

states computed for the same values of L_x and L_z, though turbulence is not always sustained depending on initial conditions in the case of the closed circles at lower Reynolds numbers close to Re_{SN}. The wall shear rate and therefore h^+ (and L_z^+) for the lower-branch periodic solution are in good agreement with those for the minimal turbulence at $Re \gtrsim 390$, i.e. beyond the turnback Reynolds number.

The onset Reynolds number Re_{SN} of a saddle-node bifurcation, at which the periodic solution appears, is different depending on L_x and L_z. By changing the streamwise and the spanwise periods, we find that the minimum of $Re_{\text{SN}} \simeq 291$ is attained for $(L_x, L_z) \simeq (1.82\pi, 1.09\pi)$ [see the closed circles in Fig. 6 (*a*, *b*)], though it is unknown whether this minimum is global or not. At the minimum onset Reynolds number $Re_{\text{SN}} \simeq 291$ the solution has the time period $T^+ \simeq 201$, and the spatial dimensions $h^+ \simeq 27$, $(L_x^+, L_z^+) \simeq (155, 93)$ (see the closed circle in Fig. 7). The height $h^+ \simeq 27$ is comparable with the thickness of the viscous wall layer, and the spanwise period $L_z^+ \simeq 93$ is close to the widely observed spanwise spacing of near-wall streaks. We trace the solution at a slightly higher Reynolds number $Re = 300$ by changing L_x (or L_z) for $L_z = 1.09\pi$ (or $L_x = 1.82\pi$) to obtain the solid (or dashed) loop shown in Fig. 7. It can be seen that the periodic solution is confined to a certain range of the wall-parallel periods, $L_x^+ \simeq 154 - 166$ and $L_z^+ \simeq 91 - 97$. This confinement of the periodic solution is conjectured to be related with the scale selection mechanisms of near-wall coherent structures, though spatially subharmonic

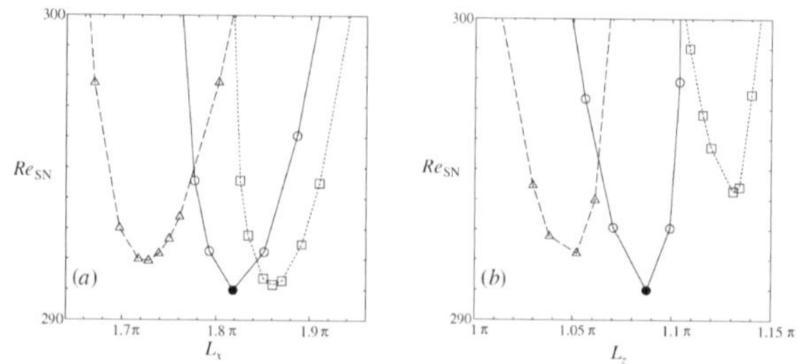

Figure 6. The onset Reynolds number Re_{SN} of saddle-node bifurcations versus (a) L_x for fixed L_z [$= 1.05\pi$ (triangles), 1.09π (circles), 1.10π (squares)]. (b) L_z for fixed L_x [$= 1.71\pi$ (triangles), 1.82π (circles), 1.95π (squares)].

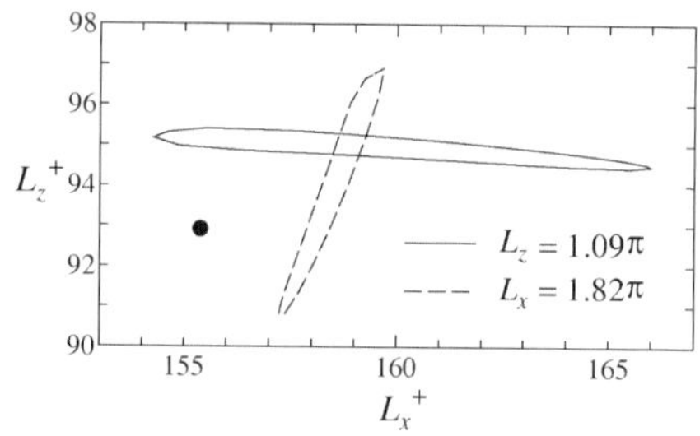

Figure 7. The streamwise period L_x^+ versus the spanwise period L_z^+. The solid (or dashed) loop is obtained by changing L_x (or L_z) for fixed L_z (or L_x) at $Re = 300$. The (L_x^+, L_z^+) for the minimum Re_{SN} ($\simeq 291$) is also shown by the closed circle.

solutions could appear as observed in a steady state of plane Couette flow [11]. At higher Reynolds numbers the upper bound of the ranges of the wall-parallel periods, L_x^+ and L_z^+, should become much larger. Consistency in the length scales between coherent structures and the structures of equilibrium solutions was also reported in plane Poiseuille [11] and Couette [11,16] systems.

The RMS velocities, u'^+, v'^+ and w'^+, of the periodic solutions at the minimum onset Reynolds number $Re_{SN} \simeq 291$ (thick lines), at $Re = 400$ on the lower branch (open symbols), and at $Re = 393$ on the upper branch (closed symbols) are plotted in Fig. 8 against the distance y^+ from the wall. The corresponding data at a higher Reynolds number $Re = 3000$ are also shown for turbulent Couette flow (thin lines) with the spatial dimensions $(L_x^+, h^+, L_z^+) = (1008, 160, 252)$. Circles and solid lines indicate the streamwise component, triangles and dotted lines the wall-normal component, and squares and dashed lines the spanwise component. The RMS velocities for $Re_{SN} \simeq 291$ are similar to those for $Re = 393$ and 400. Spatiotemporal structures visualized for different Reynolds numbers and different branches exhibit qualitatively the same regeneration cycle of the near-wall coherent structures as in Fig. 2. All the RMS velocities of the periodic solution show a qualitative agreement with those of high-Reynolds-number Couette turbulence in the buffer layer.

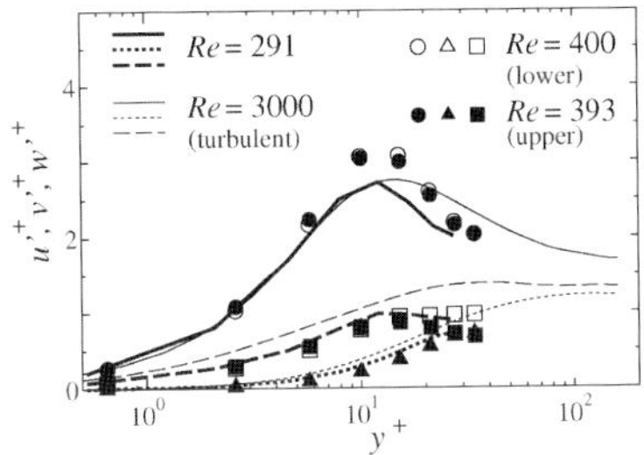

Figure 8. The RMS velocities u'^+, v'^+, w'^+ as a function of y^+.

Eigenvalues (i.e. the Floquet multipliers) for the Jacobian matrix of the Poincaré map on a fixed point represent the stability characteristics of the periodic solution to infinitesimal disturbances with the same wall-parallel periods and symmetries as the periodic solution. We compute the eigenvalues on the upper and lower branches of the solution at $Re \leq 400$ for $(L_x, L_z) = (1.755\pi, 1.2\pi)$. It is found that there are two, three or four unstable eigenvalues with modulus greater than unity and the most unstable ones are $O(10)$. We also compute the eigenvalues on the solution in the vicinity of the minimum onset point $Re_{\mathrm{SN}} \simeq 291$ for $(L_x, L_z) \simeq (1.82\pi, 1.09\pi)$ to find only one or two unstable eigenvalues. The present periodic solution is unstable, but the dimension, in phase space, of its unstable manifold could be quite low at low Reynolds numbers.

4. CONCLUDING REMARKS

In this paper the recently found periodic solution [15], which exhibits a full regeneration cycle of near-wall coherent structures in plane Couette turbulence, is reviewed and recomputed with higher accuracy. The periodic solution appears through a saddle-node bifurcation and is traceable down to a Reynolds number of $Re \approx 290$, at which the spatial dimensions of the solution are marginal in the sense that both the height $h^+ \simeq 27$ and the spanwise period $L_z^+ \simeq 93$ just fit in the viscous-wall-layer thickness and the streak spacing, respectively. This lowest value of Re is comparable with the experimentally observed nonlinear threshold Reynolds numbers, $Re \approx 320$, above which a certain kind of finite-amplitude disturbance [19] is sustained, or below which all turbulence [20] is no longer sustained in plane Couette flow. Therefore it is expected that the appearance of the periodic solution would be relevant to the transition to turbulence in plane Couette flow.

This work was partially supported by a Grant-in-Aid for Scientific Research and by Center of Excellence for Research and Education on Complex Functional Mechanical Systems from the Ministry of Education, Culture, Sports, Science and Technology, Japan.

REFERENCES

1. Prandtl L. "Uber flüssigkeitsbewegung mit kleiner reibung", *Verh. III Int. Math. Kongr.*, Heidelberg, Germany, 484–491, 1905.

2. Prandtl L. "Zur turbulenten strömung in rohren und längs platten", *Ergeb. AVA Göttingen*, **4**, 18–29, 1932.
3. Jiménez J. "The physics of wall turbulence", *Physica* A, **263**, 252–262, 1999.
4. Jiménez J, Pinelli A. "The autonomous cycle of near-wall turbulence", *J. Fluid Mech.*, **389**, 335–359, 1999.
5. Kline SJ, Reynolds WC, Schraub FA, Runstadler PW. "The structure of turbulent boundary layers", *J. Fluid Mech.*, **30**, 741–773, 1967.
6. Waleffe F. "On a self-sustaining process in shear flows", *Phys. Fluids*, **9**, 883–900, 1997.
7. Schoppa W, Hussain F. "Coherent structure generation in near-wall turbulence", *J. Fluid Mech.*, **453**, 57–108, 2002.
8. Kawahara G, Jiménez J, Uhlmann M, Pinelli A. "Linear instability of a corrugated vortex sheet — a model for streak instability", *J. Fluid Mech.*, **483**, 315–342, 2003.
9. Hamilton JM, Kim J, Waleffe F. "Regeneration mechanisms of near-wall turbulence structures", *J. Fluid Mech.*, **287**, 317–348, 1995.
10. Nagata M. "Three-dimensional finite-amplitude solutions in plane Couette flow: bifurcation from infinity", *J. Fluid Mech.*, **217**, 519–527, 1990.
11. Waleffe F. "Homotopy of exact coherent structures in plane shear flows", *Phys. Fluids*, **15**, 1517–1534, 2003.
12. Itano T, Toh S. "The dynamics of bursting process in wall turbulence", *J. Phys. Soc. Jpn.*, **70**, 703–716, 2001.
13. Waleffe F. "Exact coherent structures in channel flow", *J.Fluid Mech.*, **435**, 93–102, 2001.
14. Jiménez J, Simens MP. "Low-dimensional dynamics in a turbulent wall flow", *J. Fluid Mech.*, **435**, 81–91, 2001.
15. Kawahara G, Kida S. "Periodic motion embedded in plane Couette turbulence: regeneration cycle and burst", *J. Fluid Mech.*, **449**, 291–300, 2001.
16. Jiménez J, Kawahara G, Simens MP, Nagata M, Shiba M. "Characterization of near-wall turbulence in terms of equilibrium and 'bursting' solutions", submitted to *Phys. Fluids*, 2004.
17. Jiménez J, Moin P. "The minimal flow unit in near-wall turbulence", *J. Fluid Mech.*, **225**, 213–240, 1991.
18. Kim J, Moin P, Moser R. "Turbulence statistics in fully developed channel flow at low Reynolds number", *J. Fluid Mech.*, **177**, 133–166, 1987.
19. Dauchot O, Daviaud F. "Finite amplitude perturbation and spots growth mechanism in plane Couette flow", *Phys. Fluids*, **7**, 335–343, 1995.
20. Bottin S, Chaté H. "Statistical analysis of the transition to turbulence in plane Couette flow", *Eur. Phys. J.* B, **6**, 144–155, 1998.

SOME CLASSIC THERMAL BOUNDARY LAYER CONCEPTS RECONSIDERED

and their relation to compressible Couette flow

B.W. van Oudheusden
Delft University of Technology, Department of Aerospace Engineering, Kluyverweg 1, 2629 HS Delft, The Netherlands, B.W.vanOudheusden@LR.TUDelft.nl

Abstract: Some classic analytical concepts regarding thermal effects in boundary layers are reconsidered, in their application to Couette flow as well as external boundary layers. Particular attention is given to the role of the Prandtl number in this.

Key words: recovery factor, Reynolds analogy, reference temperature method.

1. INTRODUCTION

When commemorating the scientific legacy of Ludwieg Prandtl it is appropriate to consider that area of boundary layer theory where the most relevant dimensionless number is the one named after him. The Prandtl number, $\Pr = c_p\,\mu/k$, expresses the ratio of mechanical to thermal diffusion and plays a dominant role in thermal boundary layer effects [1-2]. Some classic analytical concepts can be recognised in this field, like the Crocco-Busemann relation, the Reynolds analogy, the recovery factor and the reference-temperature method. The objective of the present discussion is to reconsider the background of these concepts, assisted by a supporting analysis of compressible Couette flow, as well as looking at the role of the Prandtl number in their extension to more general boundary layer conditions.

Most of these concepts follow from the analogy between momentum and heat transport by convection and diffusion, the latter being characterised by the Prandtl number. For steady two-dimensional laminar boundary layer flow the momentum and energy equations in terms of velocity u and enthalpy h, are given in a Cartesian coordinate frame by:

G.E.A. Meier and K.R. Sreenivasan (eds.), IUTAM Symposium on One Hundred Years of Boundary Layer Research, 425-434,

$$\rho \frac{Du}{Dt} = -\frac{dp}{dx} + \frac{\partial}{\partial y}\left(\mu \frac{\partial u}{\partial y}\right), \quad \rho \frac{Dh}{Dt} = u\frac{dp}{dx} + \mu\left(\frac{\partial u}{\partial y}\right)^2 + \frac{\partial}{\partial y}\left(\frac{\mu}{\Pr}\frac{\partial h}{\partial y}\right) \quad (1)$$

For analytical convenience, the specific heat $c_p = dh/dT$ is assumed constant, while the viscosity-temperature relation is taken as $\mu \sim T$.

1.1 The relation between temperature and velocity

The complete analogy between the mechanical and thermal diffusion in case that $\Pr = 1$ allows the temperature to be written in terms of the velocity:

$$T = T_w - (T_w - T_{aw})(u/u_e) - (T_{aw} - T_e)(u/u_e)^2 \quad (2)$$

where T_w and T_e are the wall and freestream temperatures, while T_{aw} is the adiabatic wall temperature, under this condition equal to the freestream total temperature T_0. This result, known as the 'Crocco integral' or 'Crocco-Busemann relation' [3-4] applies for (1) zero heat transfer (adiabatic wall, the second term vanishes as $T_w = T_{aw}$) or (2) with heat transfer if the wall is isothermal and pressure gradient zero. In direct relation to this, the recovery factor r and the Reynolds analogy factor s are obtained as:

$$r = \frac{T_{aw} - T_e}{T_0 - T_e} = 1, \qquad s = \frac{q_w / \rho u_e c_p (T_w - T_{aw})}{\tau_w / \rho u_e^2} = 1 \quad (3)$$

Based on numerical assessment [4], the common approach for $\Pr \neq 1$ is that these results still hold in approximation, with $r \approx \Pr^{1/2}$ and $s \approx \Pr^{-2/3}$.

1.2 The reference temperature method

The reference temperature method is a convenient engineering tool to approximate the effect of compressibility on technically relevant properties, like skin friction and heat transfer. The approach implies that reasonable predictions can be obtained from incompressible flow relations when fluid properties are evaluated at an appropriate 'reference temperature', T^*. The method was originally developed as a data fit for the laminar flat-plate flow of a perfect gas [6-7], but was later extended to hypersonic flow conditions by including real-gas effects [8-9]. The reference temperature method was introduced by Rubesin and Johnson [6] who expressed T^* as:

$$T^* = T_e + 0.58(T_w - T_e) + 0.16(T_0 - T_e) \quad (4)$$

The approach was further explored notably by Eckert [7] who replaced the total temperate T_0 by the adiabatic wall temperature T_{aw} to obtain:

$$T^* = T_e + 0.5(T_w - T_e) + 0.22(T_{aw} - T_e) \tag{5}$$

Although initially derived only as a data correlation, Dorrance [10] provided theoretical basis to the concept, by showing that for the flat-plate flow the reference temperature can be obtained, in an approximate sense, as a velocity-average of the temperature over the boundary layer:

$$T^* \approx \frac{1}{u_e} \int_0^{u_e} T\, du \approx T_e + \tfrac{1}{2}(T_w - T_e) + \tfrac{1}{6}(T_{aw} - T_e) . \tag{6}$$

The final expression results after substituting Eq.(2). The agreement between Eqs.(5) and (6) may serve to justify the reference temperature concept, although Dorrance gave no quantitative assessment of the accuracy of his analysis nor of the approximations involved.

In addition, Herwig [11] investigated the effect of compressibility and heat transfer by means of a first-order perturbation of the constant-property solution, from which the reference temperature may be formally obtained as:

$$T^* = T_e + a_1(\mathrm{Pr}) \cdot (T_w - T_e) + a_2(\mathrm{Pr}) \cdot (T_0 - T_e) . \tag{7}$$

Note that this result does not necessarily favour $T_0 - T_e$ over $T_{aw} - T_e$ as the proper temperature difference representative of frictional heating effects, as their ratio is equal to the recovery factor r, which is a function of Pr that can be absorbed in the coefficient a_2.

2. COMPRESSIBLE COUETTE FLOW

The Couette flow resembles boundary layer flow in quite some respect, as the moving plate performs a similar shearing action as the external flow, while the stationary plate acts like the boundary layer surface. For Couette flow the governing equations can be integrated analytically, yielding that the temperature-velocity relation of Eq.(2) applies exactly for an arbitrary constant value of Pr. Under more general conditions with variable Pr, the shear stress, heat transfer and adiabatic wall temperature are found to be:

$$\tau_w = \mu^* \frac{u_e}{h}, \quad q_w = \frac{\tau_w / u_e}{\mathrm{Pr}^*} c_p (T_w - T_{aw}), \quad T_{aw} = T_e + r \frac{u_e^2}{2c_p} \tag{8}$$

with h the plate distance and where the effective values of viscosity, Prandtl number and recovery factor are obtained as:

$$\mu^* = \frac{1}{u_e}\int_0^{u_e} \mu\, du\,, \qquad \Pr^* = \frac{1}{u_e}\int_0^{u_e} \Pr du\,, \qquad r = \frac{2}{u_e^2}\int_0^{u_e} \Pr u du\,. \tag{9}$$

For constant Pr, hence, $r = \Pr$ and $s = \Pr^{-1}$, while for variable properties, in approximation, $\mu^* = \mu(T^*)$, $\Pr^* = \Pr(T^*)$ and $r = \Pr(T^{**})$, where:

$$T^* = \frac{1}{u_e}\int_0^{u_e} T\, du\,, \quad T^{**} = \frac{2}{u_e^2}\int_0^{u_e} T\, u du\,. \tag{10}$$

This confirms, for the Couette flow case, the velocity-averaging concept proposed by Dorrance. Interestingly, a different reference temperature (T^{**}) is found to apply for the recovery factor, although this may be of less direct relevance as the effect of temperature on Prandtl number is small in gases.

3. COMPRESSIBLE BOUNDARY LAYER FLOW

Compressible boundary layers are conveniently analysed under the classic coordinate transformation $\eta(x,y)$ and $\xi(x)$, and introducing a stream function f and enthalpy function θ, according to:

$$\frac{d\xi}{dx} = \rho_e \mu_e u_e\,; \quad \frac{\partial \eta}{\partial y} = \rho \frac{u_e}{\sqrt{2\xi}}\,; \quad f'(\eta,\xi) = \frac{u}{u_e}\,; \; \theta(\eta,\xi) = \frac{T - T_e}{T_0 - T_e}\,; \tag{11}$$

where a prime indicates a derivative with respect to η. Under the above transformation and assuming constant Pr and an adiabatic external flow (constant total enthalpy H_e), the momentum and energy equations become:

$$\begin{aligned} &(Cf'')' + ff'' + \beta\left(\frac{H}{H_e} - f'^2\right) = 2\xi\left(f'\frac{\partial f'}{\partial \xi} - f''\frac{\partial f}{\partial \xi}\right) \\ &\frac{(C\theta')'}{\Pr} + f\theta' - 2\beta f'\theta + 2Cf''^2 = 2\xi\left(f'\frac{\partial \theta}{\partial \xi} - \theta'\frac{\partial f}{\partial \xi}\right) \end{aligned} \tag{12}$$

with: $$C = \frac{\mu\rho}{\mu_e \rho_e}\,; \quad \beta = \frac{2\xi}{M_e}\frac{dM_e}{d\xi} \tag{13}$$

3.1 Effect of Prandtl number on thermal recovery

For the adiabatic wall case the relation between temperature and velocity expressed by Eq.(2) is exactly valid, irrespective of pressure gradient and compressibility, provided that Pr = 1. This suggests that for Pr different from but near 1, the effect of Pr can be studied from a perturbation analysis [12]. Assuming incompressible flow, the solution for f is independent of Pr, while for the temperature function we introduce an asymptotic expansion with respect to the perturbation parameter $\varepsilon = \mathrm{Pr} \quad 1$. Whereas the solutions for the higher order terms are problem-specific, analytical solutions valid under general conditions may be obtained for the first two terms [12]:

$$\theta_0 = 1 - f'^2 ; \quad \theta_1 = \frac{1}{2}(1 - f'^2) - f f'' \tag{14}$$

Numerical results for the coefficients of the series expansion of the recovery factor are given in Tab.1 for some self-similar boundary layers. The results reveal the small effect of the pressure-gradient parameter β on the recovery factor, as well as the good agreement with the classic approximation $r \quad \mathrm{Pr}^{1/2}$.

Table 1. Recovery factor coefficients ($r = c_0 + c_1\varepsilon + c_2\varepsilon^2 + ...$) for constant-property self-similar boundary layer solutions.

	$f''(0)$	c_0	c_1	c_2	c_3	c_4
$\beta = -0.1988$	0	1.0000	0.5000	-0.1331	0.0641	-0.0389
$\beta = 0$	0.4696	1.0000	0.5000	-0.1345	0.0655	-0.0399
$\beta = 1$	1.2326	1.0000	0.5000	-0.1349	0.0661	-0.0404
$r = \mathrm{Pr}^{1/2}$		1	0.5	-0.125	0.0625	-0.0391

Combination of the first-order perturbation results given in Eq.(14) allows a rational and asymptotically complete extension of the Crocco-Busemann relation for an adiabatic wall and $\mathrm{Pr} \neq 1$ to be formulated as:

$$T = T_{aw} - (T_{aw} - T_e)(f')^2 + (T_0 - T_{aw})2 f f'' \tag{15}$$

In comparison to the 'classic' temperature-velocity relation, Eq.(2), it can be observed that indeed a local recovery effect is present (the second term), but the last term is missing there. This term describes the redistribution of energy 'released' by the incomplete recovery in case $\mathrm{Pr} \neq 1$, in direct consequence of the integral conservation of energy. As shown in Fig. 1 for the flat plate flow at $\mathrm{Pr} = 0.7$, the extended relation provides an improved description of the temperature profile, in particular the overshoot in total temperature near the edge of the boundary layer. The effects of compressibility on these results have been discussed elsewhere [13].

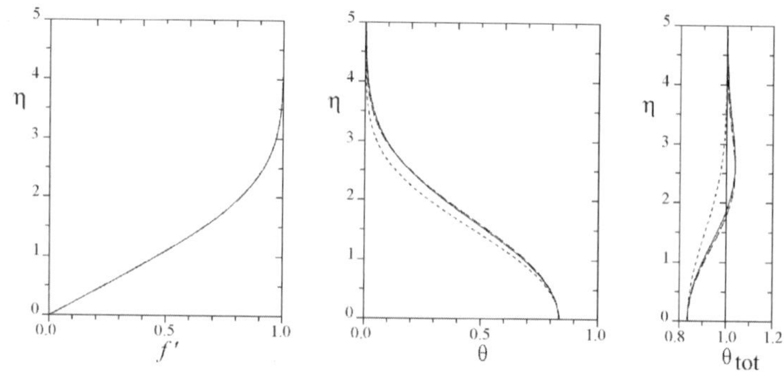

Figure 1. Boundary layer profiles of velocity, temperature and total temperature ($\theta_{tot} = \theta + f'^2$) for incompressible flat-plate flow (Pr = 0.7); solid line: numerical solution; dotted line: classic Crocco integral, Eq.(2); dashed line: extended Crocco integral, Eq.(15).

3.2 Assessment of the reference temperature concept

Returning to the self-similar solution of the flat-plate boundary layer ($\beta = 0$) under compressible flow conditions, the shear stress is obtained from the transformed solution properties as:

$$\tau_w = u_e^{\,2}\sqrt{\frac{\rho_e \mu_e}{2u_e x}}\, C_w f''(0)\,; \qquad C_w = C(0) \tag{16}$$

The reference temperature concept now implies that the shear stress can be obtained from using constant values ρ^* and μ^*, hence:

$$C_w f''(0) = C^* f^{*\prime\prime}(0) = \sqrt{C^*}\, f_0''(0) \tag{17}$$

where f^* and f_0 are the solutions for $C = C^*$ and $C = 1$ (incompressible flow), respectively. This allows exact reference temperatures to be extracted from the numerical solution data [14]. Figure 1 shows the computed reference temperature, for (left) an adiabatic wall and (right) a cooled wall with $T_w = T_e$, with constant Prandtl number. The results are given in the form of $(T^*\!-\!T_e)/(T_{aw}\!-\!T_e)$, which should be constant according to theory, Eqs.(5) and (6). For the Couette flow the variation with Mach number is relatively small (3% variation for $0 < \mathrm{M} < 10$) and due only to the nonlinearity of the viscosity-temperature relation used. For the flat plate flow the variation is much larger (ca. 10% over this Mach number range). As Eq.(2) is still exact when Pr = 1, this indicates a more approximate validity of the velocity-averaging concept in boundary layer flow. The results at low Mach number are independent of the viscosity law, but a marked influence of the Prandtl number is noted. At low Mach number the result for the adiabatic wall agrees well with

Eckert's fit $(T^*-T_e)/(T_{aw}-T_e) = 0.72$ at Pr = 0.7, while for Pr = 1 it is closer to Dorrance's prediction $(T^*-T_e)/(T_{aw}-T_e) = 2/3$.

This shift can be explained from the Prandtl number effect on the velocity-temperature relation. Whereas in Couette flow Eq.(2) is valid for arbitrary Pr, it has to be modified for boundary layer flow when $\mathrm{Pr} \neq 1$, in accordance to section 3.1. When the third term of Eq.(15) is added to Eq.(2), the velocity-averaged reference temperature becomes:

$$T^* = T_e + \tfrac{1}{2}(T_w - T_e) + \tfrac{1}{6}(T_{aw} - T_e) + a_3(T_0 - T_{aw}) \qquad (18)$$

where $a_3 = f_0''(0) = 0.22$ is obtained from averaging the $2 f_0 f_0''$ term. This addition indeed gives an accurate prediction of the Prandtl number effect on the level shift in Fig. 1 (from a value of 0.67 for Pr = 1 to 0.71 for Pr = 0.7).

It may be noted that for the cooled wall (Fig. 2 right) the discrepancies between numerical results and theoretical predictions are much larger.

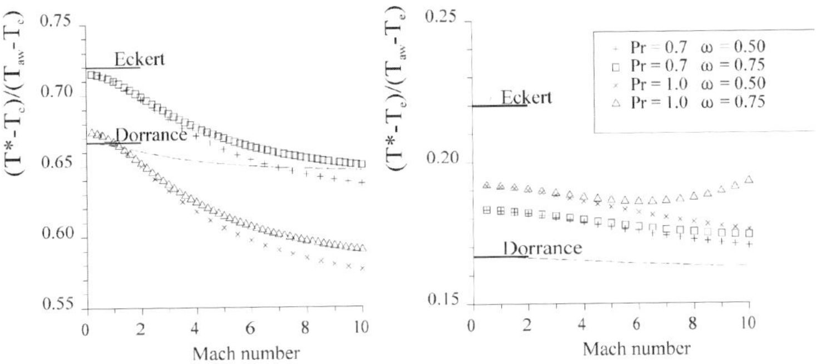

Figure 2. Reference temperature calculations for adiabatic wall (left) and cooled wall (right); symbols: flat plate boundary layer; dashed curve: Couette flow (Pr = 1, ω = 0.75).

Different reference temperature expressions are compared in Table 2, in terms of the coefficients a_1 and a_2 introduced in Eq.(7), as well as the actual predictions for the two thermal cases considered here. Apart from the three classic relations of Eqs.(4)-(6), the table contains the result of Eq.(18), labeled as 'corrected Dorrance', and of the perturbation analysis of the next section, which represent the 'exact' results for small Mach number (cf. Herwig [11]).

It is observed that none of the proposed correlations contain the Pr effect on the heat transfer component a_1, which explains why all results are rather inaccurate for the cooled wall. For the adiabatic wall case the method of Eq.(18) performs best, with an overall relative error of about 1 per cent.

Table 2. Comparison of reference temperature expressions.

Pr = 1 (r = 1)	a_1	a_2	$(T^*-T_e)/(T_0-T_e)$ adiabatic wall		$(T^*-T_e)/(T_0-T_e)$ cooled wall	
perturbation analysis	0.4823	0.1918	0.6741		0.1918	
Rubesin & Johnsson	0.58	0.16	0.74	(+0.066)	0.16	(-0.032)
Eckert	0.50	0.22	0.72	(+0.046)	0.22	(+0.028)
Dorrance	0.50	0.167	0.667	(-0.007)	0.167	(-0.025)
corrected Dorrance	0.50	0.167	0.667	(-0.007)	0.167	(-0.025)

Pr = 0.7 (r = 0.8357)	a_1	a_2	$(T^*-T_e)/(T_0-T_e)$ adiabatic wall		$(T^*-T_e)/(T_0-T_e)$ cooled wall	
perturbation analysis	0.5320	0.1532	0.5978		0.1532	
Rubesin & Johnsson	0.58	0.16	0.645	(+0.047)	0.16	(+0.007)
Eckert	0.50	0.184	0.602	(+0.004)	0.184	(+0.031)
Dorrance	0.50	0.139	0.557	(-0.041)	0.139	(-0.014)
corrected Dorrance	0.50	0.175	0.593	(-0.005)	0.175	(+0.022)

Closer investigation of the expression for C, through which the compressibility effect enters in the equations, reveals that an effective perturbation parameter for small Mach number can be defined as: $\varepsilon = (\omega - 1)(T_0 - T_e)/T_e$, yielding as the leading-order expression for the reference temperature [14]:

$$\theta_0^* = 2\left(\theta_0(0) + \frac{f_1''(0)}{f_0''(0)}\right) \tag{19}$$

with the functions that feature in the expansion satisfying:

$$\begin{aligned} & f_0''' + f_0 f_0'' = 0 \qquad && \frac{\theta_0''}{\Pr} + f\theta_0' = -2f_0''^2 \\ & f_1''' + f_0 f_1'' + f_1 f_0'' = -(\theta_0 f_0'')' && \end{aligned} \tag{20}$$

Inspection of the system of equations reveals that $\theta_0{}^*$ depends only on Pr and not on ω, which agrees with the computations. Also, because of linearity the effect of heat transfer and frictional heating can be superimposed, which confirms the structure of Eq.(7). Numerical results of this perturbation analysis can be found in Table 3.

The velocity-averaging concept of the reference temperature, found justified for the Couette flow, was originally derived by Dorrance starting from formal integration of the momentum equation, see Eq.(12a), yielding:

$$f''(\eta) = \frac{1}{C(\eta)} \exp\left(-\int_0^\eta \frac{f}{C} d\eta'\right) \left[\int_0^\infty \frac{1}{C} \exp\left(-\int_0^\eta \frac{f}{C} d\eta'\right) d\eta\right]^{-1} \tag{21}$$

Evaluation of Eq.(17) is then to deliver the velocity average procedure:

$$\frac{1}{C^*}=\int_0^\infty \frac{1}{C(\eta)}\left[\frac{\exp(-\int_0^\eta \frac{f}{C}d\eta)}{\int_0^\infty \exp(-\int_0^\eta \frac{f^*}{C^*}d\eta')d\eta}\right]d\eta \approx \int_0^\infty \frac{1}{C(\eta)}f''\,d\eta=\int_0^1 \frac{1}{C(\eta)}df' \tag{22}$$

The approximation involved by replacing the bracketed term by f'' was not further commented on by Dorrance. Comparison to Eq.(21) reveals that this estimate can be justified when $f \approx f^*$ and $C \approx C^*$. This suggests the approach to be exact for the limit of zero Mach number, when $C = C^* = 1$ and $f = f^* = f_0$. However, the direct procedure to extract θ^* then becomes singular and proper expansion yields [14]:

$$\theta_0^* = \int_0^\infty \left(\theta_0 + \frac{1}{2}\theta_0^* - g\right) f_0''\,d\eta = \int_0^1 \theta_0 df' + \int_0^1 (\theta_0 - 2g)df' \tag{23}$$

$$\text{where: } \quad g(\eta) = \int_0^\eta (f_0\theta_0 + f_1)\,d\eta$$

As can be seen, Eq.(23) contains a second term in addition to the proper velocity average. The computed values of the two separate terms have been included in Tab.3. The 'correction term' is found to be relatively small (ca. 1%) for the adiabatic wall case, but significantly larger in the presence of heat transfer (ca. 15% for the cooled wall). The explanation for this outcome is that for the approximation made in Eq.(22) to hold, the temperature variation across the boundary layer needs to be small in *relative* sense, which condition is best satisfied for the adiabatic wall.

Table 3. Reference temperature calculations: asymptotic analysis results.

adiabatic wall	$f_0''(0)$	$f_1''(0)$	$\theta_0(0)$	θ_0^*	θ_0^*/r	velocity average	correction term
Pr = 1	0.4696	-0.3113	1.0000	0.6741	0.6741	0.6667	0.0074
Pr = 0.7	0.4696	-0.2521	0.8357	0.5978	0.7154	0.5894	0.0085
Pr = 0.5	0.4696	-0.2065	0.7043	0.5291	0.7512	0.5211	0.0080

cooled wall	$f_0''(0)$	$f_1''(0)$	$\theta_0(0)$	θ_0^*	θ_0^*/r	velocity average	correction term
Pr = 1	0.4696	0.0450	0	0.1918	0.1918	0.1667	0.0251
Pr = 0.7	0.4696	-0.2521	0	0.1532	0.1833	0.1341	0.0191
Pr = 0.5	0.4696	-0.2065	0	0.1223	0.1736	0.1079	0.0144

4. CONCLUSIONS

Some classic thermal boundary layer concepts were reconsidered, with particular attention to the influence of the Prandtl number. It was found that for Couette flow the Crocco integral relating temperature and velocity is justified for arbitrary Prandtl number. Additional considerations have to be made for boundary layer flow. A proper extension of the Crocco integral was given for the adiabatic wall case. Secondly, the velocity-average concept of the reference temperature as suggested by Dorrance was shown to apply indeed exactly in the Couette flow, but needs modification for boundary layer flow, for which it was originally proposed. The additional effect is small for adiabatic wall flow, although a correction is needed to properly account for the Prandtl number effect on the velocity-temperature relation. In the presence of heat transfer, however, the required correction is more significant.

REFERENCES

1. White FM. *Viscous Fluid Flow*, 2nd ed., McGraw-Hill, 1991.
2. Schlichting H, Gersten K. *Boundary Layer Theory*, 8th ed., Springer, 2000.
3. Crocco L. "Sulla trasmissione del calore da una lamina piana a un fluido scorrente ad alta velocita", *L'Aerotecnica* **12**, pp. 181-197, 1932.
4. Busemann A. "Gasströmung mit laminarer Grenzschicht entlang einer Platte", *Z.A.M.M.*, **15**, pp. 23-25, 1935.
5. Pohlhausen E. "Der Wärmeaustausch zwischen festen Körpern und Flüssigkeiten mit kleiner Reibung und kleiner Wärmeleitung", *Z.A.M.M.*, **1**, pp. 115-121, 1921.
6. Rubesin MW, Johnson HA. "A critical review of skin-friction and heat-transfer solutions of the laminar boundary layer of a flat plate", *Trans. ASME*, **71**, pp. 383-388, 1949.
7. Eckert ERG. "Engineering relations for heat transfer and friction in high-velocity laminar and turbulent boundary-layer flow over surfaces with constant pressure and temperature", *Trans. ASME*, **78**, pp. 1273-1283, 1956.
8. Eckert ERG, Tewfik OE. "Use of reference enthalpy in specifying the laminar heat-transfer distribution around blunt bodies in dissociated air", *J. Aerosp. Sci.*, **27**, pp. 464-466, 1960.
9. Ott JD, Anderson JD. "Effects of nonequilibrium chemistry on the reference temperature method and Reynolds analogy", *J. Thermophys. Heat Tr.*, **8**, pp. 381-384, 1994.
10. Dorrance WH. *Viscous Hypersonic Flow*, McGraw-Hill, pp. 134-140, 1962.
11. Herwig H. "An asymptotic approach to compressible boundary-layer flow", *Int. J. Heat Mass Transfer*, **30**, pp. 59-68, 1987.
12. Van Oudheusden BW. "A complete Crocco integral for two-dimensional laminar boundary layer flow over an adiabatic wall for Prandtl numbers near unity", *J. Fluid Mech.*, **353**, pp. 313-330, 1997.
13. Van Oudheusden BW. "Compressibility effects on the extended Crocco relation and the thermal recovery factor in laminar boundary layer flow", *J. Fluids Eng.*, **126**, pp. 32-41, 2004.
14. Van Oudheusden BW. "On the justification of the reference temperature method in compressible boundary layer flow", *BAIL 2004*, 5-9 July 2004, Toulouse, France.

VORTICITY IN FLOW FIELDS
in relation to Prandtl's work and subsequent developments

Tsutomu Kambe
IDS, Higashi-yama 2-11-3, Meguro-ku, Tokyo 153-0043, Japan
e-mail:*kambe@gate01.com*

Abstract: One of the important properties of boundary layer is, from the point of view of *Strömungslehre*, that it pumps vorticity into the flow field. Vorticity in fluid flows plays diverse roles and produces tremendous variety of flow phenomena. We consider four aspects of vorticity in fluid flows: (A) Kinematical aspect of a vortex sheet, (B) mechanical aspect of hydrodynamic impulse of a vortex system, (C) dynamical aspect: vorticity dynamics, excitation of acoustic waves and formation of dissipative structure in turbulence, and (D) gauge field associated with local rotational symmetry.

Key words: vorticity, impulse, acoustic source, dissipative structure, gauge field.

1. INTRODUCTION

Boundary layer is a transition layer of velocity adjacent to a solid surface, and it is the place where vorticity is created. Boundary layer can be separated from the wall. Separated layer with vorticity is a common source of the vorticity in flow fields. All these essential properties of the boundary layer were established in the celebrated paper [1] of Prandtl in 1904.

Given a velocity field $\boldsymbol{v}(\boldsymbol{x})$, the vorticity $\boldsymbol{\omega}$ is defined by $\boldsymbol{\omega} = \mathrm{rot}\, \boldsymbol{v}$, which is a measure of local rotation of a fluid element. From the point of view of *Strömungslehre*, one of the important characteristics of the boundary layer is that it pumps vorticity into the flow field. Prandtl explained in the same paper [1] that the boundary layer is prone to be separated from the solid wall when pressure increases along the direction of flow, and showed some experimental evidence in the case of flows around a circular cylinder. Along

G.E.A. Meier and K.R. Sreenivasan (eds.), IUTAM Symposium on One Hundred Years of Boundary Layer Research, 435-444,

a cylinder in a free stream, the pressure increases over the rear part of the cylinder surface. The fluid decelerated by the viscosity in the boundary layer is forced to separate from the wall by the adverse pressure gradient under combined action of the forward free stream.

By the separation, the vorticity $\boldsymbol{\omega}$ in the boundary layer is transported to the interior field ([**2**]: II §6 and III §6). Separation of boundary layer is most common source of vortices in fluid flows of uniform density at a high Reynolds number (under a conseravtive force), where there is no interior mechanism of creation of vortices. In §4.1, we will consider some mechanisms of creation of vorticity. The separation occurs most effectively at a sharp edge of a body.

Fluid flows with non-vanishing vorticity are called *rotational* flows. There are an infinite variety of rotational flows. In fact, given a simply connected bounded domain of flows together with boundary conditions for the velocity, an *irrotational* flow is determined uniquely. However, rotational flow is not unique, and one can conceive all sorts of complex flows, depending on the vorticity distribution under the same boundary conditions.

In this review article, we are going to consider various roles of vorticity in fluid flows. (A) Kinematical aspect: a vortex sheet, *i.e.* a thin vorticity layer sandwiched between two irrotational flows, is a transition layer of discontinuity in velocity. This is described as a review of Prandtl's work in §2. (B) Mechanical aspect: A system of vorticity (i.e. a certain distribution of vorticity) possesses a certain amount of momentum (and also some energy). Therefore, a body shedding a set of vortices from its surface is subject to a reaction force, which is described in §3. (C) Dynamical aspect: (i) Creation of vorticity is considered on the basis of the evolution equation of vorticity, (ii) nonlinear interaction between $\boldsymbol{\omega}$ and $\boldsymbol{v}$ excites density waves, resulting in generation of acoustic waves, and (iii) the vorticity is an agent forming dissipative structures in turbulence which are visualized as fine-scale slender objects characterized with high level of vorticity magnitude. (D) Gauge field: Vorticity is regarded as a gauge field, which is defined in the variational formulation by requiring that the equation of motion should be invariant with respect to local rotational gauge transformation.

2. KINEMATICAL ASPECT (A review of Prandtl's work)

There are two different ways nominally, in which vortices are brought to interior of flows of a fluid of small viscosity. First, when a fluid flows round a bluff body such as a circular cylinder, the boundary layer separates at a certain point on the body surface and penetrates into the interior of fluid. Secondly, when the fluid flows around a sharp edge, the streams along the two sidewalls of the edge separate at the edge. This forms a surface of separation, which is a discontinuity surface between two velocities v_1 and v_2 on the two sides. On account of viscosity, this separation is regarded as merging of two boundary layers on both sides of the edge, and the velocity profile within the layer takes a form of a function *tanh* expressing transition from one velocity v_1 to another v_2. This layer is obviously rotational, i.e. it has nonzero vorticity. In the limit of vanishing viscosity, this layer tends to a surface, called a *vortex sheet*. Thus, the thin vorticity layer (sandwiched between two irrotational flows) represents a transition layer of *velocity discontinuity*.

Two conditions must always be satisfied on the surface of discontinuity. First, this surface must be a material surface consisting of the same fluid particle. Secondly, the pressure must be continuous across the vortex sheet. Separation at a sharp edge can be interpreted as the same mechanism that is working over the surface of a circular cylinder. For, a sharp edge can be regarded as a limiting surface when one of the radii of curvature of the wall surface becomes very small ([**3**]: §4.8). The flow along the edge will encounter a sudden increase of pressure after passing by the edge.

The vortex sheet (the pressure being continuous across it) is unstable and tends to rollup into a sequence of eddies, called the Kelvin-Helmholtz instability [**6**]. This occurs as a result of imbalance of pressure across the sheet when small wavy perturbations are imposed ([**2**], II 6; [**4**], §94).

Starting vortex is another well-known example, which is shed from a sharp trailing edge of an aerofoil driven impulsively ([**4**], §93). After a short time, a vortex is left behind, and the aerofoil aquires a circulation round itself which is equal and opposite to that of the departed vortex.

3. MECHANICAL ASPECT

Suppose that a vorticity field is given by $\boldsymbol{\omega}(\boldsymbol{x}, t)$ at a spatial point $\boldsymbol{x}$ at a time t. The vorticity field is characterized by a hydrodynamic impulse when the fluid density ρ is assumed constant. The impulse $\boldsymbol{P}$ is defined by

$$\boldsymbol{P} = A_n \rho \int \boldsymbol{x} \times \boldsymbol{\omega}(\boldsymbol{x}, t)\, \mathrm{d}^n \boldsymbol{x}, \tag{1}$$

where $A_n = 1$ or $1/2$ according as the $\boldsymbol{\omega}$ distribution is 2D ($n = 2$) or 3D ($n = 3$), respectively. The impulse $\boldsymbol{P}$ is interpreted by the impulsive force $\boldsymbol{f}(\boldsymbol{x})\,\delta(t - t_0)$ necessary to generate instantaneously the motion from rest at $t = t_0$. It can be shown that $\boldsymbol{P} = \int \boldsymbol{f}(\boldsymbol{x})\, \mathrm{d}^n \boldsymbol{x}$ [**6**]. The impulse represents an effective total momentum, because $\boldsymbol{P}$ satisfies the following equation:

$$\frac{\mathrm{d}}{\mathrm{d}t} \boldsymbol{P} = \int \boldsymbol{F}_r\, \mathrm{d}^n \boldsymbol{x},$$

where $\boldsymbol{F}_r$ denotes the non-conservative body force, i.e. $\text{rot}\boldsymbol{F}_r \neq 0$ (viscoisty has no effect on the rate of change of $\boldsymbol{P}$ in unbounded fluid flows [**6**]). Hence, if the applied force has a potential (then $\boldsymbol{F}_r = 0$), the $\boldsymbol{P}$ is invariant.

Figure 1 is taken from the book [**5**] (Plate 22, Fig.55). This is a most impressive photograph, illustrating the impulse of a vortex system. The vortices are shed by an aerofoil that has been started impulsively from rest to a steady motion, and stopped suddenly shortly after. During the motion, the aerofoil gained a lift, *i.e.* an upward momentum. As a reaction, the fluid acquired a downward momentum, which is represented by the vortex pattern called a vortex pair, regarded as an object carrying some amount of downward momentum. Assuming that the flow field of Fig.1 is composed of two 2D vortices, and denoting the strength of the right vortex with $+\Gamma$ and the left by $-\Gamma$ and their separation distance with d, the magnitude of impulse of the two-vortex system is given by $P = \rho\Gamma d$ from Eq.(1), which is directed downward.

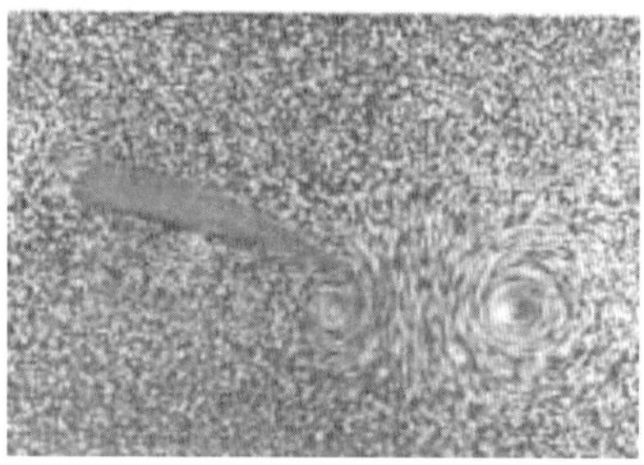

Figure 1. A vortex pair shed by an aerofoil [**5**].

Figure 2 shows a vortex ring generated by a shock wave (a circular arc on the right) emerging from a nozzle (a vertical dark shade on the left end) of circular cross-section. The generation mechanism is attributed to the impulse of the shock wave coming out impulsively to the open space. Denoting the vortex strength with Γ and its ring radius with r, and assuming axisymmetry, magnitude of the impulse of a circular vortex ring defined with Eq.(1) is given by $P = \pi\rho\Gamma r^2$, which is directed toward right.

A solid body shedding a set of vorticities from its surface is subject to a reaction force. If the reaction force is perpendicular to the direction of motion of the body, it is felt as a lift (or lateral) force. If the reaction is in the opposing direction of body's motion, then it is a drag. Drag of a bluff body can be interpreted partly by this sort of vortex drag [**8**]. On the other hand, if the reaction is in the same direction as the body's motion, then it is felt as a thrust. Animals are wise enoug to take advantage of the reaction forces variously.[1] A pair of tip vortices leaving a finite wing in steady rectilinear motion can be interpreted in this context ([**5**]: Chap.VI, C).

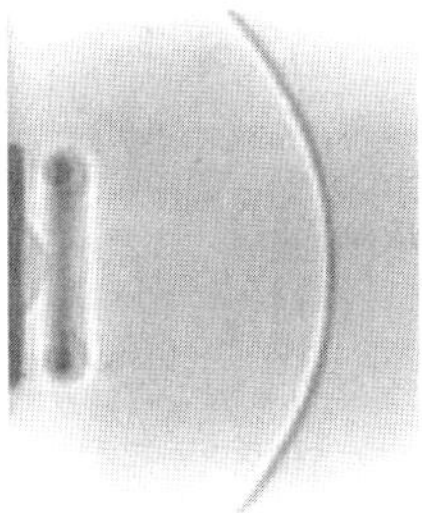

Figure 2. A vortex ring generated by a shock impulse [**7**].

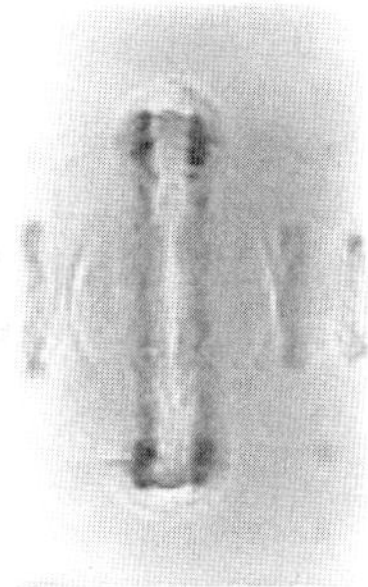

Figure 3. Shadowgraph of head-on collision of two vortex rings [**15**].

4. DYNAMICAL ASPECT

4.1 Creation of vorticity

Euler's equation of motion for a compressible ideal fluid moving under an external force $\boldsymbol{F}$ per unit mass is

$$\partial_t \boldsymbol{v} + \boldsymbol{\omega} \times \boldsymbol{v} + \nabla(\tfrac{1}{2}\, v^2) = \rho^{-1}\, \nabla p + \boldsymbol{F}, \tag{2}$$

where p is the pressure (the density ρ is not necessarily constant). Taking rot of (2), we obtain an evolution equation for $\boldsymbol{\omega} = \operatorname{rot} \boldsymbol{v}$ (see *e.g.* [**6**]):

$$\partial_t \boldsymbol{\omega} + \operatorname{rot}(\boldsymbol{\omega} \times \boldsymbol{v}) = \rho^{-2} \nabla \rho \times \nabla p + \operatorname{rot} \boldsymbol{F}_r, \tag{3}$$

where $\boldsymbol{F}_r$ is the non-conservative part of $\boldsymbol{F}$. If the fluid is *barotropic*, *i.e.* $p = p(\rho)$, then $\nabla \rho \times \nabla p = p'(\rho)\, \nabla \rho \times \nabla \rho = 0$. In addition if the force is conservative, $\boldsymbol{F}_r = 0$. Then, the right hand side vanishes and we obtain

$$\partial_t \boldsymbol{\omega} + \operatorname{rot}(\boldsymbol{\omega} \times \boldsymbol{v}) = 0. \tag{4}$$

Based on this equation, Helmholtz's three laws of vortex motion are derived [**4, 6**]: (i) Persistence of irrotationality, (ii) material line remains a vortex

[1] *e.g.* Taylor GK, Nudds RL & Thomas ALR. "Flying and swimming animals cruise at a Strouhal number tuned for high power efficiency", *Nature* **425** (16 Oct 2003), 707-711.

line, and (iii) strength of an infinitely thin vortex tube is invariant when the vortex moves.[2] Furthermore, the Kelvin's circulation theorem is also derived by using (2) with $p = p(\rho)$ and $\boldsymbol{F} = \mathrm{grad}\,\Psi$ (conservative body force with a potential Ψ). The equation (4) describes essentially that the vorticity $\boldsymbol{\omega}$ is *frozen* to the fluid flow of velocity $\boldsymbol{v}$, and that no vorticity is created in the evolution governed by (4).

Putting it the other way, the right hand side of Eq.(3) states that the vorticity is created in two ways. The pressure p in general depends on two thermodynamic variables $p = p(\rho, s)$ (say), where s is the entropy per unit mass, and $\nabla\rho\times\nabla p \neq 0$. Then, the vorticity is created by the first term. This is called the *baroclinic* effect. Bjerkness [**9**] gave a geometrical interpretation of the creation of circulation by this term [**4**, §85]. In addition, if the force is non-conservative (then $\boldsymbol{F}_r \neq 0$), the second term also can generate vorticity.

A body moving relative to fluid can be replaced kinematically by a distribution of image vorticity within the body. In steady motion, the distribution of vorticity is fixed relative to the body and is referred to as *bound* vorticity. It does not in general satisfy the Helmholtz laws. Vorticity satisfying the Helmholtz laws is referred to as *free* vorticity. A typical example of the bound vortex is the Prandtl's lifting vortex of strength Γ in a stream of velocity U and density ρ. The lift L acting on the bound vortex per unit length is given by $L = \rho U\Gamma$ (Kutta (1902), Joukowsky (1906)) [**5**, §98; **6** §3.1]. Given the circulation Γ, the lift is independent of the shape of the body. A lifting vortex is not a physical reality, but a very useful concept for the theory of lift.

4.2 Exciting acoustic waves

An acoustic wave is generated by a localized rotational flow $\boldsymbol{v}$ (with a localized vorticity $\boldsymbol{\omega}$). At a low Mach number, the sound source is identified with a term of the form $\rho_0\,\mathrm{div}(\boldsymbol{\omega}\times\boldsymbol{v})$, by Powell [**10**] and Howe [**11**], where ρ_0 is the undisturbed fluid density. Thus, the wave equation for the acoustic pressure p' is written approximately as

$$c^{-2}\,\partial_t^2 p' - \nabla^2 p' = \rho_0\,\mathrm{div}(\boldsymbol{\omega}\times\boldsymbol{v}), \tag{5}$$

in the limit of $M(= U/c) \to 0$, where c is the sound speed and U a representative flow velocity. This is obtained in the following way. From the fundamental conservation equations of mass, momentum and energy for flow of an ideal fluid of uniform entropy, one can derive [**14**, Appendix A]

$$\left(c^{-2}\,\partial_t^2 - \nabla^2\right)\left(h + \tfrac{1}{2}\boldsymbol{v}^2\right) = \mathrm{div}(\boldsymbol{\omega}\times\boldsymbol{v}), \quad = S(\boldsymbol{x}, t), \tag{6}$$

where h is the entalpy per unit mass. It is assumed that the source flow $\boldsymbol{v}(\boldsymbol{x}, t)$ is locallized in space and its representative Mach number M is sufficiently low. Then, the wave equation (6) can be transformed to an integral form,

[2] Helmholtz originally assumed $\mathrm{div}\,\boldsymbol{v} = 0$.

$$h(\boldsymbol{x},t) + \tfrac{1}{2}\boldsymbol{v}^2(\boldsymbol{x},t) = (1/4\pi)\int S(\boldsymbol{y},t_r)\,\mathrm{d}^3\boldsymbol{y}.$$

where the wave is observed at $\boldsymbol{x}$ and the source is located at $\boldsymbol{y}$, and $t_r = t - |\boldsymbol{x}-\boldsymbol{y}|/c$ is the retarded time. Suppose that the observation is made in such a far field as $|\boldsymbol{x}| \to \infty$ and $h + \frac{1}{2}\boldsymbol{v}^2 \to p'/\rho_0$ since $|\boldsymbol{v}|$ is $O(|\boldsymbol{x}|^{-3})$ and $h' = p'/\rho_0 + Ts' = p'/\rho_0$ (prime denotes acoustic fluctuation). Thus we obtain an approximate representation,

$$p'(\boldsymbol{x},t) = (\rho_0/4\pi)\int S(\boldsymbol{y},t_r)\,\mathrm{d}^3\boldsymbol{y}. \tag{7}$$

Thus, it is found that $p'(\boldsymbol{x},t)$ satisfies approximately the wave equation (5), when $\boldsymbol{x}$ is far from a *compact* source at $\boldsymbol{y}$. If $S(\boldsymbol{y},t) = \mathrm{div}(\boldsymbol{\omega}\times\boldsymbol{v})$ is evaluated with the incompressible vortex motion, then the error would be $O(M^2)$.

Based on (5), Möhring [**12**] succeeded in representing the acoustic pressure p' in terms of the vorticity $\boldsymbol{\omega}$ only, and gave a mathematical basis for the term, *vortex sound*. Much earlier, Obermeier [**13**] found a formula of an acoustic wave emitted by a spinning pair of two 2D vortices.

An acoustic wave radiated by head-on collision of two vortex rings was detected experimentally by Kambe and Minota [**14**], using a pair of vortex rings generated as in Fig.2. Figure 3 shows a shadowgraph [**15**] at the time of head-on collision of two vortex rings whose velocity were much larger than the acoustic experiment [**14**]. Two vertical dark columns are the colliding vortices, and short arcs bridging them (bright-and-dark double layers) are shocklets. Visible wave patterns are weak shocks.

4.3 Dissipative structure in turbulence

In homogeneous turbulence, average rate of dissipation ε of kinetic energy $\langle \boldsymbol{v}^2\rangle/2$ per unit mass is proportional to the average squared vorticity $\langle \boldsymbol{\omega}^2\rangle$,

$$\varepsilon = -\frac{\mathrm{d}}{\mathrm{d}t}\,\tfrac{1}{2}\,\langle \boldsymbol{v}^2\rangle = \nu\,\langle \boldsymbol{\omega}^2\rangle, \tag{8}$$

for an incompressible fluid, where ν is the kinematic viscosity, and $\langle\cdot\rangle$ denotes ensemble average.[3] This average equation can be derived from (2) by assuming $\mathrm{div}\,\boldsymbol{v} = 0$, taking scalar product with $\boldsymbol{v}$, and replacing $\boldsymbol{F}$ with the viscous force $\boldsymbol{f}_v$. Taking average over a large volume V, integrating by parts, and omitting integrated terms over bounding surface, we obtain (8). [**16**]

[3] In [**1**], Prandtl wrote that the viscous force is expressed as $\boldsymbol{f}_v = \nu\nabla^2\boldsymbol{v} = -\nu\,\mathrm{rot}\,\boldsymbol{\omega}$ if $\mathrm{div}\,\boldsymbol{v} = 0$, hence that the vorticity $\boldsymbol{\omega}$ gets involved with the viscous diffusion. But, it is noted only that $\boldsymbol{f}_v = 0$ for $\boldsymbol{\omega} = 0$, *i.e.* irrotational flow is possible for arbitrary viscosity.

In turbulence, there exists a certain straining mechanism by which vortex lines are stretched on the average. This is related to the *negative* value of the skewness S of longitudinal derivative $\partial u/\partial x$ [**16**, §8], where u is the velocity component along the x-axis, and the skewness S is defined by normalized statistical average of $(\partial u/\partial x)^3$. Non-zero value of S implies that the statistics is *non-Gaussian*, and that there exists structures in turbulence, which is often called the *intermittency* [**17**]. As a thin vortex tube is stretched, its vorticity increases, and dissipation is enhanced around it in accordance with (8).

According to the theory of energy cascade of fully developed turbulence, energy is dissipated at scales of smallest eddies of order $\eta = (\nu^3/\varepsilon)^{1/4}$, called the Kolmogorov's dissipation scale. In computer simulations, the dissipative structures in turbulence are visualized as fine-scale slender objects with high level of vorticity magnitude often called *worms*.

Figure 4 shows snapshot of vorticity field [**18**] obtained by a direct numerical simulation of incompressible turbulence in a periodic box with grid points 2048^3 carried out on the *Earth Simulator*, the largest parallel computer in operation.

(a) (b)

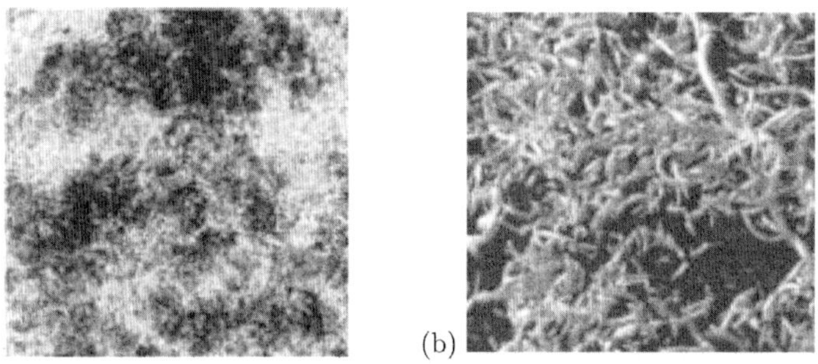

Figure 4. High-vorticity isosurfaces obtained by DNS of 2048^3 grid points with $\nu = 4.4 \times 10^{-5}$, $\eta = 1.05 \times 10^{-3}$, the Taylor-microscale Reynolds number $R_\lambda = 732$, and the integral scale $L = 1.23$. (a) Length of a side is $2992\,\eta$, and (b) 8 times enlargement of (a), *i.e.* length of a side is $374\,\eta$, the area being 1/64 of that of (a). The isosurfaces are defined by $|\boldsymbol{\omega}| = \langle \boldsymbol{\omega} \rangle + 4\sigma$ where σ is the standard deviation of the magnitude $|\boldsymbol{\omega}|$. [**18**].

5. GAUGE FIELD

Fluid mechanics is considered as a field theory of *mass flow* in Newtonian mechanics. In the theory of gauge fields, a guiding principle is that laws of physics should be expressed in a form that is independent of any particular coordinate system. The Lagrangian of fluid flow is defined in such a way as having an invariance under Galilei transformation. Next, a gauge principle is applied to the Lagrangian, requiring it to have *symmetry*, *i.e.* the gauge invariance. In regard to the fluid flows, relevant symmetry groups are *translation group* and *rotation group* [**19**].

According to the gauge principle, time derivative of velocity is given by the covariant derivative $\nabla_t \boldsymbol{v}$ in the following form:

$$\nabla_t \boldsymbol{v} = \partial_t \boldsymbol{v} + \mathrm{grad}(\boldsymbol{v}^2/2) + \Omega \boldsymbol{v}, \qquad (\Omega \boldsymbol{v})_i = \Omega_{ij} v_j \tag{9}$$

where Ω is a linear operator called a *gauge field*, assumed to vanish in irrotational flows.

For *irrotational* flows in which $\boldsymbol{v} = \boldsymbol{v}_{\mathrm{p}} = \mathrm{grad}\,\phi$ (ϕ: a velocity potential), first two terms represent a time derivative which is invariant under *local translational* gauge transformation, and the expression (9) reduces to

$$\nabla_t \boldsymbol{v}_{\mathrm{p}} = \partial_t \boldsymbol{v}_{\mathrm{p}} + \mathrm{grad}(\boldsymbol{v}_{\mathrm{p}}^2/2) = \partial_t \boldsymbol{v}_{\mathrm{p}} + (\boldsymbol{v}_{\mathrm{p}} \cdot \nabla)\boldsymbol{v}_{\mathrm{p}}. \tag{10}$$

since $\frac{1}{2}\,\partial_k(\boldsymbol{v}_{\mathrm{p}})^2 = (\partial_i\phi)\partial_k(\partial_i\phi) = (\partial_i\phi)\partial_i(\partial_k\phi) = (\boldsymbol{v}_{\mathrm{p}} \cdot \nabla)(\boldsymbol{v}_{\mathrm{p}})_k$. There exist some liquids, in which composing particles are equivalent and indistinguishable. Local rotation of such a fluid may not be captured because local rotaion (if any) should make no difference. Therefore the flow should be inevitably *irrotational.* It is known that super-fluid flows such as He^4 (a boson) or a Bose-Einstein condensate are irrotational.

This is not the case when we consider motions of a fluid composed of distinguishable particles such as an ordinary fluid. Local rotation is distinguishable and flows are *rotational* in general. Then, it is required that the equation of fluid flows should be invariant under *rotational* gauge transformation, and that $\nabla_t \boldsymbol{v}$ is invariant with respect to Galilei transformation. From this requirement, we find [**20**] that the gauge term must be of the form $\Omega_{ij} = -\epsilon_{ijk}\omega_k$, where ω_k is the k-th component of the vorticity $\boldsymbol{\omega}$. Hence, we obtain

$$(\Omega \boldsymbol{v})_i = -\epsilon_{ijk}\omega_k v_j = (\boldsymbol{\omega} \times \boldsymbol{v})_i.$$

Thus, the vorticity $\boldsymbol{\omega}$ is found to be the gauge field, and the covariant derivative (9) is given by the material derivative, *i.e.* the *Lagrange derivative*,

$$\nabla_t \boldsymbol{v} = \partial_t \boldsymbol{v} + \mathrm{grad}(\boldsymbol{v}^2/2) + \boldsymbol{\omega} \times \boldsymbol{v} = \partial_t \boldsymbol{v} + (\boldsymbol{v} \cdot \nabla)\boldsymbol{v}, \tag{11}$$

by using a well-known vector identity.

This *gauge-theoretic* formulation provides a theoretical ground to the physical analogy between the aeroacoustic interaction associated with vortices and the interaction of electron and electromagnetic-field. In the latter problem, the electromagnetic field is the gauge field.

REFERENCES

1. Ludwig Prandtl. "Über Flüssigkeitbewegung bei sehr kleiner Reibung", *Verhandlungen des III. Internationalen Mathematiker-Kongress* (Heidelberg 1904), pp. 484-491, 1905.
2. Prandtl L. *Abriss der Strömungslehre*, Friedr.Vieweg & Sohn, Braunschweig, 1931.
3. Prandtl L. *Führer durch die Strömungslehre*, Friedr.Vieweg & Sohn, Braunschweig, 1965.
4. Prandtl L and Tietjens OG. *Fundamentals of Hydro- and Aeromechanics*, McGraw-Hill. (Translated from the German edition, Springer, 1931.), 1934.
5. Prandtl L and Tietjens OG. *Applied Hydro- and Aeromechanics*, McGraw-Hill. (Translated from the German edition, Springer, 1931.), 1934, [Dover].
6. Saffman PG. *Vortex Dynamics*, Cambridge Univ. Press, 1992.
7. Minota T. "Interaction of a shock wave with a high-speed vortex ring", *Fluid Dyn. Res.* **12**, pp. 335-342, 1993.
8. von Karman T. *Göttingen Nachrichten*, Göttingen, pp. 547-556, 1912.
9. Bjerkness V. "Lectures on hydrodynamic forces acting at a distance" (German), Leipzig, 1900-1902.
10. Powell A. "Theory of vortex sound", *J. Acoust. Soc. Am.* **36**, 177-195, 1964.
11. Howe MS. "Contributions to the theory of aerodynamic sound", *J. Fluid Mech.* **71**, pp. 625-973, 1975.
12. Möhring W. "On vortex sound at low Mach number", *J. Fluid Mech.* **85**, pp. 685-691, 1978.
13. Obermeier F. "Berechnung aerodynamisch erzeugter Shallfelder mittels der Methode der Matched Asymptotic Expansions", *Acustica* **18**, 238-240, 1967.
14. Kambe T and Minota T. "Acoustic wave radiated by head-on collision of two vortex rings", *Proc. R. Soc. Lond. A* **386**, pp. 277-308, 1983.
15. Minota T, Nishida M and Lee MG. "Head-on collision of two compressible vortex rings", *Fluid Dyn. Res.* **22**, pp. 43-60, 1998.
16. Batchelor GK. *The Theory of homogeneous turbulence*, Cambridge Univ. Press, 1953.
17. Kambe T. "Statistical laws governed by vortex structures in fully developed turbulence", in *IUTAM Symposium on Geometry and Statistics of Turbulence* (eds: Kambe, Nakano & Miyauchi, held in 1999), Kluwer Academic Pub., Dordrecht, pp. 117-126, 2001.
18. Yokokawa M, Itakura K, Uno A, Ishihara T, Kaneda Y. "16.4TFlops direct numerical simulation of turbulence by a Fourier spectral method on the Earth Simulator", *Proc. IEEE/ACM SC2002 Conf.*, Baltimore, 2002 (http://www.sc-2002.org/paperpdfs/pap.pap273.pdf);
    Kaneda Y, Ishihara T, Yokokawa M, Itakura K, Uno A. "Energy dissipation rate and energy spectrum in high resolution direct numerical simulations of turbulence in a periodic box", *Physics of Fluids*, **15** (2), pp. L21-L24, 2003.
19. Kambe T. *Geometrical Theory of Dynamical Systems and Fluid Flows*, Ch.7, World Scientific Publishing Co., 2004;
    Kambe T. "Gauge principle and variational formulation of ideal fluid flows", *Preprint* TK2004 (http://hw001.gate01.com/~kambe).
20. Kambe T. "Gauge principle and variational formulation for flows of an ideal fluid", *Acta Mechanica Sinica* **19**, pp. 437-452, 2003;
    Kambe T. "Gauge principle for flows of an ideal fluid", *Fluid Dyn. Res.* **32**, pp. 193-199, 2003.

AN EXPERIMENTAL INVESTIGATION OF THE BRINKMAN LAYER THICKNESS AT A FLUID-POROUS INTERFACE

Afshin Goharzadeh[1], Arash Saidi[1], Dianchang Wang[1], Wolfgang Merzkirch[2] and Arzhang Khalili[1, 3]
[1]*Max Planck Institute for Marine Microbiology*
Celsiusstr. 1, 28359 Bremen, Germany
[2]*Institute of Fluid Mechanics, University of Essen, D-45117 Essen, Germany*
[3]*International University Bremen, Campus Ring 1, 28725 Bremen, Germany*
E-mail: akhalili@mpi-bremen.de

Abstract: In the present study, we characterized the flow in the vicinity of a permeable interface using particle image velocimetry (PIV) and refractive index matching (RIM). The porous layer is exposed to a laminar flow in the overlying fluid layer, and the velocity field in the interfacial region is measured. The averaged velocities of fluid particles decrease continuously when moving from the fluid layer into the porous one. We measure the thickness of the transition layer, also known as Brinkman layer, defined as the depth below the porous interface within which the velocity decreases to a constant value. It was observed that the thickness of Brinkman layer is of order of the grain diameter, and hence, much larger than the square root of permeability predicted by previous theoretical studies. A new theoretical approach based on depth-dependent permeability and effective viscosity is performed, that agreed well with experimental results.

Key words: Brinkman layer, permeable interface, fluid-porous layer.

1. INTRODUCTION

The benthic boundary layer (BBL) refers to an interfacial thin region containing a tiny portion of the water column as well as the sediment at the bottom of lakes, seas and oceans. It plays a vital role in adjusting the biological, chemical and physical processes in the water layer and in the underlying seabed. Therefore, the physics of fluid flow at permeable interfaces has become one of the major key factors in understanding several practical problems in the field of environmental and engineering sciences. An example for this is the microbiology of seabeds where near-bed transport of solute and suspended particles need to be understood.

When the seabed is made of permeable sediments, the hydrodynamic effects might even be more pronounced than other involved processes (e.g., diffusion). The lower portion of the BBL that contains the sediment is referred to as Brinkman layer (BL) or transition layer, in which the velocity of fluid particles decreases drastically, until it reaches an averaged constant velocity that is predicted by Darcy equation. Generally, the mathematical

G.E.A. Meier and K.R. Sreenivasan (eds.), IUTAM Symposium on One Hundred Years of Boundary Layer Research, 445-454,

term expressing the BL is believed to act as a transition between the Navier-Stokes (or Stokes) equations and the Darcy equation. Hence, determination of BL thickness (δ) has been given due attention in the past decades, for the knowledge of δ and the corresponding hydrodynamic boundary conditions would enormously simplify interfacial transport modeling. The Brinkman equation **[1]** has initiated many subsequent studies **[2-6]** which aimed at clarifying the length scale of BL. While matching the Stokes flow in the fluid layer to the Darcy flow in the porous substrate, Neale and Nader **[3]** found that the transition layer thickness was of order of $\sqrt{k}$, with k denoting the bulk permeability of the porous medium. It should be noted that these workers took constant input parameters for the permeability and effective viscosity appearing in the Brinkman equation. However, a clear experimental evidence for this finding does not exist in the literature **[7]**.

Recently, Gupte and Advani **[8]** performed a local velocity measurement near a permeable interface in a Hele-Shaw cell using Laser Doppler Anemometry. Although this experiment gives the first hint for the existence of a larger transition layer thickness than $\sqrt{k}$, this result may not be considered as universal due to the limitations of the Hele-Shaw cell.

In the present study, these types of limitations were avoided by using a general setup with sufficient dimensions allowing the generation of clear interfacial flows and their measurements at the pore scale.

The experimental setup consists of a rectangular horizontal open channel filled with a saturated porous layer and its overlying fluid layer. Visualizations and measurements have been achieved by combining refractive index matching (RIM) with particle image velocimetry (PIV) to obtain global velocity profiles, which enables us to estimate the Brinkman Layer thickness. These issues are the content of section 2. Further, in section 3, a theoretical model based on depth-dependent permeability and viscosity is presented.

2. EXPERIMENTAL ANALYSIS

2.1 The experimental setup

The experiments were conducted in a rectangular horizontal open channel of size 50 x 10 x 5 cm with side-walls made of transparent Duran glass (see Fig. 1). The central region of the extent 34 x 4 x 5 cm is filled by a random packing of mono-disperse glass beads as the porous matrix. In order to study the influence of permeability, two different sizes of transparent glass beads were used. The physical properties of porous samples are reported in Table 1. The porosity, ε, and permeability, k, were measured separately by laboratory experiments. Porosity measurements were carried out many times using high-precision weighting apparatus. For measurement of permeabilities of different samples, similar method as described by Givler and Altobelli **[9]** was used. The only difference was replacement of water by oil due to refractive index matching.

Table 1. Specifications and physical features of glass beads for different experiments.

Samples	Diameter (cm)	k (10^{-8} m^2)	ε
A	0.47	1.1 ± 0.031	0.41 ± 0.01
B	0.65	1.3 ± 0.069	0.41 ± 0.01

The porous material is embedded between two 5 x 5 cm filters to fix their position. The upper surface of the porous bed is flattened gently with a solid stamp to ensure a horizontal interface. The channel is then filled with the silicon oils mixed at the ratio reported in the next section. The height of the porous layer is kept constant in all experiments ($Hp = 4\ cm$). Using a pump with a maximum flow capacity of $5\ l/min$, the pressure gradient in the longitudinal direction was maintained uniformly. With this arrangement, in the central region a laminar flow over the porous bed was produced, that was characterized by a constant Reynolds number

$$Re = \frac{u_{max} H_f}{v} = 23 \tag{1}$$

with $u_{max} = 3.3\ cm/s$ denoting the maximum flow velocity at the fluid surface in the x-direction, $H_f = 3\ cm$ being the height of the fluid layer over the porous matrix, and v the kinematic viscosity of the fluid. The kinematic viscosity, v, was measured by an Ubbelohde capillary viscometer (Schott 52520/11) for the silicon oil mixture to $v = 42.5\ 10^{-6}\ m^2/s$. The fluid was maintained at a constant temperature $T = 23°C$ using a recirculating thermal device (TD).

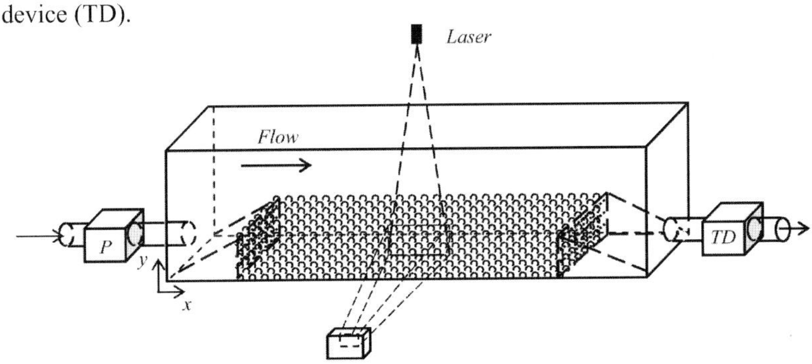

Figure 1. Experimental set-up. The pump (P) generates a recirculating flow through and above the porous layer. The temperature of fluid is kept constant using a thermal device (TD).

2.2 Visualization techniques

To access optically the interior of the channel, transparent glass beads were taken instead of the real sediments and water was replaced by a mixture of two silicon oils (Dow Corning 550 and 556) in a volume fraction of (respectively 32.2 and 67.8%) at $T = 23°C$. By doing so, the refractive indices in both layers were matched ($n \approx 1.47$), and the entire interfacial region could be visualized using particle image velocimetry (PIV). Further to use PIV, polyamide tracers (with a diameter of 22 μm), a diode laser and a CCD camera were implemented as illustrated in Fig. 1.

The entire interfacial region of the channel was scanned using a mobile mechanical. Full-frame images of 512 x 512 pixels were acquired and transferred to a computer via a frame grabber.

Each image was focused on the interface region covering portions of the fluid layer and the porous layer 25 cm downstream from the inlet of the flow.

2.3 Experimental results

Figure 2-a shows an example of a PIV image, which displays the flow near the interface at pore-scale. To obtain quantitative information of the flow, velocity vector field in the (x, y)-plane was measured (see Fig. 2b). As can be seen, the velocities in the fluid layer u_f are much higher than those inside the porous medium u_m. Consequently, in a single vector plot covering the entire interface region, the velocity vectors in the porous region would not be visible. Hence, the image (Fig. 2b) has been split into two parts in which velocity fields are presented with two different magnification factors.

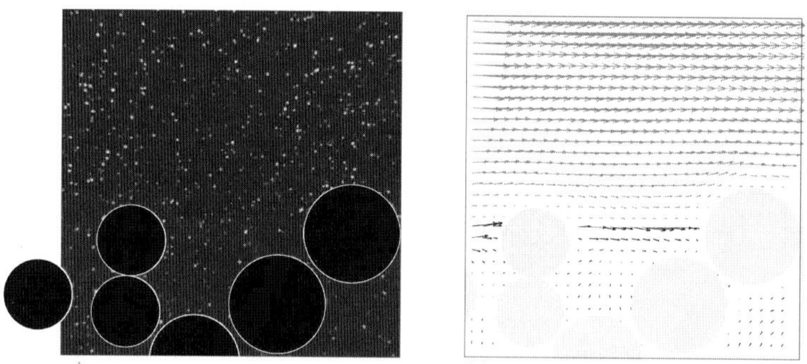

a) *b)*

Figure 2. a) Example of PIV image in (x-y) plane and b) the corresponding velocity vector plot. The field of view is 2.2 x 2.2 cm and the diameter of glass beads is 0.65 cm.

To study the velocity profile at the fluid-porous interface, in each image the horizontal velocity component is calculated by averaging the measured velocity over the fluid volume that contains both fluid and solid volume. The average is obtained using 10 neighboring sections with a small field of view (2.2 x 2.2 cm) along the x-coordinate separated by a constant distance of 1 cm. The horizontal velocity profile is plotted in figure 3. The location of the permeable interface ($y = 0$) was defined by the ordinate of the uppermost solid (glass beads) among all images. As the fluid particles approach the interface, their velocities decrease. In deeper layers inside the porous medium, the magnitude of the horizontal velocities drops drastically and is two orders of magnitude less than the velocity in the fluid region.

As can be seen from Fig. 3, the averaged velocities are continuous and decrease drastically within a layer of small extent. This layer, below the interface, represents a transition zone, and is often referred to as Brinkman layer (BL). The approximate vertical extent of the transition layer may now be obtained from these experiments, and was measured to $\delta_A \approx -0.4\ cm$ for sample A and $\delta_B \approx -0.6\ cm$ for sample B. Below these depths, the horizontal velocity components reach averaged constant value of $u_A = 25.10^{-4}\ cm/s$ and $u_B = 66.10^{-4}\ cm/s$, respectively, and correspond to the well-known Darcy velocity, u_D.

From our experimental results we can conclude that the thickness of Brinkman layer ($\delta_A \approx -0.4\ cm$ and $\delta_B \approx -0.6\ cm$) is of order of the medium grain size ($d_A = 0.47\ cm$ and $d_B = 0.65\ cm$), and hence, much larger than the square root of permeabilities ($\sqrt{k_A} = 0.01047\ cm$ and $\sqrt{k_B} = 0.0116\ cm$, respectively).

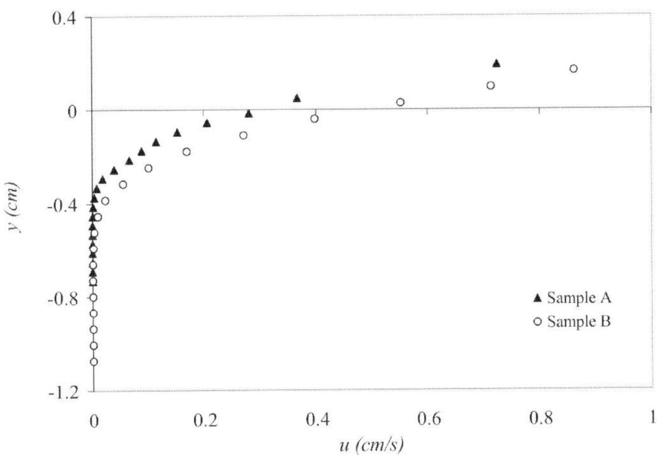

Figure 3. Vertical profile of horizontal velocity component for two different samples.

3. THEORETICAL ANALYSIS

The estimation of the thickness of the transition layer goes back to a study by Neale and Nader [3], who considered a unidirectional flow in a porous layer with its overlying fluid layer attached to solid disk from top (similar to the geometry of Beavers and Joseph [10]). By defining a transition layer thickness, δ, satisfying

$$u = 1.01\ u_D \tag{2}$$

they solved the Brinkman equation

$$\mu_{eff}\frac{d^2u}{dy^2} - \frac{\mu}{k}u = \frac{dp}{dx}, \tag{3}$$

and found an analytical expression for δ given by

$$\delta = \sqrt{k}\left(\frac{\mu_{eff}}{\mu}\right)^{1/2} ln\frac{50(h^2/k-2)}{1+\frac{h}{\sqrt{k}}\left(\frac{\mu_{eff}}{\mu}\right)^{1/2}}, \tag{4}$$

which states that δ is of order of $\sqrt{k}$. The quantities μ_{eff}, k, h and p appearing in the above equations denote, respectively, the effective viscosity, permeability, height of the fluid layer and pressure.

It should be noted that both μ_{eff} and k were kept constant in obtaining the above result.

However, as the experiments have shown, the transition layer thickness was of order of the grain diameter, and with this, much larger than $\sqrt{k}$. To address this discrepancy in more detail, we performed digital processing analysis of the PIV-images shown in Fig. 2a, and found that porosity varied with depth from *unity* in the fluid layer to its constant bulk value below a certain depth in the porous layer. Due to the fact that porosity and permeability are related to each other, it is obvious that permeability, too, must be taken to vary with depth.

It should be noted that the concept of variable physical properties has already been suggested by many researchers, and were applied to different problems. For example, Vafai [11] suggested a variable porosity model, whereas Sangani and Behl [12] varied porosity and permeability, simultaneously.

Hence, in our analysis, both the permeability and the effective viscosity are depth-dependent. As the transition layer is bounded from above by a fluid layer with an infinite permeability, and from below by a homogeneous porous medium with bulk permeability k_0, the depth-dependent permeability has to satisfy

$$\begin{aligned} &\lim_{y \to 0} k(y) = +\infty \\ &k(\delta) = k_0 \end{aligned} \tag{5}$$

Using this constraint, the following polynomial depth-dependent function was found

$$k(y) = \begin{cases} \dfrac{k_0 y^n}{2\delta^n} + \dfrac{k_0 \delta^n}{2y^n} & 0 < y < \delta \\ k_0 & \delta \le y < \infty \end{cases}, \qquad n = 4 \tag{6}$$

for the permeability change within the transition layer. Similarly, using the results of Givler and Altobelli **[9]**, for the effective viscosity within the transition layer the conditions

$$\mu_{eff}(y) = \begin{cases} \mu(1 + 6.5\dfrac{y}{\delta}) & 0 < y < \delta \\ 7.5\mu & \delta \le y < \infty \end{cases} \tag{7}$$

are implemented. Inserting Eqs. (6) and (7) into Eq. (3), one obtains a modified equation for the fluid flow in the transition layer, which, with its appropriate boundary conditions can be written as

$$\begin{aligned} &\mu_{eff}(y)\frac{d^2u}{dy^2} - \frac{\mu}{k(y)}u = \frac{dp}{dx} \\ &u(0) = u_i \\ &\left.\frac{du}{dy}\right|_{y=\delta} = 0 \end{aligned} \tag{8}$$

The above equation has been solved numerically. In the solution process, first a δ is assumed, then, the velocity profile is calculated. In the next step δ is corrected based on the velocity profile calculated from Eq. (8) using definition (2). The procedure is terminated once the convergence has been achieved. This iterative procedure converged in all cases considered independent from the choice of the initial guess for δ.

The comparison of velocity profiles using variable permeability and viscosity versus fixed ones are shown in Figs. 4 and 5. As can be depicted by the figures, the present model considerably improves the velocity prediction. To validate the numerical results, experiments with different glass beads (Sample A and B) as well as different fluid heights over the porous bed ($H_f = 3$ *cm and* $H_f = 4$ *cm*) were conducted. For all the experiments, a better match between the theoretical and experimental results could be achieved when departing from depth-dependent material properties.

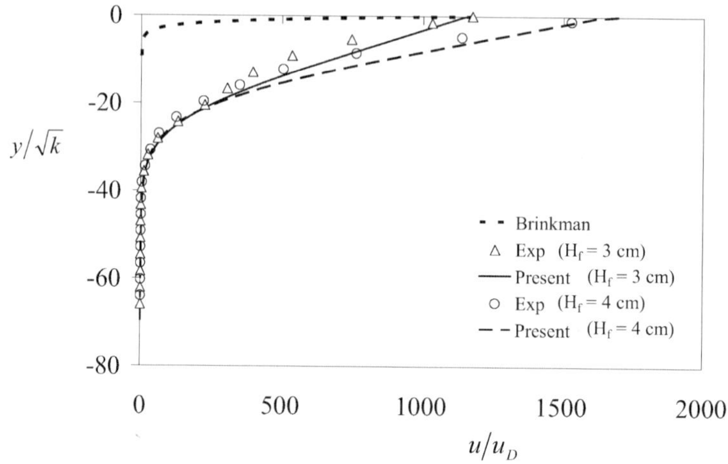

Figure 4. Vertical profile of horizontal velocity zoomed inside the porous region (Sample A, d = 0.47 cm).

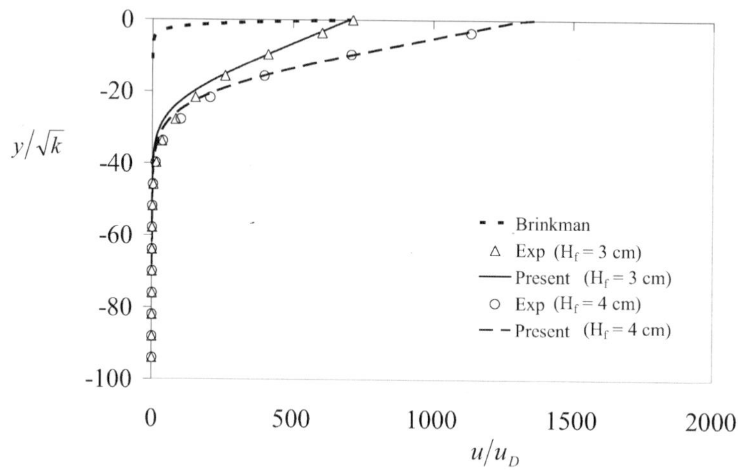

Figure 5. Vertical profile of horizontal velocity zoomed inside the porous region (Sample B, d = 0.65 cm).

A comparison of δ is made for different glass beads size (Table 2), considering the Brinkman equation with constant input parameters, Brinkman equation with depth-dependent input parameters (present model), and the experimental data. The comparison of δ depicted in the table, demonstrates that the current non-homogeneous model with depth-dependent permeability and effective viscosity predicts the thickness of the transition layer much more accurately.

Table 2. Comparison of δ between the standard Brinkman equation, the present model and the experimental data for different size of glass beads.

Method	Sample A	Sample B
Brinkman (const. k, μ_{eff})	$11.5\sqrt{k_0}$	$11.0\sqrt{k_0}$
Brinkman with variable k, μ_{eff} (Present model)	$57.4\sqrt{k_0}$	$54.8\sqrt{k_0}$
Experiment	$41.7\sqrt{k_0}$	$57.8\sqrt{k_0}$

4. CONCLUSIONS

Using the refractive index matching (RIM) technique and particle image velocimetry (PIV) we investigated a 2D flow field at the interface between a porous medium and its overlying fluid layer. With this experimental method, fine-scale velocity measurement covering small portions of fluid as well the porous layer were measured and averaged to obtain a single interfacial horizontal velocity profile for different porous samples.

From the averaged velocity profile, it was observed that the horizontal velocity profile decreases continuously when moving downward from fluid into the porous layer. The experimental data indicate clearly the existence of a transition layer, which is characterized by drastic decrease of velocity as a matching zone between the pure fluid and the Darcy region. The length scale of the transition layer was found to be of the order of grain diameter and much larger than the square root of the permeability.

When allowing the permeability and the effective viscosity to vary with depth, unlike in the work of Neale and Nader, the velocity field obtained by solving the Brinkman equation match well with those from experiments. The same is true for the Brinkman layer thickness.

REFERENCES

1. Brinkman HC. "A calculation of the viscous force exerted by a flowing fluid on a dense swarm of particles", *App. Sci. Res.,* **vol. 1**, pp. 27-34, 1947.
2. Lundgren TS. "Slow flow through stationary random beds and suspensions of spheres", *J. Fluid Mech.,* **vol. 51**, pp. 273-299, 1972.
3. Neale G, Nader W. "Practical significance of Brinkman's extension of Darcy's law: Coupled parallel flows within a channel and a bounding porous medium", *Canad. J. Chem. Eng.,* **vol. 52**, pp. 475-478, 1974.
4. Levy T, Sanchez-Palencia E. "On boundary conditions for fluid flow in porous media", *Int. J. Eng. Sci.,* **vol. 13**, pp. 923-940, 1975.
5. Vafai K, Thiyagaraja R. "Analysis of the flow and heat transfer at the interface region of a porous medium", *Int. J. Heat Mass Transfer,* **vol. 30**, pp. 1391-1405, 1987.
6. Goyeau B, Lhuillier D, Gobin D, Velarde MG. "Momentum transport at a fluid-porous interface", *Int. J. Heat Mass Transfer,* **vol. 46**, pp. 4071-4081, 2003.
7. Kaviany M. *Principles of heat transfer in porous media*, Springer-Verlag, New York, 1991.
8. Gupte SK, Advani SG. "Flow near the permeable boundary of porous medium: An experimental investigation using LDA", *Exp. Fluids,* **vol. 408**, pp. 408-422, 1997.
9. Givler RC, Altobelli SA. "A determination of the effective viscosity for the Brinkman-Forchheimer flow model", *J. Fluid Mech.*, **vol. 258**, pp. 355-370, 1994.
10. Beavers GS, Joseph DD. "Boundary conditions at a naturally permeable wall", *J. Fluid Mech.*, **vol. 30**, pp. 197-207, 1967.
11. Vafai K. "Convective flow and heat transfer in variable-porosity media", *J. Fluid Mech.* , **vol. 147**, pp. 233-259, 1984.
12. Sangani AS, Behl S. "The Planar singular solutions of Stokes and Laplace equations and their application to transport processes near porous surfaces" *Phys. Fluids*, **vol. A1**, pp. 21-37, 1989.

EXPERIMENTAL INVESTIGATIONS OF SEPARATING BOUNDARY-LAYER FLOW FROM CIRCULAR CYLINDER AT REYNOLDS NUMBERS FROM 10^5 UP TO 10^7

Three-dimensional vortex flow of a circular cylinder

Burkhard Gölling

DLR – German Aerospace Center, Institute of Aerodynamics and Flow Technology, Bunsenstraße 10, D-37073 Göttingen. E-Mail burkhard@goelling.de

Abstract: This paper gives a draft overview about the staggered processes and phenomena in boundary-layer flow separation and subsequent vortex-flow formation of a circular cylinder in cross-flow at high Reynolds numbers. In order to detect the spatial, i.e. the span-wise evolution of coherent separation structures, wall-pressure measurements and oil-flow visualizations are conducted. The unsteady separation is recognized by force measurements with a piezo-balance. The result is a description of spatial and temporal scales of shear-flow instabilities and their effects on drag and lift. For identifying the vortex-flow instabilities, an active flow control technique is applied.

Key words: Laminar-turbulent transition, circular cylinder, span-wise and stream-wise vortex flow, coherent vortex structures, macroscopic scales, flow control.

1. INTRODUCTION

The principle of boundary-layer flow separation was investigated by Ludwig Prandtl in the early part of the 20th century. He gave a first explanation, that "*turbulent flow can better work against a flow with increasing pressure gradient than a laminar flow*" which causes changes, especially, in drag **[1]**.

In this paper we present results of transitional and turbulent separated boundary-layers flows at high Reynolds numbers starting at $Re_D = 10^5$ up

G.E.A. Meier and K.R. Sreenivasan (eds.), IUTAM Symposium on One Hundred Years of Boundary Layer Research, 455-462,

to $Re_D = 10^7$. One objective is to recognize the spatial and temporal development of vortex structures in the separated shear flow in order to characterize flow topology. Another goal is to identify instabilities in the staggered transition process by application of a flow control technique.

The experimental investigations are conducted in the High-pressure wind tunnel (HDG) of German-Dutch-Windtunnel Association (DNW) located at DLR-site in Göttingen. Several measurement techniques were instrumented: unsteady force measurement by a piezo-electric balance, surface oil-flow visualization detecting the wall shear-stress, qualitatively, and unsteady wall-pressure measurements by piezo-resistive sensors. The acoustic flow control is effected by loud- speakers connected with a number of pressure chambers along the span beneath the surface of the cylinder. The cylinder model has a diameter of D = 0.06 m, length-to-diameter ratio of L/D = 10, smooth surface with an averaged roughness of $k_Z/D < 10\ \mu$.

2. LAMINAR-TURBULENT TRANSITION

The laminar-turbulent transition of the boundary-layer flow of a bluff body, especially of a circular cylinder cross-flow, is characterized by discontinuous jumps in drag and lift as well as temporal and spatial evolution of different vortex structures (see *Fig. 1*).

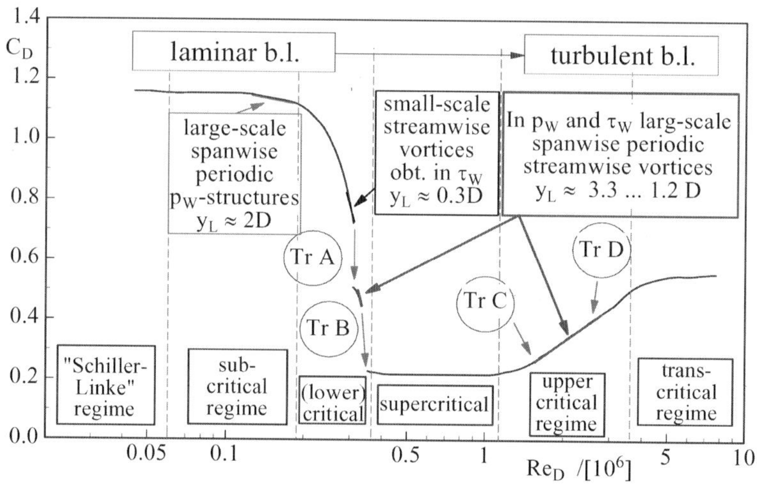

Figure 1. Drag vs. Reynolds number. Drag behavior indicates the staggered process for laminar to turbulent transition of the separating boundary-layer flow on circular cylinder.

In the ***subcritical regime*** the boundary-layer flow separates laminar and the shear flow becomes turbulent further downstream in the near wake. The laminar separation is characterized by a sharp separation line on both sides of the cylinder. In a wide Reynolds number range between $Re_D = 0.5 \cdot 10^5$ and $Re_D = 2.0 \cdot 10^5$, the mean drag is constant and the mean lift is vanished due to the stable symmetric separation of the mean flow.

By increasing the Reynolds number above $Re_D > 2.0 \cdot 10^5$, where the ***critical regime*** starts, the separating boundary-layer flow is unstable for span-wise periodic disturbances. A natural occurring span-wise length is $y_D = 2$ D, approximately, and it was recovered that a laminar separation bubble is obtained on both sides of the cylinder. How ever, the mean lift is different from zero. The drag is decreasing, rapidly, until small-scale span-wise periodic patterns will be noticed. The span-wise wave-length observed by oil-flow visualization has a value of $y_D = 0.3$ D. The following transitions called "*Transition A*" and "*Transition B*" are well-known and were studied by Schewe **[2]**, intensively. A new obtained result is that also the asymmetric separating mean flow is unstable for span-wise periodic disturbances. A span-wise length is measured to $y_D = 3.3$ D by oil-flow visualization.

The following ***supercritical regime*** is characterized by symmetric mean flow separation. Mean lift disappears and drag reaches its minimum. Comparable with the topological changes in the end of the *subcritical regime*, this originated two-dimensional separating mean flow is unstable for span-wise periodic disturbances ("*Transition C*"). In the ***upper-critical regime*** span-wise periodic patterns of wall shear-stress and wall pressure are observed which caused by stream-wise separating vortices with a span-wise length of $y_D = 3.3$ D down to $y_D = 1.2$ D with increasing Reynolds number (see *Fig. 2*). The stream-wise oriented vortices disappear if the boundary-layer flow separates turbulent from the cylinder ("*Transition D*").

Figure 2. Wall shear-stress pattern by oil-flow visualization. View on the back of the cylinder shows four and two halves span-wise cells ($y_D = 1.2$) of separating stream-wise oriented vortex structures which obtained in the *upper-critical regime*.

New results were discovered in the beginning of the ***transcritical regime***. In a small Reynolds-number range while mean drag is increasing, we obtained turbulent separation bubbles. However, at larger Re_D, the *transcritical regime* is characterized by one turbulent separation line on each side of the cylinder. The drag keeps constant or increases a little.

All reorders of the separating flow are "natural" without change of state of kinetic energy of flow molecules or any external energy input, so that all separation flow patterns must be transformable to one uniform pattern where the topology of the flow can only change under respect defined rules (see Dallmann and Schewe 1987, **[3]**).

In order to identify the instability behavior of the separating flow, we perform an acoustic, active flow control technique (see also Huerre and Monkevitz 1990 **[4]**). By applying several combined spatial and temporal wall sound-pressure disturbances, for example, $y_E = 4.4$ D and $Sr_E > 1.5$, into the separating boundary-layer, a certain flow-instability could be excited which leads to the evolution of a changed vortex shear-flow. We recognize span-wise periodic wall-pressure patterns which are comparable with patterns in the *upper-critical regime* where span-wise periodic stream-wise oriented vortices separate from the cylinder as shown in *Fig. 2*. How ever, most span-wise periodic pattern will be obtained by using high-frequency excitation, that means, the excitation frequency is, at least, one order of magnitude larger than the frequency of natural occurring cross-flow vortices (see also Gölling 2001 **[5]** and Theofilis 2003 **[6]**).

3. MAKROSCOPIC MEASURES

Especially, separated flows are connected with the evolution of coherent vortices e.g. the well-known von-Karman-vortices. The unsteady, but temporal periodic development in the near wake of bluff body flows is well described by the Strouhal number $Sr = Sr_D$ (see *Fig. 3*) – a dimensionless number given by the product of the dominant vortex separation frequency f, a defined length scale, mostly the outer diameter of the bluff body D, divided by the free-stream velocity U_0. While frequency and free-stream velocity are flow measures, the length-scale diameter of the body is a fixed datum and gives no more explanation than there is body in the flow. Therefore, another length scale should be defined which could result in clarifying of assumptions about flow behavior and instabilities.

Instead of the diameter a characteristic length scale, called feedback length L_f, should taken which is given by arc length between the time averaged separation angles via the rear side of the bluff body. In case of the separated cylinder flow we obtain a sharp constant value for the new Strouhal number $Sr(L_f)$ given by $Sr(L_f) = 1.4$ in all Reynolds number regimes where the separation of von-Karman-vortices are dominant while the development of stream-wise oriented vortices play a suborder role, i.e. in the *supercritical regime.* At these Reynolds numbers regimes or transition Reynolds numbers, where stream-wise oriented Reynolds numbers are recognized the new Strouhal number $Sr(L_f)$ differs strongly from this exact value (see *Fig. 4*).

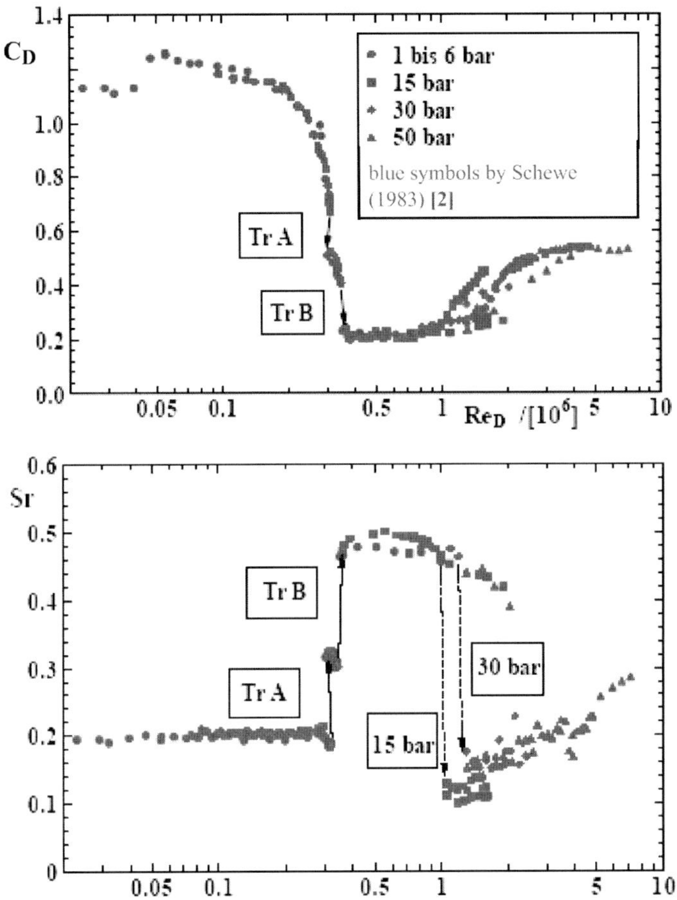

Figure 3. Experimental data of drag and (origin) Strouhal number in dependence of Reynolds numbers obtained by piezo-force measurements on a circular cylinder flow in the High Pressure Windtunnel Göttingen (DNW-HDG).

What is the physical reason why this length scale works in this manner. One explanation will be supported through a new experimental diagnostics of the separated cylinder flow by 3D-Laser Doppler Anemometry conducted by Leder and Brede 2001 **[7]**. They can shown that the periodic increasing and decreasing of $<u'^2>$ at that place where the flow separates from the cylinder surface and that place of shear-stress production $<w'^2>$ in the near wake region of the cylinder flow are time-shifted by a quart time-period. It is clearly demonstrated that in this time of a quart period the maximum of $<u'w'>$ is "walking" along this distance path, continuously, so that $<u'w'>$ is acting as a vorticity-"transporter" between the place of u'-fluctuation separation and the place of w'-fluctuation production, which can understand as the "birth place" of von-Karman-vortices.

Re_D	Sr	$\varphi_{A,oben}$ /[Grad]	$\varphi_{A,unten}$ /[Grad]	$\Delta\varphi_S$ /[Grad]	Sr_f
$1,0*10^3$	0,22	96	96	168	1,29
$5,0*10^3$	0,21	87	87	186	1,36
$1,0*10^4$	0,21	84	84	192	1,41
$5,0*10^4$	0,20	82	82	196	1,37

Re_D	Sr	$\varphi_{S,oben}$ /[Grad]	$\varphi_{S,unten}$ /[Grad]	$\Delta\varphi_S$ /[Grad]	Sr_f
$1,0*10^5$	0,20	77	77	206	1,42
$2,0*10^5$	0,21	(76+92)/2	(76+92)/2	192	1,41
$2,8*10^5$	0,22	(80+96)/2	(80+96)/2	184	1,41
$3,2*10^5$	0,18	(82+98)/2	(82+98)/2	180	1,13
$3,4*10^5$	0,30	140	(78+94)/2	134	1,40
$4,0*10^5$	0,50	140	140	80	1,40
$1,0*10^6$	0,10	ca. 130	ca. 130	100	0,35
$3,0*10^6$	0,21	85	85	190	1,39

Figure 4. Original table of experimental data (Gölling 2001, **[5]**) from left to right: Reynolds number, Strouhal number, separation angle on the top and bottom side, delta angle via rear side of the cylinder and Strouhal number with respect to the corresponding feedback length. The separation angles are obtained mainly by oil flow visualizations in dependence of Reynolds numbers on a circular cylinder flow. (see also Roshko 1961 **[9]** and Achenbach 1979 **[10]**.)

Furthermore, new explorations by Schewe 2004 **[8]** result therein that the critical Reynolds numbers for a lot of bluff-body flows, where the tremendous fall in drag occurs, have the same value by application of a defined length scale similar to the here explained feedback length which is different from the normally used length scales i.e. diameter or thickness.

In the report of Gölling 2001 **[6]** the introduction of this new Strouhal number $Sr(L_f)$ is discussed further, that the value of $Sr(L_f)$ is different from 1.4 at lower Reynolds numbers where the separated shear flow in the near-wake region of the bluff body is laminar. This supports the explanation that the evolution of von-Karman-vortices in the turbulent regime follows another instability type than in the laminar flow regime. While the evolution of von-Karman-vortices in the laminar regime is based on an absolute instability mechanism, the "birth" of cross-wise oriented vortices in the turbulent regime is based on a convective instability mechanism.

4. CONCLUSIONS

The presented experimental investigations of separating cylinder cross-flow give an overview about the several staggered transitional processes. For the uncontrolled cylinder flow various span-wise coherent length-scaling stream-wise oriented vortex flows are obtained at the different Reynolds number regimes given by wall pressure distributions and oil-flow visualization patterns. By artificially excitation it could shown that the originated span-wise homogeneous, i.e. two-dimensional, and transitional or turbulent separating boundary-layer flow is most unstable for defined in dependence of Reynolds number regime span-wise periodic disturbances as well as temporal periodic, i.e. high frequent, excitation.

This leads to the assumption that local high-frequency periodic disturbances cause the evolution of small-scale stream-wise oriented vortices in the unstable, separating boundary-layer flow, especially in the *subcritical regime.* By additionally application of a span-wise periodic disturbance distribution these small-scale vortices could merge to form large-scale span-wise periodic stream-wise oriented vortices. These large-scale vortices are itself unstable with increasing Reynolds number and similar to the further development of the span-wise periodic Lambda-vortices of the flat plate the separating boundary-layer flow on a circular cylinder becomes turbulent.

General speaking, the separated cross-wise vortices originate and get its energy from the mean flow and are stable for wide ranges of Reynolds numbers. However, through disturbing the <u’w’>-transport process or through excitation of defined other, span-wise periodic more unstable modes the separating flow has to reorder. Another topology of separated, stream-wise oriented vortex flow has to develop which is an amplified mode as long as artificial energy input or span-wise shape modulation is applied.

ACKNOWLEDGEMENT

This short excerption of experimental results of my Ph.D. thesis could be presented by thankful grant of Prof. G.E.A. Meyer as a member of the IUTAM conference committee, and Dr. K.A. Bütefisch of the DLR, Institute of Aerodynamics and Flow Technology. Furthermore I thank my *doctor fathers* PD Dr. U. Dallmann and Dr. G. Schewe from DLR Göttingen, Institute of Fluid Mechanics and Institute of Aeroelastics, respectively. My thank is also guaranteed to my former colleagues, especially to Prof. V. Theofilis, Dr. H.-P. Kreplin, and Dr. F.R. Grosche.

REFERENCES

1. Prandtl L., "Bericht über neuere Turbulenzforschung. Hydraulische Probleme", *Berlin, VDI-Verlag, 1926, pp. 1-14.*
2. Schewe G., "On the force fluctuations acting on a circular cylinder in cross-flow from subcritical up to transcritical Reynolds numbers", *JFM, vol. 133, 1983, pp. 265-285.*
3. Dallmann U. and Schewe G., "On topological changes of separating flow structures at transition Reynolds numbers", *AIAA 19th Fluid Dynamics, Plasma Dynamics and Laser Conference, June 8-10, 1987, Honolulu, Hawaii, AIAA-87-1266.*
4. Huerre P., Monkewitz P.A., "Local and global instabilities in spatially developing flow", *Ann. Fluid Mech., vol. 22, 1990, pp. 473-537.*
5. Gölling B., "Experimental investigations of laminar-turbulent transition of cylinder boundary-layer flow", *DLR, Göttingen, Research report, FB 2001-14, 2001, 106 p.*
6. Theofilis V., "Advances in global linear instability analysis of nonparallel and three-dimensional flows", *Progr. in Aerospace Science., May 2003, vol. 39, pp. 249-315.*
7. Brede M. and Leder A., "On the structure of turbulence in transitional cylinder wake", *STAB Stuttgart, Notes on Numerical Fluid Mechanics NNFM, 2001, pp. 189-198.*
8. Schewe G., "Über Reynoldszahleffekte bei Profilumströmungen und deren Einfluss auf strömungsinduzierte Schwingungen", *STAB Bremen, November 2004, to appear in Notes on Numerical Fluid Mechanics, 2005.*
9. Roshko A., "Experiments on the flow past a circular cylinder at very high Reynolds numbers", *J. Fluid Mechanics, vol. 10, pp. 345-356, 1961.*
10. Achenbach, E. "Strömung und konvektiver Übergang beim Kreiszylinder und bei der Kugel." *KFA-Jul-1591, Kernforschungsanlage Jülich, 1979.*

SCALE-SEPARATION IN BOUNDARY LAYER THEORY AND STATISTICAL THEORY OF TURBULENCE

Tomomasa Tatsumi
International Institute for Advances Studies, 9-3 Kizugawadai Kizu, Kyoto 619-0225, Japan

Abstract The idea of the "boundary layer" proposed by Prandtl [1] does not only provide us with a powerful means for solving various flow problems at large Reynolds, but also furnishes us with a universal concept for dealing with the problems of the same nature in physical sciences. This is the "scale-separation" which guarantees the independence of the local motions from the global structure. In this paper, a clear parallelism is pointed out between the "boundary layer" in laminar flows on one hand and the "local equilibrium range" in turbulence on the other. The statistical characteristics of homogeneous isotropic turbulence at large Reynolds numbers are surveyed, using the recent results by Tatsumi & Yoshimura [2] covering both the *energy-containing* and the *local equilibrium* ranges, and comparative discussions are made from the view point of the scale-separation.

Key words Boundary layer, scale separation, local equilibrium, turbulence

1. SCALE-REDUCTION IN MODERN PHYSICS

It is well known that the great success in modern physics during the last century has been achieved by the reduction of the *macroscopic* observable phenomena to the *microscopic* molecular motions. The success of such *scale-reduction* crucially depends upon the large scale-difference between the two stages as to guarantee the mutual independence of them. Actually, the enormous progress in modern physical sciences including molecular biology may be accounted for as the result of such *scale reduction.*

The situation has been rather different for fluid mechanics, since the reduction of the macroscopic flows to molecular motion is too cumbersome and there is no clear scale-difference in the range of fluid motions. Moreover, the mathematical difficulty associated with the Navier-Stokes

G.E.A. Meier and K.R. Sreenivasan (eds.), IUTAM Symposium on One Hundred Years of Boundary Layer Research, 463-471,

equation of motion due to its nonlinearity and dissipativity is so serious that there was no general method available except for employing some particular solutions of the Navier-Stokes equation for practical purposes.

2. SCALE-SEPARATION IN FLUID MECHANICS

This deadlock of fluid mechanics has been broken just 100 years ago by Prandtl [1] who introduced the idea of *scale-separation* at large Reynolds numbers in fluid mechanics. He observed that fluid flow around a solid body at large Reynolds number R is composed of two distinct regions: An inviscid flow slipping over the solid body and a viscous shear-layer along the solid boundary.

The thickness of this *boundary layer* is shown to be of $O(R^{-1/2})$, so that it gives a great opportunity for scale separation for very large R. Then, employing the enlarged coordinate by the factor $R^{1/2}$ and other simplification due to large R, Prandtl has succeeded in deriving a simplified nonlinear equation from the Navier-Stokes equation and obtaining the solution which represents the velocity profile of the *boundary layer*, connecting smoothly the velocities of the outer flow and the solid boundary.

This great success of Prandtl based upon the *scale-separation* for large Reynolds numbers has provided us with a powerful method for dealing with fluid mechanics in general. Since that time, most of typical problems in fluid mechanics have been dealt with by means of *scale-separation* techniques more or less similar to Prantdl's *boundary layer* theory.

3. SCALE-SEPARATION IN TURBULENCE

On the other hand, turbulent flows, which are very common at large Reynolds numbers, have not enjoyed the merit of *scale-separation.* In the early last century the notion of the mixing length has been introduced to turbulence taking analogy with the mean free-path of the kinetic theory of gases, but this idea has proved to be of limited validity due to the lack of scale-difference between the fluctuations and the mean flow.

This stalemate of turbulence theory has been resolved by Kolmogorov [3] who first introduced the *scale-separation* for large Reynolds numbers to turbulence. He proposed to deal with small-scale components of turbulence separately from those of large-scale and assumed that the small-scale components are in a statistical equilibrium state determined by two parameters, the mean *energy-dissipation rate,*

He proposed to deal with small-scale components of turbulence separately from those of large-scale and assumed that the small-scale components are in a statistical equilibrium state determined by two parameters, the mean *energy dissipation rate,*

$$\varepsilon = \nu\Sigma_{i,j=1}^{3} \langle(\partial u_i/\partial x_j)^2\rangle \quad (1)$$ and

the *viscosity* ν(*First hypothesis*). Then, expressing the small-scale components in terms of the velocity difference at two points **x** and **x**+**r**,

$$\Delta\mathbf{u}(\mathbf{r}, t) = \mathbf{u}(\mathbf{x}+\mathbf{r}, t) - \mathbf{u}(\mathbf{x}, t), \quad (2)$$

and using dimensional analysis, he shows that the structure functions,

$$S_n(r) = \langle|\Delta\mathbf{u}(\mathbf{r}, t)|^n\rangle, \quad (3)$$

n being an integer,are expressed as

$$S_n(r) = (\varepsilon\nu)^{n/4} F_n(r/\eta), \quad (4)$$

where $\eta=(\nu^3/\varepsilon)^{1/4}$ denotes *Kolmogorov's length* representing the scale of the *local equilibrium range*.

For extremely large Reynolds numbers,he argues that there exists an *inertial subrange* where the statistics depends only upon ε but not on ν(*Second hypothesis*). Then, Eq.(4) is written as

$$S_n(r) = C_n(\varepsilon r)^{n/3}. \quad (5)$$

C_n being a constant. Eq.(5) is known as *Kolmogorov's inertial range similarity law*.

For $n = 2$, Eq.(5) gives

$$S_2(r) = \langle|\Delta\mathbf{u}(\mathbf{r}, t)|^2\rangle = C_2\varepsilon^{2/3} r^{2/3}, \quad (6)$$

that is the celebrated *Kolmogorov's* 2/3 *power law* of the velocity correlation.

Thus, Kolmogorov's theory clearly indicates the usefulness of *scale-separation* for large Reynolds numbers, which provides us with concrete similarity laws governing the small-scale components of turbulence.

4. LOCAL EQUILIBRIUM OF TURBULENCE

Kolmogorov's idea of separating the small-scale components of turbulence at *local equilibrium* from the large-scale components may be compared with Prandtl's concept of the *boundary layer* separated from the outer inviscid flow. The two parameters ε and ν governing the equilibrium of the former are considered to be the counterpart of the pressure distribution of the outer flow controlling the latter. Likewise, Kolmogorov's length $\eta = O(\nu^{3/4})$ of the local equilibrium range corresponds to the boundary layer thickness $\theta=O(\nu^{1/2})$,both vanishing in the inviscid limit or the limit of infinite Reynolds number.

Thus, a perfect analogy exists between the *small-scale components of turbulence* under a given energy-dissipation rate and the *boundary layer* under a given pressure distribution. On the other hand, if we consider the *large-scale turbulence*, the analogy with the outer inviscid flow over the boundary layer is found not to work, since both Prandtl's and Kolmogorov's theories are genuinely concerned with the smaller sides of the scale-separation. In order to obtain a unified view of turbulence covering both the large- and small-scale components, we have to develop a more systematic approach to turbulence just like the matching boundary layer theory.

5. SYSTEMATIC APPROACH TO TURBULENCE

Although Kolmogorov's theory provides us with useful information about the small-scale components of turbulence, it says nothing about the large-scale energy-containing turbulence. In order to obtain such large-scale information as well, we have to deal with the fundamental equations of turbulence.

The author has been working on this problem by making use of the "cross-independence" closure hypothesis [1], [4]. So far, statistics of homogeneous isotropic turbulence at large Reynolds numbers have been worked out covering both the large- and small-scale components, and it may be appropriate to present an overview of the theory, taking the view point of comparison of the scale-separations in laminar and turbulent flows.

6. CROSS-INDEPENDENCE HYPOTHESIS

It is well known that the complete statistical information of turbulence is provided by the probability distribution functional of the velocity field in space and time. The equation for this functional was given by Hopf [5], but no mathematical method has been available for solving this functional equation and only a few particular solutions have been obtained by Hopf [5].

A statistically equivalent formalism was given by Lundgren [6] and Monin [7] as an infinite set of equations for the joint velocity distributions at an arbitrary number of points in space-time. This set of equations is, however, indeterminate since each equation involves a new unknown as a higher-order distribution, so that the introduction of a certain hypothesis is inevitable for closing the set of equations.

The equation for the one-point velocity distribution $f(\mathbf{u}_1, t)$ is indeterminate since it includes the two-point velocity distribution $f^{(2)}(\mathbf{u}_1,\mathbf{u}_2;\mathbf{r}, t)$. The usual assumption for closing the equation is to assume the independence of the one-point distributions and set

$$f^{(2)}(\mathbf{u}_1,\mathbf{u}_2;\mathbf{r},t) = f(\mathbf{u}_1,t)\, f(\mathbf{u}_2,t). \tag{7}$$

It may be obvious that Eq.(7) gives a good approximation for large distance $|\mathbf{r}|\to\infty$ but not for the limit of small distance $|\mathbf{r}|\to 0$, for which

$$f^{(2)}(\mathbf{u}_1,\mathbf{u}_2;\mathbf{r},t) = f(\mathbf{u}_1,t)\delta(\mathbf{u}_2 - \mathbf{u}_1), \tag{8}$$

where δ is the three-dimensional delta function.

Instead of such independence, we assume the *cross-independence* of two velocities $\mathbf{u}_1$ and $\mathbf{u}_2$, or the independence of the sum $\mathbf{u}_+ = (\mathbf{u}_1 + \mathbf{u}_2)/2$ and the difference $\mathbf{u}_- = (\mathbf{u}_2 - \mathbf{u}_1)/2$. Unlike the usual independence, the cross-independence relation is shown to be valid for both limits of large separation (7) and coincidence (8),so that it is an asymptotically good approximation for large ratios of the scale-separation. This assumption is also found to be essentially equivalent to Kolmogorov's premise of the independence of small-scale components of turbulence represented by $\Delta\mathbf{u} = 2\mathbf{u}_-$ from large-scale ones.

7. VELOCITY DISTRIBUTIONS

The one-point velocity distribution $f(\mathbf{u}_1,t)$ and the two-point velocity distribution represented by the velocity-sum distribution $g_+(\mathbf{u}_+,\mathbf{r},t)$ and the velocity-difference distribution $g_-(\mathbf{u}_-,\mathbf{r},t)$ are considered. These distributions are sufficient for deriving most of the practical information of turbulence.

These distributions are obtained by solving the Lundgren equations which are closed by making use of the *cross-independence hypothesis,* assuming the self-similarity in time.

7.1 One-point velocity distribution

Inertial range

The one-point velocity distribution is expressed as

$$f(\mathbf{u}_1,t) = f_0(\mathbf{u}_1,t) \equiv (3/4\pi\varepsilon_0)^{3/2}\, t^{3/2} \exp[-(3/4\varepsilon_0)|\mathbf{u}_1|^2 t], \tag{9}$$

where $\varepsilon(t) = \varepsilon_0 t^{-2}$ denotes the *energy dissipation rate* mentioned above.

Eq.(9) represents the *inertial normal distribution* depending only upon ε and not ν. According to the homogeneity of turbulence,it is independent of the coordinate $\mathbf{x}_1$ but changes in time t. It starts from the *uniform distribution* with zero probability density at $t = 0$, evolves in time as a *normal distribution* of decreasing variance, and eventually tends to the *delta distribution* corresponding to the rest state for $t\to\infty$. During this process the *kinetic energy* decays as

$$E(t) = (1/2)\langle|\mathbf{u}(\mathbf{x},t)|^2\rangle = E_0 t^{-1}$$

$$\varepsilon(t) = -\,dE(t)/dt = E_0 t^{-2},\ E_0 = \varepsilon_0. \tag{10}$$

The one-point velocity distribution (10), which does not include the viscosity ν separately, is said to have the *inertial similarity*. It should be noted that this *inertiality* is not the same as that assumed by Kolmogorov [3] for defining the *inertial* subrange. Here the *inertiality* means that the viscosity ν only appears as the coefficient of the energy dissipation rate ε but its magnitude is left arbitrary (see Eq. (1)). On the other hand, Kolmogorov assumed ε as a constant independent of ν.

Recently, measurements of ε for wide ranges of ν have been carried out by several authors using large-scale numerical simulations, and it has been confirmed that although ε takes large values at small values of $1/\nu$ it decreases rapidly with increasing $1/\nu$ and eventually tends to a constant for very large $1/\nu$. If we call the absence of ν with its arbitrary value the *weak inertiality* and the presence of ν with zero-value the *strong inertiality*,the majority of the measured results seem to belong to the weak inertiality while only those of very large $1/\nu$ indicate the strong inertiality.

7.2 Velocity-sum distribution

Inertial range

The velocity-sum distribution $g_+(\mathbf{u}_+, \mathbf{r}, t)$ is obtained as

$$g_+(\mathbf{u}_+, \mathbf{r}, t) = g_0(\mathbf{u}_+, t) \equiv (3/2\pi\varepsilon_0)^{3/2}\, t^{3/2} \exp[-(3/2\varepsilon_0)|\mathbf{u}_+|^2 t]. \tag{11}$$

Eq.(11) gives another *inertial normal distribution* which is identical to (9) except for the parameter $\varepsilon_0/2$ instead of ε_0. For convenience, the distributions (9) and (11) may be called the *first* and *second normal distributions* N_1 and N_2, respectively.

The distribution N_2 which is independent of $\mathbf{r}$ is valid for all values of $|\mathbf{r}|$. Since, however, it does not satisfy the coincidence condition,

$$\lim_{|\mathbf{r}|\to 0} g_+(\mathbf{u}_+, \mathbf{r}, t) = f(\mathbf{u}_1, t) = N_1, \tag{12}$$

at $|\mathbf{r}| = 0$,its validity is limited to the range $|\mathbf{r}| > 0$. This means that the distribution changes discontinuously from N_2 to N_1 for $|\mathbf{r}| \to 0$.

On the other hand, taking account of the fact that the variance of the normal distribution N_2 is a half of that of N_1,we can conclude that N_2 is represented by the convolution of two independent N_1's. This clearly shows the *inertial normality* of the velocity-sum distribution

$g_+(\mathbf{u}_+, \mathbf{r}, t)$.

The inertial similarity of the distribution $g_+(\mathbf{u}_+, \mathbf{r}, t)$ in global space $|\mathbf{r}| > 0$ and its discontinuous change at $|\mathbf{r}| = 0$ shows an eminent analogy with the *boundary layer* of laminar flows at large Reynolds numbers. Learning from the success of the boundary-layer theory, let us deal with the problem using the variables based upon the *local equilibrium range* of turbulence,

Local equilibrium range

The local variables are defined in terms of Kolmogorov's scales,

$$\text{Length } \eta = (\nu^3/\varepsilon)^{1/4}, \qquad \text{Velocity } \upsilon = (\nu\varepsilon)^{1/4}, \tag{13}$$

as follows:

$$\text{Coordinate } \mathbf{x}^* = \mathbf{x}/\eta, \quad \text{Time } t^* = t/(\eta/\upsilon),$$
$$\text{Velocity } \mathbf{u}^*(\mathbf{x},t) = \mathbf{u}(\mathbf{x},t)/\upsilon. \tag{14}$$

The velocity-sum distribution $g^*_+(\mathbf{u}^*_+, \mathbf{r}^*, t^*)$ defined in the local variables is obtained as

$$g^*_+(\mathbf{u}^*_+, \mathbf{r}^*, t^*) = g^*_0(\mathbf{u}^*_+, \mathbf{r}^*, t^*)$$
$$\equiv (3/4\pi\varepsilon^*_{+0}(r^*))^{3/2} t^{*3/2} \exp[- (3/4\varepsilon^*_{+0}(r^*))|\mathbf{u}^*_+|^2 t^*], \tag{15}$$

where $\varepsilon^*_+(r^*, t^*) = \varepsilon^*_{+0}(r^*) t^{*-2}$ with $r^* = |\mathbf{r}^*|$ denotes the *self- energy dissipation rate* defined by

$$E^*_\pm(r^*, t^*) = (1/2)\langle|\mathbf{u}^*_\pm(r^*, t^*)|^2\rangle = E^*_{\pm 0}(r^*) t^{*-1}$$
$$\varepsilon^*_\pm(r^*, t^*) = -\,dE^*_\pm(r^*, t^*)/dt^* = E^*_{\pm 0}(r^*) t^{*-2}, \qquad E^*_{\pm 0}(r^*) = \varepsilon_{\pm 0}(r^*). \tag{16}$$

Eq.(15) shows that the velocity-sum distribution g^*_+ in the local equilibrium range is identical to the normal distribution (9) except for the self-energy dissipation rate $\varepsilon^*_{+0}(r^*)$ instead of the ordinary energy dissipation rate ε_0 for the latter.

Like the boundary layer, the local equilibrium range has no definite boundary for large r^*, but we may take for practical use a coordinate $r^* = r^*_L$ such that

$$\langle|\mathbf{u}^*_+(r^*_L, t^*)|^2\rangle = \langle|\mathbf{u}^*_-(r^*_L, t^*)|^2\rangle, \tag{17}$$

which leads to the conditions,

$$E^*_+(r^*_L, t^*) = E^*_-(r^*_L, t^*), \quad \varepsilon^*_+(r^*_L, t^*) = \varepsilon^*_-(r^*_L, t^*), \tag{18}$$

for the upper boundary of the local equilibrium range. Eq.(17) follows from the consideration that, in the *inertial range,* both the self-energies $\langle|\mathbf{u}^*_+|^2\rangle$ and $\langle|\mathbf{u}^*_-|^2\rangle$ are due to the energy-containing eddies of turbulence which have the same order of magnitude, while in the *local equilibrium range*, $\langle|\mathbf{u}^*_+|^2\rangle$ due to the energy-containing eddies is supposed to be dominant over $<|\mathbf{u}^*_-|^2>$ due to the small-scale eddies.

If we take the coordinate r^*_L as the upper end of the local equilibrium range it immediately follows from Eq.(17) that

$$\varepsilon^*_+(r^*_L, t^*) = \varepsilon^*_-(r^*_L, t^*) = (1/2)\varepsilon^*(t^*)\,, \qquad (19)$$

indicating that the distribution (15) exactly coincides with N_2 given by (11) at $r^* = r^*_L$.

At the lower end $r^* = 0$, it follows from the definition that $\mathbf{u}^*_+ = \mathbf{u}^*$ and $\varepsilon^*_{+0} = \varepsilon^*_0$, so that the distribution (15) exactly coincides with N_1 given by (9) at $r^* = 0$.

Thus the discontinuous change of the velocity-sum distribution $g_+(\mathbf{u}_+, \mathbf{r}, t)$ from N_2 to N_1 in the limit of $|\mathbf{r}| \to 0$ has been completely resolved by the continuous change of the *normal distribution* $g^*_+(\mathbf{u}^*_+, \mathbf{r}^*, t^*)$ with a changeable variant throughout the finite local equilibrium range.

7.3 Velocity-difference distribution

Inertial range

Concerning the *velocity-difference distribution* $g_-(\mathbf{u}_-, \mathbf{r}, t)$ in the inertial range, the situation is the same as the velocity-sum distribution $g_+(\mathbf{u}_+, \mathbf{r}, t)$ in this range. The distribution is expressed as the *second normal distribution,*

$$g_-(\mathbf{u}_-, \mathbf{r}, t) = g_0(\mathbf{u}_-, t) \equiv (3/2\pi\varepsilon_0)^{3/2}\, t^{3/2} \exp[-(3/2\varepsilon_0)|\mathbf{u}_-|^2 t]. \qquad (20)$$

Eq.(20) is identical to N_2 given by (11) with the parameter $\varepsilon_0/2$.

Like the velocity-sum distribution (11), Eq.(20) is valid at all values of $|\mathbf{r}| > 0$, but it has to change discontinuously in the limit of $|\mathbf{r}| \to 0$ in order to satisfy the coincidence condition,

$$\lim_{|\mathbf{r}| \to 0} g_-(\mathbf{u}_-, \mathbf{r}, t) = \delta(\mathbf{u}_-) \qquad (21)$$

at $|\mathbf{r}| = 0$. This discontinuous change should be resolved by dealing with the distribution in the local equilibrium range.

Local equilibrium range

In this range the equation for the *velocity-difference distribution* $g^*_-(\mathbf{u}^*_-, \mathbf{r}^*, t^*)$ becomes axisymmetric. The *lateral velocity-difference distribution* with respect to the distance $\mathbf{r}^*$ becomes nearly identical to the normal distribution with the parameter $\varepsilon^*_-(r^*, t^*)$ instead of $\varepsilon^*_+(r^*, t^*)$ for Eq.(15). Thus, it tends to the delta distribution in the limit of $r^* \to 0$ satisfying the coincidence condition (21). The *longitudinal velocity-difference distribution* becomes asymmetric with respect to its argument but otherwise similar to the lateral component. Details concerning the velocity-difference distribution in the local equilibrium range will be dealt with in a separate paper.

8. SUMMARY AND DISCUSSIONS

8.1 Physical laws of turbulence

In order to obtain universal physical laws governing turbulence the *scale-separation* according to large Reynolds numbers has been applied to the system of equations for the velocity distributions. Close similarity is noted between Prandtl's scaling of the *boundary layer* of laminar flows and Kolmogorov's scaling of the *local equilibrium range* of turbulent motions and the analogy is fully utilized in the research.

As the result, the *inertial normality* of the one- and two-point velocity distributions has been established not only in the inertial range but also in the local equilibrium range. An exceptional case is the longitudinal velocity-difference distribution but even this distribution shows only slight asymmetry on the basically Gaussian distribution.

8.2 Scope to turbulent shear flows

The simple and elegant normal laws obtained for homogeneous isotropic turbulence seem to give a bright prospect to the application of the present approach to more complex flows including *turbulent boundary layers.* Especially the validity of the fluctuation-dissipa tion theorem in homogeneous turbulence is expected to provide a wide applicability of the theorem to various turbulent fields.

REFERENCES

1. Prandtl L. "Ueber Flussigkeitsbewegung bei sehr kleiner Reibung". *Vehr. III Intern. Math. Kongr. Heidelberg,* pp.484-491, 1904.
2. Tatsumi T, Yoshimura T. "Inertial similarity of velocity distributions of homogeneous isotropic turbulence". *Fluid. Dyn. Res.* **35**, 123-158, 2004.
3. Kolmogorov AN. "The local structure of turbulence in incompressible viscous fluid for very large Reynolds numbers". *Dokl. Akad. Nauk. SSSR,* **30**, 301-305, 1941.
4. Tatsmi T. "Mathematical physics of turbulence." In Kambe T, et al (Ed) *Geometry and Statistics of Turbulence*, Kluwer Acad. Publ. pp.3-12, 2001.
5. Hopf E. "Statistical hydromechanics and functional calculus." *J. Rat. Mech. Anal.* **1**, 87-123.
6. Lundgren TS. "Distribution functions in the statistical theory of turbulence." *Phys. Fluids,* **10**, 969-975, 1967.
7. Monin AS. "Equations of turbulent motion." *PMM J. Appl. Math. Mech.* **31**, 1057-1068, 1967.

ON BOUNDARY LAYER CONTROL IN TWO-DIMENSIONAL TRANSONIC WIND TUNNEL TESTING

Bosko Rasuo
Aeronautical Department, Faculty of Mechanical Engineering, University of Belgrade, Kraljice Marije 16, 11120 Belgrade 35, Serbia and Montenegro, Tel. +381 11 3302 261, Fax. +381 11 3370 364, brasuo@mas.bg.ac.yu

Abstract: In this paper a closed-form analysis of flow in a two-dimensional transonic wind tunnel, that uses sidewall distributed suction around the model to reduce sidewall boundary-layer effects, is presented. The point of interest of this paper is the control over the boundary layer along the side walls of the transonic wind tunnels.

Key words: Boundary Layer Control, Two-Dimensional Transonic Wind Tunnel.

1. INTRODUCTION

For the successful aerodynamic designing of a new modern aircraft it is necessary to know the accurate aerodynamic characteristics of the whole aircraft, as well as of its individual constituent parts. Since there is no adequate mathematical model of turbulent flows, we cannot solve completely the problem of aerodynamic designing by computer simulation and calculation. We still have to solve many problems related to aerodynamic designing by making tests in wind tunnels. However, wind tunnel simulation is connected with many problems which cause many distortions of flow conditions around the tested models, which finally results in inaccuracy of the measured aerodynamic values. There are many reasons for that, but it is quite understandable that even the best wind tunnels cannot provide conditions for the simulation of the flows around the tested model

G.E.A. Meier and K.R. Sreenivasan (eds.), IUTAM Symposium on One Hundred Years of Boundary Layer Research, 473-482,

which would be identical to the flows in the free air. Therefore, the resolving of the problem related to the definition and elimination of the wind tunnel wall interference is a lasting task to be solved through experimental and theoretical research, either during the construction of new wind tunnels or during their exploitation. A special group of problems are related to the simulation of flows around the tested airfoil, i.e. to the provision of two-dimensional flow conditions. Paper presents the algorithm for calculating the suction of air from the working section of the wind tunnel necessary to sustain acceptable boundary layer thickness of the wind tunnel side walls, as regards successful two-dimensional wind tunnel simulation [**1-5**].

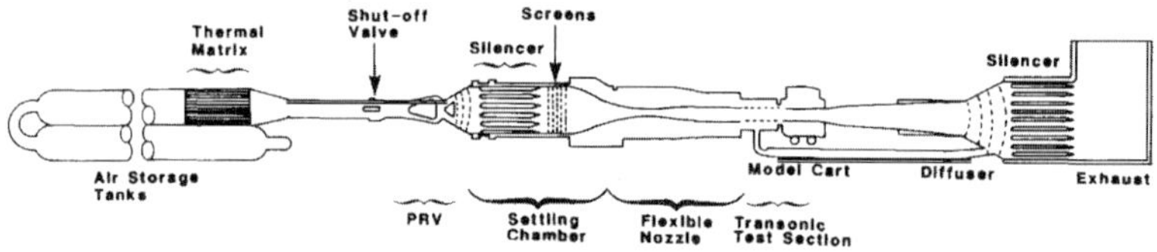

Figure 1. Schematic of the VTI s wind tunnel (PRV - Pressure Regulating Valve).

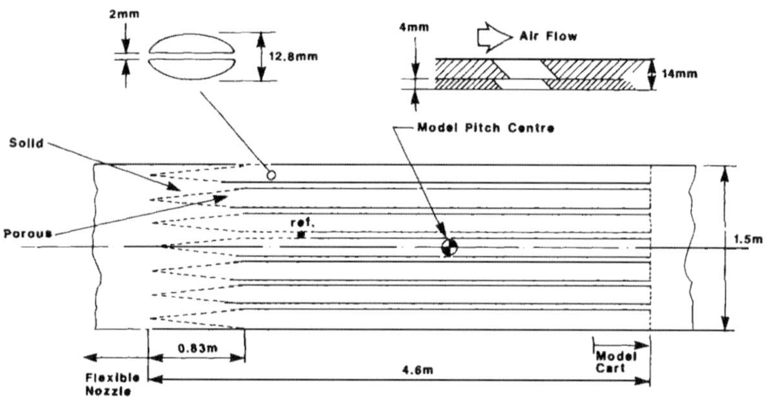

Figure 2. Schematic of test-section walls.

2. NUMERICAL AND EXPERIMENTAL APPROACHES

Some practical examples and results are given for the NACA 0012 airfoil, tested at supercritical flow conditions in perforated wall test sections of the Aeronautical Institute VTI's high Reynolds number trisonic wind tunnel, T-38 (Figure 1). The VTI's trisonic wind tunnel is a blowdown type with a two-dimensional test section, with a cross section dimensions 0.38 x 1.5 *m* with changeable perforation of walls from 0.5 to 6% (Figures 2 and 3).

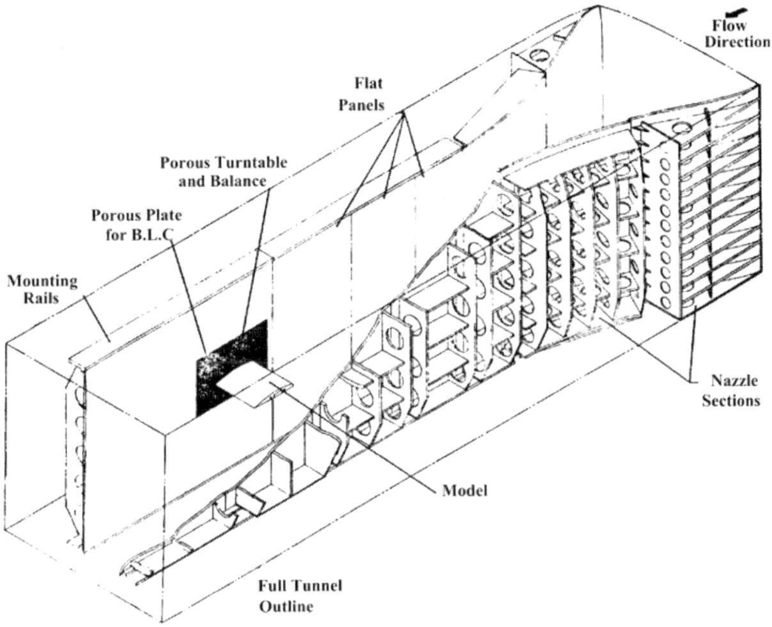

Figure 3. Two-dimensional working section.

The results of this analysis are presented in Figure 4 for NACA 0012 airfoil. They are grouped according to 21 sources of quotation. Many of these results have been achieved by the outstanding and widely known international aerodynamic institutions. For example, an analysis has been made of some old wind tunnel low speed tests made by NACA Institute (symbols 2-4), contemporary results of the NASA (1,5 and 6), the results achieved in the very good industrial facilities (10-12), detailed studies of the NPL and RAE (13-15), the results achieved by AGARD working group 04 DATA BASE (17), the results of ONERA (16-19), of the VTI and the Faculty of Mechanical Engineering (21), etc.

According to this illustration there is a great diversity in the achieved results, as a consequence of the strong influence of the Reynolds numbers effects on the test models and wind tunnels, of inadequate conditions of two-dimensional flows in the test section and the wall interference in the test section of wind tunnel. Wishing to complete this study, the analysis has been extended to the transonic speed range and it has incorporated new tests made by the VTI as well as the calculation of wall corrections made at the Faculty of Mechanical Engineering.

Experimental tests have been made in the wind tunnel T-38 with transonic two-dimensional working section. Aerodynamic coefficients have been calculated by measuring the distribution of the static pressure in 80 equally distributed tested points along the upper and lower side of NACA 0012 model with a chord of 0.254 *m*. For this measuring, the complete most modern equipment for aerodynamic measuring has been used. An additional experimental study has included the Mach test number from 0.25 to 0.8 and the Reynolds model numbers from 2 to 35 *MRe* (see Figure 4).

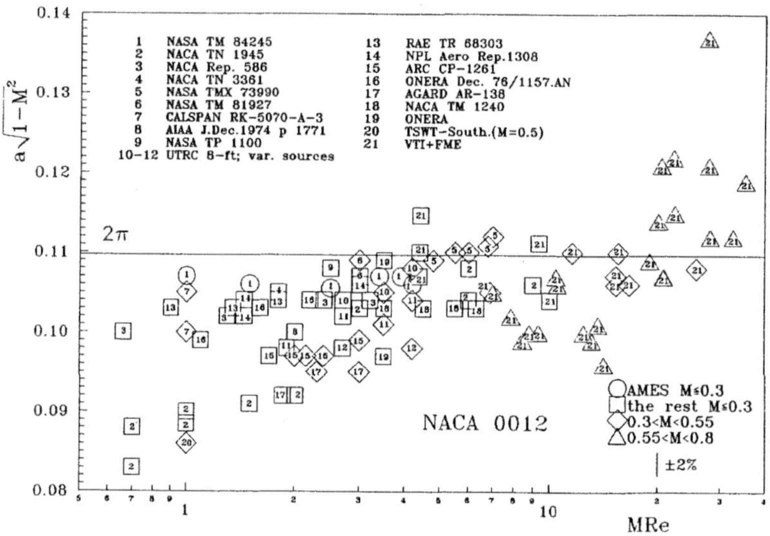

Figure 4. Lift-curve slope in function of the Reynolds number.

In order to create correct two dimensional flow conditions and uniform spanwise loading of the airfoil model, it is necessary to apply side-wall suction, i.e. the control over the boundary layer along the side walls of the wind tunnel. In the case that the control of boundary layer along the side walls is not ensured, this will certainly result in a loss of lift (and difference in drag) caused by the two basic effects of the complex flow. First, the loss of lift is caused by the decreased speed near the wall. This effect can be significantly diminished if the side-wall boundary layer is reduced to the value which is very small in comparison with the spanwise of the model. Second, the influence of the airfoil pressure range will cause nonuniform increase of boundary layer along the side walls, which will result in the creation of some tree-dimension effects in the flow around the airfoil. The separation along the side walls is also quite normal. For example, it usually occurs near a rounded leading edge (in the vicinity stagnation point), approaching the trailing edge and during the subcritical and supercritical flow, as well as in the zone of the maximum local value of pressure.

It is desirable that the quantity of the removed volume of the air through porous side walls of the wind tunnel is minimal as required for creating satisfactory conditions for two-dimensional flow. If the too much quantity of air is removed from the working section this will cause an extensive axial gradient of pressure in the wind tunnel, which will result in (buoyancy) defect in drag and in the Mach number.

The importance of the correct definition of the quantity of the removed air is evident from the ONERA tests presented in Figure 4 for its results given under point 19. The lower point is the case with inadequate suction and the upper point with right quantity of the removed air. Most frequently the removed quantity of air is expressed through the ratio of normal component of flow velocity through the wall, to the velocity of undisturbed flow (far upstream from the model) V_n/V_∞. In all tests made by the VTI which are presented in Figure 4, the velocity ratio has been within the limits V_n/V_∞ = 0.0050-0.0054 [**6-10**].

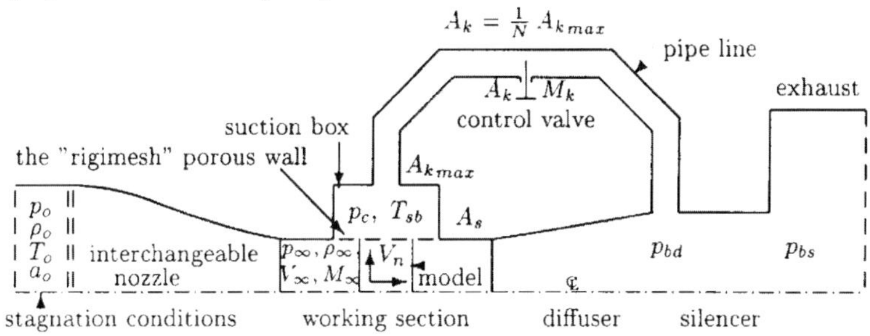

Figure 5. Two-dimensional working section.

Control of the sidewall boundary layer in the vicinity of the model is provided for by suction. The arrangement is shown in Figures 3 and 5. An area 685.88 x 156.65 *mm* for the trisonic wind tunnel T-38 is covered by compression welded multilayer woven wire sheet (stainless steel, so call "Rigimesh").

The following approximate analysis serves to calculate conditions in the system; using the usual isentropic relationships, the transonic wind tunnel working section conditions can be calculated for a stagnation values of pressure, density, and free stream Mach number (p_o, ρ_o, M_∞). Ignoring any sidewall suction outflow and any in or outflow through the top and bottom walls, the working section mass flow is

$$m_\infty = A_\infty \rho_\infty V_\infty \tag{1}$$

where A_∞ is working section area.

The sidewall suction mass flow per side is

$$m_s = A_s \rho_\infty V_n \tag{2}$$

where V_n suction velocity (normal to the wall), and A_s is suction area per sidewall

$$A_s = K_p ab \tag{3}$$

where K_p is the ratio of the open to total wall area, a and b are the sides of the porous plates for suction.

The discharge orifice area is

$$A_k = \frac{1}{N} A_{k\max} \tag{4}$$

where N is the fractional opening of discharge orifices (*1/N*, N=1, 2, 4, 8, 16, etc.), and A_{kmax} the maximum discharge orifice area.

Under steady flow conditions m_s is equal to the mass flow through the discharge orifice

$$m_s = m_k = \frac{1}{N} \rho_k V_k A_{k\max} \tag{5}$$

where ρ_k is density V_k velocity at the discharge orifice area A_k (Figure 5). After involving some isentropic relation we can write the mass flow through discharge orifice

$$m_k = \frac{\frac{1}{N} \kappa p_c M_k A_{k\max}}{a_0 \left(1 + \frac{M_k^2}{5}\right)^3} = A_s \rho_\infty V_n \tag{6}$$

where p_c is test chamber pressure and M_k the discharge orifice Mach number.

Because

$$M_k = 1 \qquad \text{for} \qquad \frac{p_c}{p_{bd}} \geq 1.892$$

$$M_k = \left\{ 5\left[\left(\frac{p_c}{p_{bd}} \right)^{2/7} - 1 \right] \right\}^{0.5} \qquad \text{for} \qquad \frac{p_c}{p_{bd}} < 1.892 \tag{7}$$

The pressure behind diffuser p_{bd} we my find from the analysis of losses in the wind tunnel [**10**]

$$p_{bd} = p_{bs} \frac{p_{bde}}{p_0} \tag{8}$$

where p_{bs} is the pressure behind silencer, and p_{bde} pressure behind diffuser at the end of the wind tunnel ran. That pressure depend on the losses in the wind tunnel and the Mach number in the working section

$$\frac{p_{bd_e}}{p_0} = 1 - \frac{\kappa K_0 M_\infty^2}{2\left(1 + \frac{\kappa - 1}{2} M_\infty^2\right)^{3.5}} \tag{9}$$

where K_0 is the losses coefficient in the wind tunnel

$$K_0 = \sum_1^n \xi_n \tag{10}$$

where ξ_n are the losses of the all parts of the wind tunnel (working section, flow screens, nozzle, dryer, diffuser, valve, silencer etc.).

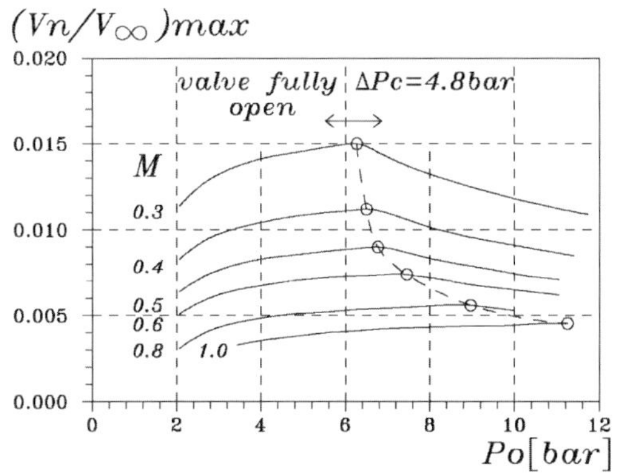

Figure 6. Maximum suction velocity versus test section total head (stagnation pressure in working section) and Mach number for the Aeronautical Institute VTI's T-38 high Reynolds number trisonic wind tunnel.

The remaining equation to close the above system represents the pressure drop across the "Rigimesh" porous plates. This can be expressed as

$$\Delta p_c = p_\infty - p_c = \frac{1}{2} K_1 \rho_\infty V_n^2 \tag{11}$$

where K_1 is the losses coefficient by the cross-flow through porous walls. The values of this coefficient we can find by experiment [**10**].

By combining previous equations we obtain fractional opening of discharge orifices

$$N = \frac{\kappa A_{k\max}}{abK_p}\left(\frac{K_1}{2}\right)^{0.5} \frac{p_c}{(p_\infty - p_c)^{0.5}} \frac{M_k}{\rho_\infty^{0.5} a_0 \left(1+\frac{M_k^2}{5}\right)^3} \tag{12}$$

This equation gives a functional links between the fractional opening of discharge orifices, the difference between the static pressures (in the working section and the suction box - test chamber) and the Mach number at the discharge orifice area (the critical section of the pipe line). This equation is possible to solve altogether with the system of equations (7), (8) and (9) by the iterative procedure and by assuming the test chamber pressure is approximately equal

$$p_c = \frac{p_\infty + p_{bd}}{n} \tag{13}$$

with the step the of iteration $n = 2, 3, 4, \ldots$. The iterative procedure is necessary to perform for the all values of the pressure ratio p_c/p_{bd}. This iterative procedure is very convenient for the calculation of the global cross-flow parameters m_s/m_∞ and V_n/V_∞ in a function of the valve fractional opening N and the pressure ratio p_c/p_{bd} and for the known stagnation conditions and the Mach number in the working section.

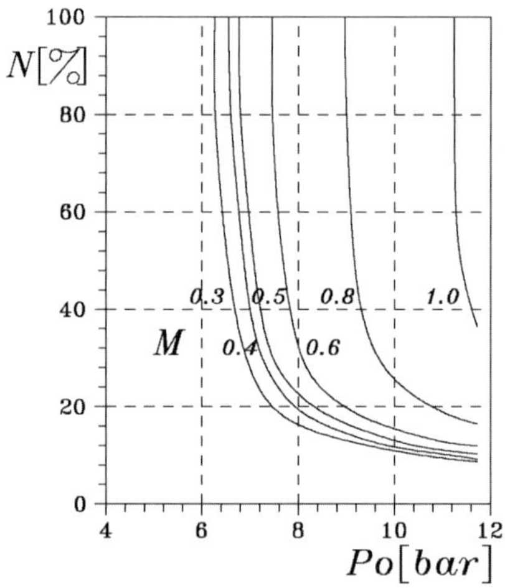

Figure 7. Suction valve maximum opening for safe operation for the Aeronautical Institute VTI's T-38 high Reynolds number trisonic wind tunnel.

In the case, that the Mach number M_∞ and the stagnation pressure P_0 are known and it is required to remove t times the sidewall boundary layer deficit mass flow, i.e.

$$m_s = t m_{\delta^*} = t \rho_\infty V_\infty \delta^* b \tag{14}$$

where b is the height of the suction area, and δ^* the boundary layer displacement thickness.

Analysis of Preston tube measurements taken just upstream of the porous sidewall plates has shown that for most test conditions the displacement thickness of the approaching boundary layer is $\delta^* = 4 \pm 0.75$ *mm* **[10]**, and hence

$$t = \frac{m_s}{m_{\delta *}} = \frac{A_s V_n}{\rho_\infty \delta^* b V_\infty} \tag{15}$$

the proportion of boundary layer deficit mass flow removal by suction, $0.6 < t < 1$ is typically.

3. RESULTS AND DISCUSSION

The maximum suction quantities that are available at any given test condition is limited by either one of two factors; the strength limitations of the side structure supporting the porous panels and the pressure difference available with the discharge orifices fully open. It is considered unsafe to exceed a pressure drop across the porous panels of $\Delta p_c = 4.8$ *bar*. For the VTI's trisonic wind tunnel T-38, the maximum suction quantities for the load limit and with the discharge orifices fully open ($N=100\%$) are given in Figure 6. The dashed line separate the region of valve fully open and the region of the maximal pressure drop across the porous panels of $\Delta p_c = 4.8$ *bar* for a maximum available suction. The maximum discharge orifice openings and suction quantities available at this maximum wall loading are given in Figure 7.

4. CONCLUSION

The establishment of exact two-dimensional conditions of flow in wind tunnels is a very difficult problem. This is evident for wind tunnels of all types and scales. In order to create correct two dimensional flow conditions and uniform spanwise loading of the tested airfoil model, it is necessary to apply side-wall suction, i.e. the control over the boundary layer along the side walls of the wind tunnel.

In this paper the model problem that is treated involves a flat plate airfoil in a transonic wind tunnel with a suction sidewall porous panel shaped to permit an analytic solution. This solution shows that the lift coefficient for airfoil depends explicitly on the porosity parameter (or the losses coefficient) of the suction porous panel, the fractional opening of discharge orifices and implicitly on the suction pressure differential (between the working section pressure and the suction box pressure). For a given sidewall displacement thickness, the lift coefficient for airfoil increases as the suction-porous panel porosity decreases.

REFERENCES

1. Prandtl L, Tietjens O.G. *Applied Hydro- and Aeromechanics*, Dover Publications, New York, 1957.
2. Göthert B. *Transonic Wind Tunnel Testing*, AGARDograph 49, Pergamon Press, New York, 1961.
3. Lachmann G.V. (Ed). *Boundary Layer and Flow Control, its Principles and Application*, Pergamon Press, New York, 1961.
4. Elfstrom G.M, Medved B, Rainbird W.J. "Wave Cancellation Properties of a Splitter-Plate Porous Wall Configuration", *Journal of Aircraft*, **vol. 26**, **no. 10**, pp. 920-924, 1989.
5. Ewald B.F.R. (Ed). *Wind Tunnel Wall Corrections*, AGARD-AG-336, RTO/NATO, 1998.
6. Rasuo B. "On Sidewall Boundary Layer Effects in Two-Dimensional Subsonic and Transonic Wind Tunnels", *ZAMM*, **vol. 81**, pp. 935-936, 2001.
7. Rasuo B. "On Results' Accuracy at Two-Dimensional Transonic Wind Tunnel Testing", *PAMM*, **vol. 2**, pp. 306-307, 2003.
8. Rasuo B. "An Experimental and Theoretical Study of Reynolds and Mach Number Effects at Two-Dimensional Wind Tunnel Testing", *AIAA/SAE Paper* No. 2000-01-5510, Washington, D.C. USA, 2000.
9. Rasuo B. "On Solving Boundary Value Problems in Fluid Mechanics by Fourier's Method: Wall Interference of Transonic Wind Tunnels", In: *Analysis and Simulation of Multifield Problems*, Wendland W.L, and Efendiev M. (eds), Berlin, Springer Verlag, pp. 317-322, 2003.
10. Rasuo B. *Two-dimensional Transonic Wind Tunnel Wall Interference*, Monographical Booklets in Applied & Computer Mathematics, MB-32/PAMM, Budapest, Technical University of Budapest, 2003.

THEORY OF BOUNDARY LAYER INSTABILITY: PARTICLE OR WAVE?

Ka-Kheng Tan
Department of Chemical and Environmental Engineering, Universiti Putra Malaysia, 43400 UPM, Serdang, Selangor, Malaysia
kakheng@netscape.com or tankk@eng.upm.edu.my.
Fax: 603 8656 7200

Abstract: No stability analyses of a growing boundary layer provides a unique solution to the position of the onset of instability because the point of instability is not known *a priori*, and the underlying equations may be normalized by any thickness of the boundary layer. Tollmien [1] has admitted in the first theoretical study of boundary-layer instability that his choice of and the definition of the displacement thickness δ_l is fundamentally unscientific and problematic. Therefore it is not surprising that the value of the conventional theoretical Reynolds number $\left(\mathrm{Re}_{\delta_1}\right)$, based as it is on an arbitrarily-defined displacement thickness, has failed to give a theoretical value of longitudinal Reynolds number (Re_X), that agrees with experimental observations. It is found in this study that the onset of instability can be predicted from a differential Reynolds number Re_y, defined as $y^2(\partial u/\partial y)/\nu$, whose maximum value for the case of the Blasius velocity profile is located at $y_c = 2.95\sqrt{\nu x/U_\infty}$ or $= 0.6\delta$. The maximum Reynolds number at this critical depth is found to be $\mathrm{Re}_{y_c} = 1.454 U_\infty \sqrt{\nu x/U_\infty}/\nu = 1.454\sqrt{\mathrm{Re}_{x_c}}$, and shows a theoretical critical value of 681 and a corresponding critical value of Re_X of 219 000. This has been verified by a plot of $\bar{c}_f$ against Re_x that shows a departure from Blasius flow at a critical Re_x = 210 000 and $\bar{c}_f = 0.0029$ for several experiments and a numerical study. Thus the onset of instability is marked by the deviation from Newton's law of friction, which is followed by the onset of convection of momentum that characterizes the transition to turbulence. This particle approach to boundary layer instability is superior to the so-called wave-amplification theory and its verification experiments of artificial wave amplifications
Key words: boundary layer, displacement thickness, Reynolds number and instability.

1. INTRODUCTION

Prandtl [2] discovered the boundary layer early last century, although Lord Rayleigh [3] was the first to investigate the stability of laminar flow, albeit both did not provide a criterion of instability. It is well-established that all boundary layer flows are inherently unstable if the plates are sufficiently long [4]. However, to date the theory of the boundary-layer instability on a flat plate is not in full agreement with experimental observations, particularly in the position and the origin of the onset of instability in laminar flow, in that the theoretical value of critical Reynolds number $\mathrm{Re}_{\delta_1}\left(= U_\infty \delta_1/\nu\right)$ of 520 from stability analyses [5] and [6] can not provide a correct critical Re_x, which also

G.E.A. Meier and K.R. Sreenivasan (eds.), IUTAM Symposium on One Hundred Years of Boundary Layer Research, 483-494,

cannot be determined unambiguously in wave-amplification experiments [7] and [**8**]. The cause of the discrepancy between theory and experiments may be traced to the first theoretical study of [**1**], who recognized that his approach to determine the interface ("Grenzschichtdicke") thickness, now known as displacement thickness $\delta_1 = 1.72\sqrt{\nu x/U_\infty}$, from the basic fundamental science is unscientific and flawed. He has deterministically defined the displacement thickness from a flow balance to delineate the portion of the flow stream in the boundary layer that would have a free stream velocity U_∞, i.e. $U_\infty \delta_1 = \int_{y=0}^{\infty} (U_\infty - u) dy$. The integral merely represents the imaginary reduced or displaced flow due to the effect of friction; thus, the displacement thickness is fundamentally irrational and problematic as it is not the point of instability. He also discovered that the value of transition Re_x of 300 000 from experiments of Hansen [**9**] yielded a large Re_{δ_1} of 950, which is more than double his theoretical value of 420. Conversely his theoretical value of 420 would yield a critical Re_x of only 59, 600. Later [7] repeated this calculation with the more accurate value of $\mathrm{Re}_{\delta_1} = 520$ and also reached the same conclusion. It is clear that the so-called displacement thickness is merely a mathematical convenience for providing a sensible scaling length for the definition of a Reynolds number in any stability analysis, it later became a convention that defied critical review. Indeed, all stability analyses, be it linear stability analysis (LSA), parabolized stability equations (PSE) or direct numerical simulations (DNS), do not yield a unique solution as the critical point of instability is not known *a priori;* hence, they may be normalized by any arbitrary length of the boundary layer.

Tollmien's [**1**] failure to obtain a correct critical Re_x from his theory is perhaps explainable since the formulation of the Reynolds number Re_{δ_1} hints at a linear velocity gradient whereas the velocity profile is non-linear. The linear form of Reynolds number $\mathrm{Re}_{\delta_1} = U_\infty \delta_1 / \nu$ implied a linear velocity gradient $du/dy = U_\infty / \delta_1$, which is roughly twice that of the Blasius velocity profile at $\delta_1 = 1.72\sqrt{\nu x/U_\infty}$, where $du/dy \approx 0.5 U_\infty / \delta_1$. Consequently the resultant critical Re_x derived from Re_{δ_1} based on $du/dy = U_\infty / \delta_1$ with a theoretical critical value of 520 will be substantially less than the expected theoretical value of Re_x. Saric [**10**] was critical of the inadequacy of δ_l, however, his attempt to use a reference thickness δ_r in defining a $\mathrm{Re}_{\delta_r} = \sqrt{\mathrm{Re}_x}$ is also equally futile, if not adding to more confusion, since it results in an unusually small value of $\mathrm{Re}_{\delta_r} = 520/1.72 = 302$. Lim [**11**] has calculated with a simplified linear velocity profile a value of Re_{δ_1} of 519, which is close to the commonly accepted value of 520. It is thus possible to define a Re_{δ_1}

based on the linear velocity gradient of $du/dy \approx 0.5U_\infty/\delta_1$, that is $Re_{\delta_1} = 0.5U_\infty\delta_1/\nu$, and arrive at a correct critical Re_x of about 216 000, based on $Re_{\delta_1} \cong 400$ to account for the leading-edge effect (see later). However, the agreement is fortuitous since δ_l is not the point of instability. This may be solved with the theory of transient instability of Tan and Thorpe [**12-16**], who had successfully predicted the onset of transient buoyancy, Marangoni and momentum instability induced by unsteady-state diffusions, which are characterized by nonlinear temperature, concentration and velocity profiles respectively. The onsets of these instabilities were easily predicted accurately with their newly defined transient Rayleigh numbers, transient Marangoni and transient Taylor numbers respectively that incorporated local temperature, concentration and velocity gradients. Tan [**16**] had systematically derived a unified transient instability theory from a comprehensive study of this broad class of instability, it will be applied to the present study of boundary layer instability.

Hinze's [**17**] detailed review of the mechanism of instability and transition revealed that the experimental velocity profile showed a rounded kink at a point approximately $y = 0.6\delta$, i.e. $3\sqrt{\nu x/U_\infty}$ in the boundary layer. Schlichting [**18**] and Kurtz and Crandall [**19**], Kaplan [**20**] and [**5**] found in their calculations and Schubauer and Skramstad [**21**] in their experiments that the perturbation velocity underwent 180° phase shift at $y = 3\sqrt{\nu x/U_\infty}$ in the region of "transition". Klebanov and Tidstrom [**22**] and later Kovasznay *et al.* [**23**] found that breakdown of local flow occurred at $y = 0.6\delta$. Perhaps the best measurement of $y_c = 0.6\delta$ was provided by the experiments of Klingmann *et al.* [**24**] and Swearingen and Blackwelder [**25**], when the filamentous convection plumes first appeared. Thus, the point of instability should be characterized by this critical depth and not by the so-called displacement thickness.

The mechanism of the onset of instability in flows is generally known to be caused by the local in-equilibrium between the inertial and viscous forces, which can be discerned from the momentum equation for a boundary layer with zero-pressure gradient and $du/dx = 0$: $v(du/dy) = \nu(d^2u/dy^2)$. If viscosity is the cause of instability in flows, then the friction drag at the point of onset of instability will rise abruptly. The defining test for the onset of instability is then the determination of the point of deviation from the theoretical friction drag coefficient for Blasius flows given by $\bar{c}_f = 1.328Re_x^{-1/2}$. Comprehensive measurements of friction drag of Hislop [**26**] and Liepmann and Dhawan [**27**] showed a departure from laminar flow at a critical Re_x of about 210 000 and $\bar{c}_f = 0.0029$, Figure 1. This has also been observed for a boundary layer on a slightly-curved plate by [25] at critical $Re_x \sim 210\,000$ and $\bar{c}_f = 0.0029$, which have been successfully simulated by Lee and Liu [28].

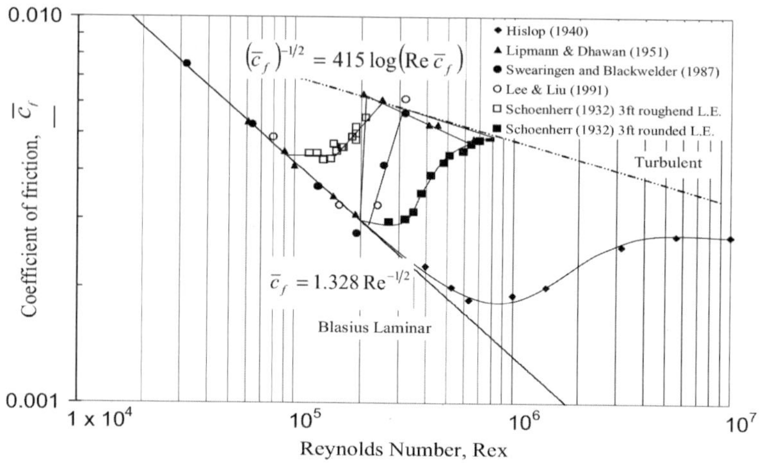

Figure 1. Onset of instability caused by momentum diffusion on a flat plate.

Historically the $\bar{c}_f$ in the transition regime for flat plate has not been measured accurately primarily because most past researchers were not aware of the emergence of the slow and fast streaks at the onset of instability, which will entail the measurement of the average friction drag in the spanwise direction. The local values of du/dy of slow and fast streaks measured by [25] and calculated by [28] appear to be a first attempt to quantify the effect of friction drag on instability. We have computed the span-wise average $\bar{c}_f$ from their data and accurately determined the onset of instability as shown in Figure 1. All these experiments of low disturbance flows support the theory that the onset of instability occur at a critical value of Re_x of 210 000. Experiments of Shoenherr [29] with a flat plate moving in quiescent water also showed that the point of onset of instability is located at $Re_x \sim 200\,000$, and roughness significantly reduced it to about 130 000. Surprisingly this simple conclusive verification of the boundary-layer theory has not been recognized, nor was there any serious attempt in addressing [1] failure to determine a correct value of critical Re_x.

The onset of instability is clearly delineated by the departure from Newton's [30] law of friction or momentum diffusion, and the ensuing onset of convection of momentum is the well-known transition to turbulence. It is interesting to note that G. I. Taylor had, thirteen years after the first publication of his unprecedented theory and experiments

[**31**] for steady-state Couette flow instability, felt compelled to measure the friction drag between the cylinders as the only direct test for his theory. He [**32**] accurately detected the onset of instability at a critical $\overline{c}_{\mathrm{f}} = 0.0126$ with a plot of the friction drag versus Reynolds number (Re_d) when the inner cylinder is moving. The corresponding critical Taylor number $\left(Ta = U_i^2 d^3 / \nu^2 R_i\right)$ can now be calculated as 1650, which is close to his theoretical value of 1706. His experiments confirm that the onset of instability in flows is the cause of increased friction, and not the amplification of waves. Recently [**16**] had shown that transient Couette instability for an impulsively started cylinder in a large gap, Chen and Christensen [**33**] and Taneda [**34**] is analogous to the onset of instability in a deep tank of fluid heated from below, where thin narrow plumes first appeared and later developed into mushroom-shaped convective plumes. They defined a new transient Taylor number $Ta = z^5 \left(\partial u / \partial z\right)^2 / \nu^2 R_i$ that incorporated the local velocity gradient and the penetration depth and determined the maximum transient Taylor number, $Ta_{\max} = 1.461 U_i^2 \left(\sqrt{\nu t}\right)^3 / \nu^2 R_i$. They showed that the point of instability for the experiments of Kirchner and Chen [**35**] occurred at a transient Taylor number of 1100, which is similar in magnitude to that of a critical Rayleigh number for thermal instability in a fluid layer bounded by a free and a solid surface. The experimental critical dimensionless wavenumber of 3.05 also agrees well with the theoretical value of 2.9. These Couette instability studies seem to confirm that the stability analyses of 2D laminar flows can only provide the criterion of the onset of instability.

Experiments of wave amplifications first performed by [**21**] suffered from a serious drawback, in that they could not detect the very small perturbation, which was replaced by artificial disturbance generated by electromagnetically-vibrated ribbon. Indeed Saric's [**36**] comprehensive review had confirmed that natural disturbances in the free stream were too small for detection. Moreover, the artificial wave experiments measured only a value of Re_{δ_1} = 400, which was lower than the theoretical value of 520. The discrepancy has been found to be caused by the leading edge. Therefore, the onset of instability on a flat plate will inevitably occur at a critical value of Re_{δ_1} approximately 400 [**37**]. Indeed, the purported experiments of flow instability have been reduced to the investigation of the influence of fluid flow on an artificial wave. Strictly conventional perturbation theory only traces the initial growth of the perturbation up to the point of instability, while the wave equations of perturbed variables merely provide a method to determine the minimum value of Reynolds number that locates the position of the point of instability. Fundamentally, the artificial waves that have substituted the undetectable elusive waves of natural flows are not identical to the hypothetical waves of the stability theory. Thus, the onset of instability in natural flows may not necessarily be caused by wave amplifications. All perturbation analyses of thermal instability Lord Rayleigh [**38**] and Pearson [**39**] and Taylor instability [**31**] also employed wave equations with a wave frequency, but they never relied on the measurements of the wave amplifications to verify the theory. Rather, the heat flux Schmidt and Milverton [**40**] and Silveston [**41**] and friction drag [**32**] respectively were measured to detect the onset of instability at a critical Rayleigh number and Taylor number.

In summary the absence of a unique solution of the OSE, PSE and DNS implies the impossibility of determining the point of onset of instability unambiguously, both theoretically and experimentally. The $\bar{c}_f - \mathrm{Re}_x$ plot employing Newton's law of friction seems to be the only correct and reliable method in determining the position of instability. This paper will propose a new theory of the onset of instability with the definition of a new spatially transient Reynolds number that will determine the point of onset of instability and the maximum transient Reynolds number, from which the theoretical value of Re_{x_c} can be determined accurately.

2. SPATIALLY TRANSIENT REYNOLDS NUMBER

The flow of fluid on a flat plate experiences progressive retardation due to the viscous shear of the fluid. At a critical distance downstream the momentary imbalance between the local inertial force and the frictional shear may result in local instability. The position of the point of instability may be simply determined from a maximum transient Reynolds number. The mathematics for determining the onset of instability is similar to that developed for the onset of instability induced by an impulsively-started rotating cylinder [**16**]. The rapidly changing velocity gradients $(\partial u/\partial y)$ in the wall region is the primary driving force of instability, which is readily seen in the momentum equation for a boundary layer with zero pressure gradient and $du/dx = 0$: $v(du/dy) = \nu(d^2u/dy^2)$. The critical point may be determined by defining a new spatially transient Reynolds number in terms of the distance from the plate y and the varying local velocity gradient as:

$$\mathrm{Re}_y = \frac{y^2}{\nu}\frac{\partial u}{\partial y} \tag{1}$$

The local velocity gradient in the laminar boundary layer is given by the Blasius velocity profile [**42**] and substituting it in Equation

$$\mathrm{Re}_y = \frac{U_\infty y^2}{\sqrt{\nu^3 x/U_\infty}} f''(\eta) \quad (1),$$

becomes (2)

$$= \frac{U_\infty \sqrt{\nu x/U_\infty}}{\nu} \eta^2 f''(\eta)$$

where $\eta = y\sqrt{U_\infty/\nu x}$, $f(\eta) = \psi/\sqrt{U_\infty x\nu}$ and $u = U_\infty f'(\eta)$. The values of η and f'' are provided by table of Blasius functions [7]. The maximum value of the transient Reynolds number can be found by plotting Re_y against η as shown in Figure 2, which shows the maximum Re_y at

$$\eta_c = 2.95 \quad (3)$$

that gives the critical depth at $y_c = 2.95\sqrt{\nu x/U_\infty}$. This agrees remarkably well with theoretical calculations and experimental observations discussed previously, that instability originates at the point $y_c = 3\sqrt{\nu x/U_\infty} = 0.6\delta$. This is the critical depth required to normalize the stability equations that may yield the point of instability unambiguously.

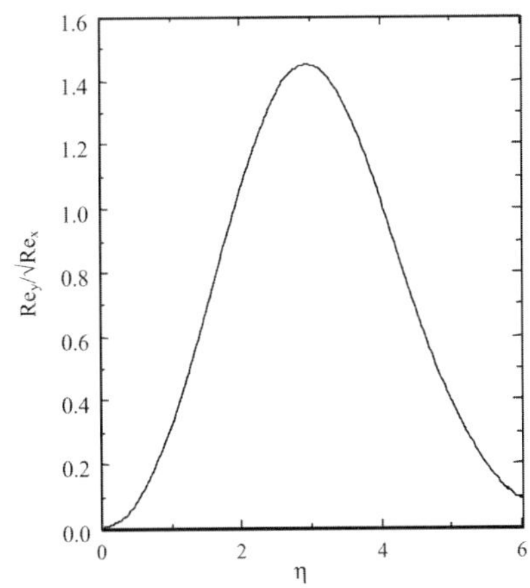

Figure 2. Variation of Reynolds number in the boundary layer

The maximum transient Reynolds number at the critical depth is easily determined by substituting $\eta_c = 2.95$ and the value of $f''(\eta) = 0.1614$ at $\eta_c = 2.95$ in Equation (2) as:

$$\mathrm{Re}_{y_c} = 1.454\sqrt{\mathrm{Re}_{x_c}} \tag{4}$$

The formulation of the new spatially transient Reynolds number is reasonable as the fluid flows with a mean velocity of nearly $0.5U_\infty$ through the critical depth $y_c = 3\sqrt{\nu x/U_\infty}$, and thus the Reynolds number is $= 0.5U_\infty \times 3\sqrt{\nu x/U_\infty}/\nu = 1.5U_\infty\sqrt{\nu x/U_\infty}/\nu$ or $1.5\sqrt{\mathrm{Re}_x}$. The variation of Reynolds numbers with the depth of boundary layer is shown in Figure 2, which has a peak value at $\eta_c = 2.95$ and drops sharply thereafter. Reynolds numbers after the peak value of the onset of instability have no more meaningful values because the laminar Blasius velocity profile would have broken down and convection commenced.

There is no known theoretical value of Reynolds number based on this critical depth y_c. However, Hinze [**17**] has shown that Reynolds number at $\delta = 5\sqrt{\nu x/U_\infty}$ can be extrapolated linearly from Reynolds number at the displacement thickness from linear stability analysis, since the Reynolds number is a function of the thickness of the boundary layer. Therefore, the value of critical Reynolds number at the critical depth y_c for $\delta_e = 1.454\sqrt{\nu x/U_\infty}$ may be extrapolated linearly from $\mathrm{Re}_{\delta_1} = 0.854\sqrt{\mathrm{Re}_x}$ at $\delta_1 = 1.72\sqrt{\nu x/U_\infty}$ with the theoretical value of $\mathrm{Re}_{\delta_1} = 400$, and is found to be $\mathrm{Re}_{y_c} = 681$. The critical Reynolds number for the horizontal distance x can now be found from $\mathrm{Re}_{x_c} = (\mathrm{Re}_{\delta_e}/1.454)^2$ as 219 000, which is close to the experimental value of 210, 000 as shown in Figure 1. This same critical value of Re_x may also be easily obtained for $\mathrm{Re}_{\delta 1} = 400$ from $\mathrm{Re}_{x_c} = (400/0.854)^2 = 219\,000$. More accurate measurements of [**26**], [**27**], [**25**] and numerical calculations of [**28**] showed that the leading-edge effect will lead to a $\mathrm{Re}_{x_c} = 210\,000$.

The experimental value of critical $\mathrm{Re}_{x_c} = 210\,000$ from Figure 1 is easily shown by $\mathrm{Re}_{y_c} = 1.454\sqrt{\mathrm{Re}_{x_c}}$, to give the value of maximum transient Reynolds number or Re_{y_c} of 666, which is close to the linearly extrapolated theoretical value of Re_{y_c} of 681. The value of Re_{y_c} of 666 results in a Re_{δ_1} of 391, which is close to the experimental value of Ross *et al.* [**42**], Strazisar *et al.* [**43**] and [**5**].

The critical distance for the onset of instability can be found from Equation (4) with a critical Reynolds number of 210 000 as $x_c = 2.10\times10^5 \nu/U_\infty$. For Blasius flows the theoretical critical $\bar{c}_f$ at $\mathrm{Re}_{x_c} = 210\,000$ is found to be 0.0029 from $\bar{c}_f = 1.328\,\mathrm{Re}^{-1/2}$. There thus exists a critical shear that causes the onset of instability, and is given by $\tau_{0_c} = 0.0029\left(0.5\rho U_\infty^2\right) = 0.00145\rho U_\infty^2$, which provides a very simple and quick calculation of the critical shear for a fluid with a known velocity. The critical dimensionless wavenumber based on the critical depth is found to be $\tilde{a}_c = a_c y_c = 0.516$, since the one normalized with the displacement thickness is $\tilde{a}_c = a_c \delta_1 = 0.301$ [**5**]. The critical horizontal wavelength of the plume is therefore $\lambda_c = 2\pi y_c/\tilde{a}_c = 12.18 y_c = 35.9\sqrt{\nu x_c/U_\infty}$, which may be expressed in terms of the critical Reynolds number of 219 000 as $\lambda_c = 35.9\,\mathrm{Re}_{x_c}^{\;1/2}\,\nu/U_\infty = 16450\nu/U_\infty$, which shows that the λ_c is dependent only on the free-stream velocity.

3. CONCLUSIONS

It is found that all stability analyses that have been normalized by arbitrary length scale cannot provide a unique solution for the position of instability. This is because one does not know *a priori* nor *a posteiori* from these analyses the critical depth within the boundary layer at which instability originates. The origin of instability is found to be located at a critical depth $y_c = 3\sqrt{\nu x/U_\infty}$ or $= 0.6\delta$ at a critical horizontal distance from the leading edge $x_c = 2.19\times10^5 \nu/U_\infty$. The critical depth may be used for the normalization of the stability equations so as to yield a unique solution that may predict the onset of instability unambiguously. This has provided a rational and scientific basis for determining the critical depth of the boundary layer where instability originates.

The new differential form of Reynolds number can successfully predict the onset of instability at the critical point by Equation (4), i.e. $\mathrm{Re}_{y_c} = 1.454\sqrt{\mathrm{Re}_{x_c}}$. The theoretical values of Reynolds number at the onset of instability are found to be $\mathrm{Re}_{y_c} = 681$ and Re_{x_c} = 219 000, which are in good agreement with the average experimental values of 670 and 210 000, respectively. The onsets of instability in several studies were found to occur at a critical friction drag coefficient of 0.0029, which agrees very well with the theoretical value of 0.0029. The maximum $\bar{c}_f$ at the point of transition to turbulence is about 0.006, which is about twice the value of critical $\bar{c}_f$ for the onset of instability.

REFERENCES

1. Tollmien, W. "Uber die entstehung der Turrbulenz", *I. Mitt. Nachr. Fes. Wiss. Gottingen, Math. Phys. Klasse*, pp. 21-44, 1929; English translations in NACA TM 609, 1931.

2. Prandtl, L. "Über Flüssigkeitsbewegung bei sehr kleiner Reibung, or Fluid motion with very small friction", *Proc. Third Intern. Math. Congr.*, Heidelberg, 1904, pp. 484-491, 1904. Reprinted in: NACA TM 452 (1928)

3. Lord Rayleigh. "On the stability or instability of certain fluid motions", *Proc. London Mathematics Society*, **11**, pp. 57-70, 1880.

4. Drazin, P. G. Personal communications by e-mail 5th July 2000. "So the critical value of the Re gained from the OSE tells more of *where* the boundary layer becomes unstable, rather than indicating *whether* it is stable or unstable. In fact *all* boundary layers on plates with large chords length are unstable."

5. Jordinson, R. "The flat plate boundary layer. Part 1. Numerical integration of Orr-Sommerfeld equation", *Jounal of Fluid Mechanics*, **43**, 801, 1970.

6. Gaster, M. "On the effects of boundary layer growth on flow stability", *Journal of Fluid Mechanics*, **43**, pp. 813-818, 1974.

7. Schlichting, H. "Boundary-Layer Theory" 1979. (translated by J. Kestin. McGraw-Hill, 7th ed., chap. V and XVII). Also 8th ed. revised by Gersten, K., 2000.

8. Saric, W. S., White, E. B., & Reed, H. L. "Boundary layer 'receptivity to free stream disturbances and its role in transitions", *Invited paper at 30th AIAA Fluid Dynamics Conference*, 28 June – 1st July 1999, Norfolk, VA, American Institute of Aeronautics and Astronautics, 99-3788, 1999.

9. Hansen, M. "Die Geschwindigkeitsverteilung in der Grenzschicht an einer eingetauchten Platte", ZAMM **8**, pp. 185-199, 1928. (NACA TM 585, 1930).

10. Saric, W. S. "Skin fiction drag reduction", AGARD Report 786, 1992.

11. Lim, C. W. (2003) personal communications. Please refer to: C.W. Lim, The stability of flow over periodically supported plates, M. Eng. thesis, National University of Singapore, November, 1991, and K.S. Yeo and C.W. Lim, The stability of flow over periodically supported plates – Potential flow, *J. Fluid Structure*, 8, 331-354, 1994.

12. Tan, K. K. & Thorpe, R. B. "The onset of convection caused by buoyancy during transient heat conduction in deep fluids", *Chemical Engineering Science*, **51**, 17, 4127, 1996.

13. Tan, K. K. & Thorpe, R. B. "The onset of convection driven by buoyancy caused by transient heat conduction: Part I. Transient Rayleigh numbers", *Chemical Engineering Science*, **54,** 225, 1999a.

14. Tan, K. K. & Thorpe, R. B. "The onset of convection driven by buoyancy caused by transient heat conduction: Part II: The Sizes of Plumes", *Chemical Engineering Science,* **54,** 239, 1999b.

15. Tan, K. K. & Thorpe, R. B. "On convection driven by surface tension caused by transient cooling", *Chemical Engineering Science*, **54**, 775, 1999c.

16. Tan, K. K. & Thorpe, R. B. "Transient Instability Of The Flow Induced By An Impulsively Started Rotating Cylinder", *Chemical Engineering Science*, **58**, pp. 149-156, 2003.

17. Hinze, J. O. (1975). *Turbulence*, 2nd ed. (McGraw-Hill, New York,).

18. Schlichting, H. "Amplitudenverteilung und Energiebilanz der kleinen storungen bei der Plattenstromung" *Nachr. Ges. wiss. Gottingen, Mathematics Physics*, Klasse, Fachgruppe I, 1. pp. 47-78, 1935.

19. Kurtz, E. F., & Crandall, S. H. "Computer-aided analysis of hydrodynamic stability", *Journal of Mathematics Physics*, 44, 264, 1962.

20. Kaplan, R. E. (1964). "The stability of laminar incompressible boundary layers in the presence of compliant boundaries" (Ph.D. Thesis, Massachusetts Institute of Technology, Aero-Elastic and Structures Research Laboratory, ASRL TR 116-1)

21. Schubauer, G. B., & Skramstad, H. K. "Laminar boundary layer oscillations and stability of laminar flow", *Journal of Aeronautical Science,* **14**, pp. 69-78, 1947. Also Laminar boundary layer oscillations on a flat plate. *NACA Report*, 909.

22. Klebanov, P. S., Tidstrom, K. D. & Sargent, L. M. "The three-dimensional nature of boundary layer instability", *Journal of Fluid Mechanics*, **40**, pp. 39-47, 1962.

23. Kovasznay, L. S. G., Komoda, H. & Vsudeva, B. R. "Detailed flow field in transition", *Proc. of the 1962 Heat transfer and Fluid Mechanics Institute* (Stanford University Press, Stanford, California) pp. 1-26, 1962.

24. Klingmann, B. G. B., Boiko, A. V., Westin, K. J. A., Kozlov, V. V. & Alfredsson. "Experiments on the stability of Tollmien-Schlichting waves", *European Journal Mechanics*, B/Fluids, 12, no. **4**, pp. 493-514, 1993.

25. Swearingen, J. D. & Blackwelder, R. F. "The growth and breakdown of streamwise vortices in the presence of a wall", *Journal of Fluid Mechanics*, **182**, pp. 255-290, 1987.

26. Hislop, G. S. (1940). "The transition of a laminar boundary layer in a wind tunnel", PhD thesis, Department of Engineering, University of Cambridge.

27. Liepmann, H.W. and Dhawan, S. "Direct measurements of skin friction in low-speed and high-speed flow", *Proc. First US Nat. Congr. Appl. Mech.* 869, 1951.

28. Lee, K. & Liu, J. T. C. "On the growth of mushroom-like structures in nonlinear spatially developing Goertler vortex flow", *Physics of Fluids A*, **4**, pp. 95-103, 1992.

29. Schoenherr, K. E. "Resistance of Flat surfaces moving through a fluid", *Trans. Society. Nav. Arch. And Mar. Eng.,* 40, 279, 1932. (See Emmons, 1951)

30. Newton, I. (1687). Principia Mathematica, Book II, Section IX: The circular motion of fluids.

31. Taylor, G. I. "Stability of a viscous liquid contained between two rotating cylinders", *Phil. Trans. Roy. Society,* (London) A223, 289, 1923.

32. Taylor, G. I. "Fluid friction between two rotating cylinders", *Proc. Roy. Soc.*, A157, 546-64 and 565-78, 1936.

33. Chen, C. F. & Christensen, D. K. "Stability of flow induced by an impulsively started rotating cylinder" *Physics of Fluid,* 10, 8, 1845, 1967.

34. Taneda, S. "Visual study of unsteady separated flows around bodies", *Prog. Aerospace Science*, **17**, pp. 287, 1977. Also in *An Album of Fluid Motion*, (edited by M. van Dyke, The Parabolic Press, 1982).

35. Kirchner, R. P. & Chen, C. F. "Stability of time-dependent rotational Coutte flow. Part 1. Experimental investigation", *Journal of Fluid Mechanics,* 40, 39, 1970.

36. Saric, W.S. "Progress in transition modeling", *AGARD Report* 793, 1993.

37. Herbert, T. "The dimensional phenomena in the transitional flat-plate boundary layer", *American Institute of Aeronautics and Astronautics*, 85-0489, 1985.

38. Lord Rayleigh. "On convective currents in a horizontal layer of fluid when the higher temperature is on the under side" *Phil. Magazine.* 32, 529, 1916.

39. Pearson, J. R. A. "On convection cells induced by surface tension", *J. Fluid Mech.*, 4, 489, 1958.

40. Schmidt, R. J., & Milverton, S. W. "On the instability of a fluid when heated from below", *Proc. Ray. Soc.*, A 152, 586, 1935.

41. Silveston, P. L. "Wärmedurchgang in Waagerechten Flüssigkeitsschichten Forch", *Gebiete Ingenieurwes*. **24**, pp. 29-32 & pp. 59-69, 1958.

42. Ross, J. A., Barnes, F. H., Burns, J. G., & Ross, M. A. S. "The flat plate boundary layer. Part 3. Comparison of theory with experiment", *Journal of Fluid Mechanics*, 43, 819, 1970.

43. Strazisar, A. J., Prahc, J. M. & Reshotko, E. "Experimental study of the study of heated boundary layers in waters". *FTAS/TR* -75 -113, 1976. Case Western University, Department Fluid Thermal Aerospace Science.

Mechanics

SOLID MECHANICS AND ITS APPLICATIONS

Series Editor: G.M.L. Gladwell

Aims and Scope of the Series

The fundamental questions arising in mechanics are: *Why?*, *How?*, and *How much?* The aim of this series is to provide lucid accounts written by authoritative researchers giving vision and insight in answering these questions on the subject of mechanics as it relates to solids. The scope of the series covers the entire spectrum of solid mechanics. Thus it includes the foundation of mechanics; variational formulations; computational mechanics; statics, kinematics and dynamics of rigid and elastic bodies; vibrations of solids and structures; dynamical systems and chaos; the theories of elasticity, plasticity and viscoelasticity; composite materials; rods, beams, shells and membranes; structural control and stability; soils, rocks and geomechanics; fracture; tribology; experimental mechanics; biomechanics and machine design.

1. R.T. Haftka, Z. Gürdal and M.P. Kamat: *Elements of Structural Optimization*. 2nd rev.ed., 1990 ISBN 0-7923-0608-2
2. J.J. Kalker: *Three-Dimensional Elastic Bodies in Rolling Contact*. 1990 ISBN 0-7923-0712-7
3. P. Karasudhi: *Foundations of Solid Mechanics*. 1991 ISBN 0-7923-0772-0
4. *Not published*
5. *Not published.*
6. J.F. Doyle: *Static and Dynamic Analysis of Structures*. With an Emphasis on Mechanics and Computer Matrix Methods. 1991 ISBN 0-7923-1124-8; Pb 0-7923-1208-2
7. O.O. Ochoa and J.N. Reddy: *Finite Element Analysis of Composite Laminates*. ISBN 0-7923-1125-6
8. M.H. Aliabadi and D.P. Rooke: *Numerical Fracture Mechanics*. ISBN 0-7923-1175-2
9. J. Angeles and C.S. López-Cajún: *Optimization of Cam Mechanisms*. 1991 ISBN 0-7923-1355-0
10. D.E. Grierson, A. Franchi and P. Riva (eds.): *Progress in Structural Engineering*. 1991 ISBN 0-7923-1396-8
11. R.T. Haftka and Z. Gürdal: *Elements of Structural Optimization*. 3rd rev. and exp. ed. 1992 ISBN 0-7923-1504-9; Pb 0-7923-1505-7
12. J.R. Barber: *Elasticity*. 1992 ISBN 0-7923-1609-6; Pb 0-7923-1610-X
13. H.S. Tzou and G.L. Anderson (eds.): *Intelligent Structural Systems*. 1992 ISBN 0-7923-1920-6
14. E.E. Gdoutos: *Fracture Mechanics*. An Introduction. 1993 ISBN 0-7923-1932-X
15. J.P. Ward: *Solid Mechanics*. An Introduction. 1992 ISBN 0-7923-1949-4
16. M. Farshad: *Design and Analysis of Shell Structures*. 1992 ISBN 0-7923-1950-8
17. H.S. Tzou and T. Fukuda (eds.): *Precision Sensors, Actuators and Systems*. 1992 ISBN 0-7923-2015-8
18. J.R. Vinson: *The Behavior of Shells Composed of Isotropic and Composite Materials*. 1993 ISBN 0-7923-2113-8
19. H.S. Tzou: *Piezoelectric Shells*. Distributed Sensing and Control of Continua. 1993 ISBN 0-7923-2186-3
20. W. Schiehlen (ed.): *Advanced Multibody System Dynamics*. Simulation and Software Tools. 1993 ISBN 0-7923-2192-8
21. C.-W. Lee: *Vibration Analysis of Rotors*. 1993 ISBN 0-7923-2300-9
22. D.R. Smith: *An Introduction to Continuum Mechanics*. 1993 ISBN 0-7923-2454-4
23. G.M.L. Gladwell: *Inverse Problems in Scattering*. An Introduction. 1993 ISBN 0-7923-2478-1

Mechanics

SOLID MECHANICS AND ITS APPLICATIONS

Series Editor: G.M.L. Gladwell

24. G. Prathap: *The Finite Element Method in Structural Mechanics.* 1993 ISBN 0-7923-2492-7
25. J. Herskovits (ed.): *Advances in Structural Optimization.* 1995 ISBN 0-7923-2510-9
26. M.A. González-Palacios and J. Angeles: *Cam Synthesis.* 1993 ISBN 0-7923-2536-2
27. W.S. Hall: *The Boundary Element Method.* 1993 ISBN 0-7923-2580-X
28. J. Angeles, G. Hommel and P. Kovács (eds.): *Computational Kinematics.* 1993 ISBN 0-7923-2585-0
29. A. Curnier: *Computational Methods in Solid Mechanics.* 1994 ISBN 0-7923-2761-6
30. D.A. Hills and D. Nowell: *Mechanics of Fretting Fatigue.* 1994 ISBN 0-7923-2866-3
31. B. Tabarrok and F.P.J. Rimrott: *Variational Methods and Complementary Formulations in Dynamics.* 1994 ISBN 0-7923-2923-6
32. E.H. Dowell (ed.), E.F. Crawley, H.C. Curtiss Jr., D.A. Peters, R. H. Scanlan and F. Sisto: *A Modern Course in Aeroelasticity.* Third Revised and Enlarged Edition. 1995 ISBN 0-7923-2788-8; Pb: 0-7923-2789-6
33. A. Preumont: *Random Vibration and Spectral Analysis.* 1994 ISBN 0-7923-3036-6
34. J.N. Reddy (ed.): *Mechanics of Composite Materials.* Selected works of Nicholas J. Pagano. 1994 ISBN 0-7923-3041-2
35. A.P.S. Selvadurai (ed.): *Mechanics of Poroelastic Media.* 1996 ISBN 0-7923-3329-2
36. Z. Mróz, D. Weichert, S. Dorosz (eds.): *Inelastic Behaviour of Structures under Variable Loads.* 1995 ISBN 0-7923-3397-7
37. R. Pyrz (ed.): *IUTAM Symposium on Microstructure-Property Interactions in Composite Materials.* Proceedings of the IUTAM Symposium held in Aalborg, Denmark. 1995 ISBN 0-7923-3427-2
38. M.I. Friswell and J.E. Mottershead: *Finite Element Model Updating in Structural Dynamics.* 1995 ISBN 0-7923-3431-0
39. D.F. Parker and A.H. England (eds.): *IUTAM Symposium on Anisotropy, Inhomogeneity and Nonlinearity in Solid Mechanics.* Proceedings of the IUTAM Symposium held in Nottingham, U.K. 1995 ISBN 0-7923-3594-5
40. J.-P. Merlet and B. Ravani (eds.): *Computational Kinematics '95.* 1995 ISBN 0-7923-3673-9
41. L.P. Lebedev, I.I. Vorovich and G.M.L. Gladwell: *Functional Analysis.* Applications in Mechanics and Inverse Problems. 1996 ISBN 0-7923-3849-9
42. J. Menčik: *Mechanics of Components with Treated or Coated Surfaces.* 1996 ISBN 0-7923-3700-X
43. D. Bestle and W. Schiehlen (eds.): *IUTAM Symposium on Optimization of Mechanical Systems.* Proceedings of the IUTAM Symposium held in Stuttgart, Germany. 1996 ISBN 0-7923-3830-8
44. D.A. Hills, P.A. Kelly, D.N. Dai and A.M. Korsunsky: *Solution of Crack Problems.* The Distributed Dislocation Technique. 1996 ISBN 0-7923-3848-0
45. V.A. Squire, R.J. Hosking, A.D. Kerr and P.J. Langhorne: *Moving Loads on Ice Plates.* 1996 ISBN 0-7923-3953-3
46. A. Pineau and A. Zaoui (eds.): *IUTAM Symposium on Micromechanics of Plasticity and Damage of Multiphase Materials.* Proceedings of the IUTAM Symposium held in Sèvres, Paris, France. 1996 ISBN 0-7923-4188-0
47. A. Naess and S. Krenk (eds.): *IUTAM Symposium on Advances in Nonlinear Stochastic Mechanics.* Proceedings of the IUTAM Symposium held in Trondheim, Norway. 1996 ISBN 0-7923-4193-7
48. D. Ieşan and A. Scalia: *Thermoelastic Deformations.* 1996 ISBN 0-7923-4230-5

Mechanics

SOLID MECHANICS AND ITS APPLICATIONS

Series Editor: G.M.L. Gladwell

49. J.R. Willis (ed.): *IUTAM Symposium on Nonlinear Analysis of Fracture*. Proceedings of the IUTAM Symposium held in Cambridge, U.K. 1997 ISBN 0-7923-4378-6
50. A. Preumont: *Vibration Control of Active Structures*. An Introduction. 1997 ISBN 0-7923-4392-1
51. G.P. Cherepanov: *Methods of Fracture Mechanics: Solid Matter Physics*. 1997 ISBN 0-7923-4408-1
52. D.H. van Campen (ed.): *IUTAM Symposium on Interaction between Dynamics and Control in Advanced Mechanical Systems*. Proceedings of the IUTAM Symposium held in Eindhoven, The Netherlands. 1997 ISBN 0-7923-4429-4
53. N.A. Fleck and A.C.F. Cocks (eds.): *IUTAM Symposium on Mechanics of Granular and Porous Materials*. Proceedings of the IUTAM Symposium held in Cambridge, U.K. 1997 ISBN 0-7923-4553-3
54. J. Roorda and N.K. Srivastava (eds.): *Trends in Structural Mechanics*. Theory, Practice, Education. 1997 ISBN 0-7923-4603-3
55. Yu.A. Mitropolskii and N. Van Dao: *Applied Asymptotic Methods in Nonlinear Oscillations*. 1997 ISBN 0-7923-4605-X
56. C. Guedes Soares (ed.): *Probabilistic Methods for Structural Design*. 1997 ISBN 0-7923-4670-X
57. D. François, A. Pineau and A. Zaoui: *Mechanical Behaviour of Materials*. Volume I: Elasticity and Plasticity. 1998 ISBN 0-7923-4894-X
58. D. François, A. Pineau and A. Zaoui: *Mechanical Behaviour of Materials*. Volume II: Viscoplasticity, Damage, Fracture and Contact Mechanics. 1998 ISBN 0-7923-4895-8
59. L.T. Tenek and J. Argyris: *Finite Element Analysis for Composite Structures*. 1998 ISBN 0-7923-4899-0
60. Y.A. Bahei-El-Din and G.J. Dvorak (eds.): *IUTAM Symposium on Transformation Problems in Composite and Active Materials*. Proceedings of the IUTAM Symposium held in Cairo, Egypt. 1998 ISBN 0-7923-5122-3
61. I.G. Goryacheva: *Contact Mechanics in Tribology*. 1998 ISBN 0-7923-5257-2
62. O.T. Bruhns and E. Stein (eds.): *IUTAM Symposium on Micro- and Macrostructural Aspects of Thermoplasticity*. Proceedings of the IUTAM Symposium held in Bochum, Germany. 1999 ISBN 0-7923-5265-3
63. F.C. Moon: *IUTAM Symposium on New Applications of Nonlinear and Chaotic Dynamics in Mechanics*. Proceedings of the IUTAM Symposium held in Ithaca, NY, USA. 1998 ISBN 0-7923-5276-9
64. R. Wang: *IUTAM Symposium on Rheology of Bodies with Defects*. Proceedings of the IUTAM Symposium held in Beijing, China. 1999 ISBN 0-7923-5297-1
65. Yu.I. Dimitrienko: *Thermomechanics of Composites under High Temperatures*. 1999 ISBN 0-7923-4899-0
66. P. Argoul, M. Frémond and Q.S. Nguyen (eds.): *IUTAM Symposium on Variations of Domains and Free-Boundary Problems in Solid Mechanics*. Proceedings of the IUTAM Symposium held in Paris, France. 1999 ISBN 0-7923-5450-8
67. F.J. Fahy and W.G. Price (eds.): *IUTAM Symposium on Statistical Energy Analysis*. Proceedings of the IUTAM Symposium held in Southampton, U.K. 1999 ISBN 0-7923-5457-5
68. H.A. Mang and F.G. Rammerstorfer (eds.): *IUTAM Symposium on Discretization Methods in Structural Mechanics*. Proceedings of the IUTAM Symposium held in Vienna, Austria. 1999 ISBN 0-7923-5591-1

Mechanics

SOLID MECHANICS AND ITS APPLICATIONS

Series Editor: G.M.L. Gladwell

69. P. Pedersen and M.P. Bendsøe (eds.): *IUTAM Symposium on Synthesis in Bio Solid Mechanics*. Proceedings of the IUTAM Symposium held in Copenhagen, Denmark. 1999 ISBN 0-7923-5615-2
70. S.K. Agrawal and B.C. Fabien: *Optimization of Dynamic Systems*. 1999 ISBN 0-7923-5681-0
71. A. Carpinteri: *Nonlinear Crack Models for Nonmetallic Materials*. 1999 ISBN 0-7923-5750-7
72. F. Pfeifer (ed.): *IUTAM Symposium on Unilateral Multibody Contacts*. Proceedings of the IUTAM Symposium held in Munich, Germany. 1999 ISBN 0-7923-6030-3
73. E. Lavendelis and M. Zakrzhevsky (eds.): *IUTAM/IFToMM Symposium on Synthesis of Nonlinear Dynamical Systems*. Proceedings of the IUTAM/IFToMM Symposium held in Riga, Latvia. 2000 ISBN 0-7923-6106-7
74. J.-P. Merlet: *Parallel Robots*. 2000 ISBN 0-7923-6308-6
75. J.T. Pindera: *Techniques of Tomographic Isodyne Stress Analysis*. 2000 ISBN 0-7923-6388-4
76. G.A. Maugin, R. Drouot and F. Sidoroff (eds.): *Continuum Thermomechanics*. The Art and Science of Modelling Material Behaviour. 2000 ISBN 0-7923-6407-4
77. N. Van Dao and E.J. Kreuzer (eds.): *IUTAM Symposium on Recent Developments in Non-linear Oscillations of Mechanical Systems*. 2000 ISBN 0-7923-6470-8
78. S.D. Akbarov and A.N. Guz: *Mechanics of Curved Composites*. 2000 ISBN 0-7923-6477-5
79. M.B. Rubin: *Cosserat Theories: Shells, Rods and Points*. 2000 ISBN 0-7923-6489-9
80. S. Pellegrino and S.D. Guest (eds.): *IUTAM-IASS Symposium on Deployable Structures: Theory and Applications*. Proceedings of the IUTAM-IASS Symposium held in Cambridge, U.K., 6–9 September 1998. 2000 ISBN 0-7923-6516-X
81. A.D. Rosato and D.L. Blackmore (eds.): *IUTAM Symposium on Segregation in Granular Flows*. Proceedings of the IUTAM Symposium held in Cape May, NJ, U.S.A., June 5–10, 1999. 2000 ISBN 0-7923-6547-X
82. A. Lagarde (ed.): *IUTAM Symposium on Advanced Optical Methods and Applications in Solid Mechanics*. Proceedings of the IUTAM Symposium held in Futuroscope, Poitiers, France, August 31–September 4, 1998. 2000 ISBN 0-7923-6604-2
83. D. Weichert and G. Maier (eds.): *Inelastic Analysis of Structures under Variable Loads*. Theory and Engineering Applications. 2000 ISBN 0-7923-6645-X
84. T.-J. Chuang and J.W. Rudnicki (eds.): *Multiscale Deformation and Fracture in Materials and Structures*. The James R. Rice 60th Anniversary Volume. 2001 ISBN 0-7923-6718-9
85. S. Narayanan and R.N. Iyengar (eds.): *IUTAM Symposium on Nonlinearity and Stochastic Structural Dynamics*. Proceedings of the IUTAM Symposium held in Madras, Chennai, India, 4–8 January 1999 ISBN 0-7923-6733-2
86. S. Murakami and N. Ohno (eds.): *IUTAM Symposium on Creep in Structures*. Proceedings of the IUTAM Symposium held in Nagoya, Japan, 3-7 April 2000. 2001 ISBN 0-7923-6737-5
87. W. Ehlers (ed.): *IUTAM Symposium on Theoretical and Numerical Methods in Continuum Mechanics of Porous Materials*. Proceedings of the IUTAM Symposium held at the University of Stuttgart, Germany, September 5-10, 1999. 2001 ISBN 0-7923-6766-9
88. D. Durban, D. Givoli and J.G. Simmonds (eds.): *Advances in the Mechanis of Plates and Shells The Avinoam Libai Anniversary Volume*. 2001 ISBN 0-7923-6785-5
89. U. Gabbert and H.-S. Tzou (eds.): *IUTAM Symposium on Smart Structures and Structonic Systems*. Proceedings of the IUTAM Symposium held in Magdeburg, Germany, 26–29 September 2000. 2001 ISBN 0-7923-6968-8

Mechanics

SOLID MECHANICS AND ITS APPLICATIONS

Series Editor: G.M.L. Gladwell

90. Y. Ivanov, V. Cheshkov and M. Natova: *Polymer Composite Materials – Interface Phenomena & Processes*. 2001 ISBN 0-7923-7008-2
91. R.C. McPhedran, L.C. Botten and N.A. Nicorovici (eds.): *IUTAM Symposium on Mechanical and Electromagnetic Waves in Structured Media*. Proceedings of the IUTAM Symposium held in Sydney, NSW, Australia, 18-22 Januari 1999. 2001 ISBN 0-7923-7038-4
92. D.A. Sotiropoulos (ed.): *IUTAM Symposium on Mechanical Waves for Composite Structures Characterization*. Proceedings of the IUTAM Symposium held in Chania, Crete, Greece, June 14-17, 2000. 2001 ISBN 0-7923-7164-X
93. V.M. Alexandrov and D.A. Pozharskii: *Three-Dimensional Contact Problems*. 2001 ISBN 0-7923-7165-8
94. J.P. Dempsey and H.H. Shen (eds.): *IUTAM Symposium on Scaling Laws in Ice Mechanics and Ice Dynamics*. Proceedings of the IUTAM Symposium held in Fairbanks, Alaska, U.S.A., 13-16 June 2000. 2001 ISBN 1-4020-0171-1
95. U. Kirsch: *Design-Oriented Analysis of Structures*. A Unified Approach. 2002 ISBN 1-4020-0443-5
96. A. Preumont: *Vibration Control of Active Structures*. An Introduction (2nd Edition). 2002 ISBN 1-4020-0496-6
97. B.L. Karihaloo (ed.): *IUTAM Symposium on Analytical and Computational Fracture Mechanics of Non-Homogeneous Materials*. Proceedings of the IUTAM Symposium held in Cardiff, U.K., 18-22 June 2001. 2002 ISBN 1-4020-0510-5
98. S.M. Han and H. Benaroya: *Nonlinear and Stochastic Dynamics of Compliant Offshore Structures*. 2002 ISBN 1-4020-0573-3
99. A.M. Linkov: *Boundary Integral Equations in Elasticity Theory*. 2002 ISBN 1-4020-0574-1
100. L.P. Lebedev, I.I. Vorovich and G.M.L. Gladwell: *Functional Analysis*. Applications in Mechanics and Inverse Problems (2nd Edition). 2002 ISBN 1-4020-0667-5; Pb: 1-4020-0756-6
101. Q.P. Sun (ed.): *IUTAM Symposium on Mechanics of Martensitic Phase Transformation in Solids*. Proceedings of the IUTAM Symposium held in Hong Kong, China, 11-15 June 2001. 2002 ISBN 1-4020-0741-8
102. M.L. Munjal (ed.): *IUTAM Symposium on Designing for Quietness*. Proceedings of the IUTAM Symposium held in Bangkok, India, 12-14 December 2000. 2002 ISBN 1-4020-0765-5
103. J.A.C. Martins and M.D.P. Monteiro Marques (eds.): *Contact Mechanics*. Proceedings of the 3rd Contact Mechanics International Symposium, Praia da Consolação, Peniche, Portugal, 17-21 June 2001. 2002 ISBN 1-4020-0811-2
104. H.R. Drew and S. Pellegrino (eds.): *New Approaches to Structural Mechanics, Shells and Biological Structures*. 2002 ISBN 1-4020-0862-7
105. J.R. Vinson and R.L. Sierakowski: *The Behavior of Structures Composed of Composite Materials*. Second Edition. 2002 ISBN 1-4020-0904-6
106. Not yet published.
107. J.R. Barber: *Elasticity*. Second Edition. 2002 ISBN Hb 1-4020-0964-X; Pb 1-4020-0966-6
108. C. Miehe (ed.): *IUTAM Symposium on Computational Mechanics of Solid Materials at Large Strains*. Proceedings of the IUTAM Symposium held in Stuttgart, Germany, 20-24 August 2001. 2003 ISBN 1-4020-1170-9

Mechanics

SOLID MECHANICS AND ITS APPLICATIONS

Series Editor: G.M.L. Gladwell

109. P. Ståhle and K.G. Sundin (eds.): *IUTAM Symposium on Field Analyses for Determination of Material Parameters – Experimental and Numerical Aspects.* Proceedings of the IUTAM Symposium held in Abisko National Park, Kiruna, Sweden, July 31 – August 4, 2000. 2003 ISBN 1-4020-1283-7
110. N. Sri Namachchivaya and Y.K. Lin (eds.): *IUTAM Symposium on Nonlinear Stochastic Dynamics.* Proceedings of the IUTAM Symposium held in Monticello, IL, USA, 26 – 30 August, 2000. 2003 ISBN 1-4020-1471-6
111. H. Sobieckzky (ed.): *IUTAM Symposium Transsonicum IV.* Proceedings of the IUTAM Symposium held in Göttingen, Germany, 2–6 September 2002, 2003 ISBN 1-4020-1608-5
112. J.-C. Samin and P. Fisette: *Symbolic Modeling of Multibody Systems.* 2003 ISBN 1-4020-1629-8
113. A.B. Movchan (ed.): *IUTAM Symposium on Asymptotics, Singularities and Homogenisation in Problems of Mechanics.* Proceedings of the IUTAM Symposium held in Liverpool, United Kingdom, 8-11 July 2002. 2003 ISBN 1-4020-1780-4
114. S. Ahzi, M. Cherkaoui, M.A. Khaleel, H.M. Zbib, M.A. Zikry and B. LaMatina (eds.): *IUTAM Symposium on Multiscale Modeling and Characterization of Elastic-Inelastic Behavior of Engineering Materials.* Proceedings of the IUTAM Symposium held in Marrakech, Morocco, 20-25 October 2002. 2004 ISBN 1-4020-1861-4
115. H. Kitagawa and Y. Shibutani (eds.): *IUTAM Symposium on Mesoscopic Dynamics of Fracture Process and Materials Strength.* Proceedings of the IUTAM Symposium held in Osaka, Japan, 6-11 July 2003. Volume in celebration of Professor Kitagawa's retirement. 2004 ISBN 1-4020-2037-6
116. E.H. Dowell, R.L. Clark, D. Cox, H.C. Curtiss, Jr., K.C. Hall, D.A. Peters, R.H. Scanlan, E. Simiu, F. Sisto and D. Tang: *A Modern Course in Aeroelasticity.* 4th Edition, 2004 ISBN 1-4020-2039-2
117. T. Burczyński and A. Osyczka (eds.): *IUTAM Symposium on Evolutionary Methods in Mechanics.* Proceedings of the IUTAM Symposium held in Cracow, Poland, 24-27 September 2002. 2004 ISBN 1-4020-2266-2
118. D. Ieşan: *Thermoelastic Models of Continua.* 2004 ISBN 1-4020-2309-X
119. G.M.L. Gladwell: *Inverse Problems in Vibration.* Second Edition. 2004 ISBN 1-4020-2670-6
120. J.R. Vinson: *Plate and Panel Structures of Isotropic, Composite and Piezoelectric Materials, Including Sandwich Construction.* 2005 ISBN 1-4020-3110-6
121. *Forthcoming*
122. G. Rega and F. Vestroni (eds.): *IUTAM Symposium on Chaotic Dynamics and Control of Systems and Processes in Mechanics.* Proceedings of the IUTAM Symposium held in Rome, Italy, 8–13 June 2003. 2005 ISBN 1-4020-3267-6
123. E.E. Gdoutos: *Fracture Mechanics. An Introduction. 2nd edition.* 2005 ISBN 1-4020-3267-6
124. M.D. Gilchrist (ed.): *IUTAM Symposium on Impact Biomechanics from Fundamental Insights to Applications.* 2005 ISBN 1-4020-3795-3
125. J.M. Huyghe, P.A.C. Raats and S. C. Cowin (eds.): *IUTAM Symposium on Physicochemical and Electromechanical Interactions in Porous Media.* 2005 ISBN 1-4020-3864-X
126. H. Ding, W. Chen and L. Zhang: *Elasticity of Transversely Isotropic Materials.* 2005 ISBN 1-4020-4033-4
127. W. Yang (ed): *IUTAM Symposium on Mechanics and Reliability of Actuating Materials.* Proceedings of the IUTAM Symposium held in Beijing, China, 1–3 September 2004. 2005 ISBN 1-4020-4131-6

Mechanics

SOLID MECHANICS AND ITS APPLICATIONS

Series Editor: G.M.L. Gladwell

128. J.-P. Merlet: *Parallel Robots*. 2006 ISBN 1-4020-4132-2
129. G.E.A. Meier and K.R. Sreenivasan (eds.): *IUTAM Symposium on One Hundred Years of Boundary Layer Research*. Proceedings of the IUTAM Symposium held at DLR-Göttingen, Germany, August 12–14, 2004. 2006 ISBN 1-4020-4149-7
130. H. Ulbrich and W. Günthner (eds.): *IUTAM Symposium on Vibration Control of Nonlinear Mechanisms and Structures*. 2006 ISBN 1-4020-4160-8
131. L. Librescu and O. Song: *Thin-Walled Composite Beams*. Theory and Application. 2006 ISBN 1-4020-3457-1
132. G. Ben-Dor, A. Dubinsky and T. Elperin: *Applied High-Speed Plate Penetration Dynamics*. 2006 ISBN 1-4020-3452-0
133. X. Markenscoff and A. Gupta (eds.): *Collected Works of J. D. Eshelby*. Mechanics and Defects and Heterogeneities. 2006 ISBN 1-4020-4416-X
134. R.W. Snidle and H.P. Evans (eds.): *IUTAM Symposium on Elastohydrodynamics and Microelastohydrodynamics*. Proceedings of the IUTAM Symposium held in Cardiff, UK, 1–3 September, 2004. 2006 ISBN 1-4020-4532-8
135. T. Sadowski (ed.): *IUTAM Symposium on Multiscale Modelling of Damage and Fracture Processes in Composite Materials*. Proceedings of the IUTAM Symposium held in Kazimierz Dolny, Poland, 23–27 May 2005. 2006 ISBN 1-4020-4565-4
136. A. Preumont: *Mechatronics*. Dynamics of Electromechanical and Piezoelectric Systems. 2006 ISBN 1-4020-4695-2
137. M.P. Bendsøe, N. Olhoff and O. Sigmund (eds.): *IUTAM Symposium on Topological Design Optimization of Structures, Machines and Materials*. Status and Perspectives. 2006 ISBN 1-4020-4729-0
138. A. Klarbring: *Models of Mechanics*. 2006 ISBN 1-4020-4834-3
139. H.D. Bui: *Fracture Mechanics*. Inverse Problems and Solutions. 2006 ISBN 1-4020-4836-X
140. M. Pandey, W.-C. Xie and L. Xu (eds.): *Advances in Engineering Structures, Mechanics and Construction*. Proceedings of an International Conference on Advances in Engineering Structures, Mechanics & Construction, held in Waterloo, Ontario, Canada, May 14–17, 2006. 2006 ISBN 1-4020-4890-4
141. G.Q. Zhang, W.D. van Driel and X. J. Fan: *Mechanics of Microelectronics*. 2006 ISBN 1-4020-4934-X
142. Q.P. Sun and P. Tong (eds.): *IUTAM Symposium on Size Effects on Material and Structural Behavior at Micron- and Nano-Scales*. Proceedings of the IUTAM Symposium held in Hong Kong, China, 31 May–4 June, 2004. 2006 ISBN 1-4020-4945-5